FRENCH–ENGLISH

HORTICULTURAL DICTIONARY

with English–French Index

by

D.O'D. Bourke, MA, Dip Agric, FI Hort, FLS
former Director, CAB International Bureau of Horticulture and Plantation Crops

CAB INTERNATIONAL

CAB INTERNATIONAL
Wallingford
Oxon OX10 8DE
UK

Tel: +44 (0)1491 832111
Fax: +44 (0)1491 833508
E-mail: cabi@cabi.org

CAB INTERNATIONAL
198 Madison Avenue
New York, NY 10016-4341
USA

Tel: +1 212 726 6490
Fax: +1 212 686 7993
E-mail: cabi-nao@cabi.org

A catalogue record for this book is available from the British Library, London, UK

First edition published in 1974
This edition published 1989, reprinted 1997

ISBN 0 85198 626 9

Printed and bound in the UK by Antony Rowe Ltd, Chippenham

For Shirley, Theresa, Patrick,
Isabelle and Andrew

Preface

The aim of this dictionary is to make it possible for readers whose language is English (or who know English) to discover the meaning of words they may meet in horticultural literature written in the French language. It has been compiled from the widest possible range of horticultural literature, scientific and popular, temperate and tropical, modern and not-so-modern, emanating mostly from France and French Overseas Territories, but also from Belgium, Canada, Mauritius, Switzerland and francophone African countries. The term 'horticulture' has also been interpreted in its widest sense, to include not only the growing of fruit, flowers and vegetables but also plantation crops, ornamental arboriculture, parks and wild gardens, turf, medicinal plants, minor industrial crops and edible fungi, together with selected attendant disorders, diseases, pests and weeds.

The plan of the dictionary has been kept as simple as possible. Indications have been inserted only where there is a risk of confusion. Nouns are not usually indicated as such unless a distinction has to be made between singular and plural or unless they are grammatical forms of identical words which are also adjectives. The genders of nouns are given throughout, as are feminine forms of adjectives, plurals of certain compound words and most plurals which do not end in 's'. Explanations of terms have been provided where appropriate and English names where possible. Where two scientific names are still in common use both have been given.

In the same way that no dictionary is ever complete, so no modern dictionary in a major language is ever entirely original. The lexicographer must always rely on the work of others to confirm or expand his ideas. A list of dictionaries and works consulted appears below. To my wife, who looked up many references for me, I offer my special thanks. In addition, many friends and colleagues have most helpfully given me the benefit of their experience, knowledge and advice. It is with much pleasure that I acknowledge my debt to: Mr. G. E. Tidbury, Director, Commonwealth Bureau of Horticulture and Plantation Crops, for including the dictionary in his publications list and for valuable suggestions about its scope; Miss M. H. Blackler, for data on pome and stone fruits; Miss J. E. K. Cox, for many details on ornamentals and other horticultural matters; Mrs. T. A. L. Crowhurst, for typing out all my notes and anticipating many corrections; Mr. R. J. Garner, for grafting terms; Miss E. E. Hardman, for information on berries and grapes; MM. J.-C. Lamontagne and J. Laurent (on the other side of the Channel), for explanations of puzzling terms; Mr. W. L. Millen, for spraying terms; Mr. A. P. Preston, for horticultural terms; Professor J. W. Purseglove, for many details on tropical crops; Miss J. P. Rowe-Dutton, for checking the typescript and proofs and making many relevant suggestions; and Mr. T. E. Walter, for tea pruning terms. I owe a special debt to all members of the Bureau who have ever been concerned with the abstracting of French horticultural literature, including the late Mr. J. Dundas who compiled French vocabularies for internal use. This list is not complete, however, as many others, too numerous to list individually, have helped to guide me to sources of information. I wish to take this opportunity of thanking them all, while stressing that they are in no way responsible for any errors in the text.

Preface to the Second Edition

The advance of horticultural technology since the publication of this dictionary has led to the appearance of new words and to the adaptation of existing words to new techniques. Words have been added which deal with the environment, greenhouse construction, the use of plastics, pollution and tissue culture. Additions have also been made in the fields of African and Canadian usage, chromatography, marketing, medicinal plants, pedology, plant physiology and statistics. Brief descriptions of many lesser-known horticultural plants have been included, together with their English names whenever possible.

Some of the works consulted

Le Bon Jardinier. Nouvelle Encyclopédie Horticole. La Maison Rustique. 1947.
Cassell's New French Dictionary. French–English, English–French. 1965.
Chamber's Twentieth Century Dictionary. 1952.
Diccionario de Plantas Agrícolas. Enrique Sánchez-Monge y Parellada. Madrid. 1981.
Dictionary of Botany. G. Usher. Constable. 1966.
Dictionary of Flowering Plants and Ferns. J. C. Willis. (Eighth Edition, revised by H. K. Airy Shaw). Cambridge. 1973.
Dictionary of the Fungi. Ainsworth and Bisby. (Sixth Edition). C.A.B. 1971.
Dictionary of Gardening. Royal Horticultural Society. Oxford. 1951.
Dictionnaire d'Agriculture et des Sciences Annexes. La Maison Rustique. 1977.
Dictionnaire des Difficultés de la Langue Française. Larousse. 1956.
Dictionnaire Moderne Larousse. Français–Anglais, Anglais–Français. 1960.
Duden Français. Dictionnaire en Images. Harraps. 1962.
Elsevier's Dictionary of Horticulture. 1970.
French–English Science Dictionary. L. De Vries. McGraw-Hill. 1951.
Gartenbautechnik I. Mehrsprachen–Bildwörterbuch. H. Steinmetz. West Germany. 1972.
Grand Larousse Encyclopédique en Dix Volumes. 1960. Supplément. 1969.
A Handbook of Agricultural Terms. J. W. A. Newhouse. 1973.
Harrap's New Standard French and English Dictionary. 1972. Part I. French–English.
Harrap's Standard French and English Dictionary. 1956. Part I. French–English. 1962. Part II. English–French.
Harrap's Visual French–English Dictionary. Editions Québec/Amérique Inc. 1987.
Henderson's Dictionary of Biological Terms. Oliver and Boyd. 1963.
Langenscheidt's Standard Dictionary of the French and English Languages. Hodder and Stoughton. 1968.
Larousse. Trois Volumes en Couleurs. 1965.
Lexique de la Vigne et du Vin. Office International du Vin. 1963.
McGraw-Hill Dictionary of Scientific and Technical Terms. Second Edition. 1978.
Nouveau Larousse Universel. 1948.
Oxford Universal Dictionary. (Illustrated). 1970.
Petit Larousse. 1966.
Sanders' Encyclopaedia of Gardening. 1971.
Shorter Oxford English Dictionary. 1956.
Webster's Third New International Dictionary of the English Language. 1966.

INDICATIONS

A in former use
a. (*a*.) adjective
Adm. Administration
adv. adverb
adv. *phr*. adverbial phrase
Afr. Africa
Ag. Agriculture
AH Animal husbandry
AM Agricultural/horticultural machinery
Ap. Apiculture
Arb. Arboriculture
Ban. Banana culture
Biol. Biology
Bot. Botany
(*Cacao*) Cocoa culture
(*Canada*)
Chem. Chemistry
Cit. Citrus culture
(*Coconut*) Coconut culture
Coff. Coffee culture
Coll. Colloquial
Comm. Commercial
(*Date*) Date culture
Ent. Entomology
f. (*f*.) feminine
Fung. Fungi
Hort. Horticulture
inv. invariable
Irrig. Irrigation
LC Locust control
m. (*m*.) masculine
Mush. Mushroom culture
n. (*n*.) noun
num. numerical
OP Oil palm culture
PB Plant breeding
PC Pest control and pesticides
PD Plant diseases and disorders
Ped. Pedology
Physics
Physiol. Plant physiology
(*Pineapple*) Pineapple culture
pl. plural
Plant. Plantations
prep. preposition
prep. *phr*. prepositional phrase
pron. pronoun
q.*v*. (*q*.*v*.) which see
Rubb. Rubber culture
SC Soil conservation
sp., spp. species
Stat. Statistics
Sug. Sugar cane culture
Tea (*Tea*) Tea culture
(*USA*) American usage
v. verb
var. variety
v.*i*. intransitive verb
Vit. Viticulture
v.*p*. pronominal verb
v.*t*. transitive verb
(*West Indies*)

A

abaca *m.* Manila hemp (*Musa textilis*) (= **chanvre de Manille**)
abaissé,-e *a.* lowered, lessened
abaissement *m.* lowering, reduction; **a. significatif** significant reduction
abâtardi,-e *a.* degenerate
abâtardissement *m.* degeneration
abattage *m. Arb.* felling
abattement *m.* = **abattage** (*q.v.*)
abatteur *m. Arb.* person who fells trees
abattis *m. Arb.* felling, clearing
abattre *v.t.* to knock down, to throw down; to fell, to cut, to clear (trees)
abattu,-e *a.* knocked down, thrown down; felled, cut down (of trees)
abeille *f.* bee, honey-bee; **a. coupeuse de feuilles** leaf-cutter bee (*Megachile* sp.); **a. mère** queen bee; **a. mâle** drone; **a. ouvrière** worker; **a. mécanique** artificial bee, electric vibrator
abeiller, abeillier,-ère *a.* apiarian, relating to bees; **industrie abeillère** bee-keeping industry
abelmosch *m.* musk mallow (*Hibiscus abelmoschus*) (see also **ambrette**)
aberration *f.* aberration; **a. chromosomique** chromosome aberration
abies *m.* fir (*Abies* sp.) (see **sapin**)
abiotique *a.* abiotic
ablation *f.* excision, removal; **a. totale** complete removal
abonk *m.* West African ratbane (*Dichapetalum toxicarium*)
abord *m.* access, approach; *pl.* approaches
abornement *m.* marking out, delimitation, demarcation
abortif,-ive *a.* abortive; **fruit abortif** abortive fruit
abrasin *m.* tung tree (*Aleurites* sp.) (= **aleurite**)
abraxas *m.* magpie moth (*Abraxas grossulariata*)
abre *m.* crab's eye vine (*Abrus precatorius*) (= **abrus**)
abreuvage *m.* watering, drenching
abreuver *v.t.* to water, to soak, to drench, to irrigate
abri *m.* shelter, cover, protective structure; **à l'abri** sheltered, under cover; **à l'abri de la gelée** protected from frost, frostproof; **à l'abri du vent** protected from the wind; **a. arboré** shelterbelt, windbreak; **a. artificiel** artificial shelter; **a. météorologique** screen for meterological instruments; **a. A.M.P.S.** = **abri météorologique Piche simplifié** Piche screen; **a. antipucerons** aphid-free shelter; **a. plastique** plastic cover; **a. vitré** glasshouse, cloche or frame structure (see also **abri-serre**)
abriba *m.* biriba (*Rollinia pulchrinervia*)
abricot *m.* apricot (see **abricotier**); **a. du Japon** (Japanese) persimmon; **a. de Saint-Domingue** mammey apple
abricoté,-e *a.* resembling apricot, apricot-flavoured; *n. m.* preserved apricot slice
abricotier *m.* apricot tree (*Prunus armeniaca*); **a. de Briançon** Briançon apricot (*P. brigantina*) whose seeds yield **huile de marmotte**; **a. d'Afrique** African mammey apple tree (*Mammea africana*); **a. des Antilles** = **a. de Saint-Domingue** (*q.v.*); **a. du Japon** (Japanese) persimmon tree (*Diospyros kaki*); **a. de Saint-Domingue** mammey apple tree (*M. americana*); **a. de Sibérie** *Prunus sibirica*
abri-serre *m. pl.* **abris-serres** plastic greenhouse, plastic shelter, tunnel
abrité,-e *a.* sheltered, screened, protected; **endroit abrité** sheltered place
abriter *v.t.* to shelter, to screen, to protect
abrivent *m.* windbreak, matting (used as windbreak)
abroma, abrome *m. Abroma* sp., including cotton abroma (*A. augusta*)
abronia *f. Abronia* sp., sand verbena
abrotone *f.* southernwood (*Artemisia abrotanum*) (= **aurone**)
abrouti,-e *a.* browsed
abroutissement *m.* deformation of tree by browsing
abrus *m.* crab's eye vine (*Abrus precatorius*) (= **abre, jéquirity**)
abscission *f.* abscission; **a. foliaire** leaf abscission
absinthe *f.* wormwood; (**grande**) **absinthe** *Artemisia absinthium*; **a. maritime** sea wormwood (*A. maritima*); (**petite**)**absinthe** Roman wormwood (*A. pontica*)
absorbable *a.* capable of being absorbed
absorbant,-e *a.* absorbing; **poils absorbants** *Bot.* root hairs; **pouvoir absorbant** absorbing capacity

absorbé,-e *a.* absorbed
absorber *v.t.* to absorb
absorption *f.* absorption, uptake; **a. de la lumière** light absorption; **a. de substances nutritives** nutrient absorption
abutilon *m. Abutilon* sp., including several ornamental plants
acacia *m. Acacia* sp., acacia and other spp. (see also **gommier, mimosa**); **a. chenille** Sydney golden wattle (*A. longifolia*); **a. de Constantinople** pink siris (*Albizzia julibrissin*); **a. de Farnèse** sweet acacia (*Acacia farnesiana*) (see also **cassie**); **a. jaune** pea tree (*Caragana* sp.); **(faux)acacia** = **robinier** *q.v.*; **petit acacia** (*Mauritius*) *Desmanthus virgatus*
acajou *a. inv.* 1. dark auburn, mahogany, 2. *n.m.* **a. de Chine** bastard cedar (*Cedrela sinensis*); **a. du Sénégal** see **caïlcédrat**
acalypha *m. Acalypha* sp., copper-leaf
acanthacées *n. f. pl.* Acanthaceae
acanthe *f.* acanthus (*Acanthus* sp.); **a. épineuse** *A. spinosus*; **a. à feuilles molles, a. à larges feuilles** bear's breeches (*A. mollis*) (= **branche-ursine**)
acanthoïde *a. Bot.* acanthoid, spiny
acantholimon *m. Acantholimon* sp., prickly thrift
acare *m.* mite, acarid (= **acarien** *q.v.*)
acaricide *a.* acaricidal; *n. m.* acaricide
acarien *m.* mite, acarid (see also **tisserand**); **a. des bourgeons** *Cit.* bud mite (*Aceria sheldoni*); **a. brévipalpe** *Cit.* flat mite (*Brevipalpus* sp.); **a. jaune** *Rubb.* yellow (tea) mite, *Tea* broad mite (*Hemitarsonemus latus*); **a. du narcisse** bulb scale mite (of narcissus) (*Steneotarsonemus laticeps*); **a. prédateur** predacious mite (see also **typhlodrome**); **a. rouge** fruit tree red spider mite (*Panonychus ulmi*); **a. ravisseur** *Cit.* silver mite (*H. latus*); **a. à deux taches, a. tisserand** red (spider) mite (*Tetranychus urticae*) *Cit.* spider mite (*T. cinnabarinus*)
acarifuge *a.* mite-repelling; *n. m.* mite repellent
acariose *f.* acariosis, mite infestation; **a. de la vigne** *Phyllocoptes vitis*
acarpe *a. Bot.* acarpous
acarus *m.* = **acare** (*q.v.*)
acaule *a. Bot.* acauline, acaulose, stemless
accélération (*f.*) **de récolte** advancement of cropping, harvest
accéléré,-e *a.* & *n. Hort.* bolting plant, bolter
accepteur *m.* acceptor; **a. de CO_2** CO_2 acceptor; **a. d'hydrogène** hydrogen acceptor
accidenté,-e *a.* 1. uneven, rough (of ground), 2. adversely affected; **culture accidentée** injured crop, failed crop
acclimatable *a.* capable of being acclimatized
acclimatation *f.* acclimatization, acclimatation, hardening off
acclimaté,-e *a. Hort.* hardened
acclimatement *m.* acclimatization, acclimatation
acclimater *v.t.* to acclimatize, to harden off
acclimateur *m.* one who acclimatizes, acclimatizer
accolade *f.* brace, bracket (in typography), used to describe the shape of the grapevine leaf petiolar sinus
accolage *m. Hort. Vit.* tying plant(s) to stake, support, tying scion shoot to stock
accolé,-e *a.* tied, joined to
accolement *m.* joining, uniting; **a. des thylakoïdes** thylakoid stacking
accoler *v.t. Hort., Vit.* to tie plant(s) to stakes, supports, to tie scion shoot to stock
accolure *f.* tie (of straw or osier, etc.) for vines, trees, etc.
accombant,-e *a.* accumbent
accommodat *m.* accommodation, the capacity of a plant to adapt itself to environmental changes
accommodation *f.* adaptation, accommodation
accorage *m. Hort.* staking, supporting, propping
accot *m.* bank of manure or earth placed around a bed or frame to protect it from the cold (= **réchaud**)
accotement *m.* verge, verge of road, escarpment, side (of ditch, etc.)
accouple *f.* link, tie
accouplement *m.* coupling, joining; mating (of animals)
accoupler *v.t.* to couple; **s'accoupler** to mate (of animals)
accoutumance *f. PC* habituation, tolerance, resistance (to a pesticide)
accrescent,-e *a. Bot.* accrescent
accroissement *m.* growth, increase; **a. de production** yield increase
accru *m. Bot.* sucker
accu *m.* = **accumulateur**
accumulateur *m.* accumulator; **a. de manganèse** manganese accumulator
accumulation *f.* accumulation
acellulaire *a.* without cells; **extrait acellulaire** cell-free extract
acer *m. Acer* sp. (see **érable**)
acéracées *f. pl.* Aceraceae
acerbe *a.* tart, bitter
acerbité *f.* tartness, sourness
acérolier *m.* West Indian cherry, Barbados cherry (*Malpighia glabra*) (= **cerisier des Antilles**)
acétaldéhyde *f.* acetaldehyde
acétate *m.* acetate
acétique *a.* acetic
acétyl-cholinesterase *m.* acetylcholinesterase

acétylène *m.* acetylene
achaine *m.* = **akène** *q.v.*
achat *m.* buying, purchase; **a. d'impulsion** impulse buying
ache *f.* **a. (odorante)**; **a. (des marais)** wild celery (*Apium graveolens*); **a. des chiens** fool's parsley (*Aethusa cynapium*); **a. d'eau** water parsnip (*Sium latifolium*) (see also **berle**); **a. de montagne** lovage (*Levisticum officinale*) (= **livèche**), Scotch lovage (*Ligusticum scoticum*)
achillée *f.* *Achillea* sp. **a. millefeuille** milfoil, yarrow (*Achillea millefolium*) (see also **bouton d'argent**)
achimène *m.* achimenes (*Achimenes* sp.)
achyranthe *f.* *Achyranthes* sp.
aciculaire *a.* acicular, needle-shaped, slender and sharp pointed
acidanthera *m.* *Acidanthera* sp.
acide *a.* acid, acidic; *n.m.* acid; **a. abscis(s)ique** abscisic acid, ABA; **a. acétique** acetic acid; **a. alpha** alpha acid (in hops); **a. aminé** amino acid; **a. ascorbique** ascorbic acid; **a. carbonique** carbon dioxide; **a. cyanhydrique** = **a. prussique** (*q.v.*); **a. chlorhydrique** hydrochloric acid; **a. citrique** citric acid; **a. gras** fatty acid; **a. humique** humic acid; **a. phosphorique** phosphoric acid; **a. prussique** prussic acid; **a. ribonucléique** ribonucleic acid (see also **A.I.A., A.I.B., A.N.A., R.N.A.**)
acidification *f.* acidification; **a. des sols** soil acidification
acidifier *v.t.* to acidify
acidiphile *a.* *Hort.* lime-hating, acid-loving
acidité *f.* acidity, sourness; **a. titrable** titratable acidity
acidulé,-e *a.* acidulous
acineux,-euse *a.* acinose
aconit *m.* aconite; **a. napel** monkshood (*Aconitum napellus*) (= **napel**)
aconitine *f.* aconitine
aconitique *a.* aconitic
acore *m.* *Acorus* sp.; **a. odorant, a. vrai** sweet flag (*A. calamus*) (= **lis des marais**)
acotylédone, acotylédoné,-e *a.* acotyledonous
à-coup *m.* jerk, jolt, shock; sudden stoppage; **à-c. de végétation** growth stoppage
acrage *m.* (*Canada*) acreage
âcre *a.* acid, bitter
âcreté *f.* acridity
acridien,-enne *a.* relating to Acridiidae; *n.m.* acridian, (short-horned) grasshopper; **a. migrateur** locust (see also **antiacridien**)
acrocome *m.* macaw palm (*Acrocomia sclerocarpa*)
acrodinium *m.* Swan river everlasting flower (*Helipterum* sp.)
acropétale *a.* acropetal
acropète *a.* acropetal
acrostic, acrostique *m.* *Acrosticum* sp., elephant's ear fern
acrotone *a.* acrotonic
actée *f.* *Actaea* sp., including baneberry, herb Christopher (*Actaea spicata*) (= **herbe de Saint Christophe**)
actine *f.* Chinese gooseberry
actinidia (de Chine), actinidier *m.* Chinese gooseberry plant (*Actinidia chinensis*) (= **yang-tao**)
actinographe *m.* actinograph
actinomorphe *a.* actinomorphic
actinomycète *f.* actinomycete
actinotropisme *m.* *Bot.* inflexion towards light
action *f.* action; **a. capillaire** capillary action; **a. immédiate, a. de contact** *PC* contact action; **a. inhibitrice** inhibiting action
activation *f.* activation
activité *f.* activity; **a. biologique** biological activity; **a. radiculaire** root activity
aculé,-e *a.* *Bot.* aculeate
aculéiforme *a.* *Bot.* needle-shaped
acuminé-e *a.* *Bot.* acuminate
acylé,-e *a.* acylated
adansonia *f.* baobab (*Adansonia digitata*) (= **baobab**)
adaptation *f.* adaptation
additif,-ive *a.* additive
adénanthère *f.* bead tree (*Adenanthera pavonina*)
adhérence *f.* adhesion
adhérent,-e *a.* adherent, adhesive; *Bot.* adnate
adhésif,-ive *a.* adhesive, sticky; *n. m.* *PC* sticker, adhesive
adhésivité *f.* adhesiveness, stickiness
adiante *m.* maidenhair fern (*Adiantum* sp.) (see also **capillaire, cheveux de Vénus**)
adjuvant,-e *a.* & *n. m.* *PC* adjuvant, additive
adné,-e *a.* *Bot.* adnate
adonis, adonide *f.* adonis; **a. d'automne** pheasant's eye (*Adonis annua* and other spp.); **a. de printemps** yellow adonis (*A. vernalis*) (see also **goutte-de-sang**)
ados *m.* ridge; sloping south-facing or east-facing bed, bank-bed, raised bed
adsorbant *m.* adsorbent
adsorber *v.t.* to adsorb
adsorption *f.* adsorption
advection *f.* advection
adventice *a.* adventitious, casual; *n. f.* *Bot.* plant which spreads without deliberate introduction, weed (see also **mauvaise herbe**)
adventif,-ive *a.* *Bot.* adventitious; **œil adventif** adventitious bud; **racine adventive** adventitious root
æchmea, æchmée *f.* aechmea (*Aechmea* sp.)

ægopodium *m.* *Aegopodium* sp., especially ground elder (*A. podagraria*) (= **herbe aux goutteux**) (see also **égopode**)
aérateur *m.* *Hort.* spiked roller
aération *f.* aeration; **a. du sol** soil aeration
aéré,-e *a.* aerated
aérenchyme *m.* aerenchyma
aérer *v.t.* to aerate, to ventilate
aéricole *a.* *Bot.* aerial
aéride *f.* air-plant (*Aërides* sp.)
aérien,-enne *a.* aerial; **tige aérienne** aerial stem; **parties aériennes** aerial parts
aérobie *a.* aerobic
aérobiose *f.* aerobiosis
aéroponique *a.* aeroponic; **culture aéroponique** aeroponic culture
aérosol *m.* aerosol
aérotherme *m.* air heater
æschynanthe *f.* *Aeschynanthus* sp., blushwort
æsculus *m.* *Aesculus* sp., horse chestnut (see also **marronnier**)
aethionema *m.* *Aethionema* sp., candy mustard
affaissement *m.* 1. subsidence, sinking, settling. 2. wilt, decline (of plant)
affection *f.* *PD* disease, disorder
affin,-e *a.* akin, closely connected
affinité *f.* affinity, compatibility
afflux *m.* surge, enhanced flow
affouillable *a.* erodable (of channel bank etc.)
affouillement *m.* erosion, undermining (of channel bank etc.)
affranchir *v.t.* to liberate; *Hort.* **s'affranchir** to scion root
affranchissement *m.* **(du greffon)** scion rooting
affruiter *v.i.* to bear fruit, to produce fruit; *v.t.* to plant fruit trees, to garnish with fruit
affûtage *m.* sharpening
affûter *v.t.* to sharpen
agalloche *m.* agallocha, blinding tree (*Excœcaria agallocha*) (= **arbre aveuglant**)
agame, agamique *a.* agamous, agamic, asexual
agapanthe *m.* African lily (*Agapanthus* sp.) (see also **lis** 2, **tubéreuse**)
agar-agar *m.* agar-agar
agassin *m.* low bud on a vine branch which does not produce fruit
agati *m.* agati sesbania (*Sesbania grandiflora*)
agave *m.* *Agave* sp.; **a. d'Amérique** American aloe (*A. americana*)
age *m.* plough beam
agent *m.* agent; **a. causal** causal agent; **a. de conservation** preservative; **a. (de maladie) a. pathogène** *PD* pathogen; **a. de lutte biologique** biological control agent; **a. de surface** *PC* surface-acting agent, surfactant
agérate, agératum *m.* *Ageratum* sp., floss-flower
agglutinant *m.* adhesive
agitateur *m.* agitator, mixer
aglaia *m.* *Aglaia* sp. (Meliaceae)
agneau-chaste *m.* *pl.* **agneaux-chastes** chaste tree, tree of chastity (*Vitex agnus-castus*) (= **arbre au poivre, gattilier**)
agobiada *m.* *Coff.*, agobiada pruning system (= **arcure, taille en archet**)
agouti *m.* agouti (*Dasyprocta* sp.); *Afr.* cane rat (*Thryonomys swinderianus*) (= **aulacode**)
agrafe *f.* staple
agrafer *v.t.* to staple
agrafeuse *f.* stapling machine
agraire *a.* agrarian
agrégat *m.* aggregate
agrégé,-e *a.* aggregate
agrenette sarmenteuse *f.* *Desmoncus macroacanthus*
agressif,-ive *a.* aggressive; *PD* virulent
agreste *a.* rustic, uncultivated (plant), rural
agricole *a.* agricultural
agriculture *f.* agriculture
agriflamme *m.* *Hort.* flame gun
agrile *m.* *Ent.* *Agrilus* sp.; **a. du cassis** *A ribesi*
agripaume *f.* motherwort (*Leonurus cardiaca*) (= **léonure**)
agroclimatologie *f.* agroclimatology (= **climatologie agricole**)
agro-économique *a.* agro-economic
agroécosystème *m.* agroecosystem
agroforesterie *f.* agroforestry
agrologie *f.* agrology
agrologique *a.* agrological
agrométéorologie *f.* agrometeorology
agronome *m.* agronomist, (scientific) agriculturist, agricultural officer (employed by the government); **ingénieur agronome** agricultural graduate
agronomie *f.* agronomy
agronomique *a.* agronomic
agropyre *m.* *Agropyron* sp., couch grass (see also **chiendent**)
agrostide *f.* *Agrostis* sp., bent-grass; **a. jouet du vent** silky apera (*Apera (Agrostis) spica-venti*); **a. stolonifère** creeping bent-grass (*A. stolonifera*)
agrotextile *m.* plastic material which is not woven, such as polypropylene or polyester
agrume *m.* (usually *pl.*) citrus fruit, citrus tree
agrumicole *a.* citrus-growing, relating to citrus
agrumiculteur *m.* citrus grower
agrumiculture *f.* citrus growing, citriculture
A.I.A. = **(acide indole-acétique)** IAA indolyl-acetic acid
A.I.B. = **(acide indole-butyrique)** IBA indolyl-butyric acid
aiélé *m.* African elemi, incense tree (*Canarium schweinfurthii*)
aigre *a.* sour, acid, tart

aigrelet,-ette *a.* tart, slightly sour; **saveur aigrelette** tart taste
aigrelier *m.* wild service tree (*Sorbus torminalis*) (= **alisier**)
aigremoine *f.* agrimony (*Agrimonia eupatoria*)
aigrette *f. Bot.* egret, pappus
aigrin *m.* young apple or pear tree seedling; young crab apple or wild pear tree (= **égrain, égrin**)
aigu,-ë *a.* sharp, pointed
aiguille *f.* needle, *Bot.* needle; *Hort.* (fine) tender green pod (of bean)
aiguillon *m.* sting; *Bot.* prickle, thorn; **en aiguillon** thorn-like, curved
ail *m. pl.* **ails**, *Bot.* **aulx**, *Allium* sp. but especially garlic (*A. sativum*); **gousse d'ail** garlic clove; **a. blanc** white garlic (*A. neapolitanum*); **a. d'Espagne** = **rocambole** (*q.v.*); **a. doré** moly (*A. moly*) (= **moly**); **a. à fleurs comestibles** Chinese chive (*A. odorum*); **a. gigantesque** *A. giganteum*; **a. d'Orient** wild leek (*A. ampeloprasum*); **a. des vignes** crow garlic (*A. vineale*)
ailante *m. Ailanthus* sp., including tree of heaven (= **(faux) vernis du Japon**)
aile *f.* wing, pinion; *Bot.* wing (of leguminous flower)
ailé,-e *a.* winged; **fruit ailé** winged fruit
aileron *m. Sug.* aerial branching; *Vit.* small bunch forming part of cluster
air *m.* air, atmosphere; **en plein air** in the open air, outside; **a. comprimé** compressed air
aire *f.* 1. surface. 2. area; **a. foliaire** leaf surface, leaf area; **a. d'arasement de tourbière** cut-over peat bog area; **a. d'excavation de tourbière** cut-away peat bog area
airelle *f.* 1. bilberry, cranberry (*Vaccinium* sp.); **airelle, a. noire** bilberry (*V. myrtillus*); **a. bleue, a. des marais** bog bilberry (*V. uliginosum*); **a. rouge** cowberry (*V. vitis-idaea*) 2. (*Canada*) lowbush blueberry (= **bleuet de Pennsylvanie** *q.v.*); **a. en corymbe** highbush blueberry (*V. corymbosum*); **a. fausse** sourtop blueberry (*V. myrtilloides*) (see also **bleuet** 2, **canneberge, myrtille**)
aisselle *f. Bot.* (leaf) axil
ajonc *m.* gorse (*Ulex* sp.); **a. commun, a. marin** common gorse (*U. europaeus*); **a. nain** dwarf gorse (*U. nanus*) (= **genêt épineux**)
ajouré,-e *a.* perforated, pierced
ajout *m.* addition; **a. d'engrais** addition of fertilizer
ajuga *m. Ajuga* sp., bugle (= **bugle**)
ajutage *m. Irrig.* delivery pipe
akébia (*m.*) **à cinq feuilles** *Akebia quinata*
akène *m.* achene (= **achaine**)
ako *m.* cloth tree (*Antiaris africana*)
alaterne *m.* Mediterranean buckthorn (*Rhamnus alaternus*) (see also **bourdaine, nerprun**)
albarelle *f.* rough-stemmed boletus (*Boletus scaber*)
albedo, albédo *m. Cit.* albedo
alberge *f.* clingstone apricot
albergier *m.* clingstone apricot tree (see also **abricotier**)
albinisme *m.* albinism
albinos *a.* & *n. inv.* albino
albizzie *f. Albizzia* sp.
albumen *m.* albumen, endosperm (*Coconut, OP* etc.) kernel
albumine *f.* albumin
albuminé,-e *a.* **albuminous**
albumineux,-euse *a.* albuminous, albuminose
albuminoïde *a.* albuminoid
alcali *m.* alkali
alcalin,-e *a.* alkaline
alcalinité *f.* alkalinity
alcalinophile *a.* basophile, lime-loving (of plant) (= **basiphile**)
alcaloïde *a.* & *n. m.* alkaloid; **a. dimère** dimeric alkaloid
alcaloïdogenèse *f.* alkaloidogenesis
alcane *m.* alkane
alcée *f. Althaea* sp., hollyhock (see also **rose trémière**)
alchémille *f. Alchemilla* sp., including lady's mantle (*A. vulgaris*); **a. des champs, a. perce-pierre** parsley piert (*A. arvensis*)
alcool *m.* alcohol; **a. de bouche** drinking alcohol; **a. distillé (de sève)** (*Coconut*) arrack; **a. éthylique** ethyl alcohol; **a. industriel** industrial alcohol; **a. ordinaire** = **a. éthylique** (*q.v.*); **a. solidifié** metaldehyde
alcoolique *a.* alcoholic; **boisson alcoolique** alcoholic drink
aldéhyde *m.* aldehyde
aléné,-e *a.* acuminate; **en alêne** acuminate
aleurite *m.* tung tree (*Aleurites* sp.) (= **abrasin**)
aleurode *m.* whitefly (Aleyrodidae); **a. floconneux** woolly whitefly (*Aleurothrixus floccosus*); **a. des serres** greenhouse whitefly (*Trialeurodes vaporariorum*) (= **mouche blanche**)
aleurodiphage *a.* whitefly-eating
aleurone *f.* aleurone
alfa *m.* esparto grass (*Stipa tenacissima*) (see also **sparte**)
alfalfa *m.* lucerne, alfalfa; **a. tropical** Brazilian lucerne, stylo (*Stylosanthes gracilis*)
alfatier,-ère *a.* relating to esparto culture; *n.* dealer in esparto grass
algicide *a.* algicidal; *n. m.* algicide
algue *f.* alga, seaweed; **a. microscopique** microalga
algueux,-euse *a.* covered with algae

alibile *a.* nutritive
aliboufier *m.* styrax (*Styrax officinalis*); **a. du Japon** Japanese styrax (*S. japonica*)
alicante *m.* dark grape variety; wine of the same name
alignée *f.* row
alignement *m.* aligning, lining up, putting in a row
aligner *v.t.* to align, to line up, to put in a row
aligneuse *f.* windrowing machine (see also **arracheuse**)
aligoté *m.* white grape variety; white wine of the same name
aliment *m.* food, nutriment; **a. de base** staple food
alimentation *f.* feeding, nutrition; **a. en eau, a. hydrique** water supply (of crop etc.); **a. minérale** mineral nutrition
alios *m. Ped.* (**ferrugineux**) iron pan, hard pan
aliphatique *a.* aliphatic
aliquote *a.* & *n.f.* aliquot
alise *f.* fruit of service tree (**alisier**)
alisier, alizier *m.* wild service tree (*Sorbus torminalis* (= **aigrelier**); **a. blanc** whitebeam (*S. aria*) (= **allouchier**); (see also **cormier, sorbier**); **a. de Fontainebleau** *S. latifolia*
alisma, alisme *m.* water plantain (*Alisma plantago-aquatica*) (= **flûteau, plantain d'eau**)
alizari *m.* madder root
alizier *m.* = **alisier** (*q.v.*)
alkanna *f.* = **orcanette** (*q.v.*)
alkékenge *m.* bladder cherry (*Physalis alkekengi*); **a. doux, a. jaune** downy ground-cherry, dwarf Cape gooseberry (*P. pubescens*); **a. du Mexique** tomatillo, ground-cherry, cowpops (*P. angulata*); **a. coquerelle, a. du Pérou** Cape gooseberry (*P. peruviana*) (see also **coqueret**)
allamande, allamanda *f.* allamanda (*Allamanda* sp.)
allée *f.* avenue, lane, walk; *Plant.* **en allées simples** hedge planting; **en allées doubles** double rows
alléger *v.t.* to lighten; (of soil) to make lighter (by adding sand, strawy manure etc.)
allèle *m. PB* allele
allélique *a.* allelic
allélomorphe *a.* allelomorphic
allélopathie *f.* allelopathy, competition between plants of different species by means of toxic substances excreted by leaves or roots
alliacé,-e *a.* alliaceous; *n. f. pl.* **alliacées** Alliaceae (see also **pourriture**)
alliance *f.* alliance, union; *Hort.* **a. clone-sujet** (clone-) scion-rootstock union
allier *m. Hort.* = **allouchier** (*q.v.*)
allogame *a. Bot., PB* cross-fertilized, cross-pollinated
allogamie *f.* allogamy, cross-fertilization, cross-pollination
allongement *m.* lengthening, extension, elongation; **a. du jour** long day treatment; **a. de la nuit** short day treatment; **a. des rameaux** shoot elongation
allopolyploïde *a.* & *n.m.* allopolyploid
allopolyploïdie *f.* allopolyploidy
allopollen *m.* pollen from a plant other than the one under consideration but from the same species
allouchier *m.* whitebeam (*Sorbus aria*) (= **alisier blanc, allier**)
alluvial,-e *a., pl.* **-aux** alluvial
alluvion *f.* alluvial soil, deposit
alluvionnaire *a.* alluvial (see also **bourrelet**)
alocasie *f.* *Alocasia* sp.
aloès *m.* aloe (*Aloe* sp.)
aloma *m. Afr.* bitterleaf (*Vernonia amygdalina*)
alonsoa, alonzoa *f.* mask-flower (*Alonsoa* sp.)
alopécure *f.* foxtail grass (*Alopecurus* sp.) (= **vulpin**)
alpestre *a.* alpestrine, alpine
alpicole *a.* alpine (of plant)
alpin,-e *a.* alpine; **plante alpine** alpine plant, alpine
alpine *f.* rock garden plant, alpine
alpinum *m.* garden or collection of alpine plants
alpiste *m.* alpist; **a. des Canaries** Canary grass (*Phalaris canariensis*); **a. panaché** gardener's garters (*P. arundinacea variegata*); **petit alpiste** lesser Canary grass (*P. minor*)
alstonie *f.* *Alstonia* sp.
alstrœmère, alstrœmérie *f.* *Alstroemeria* sp., herb lily, Peruvian lily (see also **lis** 2)
altération *f.* change for the worse, deterioration; **a. foliaire** leaf damage; *Ped.* weathering
alternance *f.* alternation; *Hort.* **a.** (**de la fructification**) biennial bearing
alternant,-e *a.* alternating, rotating (of crops); *Hort.* biennially bearing, bearing in alternate years
alternanthera *m.* *Alternanthera* sp., joy-weed
alternariose *f.* fungal disease caused by *Alternaria* sp.; **a. des feuilles** (**de carotte**) carrot blight (*A. dauci*)
alternat *m.* = **assolement** (*q.v.*)
alterne *a.* alternate; **feuilles alternes** alternate leaves
althæa, althée *m.* *Althaea* sp.
altise *f.* flea beetle; (= **puce de terre**); **a. des bois** *Phyllotreta nemorum* (one of the turnip flea beetles); **a. du chou** *P. nigripes* (turnip flea beetle); **a. noire** *P. atra* (turnip flea beetle); **a. ondulée** *P. undulata* (turnip flea beetle); **a. à pieds noirs** = **a. du chou** (*q.v.*); **a. de la vigne** grapevine flea beetle (*Haltica*

ampelophaga, H. lythri, H. chalybea) (= **pucerotte**)
altitude *f.* altitude
alun *m.* alum; **a. ammoniacal** ammonium alum; **a. de potasse** potash alum
alvéole *m.* or *f.* alveole, alveolus
alvéolé,-e *a.* honeycombed, pitted
alysse *f.*, **alysson** *m.* *Alyssum* sp., including sweet alyssum (see also **corbeille**)
amadouvier *m.* tinder fungus (*Fomes fomentarius*)
amandaie *f.* almond grove, plantation
amande *f.* almond, nut; seed of a drupe; **a. amère** bitter almond; **a. en coque** unshelled almond; **a. d'Amérique** Brazil nut; **a. de karité** shea-nut; **a. de palme** palm kernel (see also **palmiste**); **a. de terre** chufa (*Cyperus esculentus*) (= **souchet domestique**)
amandier *m.* almond tree (*Prunus amygdalus*); **a. nain** *P. nana* (= *P. tenella*); **a. d'Amérique** Brazil nut tree (*Bertholettia excelsa*); **a. des Antilles** tropical almond, Indian almond (*Terminalia catappa*)
amandon *m.* green almond; **a. naturel** shelled almond; **a. mondé** blanched almond
amanite *f.* *Fung.* *Amanita* sp.; **a. engainée** grisette (*A. vaginata* (= **grisette**); **a. phalloïde** death cap (*A. phalloides*); **a. tue-mouches** fly agaric (*A. muscaria*) (see also **oronge**)
amarante *f.* *Amaranthus* sp., amaranth; *Afr.* African spinach (*A. hybridus* and related pot herb spp.); **a. commune** love-lies-bleeding (*A. caudatus*); **a. couchée** *A. deflexus;* **a. crête-de-coq** cockscomb (*Celosia cristata*) (= **célosie**); **a. éclatante** *A. bicolor*; **a. élégante** prince's feather (*A. hypochondriacus*); **a. fleurs en queue** = **a. commune** (*q.v.*); **a. fournaise** *A. tricolor aurea;* **a. gigantesque** *A. paniculatus* var. *speciosus*; **a. queue de renard** = **a. commune** (*q.v.*); **a. réfléchie** common amaranth (*A. retroflexus*); **a. soleil levant** Joseph's coat (*A. splendens*) (see also **épinard du Soudan, fotètè**); *a. inv.* amaranth, purplish
amarantoïde *f.* globe amaranth (*Gomphrena globosa*); **a. orange, a. orangée** *G. aurantiaca*
amarre *f.* *Vit.* etc. trellis wire anchorage
amaryllidacées *f. pl.* Amaryllidaceae
amaryllis *f.* 'amaryllis' (*Hippeastrum* sp.); **a. agréable, a. belladone, a. très belle** belladonna lily (*Amaryllis belladonna*) (= **lis belladone**); **a. croix de Saint-Jacques** Jacobaea lily (*Sprekelia formosissima*) (= **lis croix de Saint-Jacques**); **a. de Guernesey** Guernsey lily (*Nerine sarniensis*) (= **nérine**); **a. impératrice du Brésil** *Hippeastrum procerum*; **a. jaune** winter daffodil (*Sternbergia lutea*); **a. de Joséphine** *Brunsvigia josephinae*; **a. de Rouen** Barbados lily (*Hippeastrum vittatum*) and hybrids; **a. saltimbanque** *Sprekelia cybister* (= *Hippeastrum cybister*); **a. de Virginie** Atamasco lily (*Zephyranthes atamasco*)
amazonien,-enne *a. n.* (*Cacao*) Amazonian
ambérique *m.* mung bean (*Vigna* (*Phaseolus*) *mungo*)
ambiant,-e *a.* ambient
ambré,-e *a.* amber-coloured
ambrette *f.* musk mallow (*Hibiscus abelmoschus*) and its seeds (see also **gombo**)
ambrevade *f.* pigeon pea (*Cajanus cajan*) (= **pois d'Ambrevade, pois d'Angole**)
ambroise (*f.*) **à feuilles d'armoise** ragweed (*Ambrosia artemisaefolia*)
ambroisie (*f.*) (**du Mexique**) Mexican tea (*Chenopodium ambrosioides*) (= **thé du Mexique**)
ambrosie *f.* *Ambrosia* sp.; **a. maritime** oak of Cappadocia (*A. maritima*)
amélanche *f.* fruit of **amélanchier** (*q.v.*)
amélanchier *m.* *Amelanchier* sp., shadbush, including *A. alnifolia*; **a. du Canada, a. à grappes** juneberry, service berry (*A. canadensis*)
amélioration *f.* improvement, amelioration, breeding (of crops, plants etc.); **a. des cultures** crop improvement; **a. génétique** genetic improvement; **a. des plantes** plant breeding; **a. structurale du sol** soil structure improvement
améliorer *v.t.* to improve, to breed (crops, plants etc.)
aménagement *m.* planning, arranging; arrangement, disposition; **plan d'aménagement** development plan
aménager *v.t.* to plan, to arrange, to fit up
amendable *a.* capable of improvement
amendage *m.* improving, reclaiming
amendement *m.* 1. improvement, reclamation, rehabilitation. 2. (soil) amendment; **a. calcaire** lime; **a. en CO_2** CO_2 supplementation; **a. humifère, a. organique** organic manure; **a. siliceux** sand
amender *v.t.* to improve, to make better; **s'amender** (of soil) to become more fertile
amentacé,-e *a.* amentaceous; *n. f.* amentaceous plant, catkin-bearing plant
amentifère *a.* amentiferous, bearing catkins
amentiforme *a.* catkin-shaped
amer,-ère *a.* bitter
amertume *f.* bitterness
améthyste *a. inv.* amethyst (colour)
ameubli,-e *a.* loosened, friable, mellowed (of soil)

ameublir *v.t. Hort.* to loosen, to break up, to mellow (soil)
ameublissement *m.* loosening, breaking up, mellowing (of soil)
ameulonner *v.t.* to stack hay, straw etc.
amiante *m.* asbestos
amide *f.* amide
amidon *m.* starch
amincir *v.t.* to make thinner; **s'amincir** to grow thinner, to become thinner
amincissement *m.* thinning down, growing thinner, becoming thinner
amine *f.* amine
aminé,-e *a.* amino; **acides aminés** amino acids
amino-acide *m.* amino acid
amitose *f.* amitosis
ammi *m. Ammi* sp.; **a. élevé** *A. majus*
ammoniac, ammoniaque *a.* ammoniacal; **gaz ammoniac** ammonia
ammoniaque *f.* ammonia
ammonisation *f.* ammonification, ammonization, conversion of humus or fertilizer organic nitrogen into ammonium salts by bacterial action
ammonitrate *m.* calcium ammonium nitrate
ammonium *m.* ammonium
amolissement *m.* softening; **a. précoce (des pommes)** early breakdown (of apples)
amome *m. Amomum* sp. (see also **cardamome**)
amorce *f.* beginning; **a. de branche** base of branch
amorphe *a.* amorphous
amorphophalle *f. Amorphophallus* sp. (Araceae)
amortissement *m.* amortization, depreciation
amour en cage *m.* bladder cherry (*Physalis alkekengi* and related spp.) (see also **alkékenge, coqueret**)
amourette *f.* quaking grass (*Briza maxima*) (see also **brize**)
amovible *a.* removable, detachable, interchangeable (of parts of machine)
ampélographe *m.* vine scientist
ampélographie *f* vine science, description of vines
ampélographique *a.* relating to vine science
ampélologie *f.* science of grape production
ampélométrie *f.* statistical study of vine characters
ampélophage *a.* & *n. m.* applied to pests of vine
ampélopsis *m. Ampelopsis* sp., ampelopsis
amphibie *a.* amphibious
amphidiploïde *a.* amphidiploid
amphigynie *f. Fung.* amphigyny
amplexicaule *a.* amplexicaule, stem clasping (= **embrassant**)
amplexiflore *a.* clasping the flower
A.M.P.S. see **abri**
amputation *f.* amputation, cutting off
amputer *v.t.* to amputate, to cut off
amygdaline *f.* amygdalin
amylacé,-e *a.* starchy
amylase *f.* amylase
amylolytique *a.* amylolytic
amyloplaste *m.* amyloplast
ANA = acide naphtylacétique NAA, naphthylacetic acid
anabolisme *m.* anabolism
anacarde *m.* cashew fruit (see also **pomme (de) cajou, noix d'acajou**)
anacarderaie *f.* cashew plantation
anacardiacées *f. pl.* Anacardiaceae
anacardier *m.* cashew tree (*Anacardium occidentale*) (= **cajou, pommier-cajou**)
anacycle (*m.*) **tomenteux** *Anacyclus tomentosus*
anaérobie *a.* anaerobic
anaérobiose *f.* anaerobiosis
analeptique *a.* & *n.m.* analeptic
analyse *f.* analysis; **a. de corrélation** correlation analysis; **a. courante** routine analysis; **a. foliaire** foliar analysis; **a probit** probit analysis; **a. de sol** soil testing; **a. statistique** statistical analysis; **a. de (la) variance** analysis of variance
analyseur *m.* analyser; **a. infrarouge** infrared analyser
anamirte *m. Anamirta* sp., Indian berry
ananas *m.* pineapple plant and fruit (*Ananas comosus*)
anaphalis *m. Anaphalis* sp., pearl everlasting
anaphase *f.* anaphase
anastomose *f.* anastomosis
anastomoser *v.t.* to anastomose; **s'anastomoser** to anastomose
anatomie *f.* anatomy
anatrope *a. Bot.* anatropous
ancestral,-e *a., m.pl.* **-aux** ancestral
ancienneté *f.* oldness, antiquity
ancipité,-e *a.* ancipital, ancipitous
ancolie *f.* columbine (*Aquilegia* sp.); **a. des Alpes** *A. alpina*; **a. des jardins, a. hybride** garden columbine, hybrid columbine; **a. hybride de caerulea** (*A. caerulea* × *A. chrysantha*)
ancrage *m.* anchorage, anchoring, fixing; **a. pivotant** tap-rooted anchorage
ancré,-e *a.* anchored, firmly rooted (of trees)
ancrer *v.t.* to anchor
andain *m.* swath, windrow
andainage *m.* windrowing
andainer *v.t.* to windrow
andaineur *m.* **andaineuse** *f.* windrower, windrowing machine
andosol *m.* andosol
androcée *m.* androecium
androgenèse *f.* androgenesis
androgénétique *a.* androgenetic

androgyne *a.* androgynous; *n. m.* androgyne
andromède *f.* andromeda (*Andromeda* sp.); **a. à feuilles brillantes** marsh andromeda (*A. polifolia*); **a du Japon** *Pieris japonica*
andromonoïque *a.* andromonoecious
andropogon *m. Andropogon* sp.
androsace *f. Androsace* sp., rock jasmine; **a. sarmenteuse** *A. sarmentosa*
androstérile *a. PB* male sterile
androstérilité *f. PB* male sterility
anémochore *a.* anemochorous
anémogamie *f.* anemogamy, wind pollination
anémomètre *m.* anemometer, for measuring wind speed
anémone *f.* anemone; **a. aimable** *Anemone blanda*; **a. des bois** wood anemone (*A. nemorosa*); **a. éclatante** scarlet windflower (*A. fulgens*); **a. des fleuristes** poppy anemone (*A. coronaria*); **a. du Japon** Japanese anemone (*A. hupehensis* (= *A. japonica*)); **a. des jardins** anemone (*A. hortensis*); **a. hépatique** hepatica (*A. angulosa*) (= *Hepatica anemone*); **a. œil-de-paon** *A. pavonina*
anémophile *a.* anemophilous
anémophilie *f.* anemophily
anergié,-e *a.* conditioned, habituated
aneth, anet *m. Anethum* sp.; **a. (odorant)** dill (*A. graveolens*); **a. doux** fennel (= **fenouil** *q.v.*)
aneuploïde *a.* aneuploid
angélique *f.* angelica (*Angelica archangelica*); **a. épineuse** Hercules' club (*Aralia spinosa*) (= **bâton du diable, massue d'Hercule**)
angioptère, angiopteris *f. Angiopteris* sp., turnip-fern
angiosperme *a.* angiospermous
angiospermes *f.* (sometimes *m.*) *pl.* angiosperms
angle (*m.*) **d'insertion** *Hort.* crotch angle
anguillule *f.* nematode, eelworm (see also **nématode**); **a. de la betterave** beet cyst nematode (eelworm) (*Heterodera schachtii*); **a. de la carotte** carrot cyst nematode (*H. carotae*); **a. des crucifères** brassica cyst nematode (*H. cruciferae*); **a. endoparasite des racines** = **a. des racines noueuses** (*q.v.*); **a. des feuilles (du fraisier), a. du fraisier** (strawberry) leaf blotch nematode (*Aphelenchoides fragariae*); **a. libre** free-living nematode; **a. du pois** pea cyst nematode (*H. goettingiana*); **a. de la pomme de terre** potato cyst nematode (*H. rostochiensis*); **a. (parasite) des racines** root lesion nematode (*Pratylenchus* sp.); **a. des racines noueuses** rot-knot nematode (*Meloidogyne* sp.); **a. de la tige** stem eelworm, **a. des tiges et des bulbes** stem and bulb nematode (*Ditylenchus dipsaci*)
anguillulose *f.* eelworm infestation
angulaire *a.* angular
anguleux,-euse *a.* angular pointed (teeth), finely ribbed (shoot); *n. m. Vit.* shoot which is polygonal in cross-section
angurie (*f.*) **(des Antilles)** West Indian gherkin (*Cucumis anguria*) (= **concombre à épines**)
angustifolié,-e *a.* angustifoliate, narrow-leaved
angustura, angusture *f.* angostura (*Cusparia febrifuga*)
anhydre *a.* anhydrous
anhydride *m.* anhydride; **a. carbonique** carbon dioxide (= **acide carbonique**); **a. sulfureux** sulphur dioxide; **a. sulfurique** sulphur trioxide
anil *m.* = **indigotier** (*q.v.*)
animal *m., pl.* **-aux** animal; **animaux nuisibles** (animal) pests
anion *m.* anion
anis *m.*, **(vert)** aniseed (*Pimpinella anisum*; **a. étoilé** star anise (*Illicium anisatum* (= **badiane**); **a. des Vosges** = **carvi** (*q.v.*)
anisé,-e *a.* flavoured with aniseed, aniseed-scented
aniser *v.t.* to flavour with aniseed
anisopétale *a.* with unequal petals
anisophylle *a.* with unequal leaves
anisotropie *f. Stat.* anisotropy
anneau *m.* ring; **a. de croissance** growth ring; **a. rouge** (*Coconut*) red ring disease (*Rhadinaphelenchus cocophilus*)
année *f.* year; **a. de cru** *Vit.* vintage year; **a du semis** year of sowing
annélation *f.* ringing (of tree), bark-ringing
annelé,-e *a.* ringed, annulated
annexe *f. Bot.* accessory part (of structure), appendage
annuel,-elle *a.* annual; **plante annuelle** annual (plant); *n.f.* annual (plant)
annulaire *a.* annular, ring-shaped
annulation (*f.*) **des bourgeons** bud removal
anomalie *f.* anomaly, abnormality; **a. de croissance** growth anomaly
anonacées *f. pl.* Annonaceae
anone *f. Annona* sp.; **a. écailleuse** sweetsop, sugar apple (*A. squamosa*); **a. des marais** pond apple (*A. glabra*); **a. (muriquée)** soursop (*A. muricata*) (= **corossol**) (see also **chérimolier, cœur-de-bœuf, corossolier, pommier-cannelle**)
anorexigène *m. Ent.* antifeedant
anoxie *f.* anoxia, oxygen deprivation; **en anoxie** in anaerobic conditions
anoxopus *m.* carpet grass (*Anoxopus compressus*)
anoxybiose *f.* oxygen deprivation
anse *f.* handle (of basket etc.)
ansérine *f. Chenopodium* sp.; **a. amarante** red goosefoot (*C. amaranticolor*); **a. blanche** fat hen (*C. album*); **a. bon-henri** Good King

Henry (*C. bonus-henricus*) (= **épinard sauvage**); **a. quinoa** quinoa (*C. quinoa*) (= **quinoa**)

antagonisme *m.* antagonism

antagoniste *a.* antagonistic; *n.* antagonist

antaque *f.* hyacinth bean (*Dolichos lablab*) (= **pois indien**)

antennaire *f. Antennaria* sp., cat's ear

anthémis *f. Anthemis* sp., camomile; marguerite, oxeye daisy (*Chrysanthemum leucanthemum*); marguerite, Paris daisy (*C. frutescens*); **a. d'Arabie** *Cladanthus arabicus* (see also **marguerite**)

anthère *f.* anther

anthéridie *f.* antheridium

anthérifère *a.* anther-bearing

anthérozoïde *m.* antherozoid

anthèse *f.* anthesis

anthocyane *m.*, **anthocyanine** *f.* anthocyanin

anthocyané,-e *a.* containing anthocyanins

anthogène *a.* relating to factors determining flowering (= **florigène**)

anthogenèse *f.* anthogenesis

anthonome *m. Anthonomus* weevil; **a. du fraisier** strawberry blossom weevil (*Anthonomus rubi*); **a. du poirier** apple bud weevil (*A. piri*); **a. du pommier** apple blossom weevil (*A. pomorum*)

anthophage *a.* flower-eating

anthophile *a.* commonly found on flowers

anthracnose *f.* anthracnose; **a. des agrumes** *Cit.* anthracnose (*Colletotrichum gloeosporioides*); **a. des bananes** banana anthracnose (*C. musae*); **a. du caféier, a. des baies du caféier**, coffee berry disease, CBD (*C. coffeanum*); **a. du concombre, a. des cucurbitacées** cucumber anthracnose (*C. lagenarium*) (= **nuile rouge**); **a. du framboisier** cane spot of raspberry (*Elsinoë veneta*); **a. du haricot** bean anthracnose (*C. lindemuthianum*); **a. de la laitue** lettuce ring spot (*Marssonina panattoniana*); **a. du muguet** lily-of-the-valley anthracnose (*Gloeosporium convallariae*); **a. de l'oignon** onion smudge (*C. circinans*); **a. du pois** pea leaf and pod spot (*Ascochyta pisi*); **a. du rosier** rose leaf scorch (*Sphaerulina rehmiana*); **a. (maculée) de la vigne** vine anthracnose (*Elsinoë ampelina = Gloeosporium ampelophagum*) (= **rouille noire**)

anthraquinone *f.* anthraquinone

anthrisque (*f.*) **sylvestre** corn chamomile (*Anthemis arvensis*) (see also **camomille, maroute, matricaire**); **a. vulgaire** bur chervil (*A. vulgaris = A. neglecta*)

anthure *m.* flamingo flower (*Anthurium* sp.)

antiacridien,-enne *a.* **lutte antiacridienne** locust control

antiaphidien,-enne *a.* **lutte antiaphidienne** aphid control

antiappétant *m. Ent.* etc. antifeedant

antiar, antiaris *m.* upas tree (*Antiaris toxicaria*)

anti-auxine *f.* growth-inhibitor

antiaviaire *a.* anti-bird; **lutte antiaviaire** *Afr.* etc. bird control (see also **pétard**)

antibactérien,-enne *a.* bactericidal

antibiotique *a.* antibiotic; *n. m.* antibiotic

antibotrytis *a. inv.* for *Botrytis* control

antibuée *a.* anti-condensation; *n.m.* product to prevent condensation

anticapside *a.* anti-capsid; **lutte anticapside** capsid control

antichlorose *a. inv.* for chlorosis control

antichoc *a. inv.* anti-shock

anticipé,-e *a. Hort.* which develops before the normal time (either as a result of pruning or spontaneously etc.); *n. m. Hort.* feather, lateral shoot on current year's extension growth (= **rameau anticipé**) (see also **bourgeon anticipé**)

anticorps *m.* antibody

anti-coulure *a. PD* anti-coulure, for coulure prevention

anticryptogamique *a.* applied to compounds controlling cryptogams, especially fungi; **traitement anticryptogamique** fungicidal treatment (see also **antifongique**); *n. m.* fungicide

antidérapant,-e *a.* non-skidding

antidicotylé,-e *a.*, **action antidicotylée (d'un herbicide)** broad-leaved weed control action (of herbicide)

antidicotylédones *a.*, **herbicide antidicotylédones** herbicide for the control of broad-leaved weeds (see also **dicotylédonicide**)

anti-échaudure *a. PD* anti-scald

antiérosif,-ive *a.* anti-erosion

antifongique *a.* fungicidal, fungistatic (see also **anticryptogamique**)

antigel, antigelées *a. inv.*, **lutte antigel** frost protection; **réchaud antigelées** oil burner (for orchards, vineyards etc.); **antigel** *n. m.* antifreeze

antigène *m.* antigen

antigraminées *a.* for grass control; *n. m.* herbicide for grass control, grass killer

antigrêle *a.* anti-hailstone, affording protection against hail

antilépreux,-euse *a.* anti-leprosy; **plante antilépreuse** plant with anti-leprosy properties

anti-limace *a.* for slug control (see **granulé**)

anti-oiseaux *a.* anti-bird; **filet anti-oiseaux** net for bird protection

antiparasitaire *a.* anti-pest; **lutte antiparasitaire** pest control; *n. m.* pesticide

antiparasiter *v.t.* *PC* to control, to apply pesticides
anti-poussière *a.* for dust protection; **masque anti-poussière** dust mask
anti-pucerons *a. inv.*, **abri anti-pucerons** aphid-free shelter
antiputride *a.* = **antiseptique** (*q.v.*)
antirrhine *f.* *Antirrhinum* sp., snapdragon (= **gueule-de-lion, gueule-de-loup, muflier**)
antiscorbutique *a.* antiscorbutic
antiseptique *a.* & *n. m.* antiseptic
antisérum *m.* antiserum
antispasmodique *a.* & *n. m.* antispasmodic
antisporulent *m.* spore inhibitor
antisudorifique *a.* & *n. m.* antitranspirant (= **antitranspirant**)
antitavelure *a.* anti-scab (see **tavelure**)
antitranspirant,-e *a.* antitranspirant; *n. m.* antitranspirant (= **antisudorifique**)
antitumoral,-e *a.*, *m. pl.* **-aux** with anti-tumour activity; **plante antitumorale** plant with anti-tumour activity
antiverse *a.* for lodging control; **à action antiverse** with anti-lodging action
A.O.C. *Vit.* = **appellation d'origine contrôlée** (see **appellation**)
aoûté,-e *a.* ripened, lignified
aoûtement *m.* effect of a favourable temperature for ripening; ripening, lignification
aoûter *v.t.* to ripen, to lignify
apalachine *f.* Appalachian tea (*Ilex vomitoria, I. glabra*)
apate *m.* vine bostrychid beetles (*Sinoxylon sexdentatum, Schistocerus bimaculatus*) (= **bostryche**)
apérianthé,-e *a.* without perianth
apérispermé,-e *a.* without perisperm
apétale *a.* apetalous
apex *m.* apex; **a. caulinaire** stem apex
aphicide *a.* aphicidal; *n. m.* aphicide
aphide *m.* aphid, greenfly (see **aphidien**)
aphidés *m. pl.* see **aphidien**
aphididés *m. pl.* see **aphidien**
aphidien,-enne *a.* relating to aphids (see also **antiaphidien**); *n. m.* aphid, *pl.* Aphididae (see also **puceron**)
aphidiphage *a.* aphid-eating, aphidophagous; **coccinelle aphidiphage** aphid-eating ladybird
aphis *m.* aphid, greenfly (= **aphidien**)
aphylle *a.* aphyllous, leafless
api *m.*, **pomme d'api** small apple, lady apple (see also **pomme**)
apical,-e *a.*, *m. pl.* **-aux** *Bot.* apical; **méristème apical** apical meristem
apicole *a.* apiarian; **exploitation apicole** beekeeping enterprise
apiculé,-e *a.* apiculate
apiculteur *m.* apiculturist, beekeeper
apiculture *f.* apiculture, beekeeping
apion *m.* *Apion* sp., weevil; **a. du trèfle** clover seed weevil (*A. aestivum, A. apricans, A. assimile*)
apios *m.* *Apios* sp., including potato bean (*A. tuberosa*) (= **glycine tubéreuse**)
aplanir *v.t.* to level, to flatten, to smooth out
aplatir *v.t.* to make flat, to flatten
aplatissement *m.* flattening
aplomb *m.* perpendicularity, uprightness
apocarpé,-e *a.* apocarpous
apocyn, apocin *m.* *Apocynum* sp., including spreading dogbane (*A. androsaemifolium*) (= **gobe-mouche**)
apocynacées *f. pl.* Apocynaceae
apogamie *f.* apogamy
apomictique *a.* apomictic
apomixie *f.* apogamy, apomixis
aponogéton *m.* *Aponogeton* sp.; **a. à deux épis** cape pondweed (*A. distachyus*)
apoplexie *f.* *PD* apoplexy, apoplectic death; **a. de l'abricotier** apricot apoplexy; **a. parasitaire** *Vit.* vine apoplexy (caused by *Stereum hirsutum* and *Phellinus ignarius*) (= **esca**)
apozème *m.* vegetable broth, decoction
appareil *m.* apparatus, equipment, implement; **a. à dos** knapsack-type equipment; **a. d'éclairage** lighting apparatus, lighting outfit; **a. de Golgi** Golgi apparatus; **a. de pulvérisation** sprayer
appareillage *m.* 1. installation, setting up, 2. equipment, accessories
appariement *m.* 1. matching 2. coupling, pairing; **a. chromosomique** chromosome pairing
appartement *m.* flat; *Hort.* **en appartement** indoors; **plante d'appartement** indoor plant, house plant
appât *m.* bait; *PC* **a. empoisonné, a. toxique** poison bait
appauvrissement *m.* impoverishment
appellation *f.* name, term, trade name; **a. d'origine** name specifying and guaranteeing the origin of a horticultural product; **a. contrôlée, a. d'origine contrôlée** (= **A.O.C.**) *Vit.* guaranteed vintage; **sans appellation** *Vit.* non-vintage; **vin d'appellation** *Vit.* vintage wine
appel-sève *m.*, *pl.* **appels-sève** sap-drawer
appendiculé,-e *a.* appendiculate
appertisation *f.* conservation of food by heating to 80–120°C. (as devised by N. F. Appert), autoclaving
appertisé,-e *a.* heat-treated (by the Appert process), autoclaved
appétibilité *f.* palatability
application *f.* application; **a. d'engrais** fertilizer or manurial dressing; **a. en couronne** circular

dressing; **a. en couverture** top dressing; **a. fractionnée** split dressing; **a. ponctuelle** spot application

appoint *m.* added portion; **lumière d'appoint** extra lighting

apport *m.* supply, application, input; **a. d'eau** irrigation, watering; **a. (d'engrais)** *Hort.* fertilizer, manurial dressing; **a. des glaciers** glacial drift

appréciation *f.* estimate, appraisement, valuation

appressé,-e *a.* appressed, adpressed

apprimé,-e *a. Bot.* adpressed, appressed

approvisionnement *m.* provisioning, supplying; **a. régulier** regular supply; **a. en eau** water supply

âpre *a.* rough, harsh; bitter

après-vente *a.* after-sale; **service après-vente** after-sale service

âpreté *f.* roughness, harshness; bitterness

aptère *a. Ent.* apterous

aptitude *f.* aptitude; **a. à la combinaison** combining ability; **a. à l'établissement** (of plant) establishing ability

apyrène *a.* without seed, seedless

apyrénie *f.* seedlessness (of fruit)

aquaculture *f.* hydroponics, water culture

aqueux,-euse *a.* aqueous

arabette *f.* garden arabis (*Arabis caucasica*) (= **corbeille d'argent**) and other spp.); **a. des Alpes** alpine rock-cress (*A. alpina*); **a. des sables** *A. arenosa*; **a. de Thalie** Thale cress (*Arabidopsis thaliana*) (= **arabidopsis**)

arabicaculture *f. Coff.* arabica coffee growing

arabidopsis *m.* Thale cress (*Arabidopsis thaliana*) (see also **arabette**)

arabis *f.* = **arabette** *q.v.*

arable *a.* arable; **couche arable** tilth, topsoil, plough layer

aracées *f. pl.* Araceae

araignée *f.* spider, arachnid; **a. rouge, a. jaune** red (spider) mite (*Tetranychus urticae*) (also applied to *Panonychus*, *Eotetranychus* and other *Tetranychus* spp.); **a. rouge du groseillier** gooseberry bryobia (*Bryobia ribis*)

araire *m.* swing plough; **a. vigneron** vineyard plough

aralia *m. Aralia* sp., aralia; **a. des appartements** threadleaf (*Dizygotheca* sp.)

araliacées *f. pl.* Araliaceae

aramon *m. Vit.* Aramon cultivar

aranéeux,-euse *a.* araneous, cobweb-like

aratoire *a.* agricultural, relating to agriculture

araucaria *m.* **a. du Chili** Chilean pine, monkey puzzle (*Araucaria araucana*) (= *A. imbricata*) and other spp., including Norfolk Island pine (*A. excelsa*)

arboise *f.* fruit of **arbousier** (*q.v.*) (= **arbouse**)

arborescence *f.* arborescence, tree-like growth

arborescent,-e *a.* arborescent

arboretum *m.* arboretum

arboricide *m.* arboricide

arboricole *a.* arboreal; relating to trees *Bot.* arboricolous, growing on trees

arboriculteur *m.* arboriculturist, (tree and shrub) nurseryman, fruit grower

arboriculture *f.* arboriculture; **a. fruitière** fruit growing; **a. ornementale** ornamental tree growing

arboriforme *a.* shaped like a tree

arboriser *v.i.* to grow trees

arboriste *m.* tree expert

arbouse *f.* fruit of **arbousier** (*q.v.*) (= **arboise**)

arbousier *m.* strawberry tree (*Arbutus unedo*) (= **arbre aux fraises**); **a. nain** black bearberry (*Arctostaphylos alpina*)

arbre *m.* tree; **a. d'abri** nurse tree; **a. à ail** garlic tree (*Scorodophloeus zenkeri*); **a. d'alignement** avenue, roadside tree; **a. d'argent** silver tree (*Leucadendron argenteum*); **a. d'arrière cour** dooryard tree; **a. aveuglant** blinding tree (*Excoecaria agallocha*); **a. à beurre** Indian butter tree (*Madhuca butyracea*); *Afr.* shea-nut tree (*Butyrospermum paradoxum*) (= **karité**); **a. bouteille** Queensland bottle tree (*Sterculia rupestris*); **a. à caoutchouc** rubber tree; **a. à chapelets** bead tree (*Melia azedarach*); **a. à châtaigne** (*West Indies*) (seeded form of) breadfruit (*Artocarpus altilis* var. *seminifera*); **a. à cire** wax myrtle (*Myrica cerifera*); **a. aux cloches d'argent** *Halesia* sp.; **a. au corail** coral tree (*Erythrina corallodendron* and other spp.); **a. demi-tige** half-standard (= **demi-tige**); **a. à encens** frankincense (*Boswellia* sp.); **a. en espalier** espalier; **a. étalon** control tree; **a. de feu** *Acer ginnala*; **a. à feuilles persistantes** evergreen tree, evergreen; **a. feuillu** broadleaved tree; **a. forestier** forest tree; **a. aux fraises** = **arbousier** (*q.v.*); **a. franc de pied** tree on its own roots; **a. fruitier, a. à fruit** fruit tree; **a. fruitier à pépin** pome fruit tree; **a. fruitier à noyau** stone fruit tree; **a. à fût décroissant** tapering tree; **a. greffé** grafted, worked tree; **a. greffé une fois** single worked tree; **a. greffé deux fois** double-worked, topworked tree; **a. de Judée** Judas tree (*Cercis siliquastrum*) (= **gainier**); **a. à lait** cow tree (*Brosimum utile*); **a. au mastic** mastic (*Pistacia lentiscus*) (= **lentisque**); **a. (-)mère** (*m.*) mother tree, stock plant; **a. aux mouchoirs** pocket-handkerchief tree (*Davidia involucrata*); **a. nain** dwarf tree; **a. nain conduit en buisson** dwarf bush tree; **a. de neige** fringe-tree (*Chionanthus virginica*); **a. de Noël** Christmas tree (*Picea abies*); **a.**

d'ombrage shade tree; **a. d'ornement** ornamental tree; **a. à pain** breadfruit (*Artocarpus communis*); **a. à pain de l'Afrique** African breadfruit tree (*Treculia africana*); **a. permanent** permanent tree; **a. à (la) perruque, a. à perruques** wig tree, smoke tree, Venetian sumach (*Cotinus coggygria* = *Rhus cotinus*); **a. -piège** tree felled to serve as a trap for insect pests which are later destroyed; **a. pleureur** weeping tree; **a. aux pochettes** = **a. aux mouchoirs** (*q.v.*); **a. au poivre** chaste tree (*Vitex agnus-castus*) (= **agneau-chaste, gattilier**); **a. aux quarante écus** ginkgo (*Ginkgo biloba*); **a. aux quatre épices** Madagascar nutmeg (*Ravensara aromatica*); **a. pompadour** Carolina allspice (*Calycanthus floridus*); **a. riz de veau** akee (*Blighia sapida*); **a. saint** bead tree (*Melia azedarach*) (= **a. à chapelets**); **a. à semelles** owala oil tree (*Pentaclethra macrophylla*); **a. à soie** silk-cotton, kapok tree (*Ceiba, Bombax* spp.); **à de soie** pink siris (*Albizzia julibrissin*); **a. à suif** Chinese tallow tree (*Sapium sebiferum*); *Virola sebifera* (see also **virole**); **a. à suif du Gabon** false nutmeg (*Myristica kombo*); **(a.) temporaire** filler (in orchard) (= **temporaire**); **a. tige** standard; **a. à tomates** tree tomato (*Cyphomandra betacea*); **a. à Tolu, a. au baume de Tolu** Tolu balsam tree (*Myroxylon toluiferum*); **a. transplanté** transplanted tree, transplant; **a.-tuteur** (*pl.* **arbres-tuteurs**) supporting tree (for vanilla etc.); **a. type** sample tree; **a. à la vache** = **a. à lait** (*q.v.*); **a. (toujours) vert** evergreen; **a. du voyageur, a. des voyageurs** traveller's tree (*Ravenala madagascariensis*)

arbreux,-euse *a.* woody, wooded

arbrisseau *m. pl.* **-eaux** shrubby tree, shrub, bush; **a. à fruits** bush fruit tree; **plantation d'arbrisseaux** shrubbery

arbuscule *m. Fung.* arbuscule

arbuste *m.* bush, shrub; **plantation d'arbustes** shrubbery; **a. à feuillage, a. feuillu** ornamental foliage shrub; **a. à fleurs** ornamental flowering hrub; **a. à forcer** shrub for forcing; **a. à fruits décoratifs** ornamental fruit shrub; **a. d'ornement** ornamental decorative shrub; **a. aux papillons** butterfly bush (*Buddleia* sp.)

arbustif,-ive *a.* shrubby; **plantation arbustive** shrubbery

arc-boutant *m.* strut (see also **greffe**)

arceau *m. pl.* **-eaux** arch (of vault), hoop; arbor; *Hort.* arched shoot; *Vit.* fruiting vine branch bent over and tied (= **archelot, archet, archot**)

archégone *m.* archegonium

archelot *m.* see **arceau**

archet *m.* see **arceau**

archot *m.* see **arceau**

arçon *m. Vit.* 1. shoot retained at pruning, 2. = **arceau** *q.v.*

arctotis *m. Arctotis* sp.

arcure *f. Hort.* arching, bending method of training, arcure; **a. (système) Lepage** Lepage arcure; *Vit.* bending a shoot to encourage fruiting; (*Tea*) pegging

ardoise *f.* slate

ardoisé,-e *a.* slate-coloured; **fève ardoisée** (*Cacao*) badly fermented bean

are *m.* unit of metric land measure, 100 m^2 (about 119.6 square yards)

aréage *m.* land-surveying in **ares**

arec, areca *m.* areca palm (*Areca catechu*) (= **aréquier**); **noix d'arec** areca nut

arénaire *a.* living in sand

arénation *f.* covering with sand

arène *f. Ped.* **a. (granitique)** quartz sand

aréneux,-euse *a.* sandy

arénicole *a.* arenicolous, growing or living in sand

aréole *f.* areola

aréomètre *m.* hydrometer

aréquier *m.* areca palm (*Areca catechu*) (= **arec**)

arête *f.* fishbone; **en arête de poisson** in a herringbone pattern; *Bot.* awn, arista; **a. collectrice** *Irrig.* main collecting channel

arganier *m.* argan tree (*Argania sideroxylon*)

argémone *f.* argemony, prickly poppy (*Argemone grandiflora*)

argent *m. PD* silver streak virus

argenté,-e *a.* silvery; *Coff.* **pellicule argentée** silver-skin

argentine *f.* silverweed (*Potentilla anserina*) (= **potentille argentée**); Venus' navelwort (*Omphalodes linifolia*) (= **gazon blanc**)

argenture *f.* **(de la tomate)** *PD* silvering (of tomato)

argilacé,-e *a.* clayey, argillaceous

argile *f.* clay; **a. colloïdale** colloidal clay; **a. expansée** expanded clay

argileux,-euse *a.* clayey

argilière *f.* clay-pit

argilifère *a.* clay-bearing

argilisation *f.* enrichment of a soil in clay by illuviation

argot *m. Hort.* stub

argoter *v.t. Hort.* to cut off the tip of a dead branch, to snag

argousier *m.* sea buckthorn (*Hippophaë rhamnoides*) (= **a. faux nerprun, saule épineux**)

aride *a.* arid

aridité *f.* aridity

aridoculture *f.* dry culture

arille *m. Bot.* aril

arillé,-e *a. Bot.* arillate, arillated

arisème *m. Arisaema* sp.

aristé,-e *a. Bot.* aristate, awned
aristoloche *f. Aristolochia* sp., birthwort; **a. serpentaire** Virginia snakeroot (*A. serpentaria*)
aristolochiacées *f. pl.* Aristolochiaceae
armature *f.* framework, armature; **a. en fil de fer** wire framework
armérie, armeria *f. Armeria* sp., thrift; **a. maritime** sea thrift (*A. maritima*)
armillaire (*f.*) **(couleur) de miel** honey fungus (*Armillaria mellea*)
armoirie *f.* Deptford pink (*Dianthus armeria*) (see also **oeillet**)
armoise *f. Artemisia* sp., mugwort, wormwood (= **herbe de Saint-Jean**); **a. (commune)** mugwort (*Artemisia vulgaris*); **a. à santonine** sea wormwood (*A. maritima*) (see also **semen-contra**); **a. blanche** (*Mauritius*) *Leucas aspera*; **a. mauve** (*Mauritius*) *Leonurus sibiricus*
armoracia *f.* horse-radish (*Armoracia rusticana*) (= *Cochlearia armoracia*)
armure *f.* armour; **a. d'un arbre** tree-guard
ARN = **acide ribonucléique** RNA, ribonucleic acid; **ARN messager** messenger RNA
arnébie *f.* Arabian primrose (*Arnebia cornuta, Arnebia* sp.); **a. vipérine** prophet flower (*A. echioides*) (= **fleur du prophète**)
arnica, arnique *f.* arnica plant; **a. (des montagnes)** mountain tobacco (*Arnica montana*)
aroidacées *f. pl.* Araceae
arole, arolle *m.* arolla pine (*Pinus cembra*) (= **pin cembro**)
aromate *m.* aromatic, spice (see also **épice**)
aromatique *a.* aromatic
aromatiser *v.t.* to give aroma to, to aromatize
arôme *m.* aroma
aronie *f. Aronia* sp., chokeberry
arpent *m.* ancient measure of area: in Mauritius = 1.043 acres
arpentage *m.* land-surveying, field-surveying
arpenteuse *f. Ent.* looper caterpillar (= **géomètre**)
arqué,-e *a.* arched, bent over, curved; **feuille arquée** fornicate leaf
arquer *v.t.* to arch, to bend
arrachable *a.* capable of being uprooted
arrachage *m.* digging up, grubbing, lifting, pulling up, uprooting; **a. de la pépinière** lifting of nursery stock; **a. en motte** *Plant.* etc. lifting (plant) with a ball of soil; **a. à racines nues** *Plant.* etc. bare-root lifting (of plant); **matériel d'arrachage** plant-lifting equipment
arrache-fanes *m.* haulm-puller
arracher *v.t.* to lift, to pull up, to uproot; **a. à la charrue** to plough out
arrache-racine(s) *m.* hoe (for uprooting weeds)
arracheur *m.*, **arracheuse** *f.* lifter, uprooting machine; **a. de carottes** carrot harvester; **a. d'oignons** onion harvester; **a. de pommes de terre** potato digger, harvester; **a. attelée** (animal-) drawn lifter; **arracheuse-aligneuse de pommes de terre** potato harvester and windrower; **arracheuse avec élévateur** elevator digger; **arracheuse-calibreuse de céleri en branches** celery harvester-grader; **arracheuse-chargeuse, arracheuse-ramasseuse** harvester and loader
arrachis *m.* 1. uprooting; **plantes en arrachis** uprooted plants, 2. cleared ground, 3. uprooted plant(s)
arrachoir *m.* uprooting implement
arrêt (*m.*) **de croissance, a. de végétation** interruption, cessation of growth, check to growth
arrête-boeuf *m. inv.* rest-harrow (*Ononis* sp.)
arrière-action *f.*, *pl.* **arrière-actions** *Hort.* residual effect (e.g. of fertilizer)
arrière-effet *m.*, *pl.* **arrière-effets** after-effect, residual effect
arrière-fleur *f.*, *pl.* **arrière-fleurs** second flowering
arrière-goût *m.*, *pl.* **arrière-goûts** after-taste
arrière-saison *f.*, *pl.* **arrière-saisons** late autumn, back-end of year
arrimage *m.* stowing, stacking, making secure
arrimer *v.t.* to stack, to stow, to make secure
arrimeur *m.* stower, stacker
arroche *f.* 1. orach(e) (*Atriplex* sp.) (= **bonne-dame**); **a. des champs, a. étalée** common orache (*A. patula*); **a. halime** Mediterranean saltbush (*A. halimus*) (= **pourpier de mer**); **a. des jardins** garden orache, mountain spinach (*A. hortensis*) (= **belle dame, follette**). 2. **a. bon-henri** Good King Henry (*Chenopodium bonus-henricus*) (= **(ansérine) bon-henri**)
arrondi,-e *a.* rounded, round
arrosable *a.* capable of being watered
arrosage *m.* watering, sprinkling, spraying, irrigation; **a. par aspersion, a. en pluie** sprinkling, sprinkler irrigation; **a. goutte à goutte** drip irrigation
arroser *v.t.* to water; **a. par aspersion** to sprinkle; **a. au goulot** to water with the spout of the watering can; **a. en pluie** to water with a rose on the watering can, to sprinkle; **"arrosez abondamment"** "water copiously"
arroseur,-euse *n.* person who waters; watering machine, sprinkler; **arroseur automatique** automatic sprinkler; **arroseur à brise jet tournant** rotating sprinkler with intermittent jet; **arroseur fixe** fixed sprinkler; **arroseur rotatif** rotary sprinkler; **arroseur oscillant**

oscillating sprinkler; **arroseuse** (*f.*) **par aspersion** irrigation sprinkler; **a. rapide** speed sprayer
arrosoir *m.* watering can; **a. automatique** automatic sprinkler; **a. horticole** of 8–13 litre capacity; **a. de serre** 3–5 litres; **a. de dame** ½ to 3 litres
arrow-root *m.* arrowroot (*Maranta arundinacea*) and related species; **a. -r. de Tahiti** pia (*Tacca pinnatifida*); starch prepared from arrowroot rhizomes (see also **topinambour**)
arséniate *m.* arsenate; **a. de plomb** lead arsenate
arsénical,-e *a.*, *m. pl.* **-aux** arsenical; *n. m. pl. Coll.* arsenical pesticides
art (*m.*) **floral** flower arrangement, floral art
artémisia, artémisie *f. Artemisia* sp. (see also **armoise**)
artésien,-enne *a.* & *n.* artesian, from Artois; **puits artésien** artesian well
arthropode *m.* arthropod; *pl.* Arthropoda, arthropods
artichaut *m.* 1. (globe) artichoke (*Cynara scolymus*) 2. **a. d'Espagne** squash (*Cucurbita pepo*); **a. d'hiver** *Coll.* Jerusalem artichoke (= **topinambour** *q.v.*); **a. de Jerusalem** gourd, squash (= **courge, pâtisson**); **a. des toits** *Coll.* houseleek (*Sempervivum* sp.) (= **joubarbe**)
artichautière *f.* (globe) artichoke bed, field
article *m. Bot.* article
artocarpe *m. Artocarpus* sp. including breadfruit (*A. communis*) (= **arbre à pain**)
arum *m. Arum* and other spp.; **a. d'Éthiopie** arum lily (*Zantedeschia aethiopica*); **a. jaune d'Afrique** golden arum (*Z. elliotiana*); **a. maculé, a. tacheté** cuckoo pint (*A. maculatum*)
arundo *m.* reed (*Phragmites (Arundo) communis*)
asaret *m.* European wild ginger (*Asarum europaeum*)
ascendance *f.* ancestry
ascendant,-e *a. Bot.* ascending
ascidie *f. Bot.* ascidium
asclépiadacées *f. pl.* Asclepiadaceae
asclépiade *f. Asclepias* sp.
ascomycètes *m. pl.* Ascomycetes
asépale *a.* asepalous
aseptique *a.* aseptic, sterile
aseptisé,-e *a.* asepticized
asexué,-e *a.* asexual
asexuel,-elle *a.* asexual
asiminier *m.* American pawpaw (*Asimina triloba*)
asparagiculture *f.* asparagus growing, culture
asparagine *f.* asparagine
asparagus *m. Asparagus* sp., asparagus fern
aspect *m.* aspect; **a. économique** economic aspect
asperge (comestible) *f.* asparagus (*Asparagus officinalis*); **a. des fleuristes** asparagus fern (*A. plumosus, A. sprengeri*); **botte d'asperges** bundle of asparagus; **pointes d'asperges** asparagus tips
aspergement *m.* light watering or sprinkling
asperger *v.t.* to water lightly, to sprinkle
aspergeraie, aspergerie, aspergière *f.* asparagus bed, asparagus field
aspergilliforme *a. Bot.* shaped like a bottlebrush
aspergillus *m. Aspergillus* sp.
asperme *a. Bot.* aspermous, seedless
aspermie *f. Bot.* seedlessness
asperseur *m.* sprinkler; **a. à secteurs** (adjustable) rotary sprinkler
aspersion *f.* sprinkling, spraying, sprinkler irrigation; **a. atomisée** mist blowing, atomizing
aspersoir *m.* rose of watering can
aspérule *f. Asperula* sp.; **a. des champs** blue woodruff (*A. arvensis*); **a. à esquinancie** squinancywort (*A. cynanchica*); **a. odorante** sweet woodruff (*A. odorata* = *Galium odoratum*) (= **petit muguet**)
asphodèle *m. Asphodelus* sp., asphodel
asphyxie *f.* asphyxia; **a. racinaire, a. radiculaire, a. des racines** root asphyxia
aspic *m. Bot.* French lavender (*Lavandula latifolia*)
aspidistra *m.* aspidistra (*Aspidistra lurida, A. typica*)
aspirant,-e *a.* sucking; **pompe aspirante** suction pump
aspirateur *m.* suction machine
aspiration *f.* inhaling, suction
aspirer *v.t.* to inhale, to suck up, to suck in
asplenium *f. Asplenium* sp.; **a. capillaire** maidenhair spleenwort (*A. trichomanes*) (see also **doradille**)
asque *m.* ascus
assa *f.* asafoetida (gum resin from *Ferula* spp.)
assainir *v.t.* to reclaim (land), to make productive
assainissement *m.* (land) reclamation (including drainage), putting into production; **a. viral** *PD* virus elimination
assaisonnant,-e *a.* seasoning; **plantes assaisonnantes** seasoning plants
assaisonnement *m.* seasoning
assaisonner *v.t.* to season
assec *m.* period during which a pond or floodplain used for fishing or pisciculture is drained for cultivation (see also **évolage**)
assèchement *m.* drying, desiccation
assécher *v.t.* to dry, to drain, to pump dry; **s'assécher** *v.i.* to dry (up), to become dry

asseoir *v.t. Hort.* to bed, to bed down
assimilabilité *f.* assimilability, availability
assimilable *a.* assimilable, available
assimilat *m.* assimilate
assimilateur,-trice *a.* assimilating; **fonctions assimilatrices** assimilating functions; *n.* assimilator
assimilation *f.* assimilation; **a. de gaz carbonique** CO_2 assimilation
assise *f.* layer, bed, stratum; **a. cambiale, a. génératrice** *Bot.* cambium; **a. pilifère** piliferous layer
assiselage *m. Vit.* layering
association *f.* association; **a. végétale** plant association
associé,-e *a.* associated; **culture associée** mixed crop, mixture of crops
assolement *m. Hort.* cropping plan, rotation system; **a. libre** loose system; **a. triennal** three-course system (see also **rotation**)
assoler *v.t. Hort.* to rotate (crops)
aste *m. Vit.* fruiting branch
aster *m.* aster, Michaelmas daisy; **a. de Chine** China aster (*Callistephus chinensis*) (= **reine-marguerite**); **a. nain** *Aster dumosus* (see also **œil-du-Christ**)
asticot *m.* maggot
astilbe *m.* astilbe (*Astilbe* sp.), goat's beard
astragale *m. Astragalus* sp.
astringence *f.* astringency
astringent,-e *a.* & *n. m.* astringent, binding
asymétrique *a.* asymmetrical
atanga *m.* (*Gaboon*) fruit of **atangatier** (*q.v.*)
atangatier *m.* (*Gaboon*) bush butter tree (*Dacryodes edulis*) (= **safoutier**)
atavisme *m.* atavism
atelier *m.* workshop, workroom, studio; **a. d'emballage** packing shed
athérure *m. Afr.* brush-tailed porcupine (*Atherurus* sp.) (see also **porc-épic**)
atmomètre *m.* atmometer
atmosphère *f.* atmosphere; **a. artificielle, a. contrôlée** controlled atmosphere; **a. du sol** *Ped.* soil atmosphere
atoca *m.* (*Canada*) cranberry (*Vaccinium* sp.) (see also **airelle, bleuet 2, canneberge**)
atome *m.* atom; **atomes d'oxygène** oxygen atoms
atomisation *f.* (low volume) spraying
atomiser *v.t.* to atomize, to spray
atomiseur *m.* atomizer, sprayer; **a. à dos** knapsack sprayer
atractyle *f. Atractylis* sp. including *A. gummifera*
atramentaire *a.* atramental
atriplex *m. Atriplex* sp. (see also **arroche**)
atropa, atrope *f. Atropa* sp. (see **belladone**)
attachage *m.* tying
attache *f.* tie, fastener; **a. de plastique** plastic tie
attacus *m. Ent. Attacus* sp.; **a. de l'ailante** *A. cynthia*
attaque *f.* attack
attaquer *v.t.* to attack
attélabe *m.* vine weevil (*Byctiscus (Rhynchites) betulae*)
attelage *m.* attachment, coupling (of implement); **a. de remorque** coupling for trailed or drawn implement; **a. trois points** three-point attachment
attendrir *v.t.* to make tender, tenderize
attendrissement *m.* tenderization
atténué,-e *a.* attenuated, attenuate
attiéké *m.* (*Ivory Coast*) garri (processed cassava)
attier *m.* sweetsop (*Annona squamosa*); soncoya (*A. purpurea*) (see **anone**)
attractif *m. Ent.* attractant; **a. sexuel synthétique** synthetic sex attractant
attrape-mouche(s) *m.* Venus' fly-trap (*Dionaea muscipula*) (= **dionée**) and other insect-catching plants e.g. *Helicodiceros muscivorus, Arum* spp. (see also **gobe-mouche**)
aube *f.* dawn
aubépine *f.* hawthorn (*Crataegus oxyacantha* and other *Crataegus* spp.) (see also **azerolier**); **a. américaine** Mexican hawthorn, stone plum (*C. mexicana*) (= **prune de pierre**); **a parasol** willow-leaved cockspur thorn (*C. crus-galli* var. *salicifolia*) (see also **ergot-de-coq**)
aubergine *f.* aubergine, eggplant (*Solanum melongena*); **a. sauvage** *Afr.* children's tomatoes (*S. anomalum*)
aubier *m.* alburnum, sapwood
aubiner *v.t. Hort.* to heel in
aubour *m.* 1. laburnum (*Laburnum anagyroides*) (= **faux ébénier**), 2. guelder rose (*Viburnum opulus*)
aubriétie, aubriétia, aubriette *f.* aubrietia, rock cress (*Aubrieta* sp.)
aucuba *m. Aucuba* sp.; **a. du Japon** Japanese laurel (*A. japonica*)
auge *f.* trough; *Irrig.* flume, channel, bucket of Persian wheel
augelot *m. Vit.* small planting hole, trench
auget *m.* small trough; *Irrig.* bucket of water wheel; **roue à augets** bucket-wheel
augette *f.* small trough
augmentation *f.* augmentation, increase; **a. de rendement** increase in yield
au hasard *Stat.* at random
aulacode *m. Afr.* cane rat, grass cutter (*Thryonomys swinderianus*) (= *Afr.* **agouti**)
aune, aulne *m.* alder (*Alnus* sp.); **a. blanc, a. blanchâtre** grey alder (*A. incana*); **a. du Caucase** Caucasian alder (*A. subcordata*); **a. commun** common alder (*A. glutinosa*) (= **vergne**); **a. de Corse, a. à feuilles en cœur**

Italian alder (*A. cordata*); **a. glutineux = a. commun** (*q.v.*); **a. noir = bourdaine** (*q.v.*); **a. ordinaire = a. commun** (*q.v.*); **a. rouge** Oregon alder, red alder (*A. rubra*); **a. vert** green alder (*A. viridis*)
aunée, aulnée *f.* **(officinale), grande aunée** elecampane (*Inula helenium*)
auricule *f.* 1. auricle. 2. auricula (*Primula auricula*) (= **oreille d'ours, primevère auricule**)
auriculé,-e *a.* auriculate, eared
aurone, auronne *f.* southernwood (*Artemisia abrotanum*) (= **citronnelle**); **a. femelle** lavender cotton (*Santolina chamaecyparissus*) (= **santoline**)
autoanalyseur *m.* autoanalyser
auto-bouture *f.* natural propagule (e.g. of globe artichoke)
autochtone *a.* autochthonous, indigenous
autocide *a.* autocidal (see also **lutte**)
autoclavage *m.* steam sterilization
autoclave *a.* & *n. m.* steam sterilizer
autocollant,-e *a.* self-adhesive
autocompatibilité *f. PB* self-compatibility, self-fertility
auto-compatible *a. PB* self-compatible, self-fertile
autodestructeur,-trice *a.* self-destroying
autodiffusion *f.* self-diffusion
autodrainant,-e *a.* self-draining
autofécond,-e *a. PB* self-fertile, self-compatible
autofécondation *f. PB* self-fertilization, self-pollination, selfing; **autofécondations successives** inbreeding
autofécondé,-e *a. PB* selfed
autoféconder *v.t. PB* to self
autofertile *a.* self-fertile
autofertilité *f.* self-fertility
autogame *a.* self-fertile
autogamie *f.* autogamy, self-fertilization
autogreffe *f.* natural graft, autograft
autoincompatibilité *f. PB* self-incompatibility, self-sterility
autoincompatible *a.* self-incompatible, self-sterile
autolyse *f.* autolysis
automatisation *f.* automation
automatisé,-e *a.* automated, automatic; **arrosage automatisé** automatic watering
automnal,-e *a., pl.* **-aux** autumnal
automne *m.* autumn
automoteur, -trice *a.* self-propelled
autonettoyant,-e *a.* self-cleaning
auto-ombrage *m.* self-shading
auto-pollinisation *f.* self-pollination
autopollinisé,-e *a.* self-pollinated
autopolyploïde *a.* autopolyploid
autopolyploïdie *f.* autopolyploidy
autorégulant,-e *a.* self-regulating
autostérile *a.* self-sterile
autostérilité *f.* self-sterility
autotomie *f.* autotomy
autotracté,-e *a.* self-propelled
autotrophe *a.* autotrophic
autoxène *a. Fung.* autoxenous, autoecious
auvent *m.* open shed, porch roof; *Hort.* screen, matting
auxanomètre *m.* auxanometer, an instrument to measure the rate of elongation of a plant member or segment
auxiliaire *m. Ent., PC* auxiliary (beneficial) organism (predator or parasite)
auxine *f.* auxin
auxinomimétique *a.* & *n. m.* synthetic growth substance, synthetic auxin
ava *m.* Kava pepper (*Piper methysticum*)
avant-pal *m., pl.* **avant-pals** hole-making tool for soil injector
avant-printemps *m.* early spring
aveline *f.* hazel nut
avelinier *m.* hazel tree (*Corylus avellana*) (see also **coudrier, noisetier**)
averse *f.* shower; **a. dorée** *f.* golden shower (*Cassia fistula*)
avertissement *m.* warning, notice; **avertissements agricoles** agricultural warning service (see also **système**)
aviaire *a.* relating to birds; **attaque aviaire** bird attack (see also **antiaviaire**)
avion *m.* aeroplane; *PC* **a. poudreur** spraying 'plane
avitaminose *f.* vitamin deficiency
avocat *m.* avocado (pear)
avocatier *m.* avocado (pear) tree (*Persea americana*); **a. du Guatémala** coyo avocado (*P. schiedeana*)
avoine *f.* oat (*Avena sativa*); **folle avoine, a. sauvage** wild oat (*A. fatua, A. ludoviciana*); **a. animée** animated oat (*A. sterilis*); **a. à chapelet** onion couch, onion twitch (*A. elatius* var. *bulbosus*); **a. élevée** tall oat grass (*Arrhenatherum elatius*) (= **fromental**)
avorté,-e *a.* aborted
avortement *m. Bot.* non-formation, incomplete formation, abortion; **a. de l'évolution** abortive development; **a. de la fleur** *Hort.* blindness
avorter *v.i. Bot.* to develop imperfectly, to fail to ripen, to abort
axe *m. Bot.* axis, main stem, trunk; **a. central** *Hort.* main leader; **a. principal** rachis, main axis
axénique *a.* axenic
axial,-e *a., m. pl.* **-aux** axial
axile *a.* axile
axillaire *a. Bot.* axillary; **œil axillaire** axillary bud

azalée *f.* azalea; **a. à feuilles caduques** deciduous azalea; **a. de l'Inde** 'Indian' azalea *Rhododendron simsii* (= *Azalea indica*); **a. japonaise, a. du Japon** Japanese azalea (*R. obtusum*); **a. du pauvre** godetia (*Godetia grandiflora*); **a. pontique** *A. pontica* (see also **chèvrefeuille**)
azédarac, azédarach *m.* bead-tree, azedarach (*Melia azedarach*) (= **arbre à chapelets**)
azerole *f.* azarole, fruit of **azerolier**
azerolier *m.* azarole tree, Neapolitan medlar tree (*Crataegus azarolus*)
azobé *m* meni oil tree (*Lophira alata*) (= **bongossi**)
azolle *f.* *Azolla* sp.
azorelle *f.* *Azorella* sp., including balsam-bog (*A. caespitosa*)
azotate *m.* nitrate
azotation *m.* nitrogenization
azote *m.* nitrogen
azoté,-e *a.* nitrogenous; **engrais azotés** nitrogenous fertilizers; **nutrition azotée** nitrogen nutrition
azoteux,-euse *a.* nitric
azur *m.* azure, blue
azuréen,-enne *a.* of the Côte d'Azur

B

baba *m. Sug.* late tiller, sucker (= **rejet tardif, gourmand de tige, marotte**)
baboul *m.* saltbush (*Salvadora persica*)
bac *m.* 1. ferry. 2. tank, tub, trough etc; **b. d'acclimatation, b. d'acclimatement** hardening tank; **b. à évaporation** evaporation pan, tank; **b. à fleurs** window box; **b. de manutention** box, crate (for handling); **b. à plantes** tub for plants; **b. à pralin** slurry tank; **b.de récupération, b. à herbe** grass box (on mower); **b. à semis** seed tray; **b. de stockage** storage bin, crate; **b.de traitement** hot water tank
baccifère *a. Bot.* berry-bearing
bacciforme *a.* resembling a berry, berry-shaped
baccivore *a.* berry-eating
bâchage *m.* covering with plastic cover, tarpaulin etc., plastic mulching
bâche *f. Hort.* 1. (forcing) frame, frame structure; **b. froide** cold frame, 2. cover; **b. goudronnée** tarpaulin; **b. plastique** plastic cover, plastic sheet
bâché,-e *a.* covered with plastic or other cover
bacholle *f.* large wooden vessel for carrying grapes at harvest
backcross, back-cross *m. PB* backcross
bactéricide *a.* bactericidal; *n. m.* bactericide
bactérie *f.* bacterium, *pl.* bacteria; **b. acétique** acetic acid bacterium (*Acetobacter* sp.); **bactéries ammonifiantes** ammonifying bacteria; **bactéries dénitrifiantes** denitrifying bacteria; **bactéries nitrifiantes** nitrifying bacteria; **b. nodulaire** nodule bacterium
bactérien,-enne *a.* bacterial
bactériologie *f.* bacteriology
bactériophage *m.* bacteriophage
bactériose *f.* bacteriosis, bacterial disease; **b. du delphinium** delphinium black blotch (*Pseudomonas delphinii*); **b. des fleurs** bacterial canker, bacterial blossom blight (*Pseudomonas syringae*); **b. du noyer** bacterial disease of walnut (*Xanthomonas juglandis*)
bactériostatique *a.* bacteriostatic
baculovirus *m. inv.* baculovirus
bacuri *m.* bakuri (*Platonia insignis*), a tree with edible fruits, also yielding seed oil
badamier *m.*, **badamie** *f. Terminalia* sp., including tropical almond (*T. catappa*) (= **terminalier**)
badiane *f.*, **badianier** *m. Illicium* sp.; **b. (de Chine)** Chinese star anise (*I. verum*); Japanese star anise (*I. anisatum*) (= **anis étoilé**)
badigeon *m.* wash, coat of paint; **b. blanc** whitewash
badigeonnage *m.* daubing, painting, swabbing
badigeonner *v.t.* to daub, to paint, to swab
badigeonneur *m.* person who daubs, paints, swabs
bagasse *f. sug.* bagasse
baguage *m. Hort.* bark ringing
bague *f.* ring
baguenaudier *m. Colutea* sp.; **b. (commun)** bladder senna (*Colutea arborescens*) (= **faux séné**); **b. d'Ethiopie** cancer bush (*Sutherlandia frutescens*)
baguette *f.* rod, wand, stick; **b. de coudrier** (diviner's) hazel twig; *Cit.* shoot of mother-plant from which buds are taken for budding, bud-stick; *Vit.* fruit cane
bagueur *m. Hort.* knife for bark ringing
baie *f. Bot.* berry; **b. de raisin** grape; **baies fruitières** soft fruits
bail *m., pl.* **baux** lease
baillot *m. Vit.* harvesting basket
bain *m.* bath; **b. de trempage** dip (solution)
bain-marie *m., pl.* **bains-marie** water bath; **au bain-marie** in a water bath
baïonnette *f. Ban.* sword sucker
baisse (*f.*) **de rendement** decline, reduction, drop in yield
baissière *f. Ag., Hort.* depression, hollow where water collects
balai *m.* broom; **b. de l'estomac** *Coll.* spinach (= **épinard**); **b. à gazon** lawn broom, besom; **b. de jardin** garden broom; **b. métallique (à lames)** leaf rake, wire rake; **b. de sorcière** *PD* witches' broom disease; **b. de sorcière du bouleau** birch witches' broom (*Taphrina turgida*); **b. de sorcière du faux acacia** black locust brooming; **b. de sorcière de la pomme de terre** potato witches' broom disease
balance *f.* balance, scales; **b. à ressort** spring balance
balanin *m.*, **b. (des châtaignes)** chestnut weevil (*Balaninus (Curculio) elephas*); **b. (des noisettes)** hazel-nut weevil (*B. nucum*)
balanites *m.* desert date (*Balanites aegyptiaca*) (= **dattier du désert, soump**)

balanoïde *a.* resembling an acorn
balanophore *a.* acorn-bearing; *n. f. Balanophora* sp.
balata *m.* **b. franc, b. gomme, b. rouge** common balata (*Mimusops balata = Manilkara bidentata*); *n. f.* balata gum
balauste *f.* wild pomegranate flower
balaustier *m.* wild pomegranate plant (see **grenadier**)
balayage *m.* sweeping, scanning (by electron microscope etc.); scrubbing (in fruit storage)
balayures *f. pl.* sweepings
balcon *m.* balcony
baldingère *f.* reed grass (*Phalaris arundinacea*)
balisier *m. Canna* sp., including canna lily and Indian shot-plant (*C. indica*) (see also **canna**)
baliveau *m.* sapling
balle *f.* 1. ball; bale; **b. de paille** straw bale. 2. husk; *Bot.* glume
ballon rouge *m.* bladder cherry (*Physalis alkekengi*)
ballot *m.* bale; **b. de paille** straw bale
ballota, ballote *f.* **(fétide)** (black) horehound (*Ballota nigra*) (= **marrube noir**)
balonge *f.* wooden tub used for grape harvest
balsamier *m.* balsam-tree (*Copaifera* sp., *Myroxylon* sp. and others) (= **baumier**)
balsamifère *a.* balsamiferous
balsaminacées *f., pl.* Balsaminaceae
balsamine *f.* balsam (*Impatiens* sp.) (= **impatiens**)
balsamique *a.* balsamic, aromatic
balsamite *f.* costmary (*Chrysanthemum balsamita*) (= **tanaisie balsamite**)
bamboche *f. Bot.* rattan
bambou *m.* bamboo
bambouseraie *f.* bamboo grove, plantation
banane *f.* banana, plantain; **b. figue** (sweet) banana; **b. des Canaries, de Chine** Canary banana; **b. des Antilles, b. cochon, b. plantain, b. à cuire** plantain (= **plantain**); **b. corne** Horn plantain type; **b. dessert** dessert, eating banana; **bananes en mains** bananas in hands; **figue banane** banana as opposed to plantain (see also **plantain**)
bananeraie *f.* banana grove or plantation
bananier *m.* banana plant (*Musa* sp.); **b. d'Abyssinie** ensete (*Ensete ventricosa*); **b. figue, b. des sages** (sweet) banana; **b. du paradis, b. plantain** plantain; **b. textile** abaca, Manila hemp (*Musa textilis*)
bananier,-ère *a.* relating to bananas; **coopérative bananière** banana growers' co-operative; **navire bananier** banana boat
bananifère *a.* banana-producing
banc *m.* 1. bench. 2. layer, bed; **b. d'essai** test bed; **b. de gazon** turf bank
bancoulier *m.* candlenut oil tree (*Aleurites triloba*)
bandage *m.* 1. application of bandage. 2. bandage
bande *f.* 1. band, strip; **b. engluée** sticky band; **b.enherbée** grass strip; **b. piège** band to trap insects sheltering under bark etc.; **b. végétale** plant strip (to control erosion etc.); **bandes blanches** *OP* white stripe disorder. 2. band, flock; **b. primitive** *PC* (locust) hopper band
bandure *f.* pitcher plant (*Nepenthes* sp.)
bang, bangh *m.* bhang
banian *m.* banian, banyan tree (*Ficus benghalensis*)
bankanas *m. Afr. Icacina senegalensis*
banksie *f. Banksia* sp.
banque *f.* bank; **b. de données** data bank; **b. de gènes** gene bank
banquette *f.* bench terrace, dike, levee; **b. (irlandaise)** bank; platform, ledge
baobab *m.* baobab (*Adansonia digitata*)
baouléen,-enne *a. Afr.* of Baoulé
baptisie *f. Baptisia* sp., false-indigo
banquet *m.* tub, bucket; **b. à oreilles** tub with handles
baraque *f.* hut, shed
barbadine *f. Bot.* common granadilla, square-stalked passion flower (*Passiflora quadrangularis*) and its fruit (see also **grenadille, passiflore, pomme-liane**)
barbarée *f. Barbarea* sp.; **b. précoce** American cress (*B. verna = B. praecox*) (= **cresson de terre**); **b. vulgaire** winter cress, yellow rocket (*B. vulgaris*)
barbe *f. Bot.* beard; silk (of maize)
barbe de capucin *f.* variety of **chicorée** (*q.v.*)
barbe de Jupiter *f.* Jupiter's beard (*Anthyllis barba-jovis*); red valerian (*Centranthus ruber*) (= **valériane rouge**)
barbe de vieillard *f.* Spanish moss (*Tillandsia usneoides*) (= **fille de l'air**)
barbeau (*m.*) **bleu** cornflower (*Centaurea cyanus*) (= **bleuet**)
barbelé,-e *a.* barbed; **fil-de-fer barbelé** barbed wire
barbelure *f. Bot.* beard, awn
barbotine *f.* (a) mugwort (*Artemisia vulgaris*) (b) tansy (*Tanacetum vulgare*) (= **tanaisie**)
barbu,-e *a.* bearded; *Bot.* awned
barbue *f. Vit.* rooted vine, young vine about one year old
bard *m.* 1. two-man tray. 2. wheeled trolley
bardane *f.* great burdock (*Arctium lappa*); **b. comestible** edible burdock (*A. lappa* var. *edulis*); **petite bardane** lesser burdock (*A. minus*)
bark split *m.* bark split (of plum) (= **fente de l'écorce**)

barquette *f. Hort.* (boat-shaped) container
barre *f.* bar, rod, rail; **b. à boucher** rammer used to fill holes made by soil injector; **b. de coupe** cutter bar; **b. porte-outils** tool bar (see also **terre de barre**)
barre-faucheuse *f.* cutter bar (of mower)
baryum *m.* barium
basal,-e *a., m. pl.* **-aux** basal
base *f.* base, **b. de données** data base; **à la base de** at the base of; *Ped.* **b. échangeable** exchangeable base; *PC* **à base de . . .** containing . . . (as active ingredient)
baselle *f.* **(de Chine), b. à feuilles en cœur** vine spinach, Indian spinach, Malabar nightshade (*Basella rubra* = *B. cordifolia*); **b. blanche, b. rouge** varieties of *B. rubra* (see also **brède, épinard indien**)
bas-fond *m. pl.* **bas-fonds** depression, hollow, low ground
baside *m.* basidium
basidiomycète *m.* basidiomycete
basifixe *a.* basifixed
basilaire *a.* basilar
basilic *m. Hort.* (sweet) basil (*Ocimum basilicum*); **petit basilic** bush basil (*O. minimum*)
basipète *a.* basipetal
basiphile *a.* lime-loving (= **alcalinophile, basophile**)
basique *a.* basic; **pH basique** alkaline pH
basitone *a.* basitonic
basophile *a.* = **basiphile** (*q.v.*)
basse-tige *f. pl.* **basses-tiges** low-stemmed tree (about 0.50–1m high)
bassia *f. Bassia* sp.; **b. butyracée** shea-butter tree (*Butyrospermum parkii*) (= **karité**)
bassin *m.* 1. basin, bowl, pan. 2. ornamental lake, pond; **b. d'agrément** (ornamental) garden pool; *Irrig.* reservoir, drainage basin; **b. versant** catchment area
bassinage *m.* light watering, sprinkling, syringing; watering potted plants or seed boxes by standing them in a basin of water
bassiner *v.t.* to water (lightly), to sprinkle, to syringe, to damp-down; to water potted plants or seed boxes by standing them in a basin of water
bassineuse *f.* small watering can with a fine rose
baste *f.* wooden container for grape transport at harvest
bâtard,-e *a.* bastard
bâtardière *f.* nursery of grafted trees
batavia *f.* type of lettuce with uniform green curly leaves and a fairly loose heart
bâti *m.* frame, framework, structure, support
bâtiment *m.* building
bâtisse *f.* building in masonry; **b. de bois** frame building
bâton *m.* stick, rod; **b. à fouir, b. fouisseur** digging stick; *Hort.* globe artichoke set; **b. du diable** Hercules' club (*Aralia spinosa*) (= **angélique épineuse, massue d'Hercule**); **b. de Jacob** yellow asphodel, king's spear (*Asphodelus luteus* = *Asphodeline lutea*); **b. de Saint-Jacques** hollyhock (*Althaea rosea*) (= **rose trémière**); **b. d'or** wallflower (*Cheiranthus cheiri*) (= **giroflée**); **b. royal** white asphodel (*Asphodelus albus*)
battage *m.* beating, threshing (crops), ramming (earth)
battance (*f.*) **du sol** "capping", sealing, of soil
battant,-e *a. Ped.* **sol battant** "capping", sealing soil (see also **sol**); *n. m.* shutter, flap; **b. d'aération** shutter for ventilation
batte *f. Hort.* (earth) rammer, tamper (to firm soil after sowing); beater, mallet
battement *m. Ped.* "capping", sealing of soil
batteur *m.* beater; threshing drum
batteuse *f.* threshing machine, thresher
battu,-e *a.* beaten; *Ag., Hort.* threshed; *Ped.* **sol battu** "capped", sealed soil
bauhinie *f. Bauhinia* sp., including *B. purpurea* and *B. variegata*
baume *m.* balm, balsam; **b. blanc liquide** balsam of Peru (*Myroxylon pereira*); **b. cajou** cashew nut shell liquid, CNSL; **b. de Copahu** gurjum balsam (*Dipterocarpus alatus*); **b. coq, b. des jardins** costmary (*Chrysanthemum balsamita*) (= **balsamite, tanaisie balsamite**); **faux b. du Pérou** blue melilot (*Trigonella coerulea*) (= **mélilot bleu**); **b. vert** garden mint (*Mentha* sp.); **b. vert des Antilles, b. Marie** calaba balsam, tacamahac (*Calophyllum tacamahaca*) (= **calaba**)
baumier *m.* 1. balsam fir (*Abies balsamea*) (= **sapin baumier**). 2. balsam-tree (*Copaifera* sp., *Myroxylon* sp. and others); **b. du Pérou** Peru balsam tree (*Myroxylon peruiferum*)
baveur *m. Irrig.* trickler
bayoud (*m.*) **du palmier-dattier** bayoud disease of date palm (*Fusarium oxysporum* f. *albedinis*)
bayoudé,-e *a.* affected by bayoud disease
bdellium *m.* resin of *Commiphora* sp.
bec *m.* beak; **en bec de canard** (of) duck-bill (shape); *Irrig., PC* nozzle
bécabunga *m.* brooklime (*Veronica beccabunga*)
bec-de-grue *m. pl.* **becs-de-grue** common storksbill (*Erodium cicutarium*) and other spp.
bêchage *m.* digging with a spade
bêche *f.* spade; **b. automatique** mechanical spade; **b. à découper** edging knife; **b. à dents** digging fork; **profondeur de fer de bêche** spit
bêché,-e *a.* dug with a spade
bêchelon *m.* small hoe

bêcher *v.t.* to dig with a spade (see also **fraise, machine**)
bêcherie *f. Vit.* hand digging early in the season
bêche-tarière *f., pl.* **bêches-tarières** post-hole digger
bêcheton *m.* narrow spade
bêcheur,-euse *n.* person who digs with a spade
bêchoir *m.* type of hoe
bêchot *m.* small spade
bédégar, bédéguar *m.* bedeguar, rose gall (caused by *Diplolepis rosae*)
bégonia *m.* begonia (*Begonia* sp.); **b. rex** Rex begonia (*B.* × *rex-cultorum*); **b. tubéreux (hybride)** tuberous begonia (*B.* × *tuber-hybrida*); **b. à petites fleurs** *B. semperflorens*; **b. rhizomateux** rhizomatous begonia
bégoniacées *f. pl.* Begoniaceae
beige *a.* beige
bélier (*m.*) **hydraulique** hydraulic ram
belladone *f.* deadly nightshade (*Atropa belladonna*)
belle-dame *f., pl.* **belles-dames** garden orache, mountain spinach (*Atriplex hortensis*) (= **arroche des jardins, follette**) (also applied to **belladone**)
belle-de-jour *f., pl.* **belles-de-jour** *Bot.* convolvulus (*Convolvulus* sp. including *C. tricolor*), morning glory (*Ipomaea* sp.)
belle-de-nuit *f., pl.* **belles-de-nuit** *Bot.* moonflower (*Ipomaea bona-nox*), marvel of Peru, four o'clock plant (*Mirabilis jalapa*), *M. longiflora*
belle-d'onze-heures *f. pl.* **belles-d'onze-heures** star of Bethlehem (*Ornithogalum* sp.) including *O. umbellatum*, tiger flower (*Tigridia pavonia*) (= **œil-de-paon**)
belle-d'un-jour *f. pl.* **belles-d'un-jour** day lily (*Hemerocallis* sp.)
belle saison *f.* the season of warm weather, summer months
bellis *f. Bellis* sp., daisy (see also **pâquerette**)
belvédère *m. Hort.* belvedere (*Kochia scoparia*)
ben (*m.*) **(ailé)** horseradish tree (*Moringa oleifera*) (see also **huile de ben**)
bénéfice *m.* profit, gain
bénéfique *a.* beneficial; **effet bénéfique** beneficial effect
benincase *f.* wax or ash gourd (*Benincasa cerifera* = *B. hispida*)
benjoin *m.* benzoin, gum benjamin (*Styrax benzoin*)
benne *f.* 1. hamper, basket; pannier (for grapes) 2. skip, bucket (of dredger etc.); **b. basculante** tip lorry
benoîte *f.* avens (*Geum* sp.); **b. commune** herb bennett (*G. urbanum*); **b. écarlate** scarlet avens) (*G. chiloense* (*coccineum*))
bentamaré *m.* coffee senna (*Cassia occidentalis*)
benzène *m.* benzene
benzoin *m.* benzoin laurel (*Styrax benzoin*)
benzyladénine *f.* benzyladenine
bèp *m. Sterculia scaphigera*, a tropical tree with edible seeds
béquiller *v.t. Hort.* 1. to prop up (branches etc.) (see also **débéquiller**). 2. to hoe lightly
berbéridacées, berbéridées *f. pl.* Berberidaceae
berbéris *m. Berberis* sp., **b. commun** barberry (*B. vulgaris*) (= **épine-vinette, vinettier**); **b. de Darwin** Darwin's barberry (*B. darwinii*); **b. pourpre** purple barberry (*B. thunbergii atropurpurea*)
berce *f. Heracleum* sp. **(grande) b., b. commune** hogweed, cow parsnip (*Heracleum sphondylium*)
berceau *m. Hort.* trellis
béref *m. Afr.* watermelon fruit (*Citrullus vulgaris*)
bergamote *f.* bergamot orange; type of pear; **essence de bergamote** bergamot oil
bergamotier *m.* bergamot tree (*Citrus bergamia*)
bergenia *m.* elephant-leaved saxifrage (*Bergenia liguata* and other species)
berle *f. Sium* sp.; **b. des potagers** skirret (*Sium sisarum*) (see also **ache, chervis**)
berthollétie *f. Bertholletia* sp., brazil nut tree (= **noyer du Brésil**)
besier *m.* wild pear tree
besoche *f.* spade for tree-planting
besoin *m.* need, requirement; **besoins en eau** moisture requirements, irrigation requirements; **besoins en engrais** fertilizer requirements; **besoins en froid (hivernal)** (winter) chilling requirements
bétel *m.* betel (*Piper betle*)
bétoine *f.* betony (*Stachys officinalis*); **b. des montagnards** arnica (*Arnica montana*)
béton *m.* concrete
bette *f. Beta vulgaris*, beet; **b. à cardes, b. à côtes** Swiss chard (= **carde**); **b. à couper, b. poirée** spinach beet (*B. vulgaris* var. *cicla*) (see also **poirée**); **b. maritime** sea beet (*B. vulgaris* ssp. *maritima*)
betterave *f.* beet, beetroot (*Beta vulgaris*); **b. à feuillage ornemental** ornamental foliage beet; **b. longue** tap-rooted beet; **b. potagère, b. rouge** red garden beet; **b. ronde** turnip-rooted beet; **b. sucrière** sugar beet
betteravier,-ère *a.* relating to beet; *n.* beet-grower
betula *m. Betula* sp., birch (see **bouleau**)
bétuliné,-e *a.* betulaceous
beur *m. Afr.* danya (*Sclerocarya birrea*), a tree with edible fruits and medicinal uses
beurre *m.* butter; **b. de cacao** cocoa butter; **b. de dika** Dika butter (from *Irvingia* spp.); **b.**

de karité, beurre de Galam shea butter (from karité (*Butyrospermum parki*))
beurré *m.* butter-pear, beurré
biacide *a.* biacid
biacuminé,-e *a.* biacuminate, double-pointed
bibace, see **bibasse**
bibacier *m.* see **bibassier**
bibasse *f.* loquat fruit (= **nèfle du Japon**)
bibassier *m.* loquat (*Eriobotrya japonica*) (= **néflier du Japon**)
bicolore *a.* of two colours
bicorne *m.* *Martynia lutea* (= **cornaret**)
bidens *m.* *Bidens* sp.
bident *m.* *Bidens* sp.; **b. penché** nodding burmarigold, beggarticks (*B. cernuus*)
bidenté,-e *a.* bidentate
bifère *a.* biferous, flowering twice in one year; cropping twice (see also **unifère**)
bifide *a.* bifid
biflore *a.* biflorous, biflorate
bifolié,-e *a.* bifoliate
bifora (*f.*) **rayonnante** *Bifora radians*
bifurcation *f.* bifurcation, fork, branching
bifurqué,-e *a.* bifurcate, forked
bifurquer *v.t.* *v.i.* to fork, bifurcate, divide; *v.i.* **se bifurquer** to fork, bifurcate, divide
bigarade *f.* sour, Seville or bitter orange
bigaradier *m.* sour, Seville or bitter orange tree (*Citrus aurantium*) (= **citronnier bigaradier, oranger (à fruit) amer**); **b. bouquetier** perfume-producing sour orange tree; **b. chinois** myrtle leaf orange tree (*C. myrtifolia*)
bigarreau *m.* *pl.* **bigarreaux** firm-fruited sweet cherry, bigarreau
bigarreautier *m.* bigarreau cherry tree (see also **cerisier**)
bigarrure *f.* medley, mixture of colours; *PD* mosaic; **b. du haricot** bean mosaic virus; **b. de l'oignon** onion mosaic virus; **b. jaune de l'onion** onion yellow dwarf virus; **b. de la pomme de terre** potato virus Y
bigéminé,-e *a.* *Bot.* bigeminate
bignone *f.*, **b. à grandes fleurs** trumpet creeper (*Campsis (Tecoma) grandiflora*); **b. de Virginie** (*C. (Tecoma) radicans*); cross vine (*Anisostichus capreolatus* = *Bignonia capreolata*)
bignonia *m.* = **bignone** (*q.v.*)
bignoniacées *f. pl.* Bignoniaceae
bilabié,-e *a.* bilabiate
bilan *m.* balance, balance sheet; **b. carboné** carbon balance; **b. d'eau, b. hydrique** water economy, moisture balance, water regime; **b. radiatif** radiation balance
billard *m.* *Vit.* screen for removing grapes from stalks
billon *m.* ridge; *Vit.* vine plant (cut very short)
billonnage *m.* ridging; **b. cloisonné** tie ridging
billonner *v.t.* to ridge
billonneuse *f.* ridging plough, ridger (= **butteuse, charrue billonneuse**)
billot *m.* *Hort.* osier (or wooden) basket for fruit transport
bilobé,-e *a.* bilobed
biloculaire *a.* bilocular
bimarginé,-e *a.* bimarginate
binage *m.* hoeing, harrowing; **premier binage** *Afr.* first hoeing
binard *m.* wheeled trolley
bine *f.* kind of hoe
biné,-e *a.* 1. *Bot.* binate. 2. hoed, harrowed
biner *v.t.* to hoe, to harrow
binervé,-e *a.* binervate, having two veins
binet *m.* *Vit.* light plough
binette *f.* garden hoe; **b. (de Nanterre)** drag hoe
bineur *m.*, **bineuse** *f.* 1. hoer. 2. small plough (= **binot**); **bineuse rotative** rotary hoe
binoche *f.* two-tined hoe
binochon *m.* small **binoche**
binoir *m.* = **binot** (*q.v.*)
binôme *a.* & *n. m.* binomial
binot *m.* small plough; type of hoe
binotage *m.* ploughing, hoeing, with a **binot**
binoter *v.t.* to plough, to hoe, with a **binot**
biocénose *f.* biocoenosis, biological community
biocénotique *a.* biocoenotic
biochimie *f.* biochemistry
biochimique *a.* biochemical
bioclimat *m.* bioclimate
bioclimatologie *f.* bioclimatology
biocoenose = **biocénose** (*q.v.*)
biodégradable *a.* biodegradable
biodégradant,-e *a.* biodegrading
biodégradation *f.* biodegradation
biodisponibilité *f.* biological availability
bioessai *m.* bioassay
biogaz *m.* biogas
biogène *a.* biogenous
biogenèse *f.* biogenesis
bioindicateur,-trice *a.* & *n.* bioindicator
biologie *f.* biology; **b. moléculaire** molecular biology
biologique *a.* biological; **lutte biologique** biological control
biomasse *f.* biomass
biome *m.* biome
biométrie *f.* biometry, biometrics
bion *m.* sucker
bionner *v.t.* to plant **bions**
biosphère *f.* biosphere
biosynthèse *f.* biosynthesis
biosystématique *f.* biosystematics
biotique *a.* biotic
biotope *m.* biotope, animal and plant habitat, environment

biotype *m.* biotype
biovulé,-e *a.* biovulate
bioxyde (*m.*) **de carbone** carbon dioxide, CO_2; **b. de soufre** sulphur dioxide
bipare *a.* biparous
bipartit,-e, bipartite *a.* bipartite
bipenné,-e *a.* bipennate
birette *f.* large wooden rake
birone *f. Vit.* planting stick
bisannuel,-elle *a.* biennial; **plante bisannuelle** biennial (plant)
biseau *m.* chamfer, bevel; **en biseau** bevel-edged, bevelled
bisérié,-e *a.* biserial, biseriate
bissap *m. Afr.* condiment prepared from *Hibiscus sabdariffa* flowers (see **oseille de Guinée**)
bissexué,-e, bissexuel,-elle *a. Bot.* bisexual
bissoc *m.* plough with two shares
bistorte *f.* bistort (*Polygonum amphibium*)
biterné,-e *a.* biternate
bitter-pit *m. PD* bitter pit
bivalent,-e *a.* bivalent, divalent
bivalve *a.* bivalvular
bixa, bixe *f.* annatto (*Bixa orellana*) (= **rocouyer**)
bixacées *f. pl.* Bixaceae
black-rot *m. Vit.* black rot (*Guignardia bidwellii*) (= **rot noir**); **black-roté,-e** *a.* attacked by black rot
black-spot *m.* black spot of roses (*Actinonema rosae = Diplocarpon rosae*)
blanc,-che *a.* white; **blanc crème** *a. inv.* creamy white; *n. m.* white; **b. d'Espagne** whitewash; **b. (de champignon)** *Mush.* spawn; *PD* mildew and similar fungal diseases (see also **meunier, mildiou**); **b. du fraisier** strawberry mildew (*Sphaerotheca humuli*); **b. des racines** fungal root rot (*Dematophora necatrix*) (= **pourridié des racines**); **b. de la vigne** vine powdery mildew (*Uncinula necator*)
blanchâtre *a.* whitish
blanchi(s) *m.* blaze, mark (on tree)
blanchiment *m.* blanching
blanchir *v.t.* to blanch, to whiten, to make white
blaniule *m. Blaniulus* sp., millipede; **b. moucheté** spotted snake millipede (*B. guttulatus*)
blast (*m.*) **des citrus** *Cit.* black pit (*Pseudomonas syringae*)
blastophage *m.* fig wasp (*Blastophaga psenes*)
blattaire *f.* moth mullein (*Verbascum blattaria*) (= **molène blattaire, herbe aux mites**)
blé corn, wheat (*Triticum* sp.); **b. azur** lyme-grass (*Elymus arenarius*); **b. vivace** *m.* couch (*Elymus (Agropyron) repens*) (= **chiendent**)
blennocampe (*f.*) **ascendante** *Ent.* rose miner (*Cladardis elongatula = Monophadnus elongatus*); **b. descendante** *Ardis* sp.
blessé,-e *a.* wounded, injured; **fruits blessés** injured fruits
blessure *f.* wound; **b. de taille** pruning wound
blet,-ette *a.* over-ripe (of fruit)
blète, blette *f.* = **bette** (*q.v.*)
blettir *v.i.* to become over-ripe (of fruit), to blet
blettissement *m.* over-ripeness (of fruit), bletting
bleu,-e *a.* & *n. m.* blue; **b. d'aniline** aniline blue; **b. céleste, b. ciel** sky blue
bleuâtre *a.* bluish
bleuet *m.* 1. cornflower (*Centaurea cyanus*); **b. vivace** mountain knapweed (*C. montana*) (= **centaurée des Alpes**). 2. (*Canada*) blueberry (*Vaccinium* sp.) (see also **airelle, canneberge, myrtille**)
bleuetière *f.* (*Canada*) blueberry bog
bleuetier *m.* (*Canada*) blueberry plant (*Vaccinium* sp.); **b. en corymbe, b. géant** highbush blueberry (*V. corymbosum*); **b. nain, b. de Pennsylvanie** lowbush blueberry (*V. angustifolium*); **b. noir de Pennsylvanie** black-fruited blueberry (*V. angustifolium* var. *nigrum*)
bleuir *v.t.* to blue
bleuissement *m.* blueing, turning blue
bleuté,-e *a.* bluish
blocage *m.* blocking
bloc *m.* block; *Stat.* **blocs de Fisher, blocs randomisés** randomized blocks
blotch (*m.*) **fumeux** sooty blotch of apple (*Gloeodes pomigena*)
bluet *m.* = **bleuet** (*q.v.*)
bluté,-e *a.* sifted
boarmie *f.* **(des bourgeons)** *Vit.* bud boarmid (*Peribatodes (Boarmia) pedunculata, P. rhomboidaria*)
bobine *f.* bobbin, reel, spool, drum (for rope etc.)
bogue *f.* husk of chestnut
bois *m.* wood, xylem: copse, spinney; *Hort. pl.* cuttings; **b. de l'année** current year's wood; **b. de l'année précédente** previous year's wood; **b. d'un an** one-year-old lateral shoot, maiden lateral; **b. de deux ans** two-year-old lateral (branch); **b. de trois ans** three-year-old lateral (branch); **b. bouton** button bush (*Cephalanthus occidentalis*); **b. caoutchouc** *PD* rubbery wood virus; **b. capitaine** cowhage (*Malpighia urens*); **b. chardon** locust berry (*Byrsonima coriacea*) (= **merisier doré**); **b. creux** hollow wood; **b. dentelle** lace bark (*Lagetta lintearia*); **b. franc** *Vit.* shoot on previous year's wood; **b. fruitier** fruiting wood; **b. gentil** mezereon (*Daphne mezereum*) (= **mézéréon**); **b. de greffe, b. greffon** budwood, scion wood; **b. immortel** coral tree (*Erythrina corallodendron*) (= **arbre au corail**); **b. joli** = **b. gentil**

(*q.v.*); **b. mort** dead wood; **b. de nèfle** Brazilian cherry (*Eugenia brasiliensis*); **b. d'orignal** hobble bush (*Valnifolium lantanoides*); **petit bois** spinney, grove; **b. puant** *Anagyris foetida*, a shrub with medicinal uses; **b. de rajeunissement** replacement spurs (of branches); **b. de remplacement**, *Vit.* **b. de retour** renewal spur; **b. rouge dur** *PD* brown oak (*Fistulina hepatica*); **b. sacré** *Afr.* fetish grove, juju grove; **bois de Sainte-Lucie** = **cerisier de Sainte-Lucie** (*q.v.*); **b. satiné** satin-wood tree (*Chloroxylon swietenia*); **b. souple** = **b. caoutchouc** (*q.v.*); **b. strié** *PD* (apple) stem pitting virus (= **cannelures du tronc**); *PD, Vit.* rugose wood, stem pitting caused by fanleaf or leaf roll virus; **b. de taille** (1) bearing wood left after pruning, (2) wood removed at pruning; **vieux bois** old wood

boisage *m.* afforestation, tree planting: young trees, saplings

boisé,-e *a.* wooded; *n.m.* (*Canada*) wood, woodland; *Vit.* woody taste (in wine)

boisement *m.* tree-planting

boisseau *m. pl.* **-eaux** *Coll.* about 12.5 litres

boîte *f.* box; **b. distributrice (d'un semoir)** hopper (of sower); **b. de Pétri** Petri dish

boîtier *m.* case; *PC* guard (to shield crop plants from sprays)

boldo *m.* boldo (*Peumus boldus*); **feuilles de boldo** boldo leaves

bolet *m.* edible boletus (*Boletus edulis* and other spp.) (see also **cèpe, indigotier, nonette**); **b. amadouvier** *Polyporus ignarius* (see also **esca**); **b. jaune des pins** granulated boletus (*B. granulatus*)

bombage *m.* bulging of tins containing preserves

bombax *m.* *Bombax* sp., red-flowering silk cotton tree (see also **boumou**)

bombe (*f.*) **fumigène** smoke bomb, fumigating candle

bombé,-e *a.* convex, bulging

bombyx *m.* bombyx moth; **b. à bague, b. cul-brun** brown-tail moth (*Euproctis chrysorrhoea*); **b. cul-brun, b. cul doré** yellow-tail moth (*E. similis*); **b. étoilé** vapourer moth (*Orgyia antiqua*) (= **orgyie antique**); **b. (à) livrée, b. neustrien** lackey moth (*Malacosoma neustria*); **b. pudibond** pale tussock moth (*Dasychira pudibunda*)

bonbonne *f.* demijohn

bonduc *m.* Kentucky coffee tree (*Gymnocladus dioica*)

bongossi *m.* *Afr.* = **azobé** (*q.v.*)

bon-henri *m.* Good King Henry (*Chenopodium bonus-henricus*) (see also **ansérine**)

bonne-dame *f.* *Coll.* = **arroche** (*q.v.*)

bonne tenue *f.* (of plant) good growth, good habit (see also **tenue**)

bonnet carré *m.* = **bonnet de prêtre** (*q.v.*)

bonnet d'électeur *m.* scalloped summer squash, custard marrow (*Cucurbita pepo*) (= **pâtisson**)

bonnet de prêtre *m.* spindle tree (*Euonymus europaeus*) (= **fusain**)

bonnet (*m.*) **turc** kind of turban squash (*Cucurbita maxima*) (see also **courge, potiron**)

boqueteau *m. pl.* **-eaux** small wood, copse, spinney

borax *m.* borax

bord *m.* edge, border, bank (of stream etc.; **b. de la mer** seaside; **"convient au bord de la mer"** (of plant) "suitable for planting near the sea"

bordelais,-aise *a.* of the Bordeaux region; **bouillie bordelaise** Bordeaux mixture

bordure *f.* border (see also **effet**); **b. herbacée** herbaceous border

bore *m.* boron

boréal,-e *a.*, *m. pl.* **-aux** boreal, northern

borer *m.* *Ent.* borer, stem borer (see also **mineur**); **b. du chou** cabbage webworm (*Hellula undalis*); **b. ponctué de la canne à sucre** sugar cane borer (*Proceras sacchariphagus*)

borgne *a.* one-eyed; *Hort.* blind; **plant borgne** blind plant; **chou borgne** cabbage which does not heart

boriqué,-e *a.* containing boron

bornage *m.* 1. setting landmarks, marking out. 2. *Hort.* firming soil around transplanted plant

borner *v.t.* 1. to set landmarks, to mark out. 2. *Hort.* to firm soil around transplanted plant

bosquet *m.* small wood, grove

bosse *f.* bump, bruise, swelling, lump

bosselé,-e *a.* dented

bosselure *f.* dent, inequality

bostryche *m.* *Vit.* vine bostrychid beetles (*Sinoxylon sexdentatum, Schistocerus bimaculatus*) (= **apate**)

botanique *a.* botanical; *n. f.* botany

botaniquement *adv.* botanically

botaniser *v.i.* to collect plants

botaniste *m.* botanist

botillon *m.* small bunch

botrytis *m.* *Botrytis* sp., especially *B. cinerea*; disease caused by *Botrytis* sp.; **b. des semis d'oignons** onion leaf rot (*(Sclerotinia) (Botrytis) squamosa*)

botrytisé,-e *a.* infected by *Botrytis cinerea*; **moût botrytisé** must from grapes infected by *B. cinerea*

botte *f.* 1. bunch, bundle, truss; **b. de paille**

straw bale. 2. **b. en caoutchouc** rubber boot, wellington boot
bottelage *m.* bundling up, tying up, bunching
botteler *v.t.* to bundle, tie up, bunch
botteleuse *f.* machine for bundling, tying or bunching
boucage *m. Bot. Pimpinella* sp.; **b. anis** = **anis** (*q.v.*)
bouche *f.* mouth; **de bouche** for eating, dessert; **b. d'eau** *f.* hydrant
bouchon (de liège) *m.* cork; *Coll.* small sub-standard aubergine fruit
bouchure *f.* live hedge
bouclage *m.* **(des feuilles)** *PD* (leaf) curl
boucle *f.* 1. buckle. 2. loop, bow. 3. ring
boucler *v.t.* to buckle, to loop, to tie up; *v.i.* to curl, to be curled
bouclier *m.* shield; *PC* **b. (cireux)** waxy scale of scale insect
boue *f.* mud, sludge; **b. (de station) d'épuration** sewage sludge
boueux,-euse *a.* muddy, sludgy
bougainvillea, bougainvillée *f.*, **bougainvillier** *m.* bougainvillea (*Bougainvillea* sp.)
bougie *f.* candle; **b. de paraffine** paraffin wax candle
bouillie *f. Hort.* plant protection mixture; **b. bordelaise** Bordeaux mixture; **b. bourguignonne** Burgundy mixture; **b. cuprique** copper spray; **b. herbicide, b. fongicide** herbicidal, fungicidal spray, product, mixture, slurry; **b. soufrée** sulphur-copper spray; **b. sulfocalcique** lime sulphur
bouillon *m. Vit.* small bud at base of shoot
bouillon-blanc *m., pl.* **bouillons-blancs** Aaron's rod, great mullein (*Verbascum thapsus*) (= **molène**)
bouillon-noir *m., pl.* **bouillons-noirs** = **bardane** (*q.v.*)
boulbène *f.* boulbène soil (heavy soil of central Gascony, SW France); **terres de boulbène** boulbène soils
boule *f.* ball; **en boule** like a ball, round; **b. azurée** globe thistle (*Echinops ritro*)
boule (*f.*) **de neige** 1. *Hort.* snowball tree (*Viburnum opulus* var. *sterile*) (see also **obier, viorne**); **b. de neige chinoise** Chinese snowball tree (*V. macrocephalum*). 2. *Fung.* **b. de neige** field mushroom (*Agaricus campestris*); **b. de neige des bois** wood mushroom (*A. silvicola*) (see also **champignon**)
boule d'or *f., pl.* **boules d'or** globe flower (*Trollius* sp.) (see also **trolle**)
bouleau *m.* birch (*Betula* sp.); **b. (commun)** silver birch (*B. pendula*); **b. à canots** canoe birch (*B. papyrifera*); **b. merisier** cherry birch (*B. lenta*); **b. nain** dwarf birch (*B. nana*); **b. à papier** = **b. à canots** (*q.v.*); **b. pubescent** downy, white birch (*B. pubescens*); **b. verruqueux** *B. verrucoa* (= *B. pendula*)
boulet (*m.*) **draineur** mole (of mole plough)
boulingrin *m.* bowling green, small lawn, grass plot
boumou *m. Afr.* red-flowering silk cotton tree, sunset tree (*Bombax costatum*) (= **kapokier rouge**)
boungou *m. Afr.* bungu (*Ceratotheca sesamoides*)
bouquet *m.* 1. bunch (of flowers); truss (of tomatoes); **b. de mariée** bridal bouquet; **b. sec** bunch of dried flowers; **b. d'arbres** very small grove; *Hort.* **b. (de mai)** multiple bud, blossom cluster. 2. flavour, bouquet (of wine)
bouquetier *m.* maker of flower bunches: flower vase; **b. de Nice** variety of **bigaradier**
bourbe *f.* mud, mire
bourcette = **boursette** *q.v.*
bourdaine *f.* alder buckthorn (*Frangula alnus*) (= *Rhamnus frangula*) (= **aune noir**) (see also **alaterne, nerprun**)
bourde *f. Hort.* sucker
bourdon *m.* bumble-bee (*Bombus* sp.)
bourgeon *m. Bot.* bud, eye, shoot; **b. adventif** adventitious bud; **b. anticipé** *Hort.* bud giving rise to feather *Vit.* etc. lateral bud (see also **anticipé, prompt-bourgeon**); **b. axillaire** axillary bud; **b. à bois** wood bud; **b. dormant** dormant bud; **b. écailleux** scaly bud; **b. à feuilles, b. foliaire** leaf bud; **b. à fleur, b. floral** flower bud; **b. à fruit** fruit bud; **b. inflorescentiel** flower bud; **b. latent** latent bud; **b. latéral** lateral bud; **b. mixte** mixed bud; **b. nu** naked bud; **b. ordinaire** = **b. axillaire** *q.v.*; **b. principal** main bud; **b. de remplacement** replacement bud; **b. terminal** terminal bud; **sans bourgeons** budless
bourgeonnement *m.* bud burst, bud growth, bud development; **b. adventif** adventitious budding
bourgeonner *vi.* to burgeon, to put forth buds, shoots
bourguignon,-onne *a.* Burgundian; **bouillie bourguignonne** Burgundy mixture
bourrache *f.* borage (*Borago officinalis*); **petite bourrache** blue-eyed Mary (*Omphalodes verna*)
bourrage *m.* stuffing, filling up (holes etc.), packing (of soil etc.)
bourre *f.* hair, fluff, floss, fibre, fibrous husk (of coconut); **Bot.** bud down, floss; *Vit.* bud (see also **bourrillon**)
bourreau (*m.*) **des arbres** silk vine (*Periploca graeca*); climbing bittersweet (*Celastrus scandens*)

bourrelet *m.* pad, cushion, ridge of earth etc.; **b. alluvionnaire** ridge of alluvial deposit; **b. de berge** ridge on river bank; *Hort.* thickening at graft union; *Arb.* excrescence, swelling around trunk; **b. cicatriciel** callus; *Vit.* bourrelet (thickened part of pedicel)
bourrillon *m.* large bud on 2-year-old wood; *Vit.* bud at the base of a spur or branch, basal bud (see also **bourre**)
bourse *f.* purse, bag, pouch; *Bot.* **b. -à-pasteur** shepherd's purse (*Capsella bursa-pastoris*); *Hort.* bourse
boursette *f.* corn salad, lamb's lettuce (= **mâche** *q.v.*)
boursouflement *m. Cit.* etc. puffiness, puffing
boursoufler *v.t.* to puff up, to swell; **se boursoufler** to puff up, to swell
boursouflure *f.* puffiness, swelling
bouse (*f.*) **de vache** cow dung
bout *m.* extremity, end; *Sug.* **b. blanc** top
bouté,-e *a.* bluntly rounded
bouton *m. Bot.* bud, eye; *Mush.* button; **b. blanc** white bud (stage); **b. couronne** crown bud; **b. floral, b. à fleurs, b. à fruits** flower bud; **b. latéral** lateral bud; **b. de rose** rosebud; **b. rose** pink bud (stage of apple blossom); **b. terminal** terminal bud (see also **bourgeon**)
bouton d'argent *m.* sneezewort (*Achillea ptarmica*), white bachelor's buttons (*Ranunculus aconitifolius*); pearly everlasting (*Anaphalis margariacea*) (= **immortelle blanche**) and other plants
bouton d'or *m.* buttercup (*Ranunculus acris* and other spp.)
boutonnement *m.* bud burst, bud growth, bud development (= **bourgeonnement**)
boutonner *v.i. Bot.* to bud
boutonnière *f.* buttonhole; flower worn in buttonhole
bouturage *m. Hort.* 1. propagation of plants by cuttings; **b. en sec, b. ligneux** propagation by hardwood cuttings; **b. en pousse, b. par rameau herbacé** propagation by herbaceous cuttings; **b. sous brouillard** mist propagation. 2. preparation of cuttings
bouture *f.* 1. cutting, sett, slip; **b. d'oeillet** slip, piping (of carnation); *Sug.* seed piece; **b. à crossette** (mallet) heel cutting; **b. demi-herbacée** half-ripe cutting; **b. demi-ligneuse** semi-hardwood cutting; **b. de feuille** leaf cutting; **b. greffable** *Vit.* etc. (large) rootstock cutting (ready for grafting); **b. herbacée** herbaceous, leafy cutting; **b. ligneuse, b. de rameau aoûté** hardwood, ripe, winter cutting; **b. non-racinée** unrooted cutting; **b. d'œil, b. à un œil** leaf bud cutting, single bud cutting; **b.-pépinière**, *pl.* **boutures-pépinières** *Vit.* (small) rootstock cutting (rooted in the nursery, transplanted and grafted in the field); **b. de racine** root cutting; **b. racinée, b. enracinée** rooted cutting; **b. de rameau herbacé** softwood cutting; **b. simple** cutting, straight cutting; **b. à talon** heel cutting; **b. en vert** summer cutting. 2. stem sucker
bouturer *v.i. Bot., Hort.* (of plant) to produce suckers; *v.t.* to propagate plants by cuttings, to take cuttings
bouvardia *m. Bouvardia* sp.; **b. à fleurs lisses** scarlet bouvardia (*B. leiantha*)
bouvreuil *m.* bullfinch (*Pyrrhula pyrrhula*)
boval = **bowal** (*q.v.*)
bowal *m., pl.* **bowé** *Afr.* flat area with ironstone ("laterite") cap or crust at or near the surface
bowalisation *f. Afr.* "laterization", transformation into a **bowal**
boyau (*m.*) **pollinique** pollen tube
brabant *m.* type of plough; **b. (double)** one-way plough
brachyblaste *m.* brachyblast, general term for dards and short shoots
braconide *m.* braconid (wasp); **braconides, braconidés** *m. pl.* Braconidae
bractation *f. PD* cauliflower bract proliferation (= **chitoun**)
bractée *f.* bract
bractéiforme *a.* shaped like a bract
bractéole *f.* bracteole
brai *m.* pitch, tar
brancard *m.* hand-barrow
branchage *m.* branches of a tree, boughs, heap of branches
branchaison *f.* branching
branche *f.* branch; **asperges en branches** asparagus served with the stalk; *Hort.* **b. axiale** central leader; **b. charpentière, b. de charpente** primary branch, framework branch; **b. coursonne** spur system, compound spur; **b. à fruit** fruiting branch, fruiting cane; **b. fruitière** = **b. coursonne** (*q.v.*); **b. gourmande** sucker; **b. latérale** lateral branch (= **rameau latéral**); **b. maîtresse** framework branch; **b. mère** primary branch, framework branch; **b. pendante** drooping branch, swinger; **b. principale** = **b. charpentière** (*q.v.*); **b. sous-charpentière, b. sous-mère** secondary branch; **les quatre branches de l'horticulture** the four branches of horticulture (market gardening, landscape gardening, fruit growing, nursery work)
branchette *f.* branchlet, small branch
branche-ursine *f., pl.* **branches-ursines** 1. bear's breech (*Acanthus mollis*) (= **acanthe à feuilles molles**). 2. hogweed (see **berce**)
branchillon *m.* branchlet, twig

branchu,-e *a.* branched
branc-ursine *f.*, *pl.* **brancs-ursines** = **branche ursine** (*q.v.*)
brande *f.* 1. heather (*Erica* sp. and related sp.). 2. heath, heath land
bras *m. Hort.* truss (of tomato); *Vit.* primary branch
brassage *m.* stirring (of mixture etc.); **b. de l'atmosphère** air circulation; brewing (of beer)
brassicaire *a.* pertaining to cole crops; **piéride brassicaire** cabbage white butterfly (*Pieris brassicae*)
brassicole *a.* brewing; **valeur brassicole du houblon** brewing value of hops
brède *f.* pot-herb, herbaceous plant eaten cooked as for spinach; **b. d'Angola, b. gandole** Malabar nightshade (*Basella rubra*) (= **baselle, épinard indien**); **b. mafana** Para cress (*Spilanthes oleracea*); **b. chinois** (*Mauritius*) mugwort (*Artemisia vulgaris*); **b. de Malabar** (a) = **b. d'Angola** (b) Chinese spinach (*Amaranthus gangeticus*); **b. Martin** black nightshade (*Solanum nigrum*); **b. de Zanzibar** *Trichodesma zeylanicum*
brémia *m. Fung.*, *PD* lettuce downy mildew (*Bremia lactucae*)
brenner *m.* red fire disease of grapevines (*Pseudopeziza tracheiphila*) (= **rougeot parasitaire**)
brésil gazonnant *m.* joy-weed (*Alternanthera paronychioides*)
brésillet *m.* brazil wood (*Caesalpinia eclinata, Haematoxylon brasiletto*)
brette = **brède** (*q.v.*)
brévicaule *a. Bot.* short-stemmed
brèvipalpe *m. Vit. Brevipalpus* sp. mite
bride (*f.*) **d'Angola, b. gandole** = **baselle** (*q.v.*)
brignole *f.* (Brignoles) plum or prune
brillance *f.* brilliance, brightness
brillant,-e *a.* brilliant, sparkling
brimbelle *f. Coll.* bilberry
brimbellier *m. Coll.* bilberry plant (see **myrtille**)
brin *m.* shoot (of tree); **arbre de brin** tree with a single straight stem, sapling: blade (of grass, corn), sprig
brindille *f.* sprig, twig, *pl.* brushwood; *Hort.* brindille, brindle; **b. couronnée** brindle ending in a fruit bud; **b. sur bourse** bourse shoot
brindonne *f.* fruit of **brindonnier**
brindonnier *m.* kokam (*Garcinia indica*)
brinvillière *f.* pink-root (*Spigelia* sp.)
bris *m.* breaking
brise-jet *m. inv.* anti-splash nozzle
brise-mottes *m. inv.* roller for crushing clods; hand tool for same purpose (= **émietteur**)
brisure *f.* break, crack, fissure; **b. de la tige** stem breakage
brise-vent *m. inv.* shelter belt, windbreak
brix *m. Sug.* brix; **b. réfractométrique** refractometer brix
brize *f. Briza* sp., quaking grass; **b. commune** *B. media* (see also **amourette**); **petite brize** small quaking grass (*B. minor*)
broche *f. Vit.* transplanting tool
brocoli *m.* broccoli; **b. blanc** white broccoli; **b. à jets** sprouting broccoli; **b. violet** purple broccoli (= **chou brocoli**); also applied to winter cauliflower
brodé,-e *a.* netted (of melon)
bromacil *m.* bromacil
brome *m.* bromine
bromé,-e *a.* brominated
brome *m. Bromus* sp., brome grass; **b. rameux** meadow brome (*Bromus commutatus*); **b. mou** soft brome (*B. mollis*); **b. des prés** upright brome (*B. erectus*)
broméliacées *f. pl.* Bromeliaceae, bromeliads
bromélie *f. Bromelia* sp., bromeliad
bromure *m.* bromide; **b. de méthyle** methyl bromide
bronze *m.* bronze
bronzé,-e *a.* bronze, bronzed
broque *m.* = **brocoli** (*q.v.*)
brossage *m.* brushing
brosse *f.* brush; **b. métallique** metal brush; **b. à pollen** *Ap.* pollen basket
brosse-émoussoir *f.* wire brush for removing moss
brosser *v.t.* to brush
brosseur *m.*, **brosseuse** *f.* brusher, brushing machine
brosseuse-lustreuse *f.* brusher-polisher
brou *m.* husk, hull (of walnut etc.); **b. de noix** walnut cordial, walnut stain
brouette *f.* wheelbarrow
brouettée *f.* barrow load
brouetter *v.t.* to wheel, carry in a wheelbarrow
brouillard *m.* mist, fog. **b. herbicide, fongicide** herbicide, fungicide spray, drift; **bouturage, multiplication, sous brouillard** mist propagation
broussaille *f.* often *pl.* brushwood, scrub, bushes
broussin *m.* burr-knot, excrescence
broussiné,-e *a.* gnarled
broussonétie *f. Broussoneta* sp.; **b. à papier** paper mulberry (*B. papyrifera*)
brout *m.* new shoots of trees
broutille *f.*, usually *pl.* small branches, sprigs, twigs
brouture *f.* browsed new shoots
broyage *m.* crushing, pulverizing, grinding
broyer *v.t.* to crush, to pulverize, to grind
broyeur,-euse *a.* crushing, grinding; *n. m.*

crusher, grinder; **b. à marteaux** hammer mill; *Vit.* **b. à sarments** brush cutter, pruning shredder
bruchage *m.* bruchid damage
bruche *m.* (sometimes *f.*) bruchid beetle; **b. des fèves** broad bean beetle (*Bruchus rufimanus*); **b. des haricots** bean bruchid, bean weevil (*Acanthoscelides obtectus*)
brucine *f.* brucine
brugnon *m.* clingstone nectarine (see also **nectarine, pêche**)
brugnonier *m.* clingstone nectarine tree (*Prunus persica* var. *nectarina*)
bruine *f.* drizzle (see also **crachin**)
brûlage *m.* burning (of sugar cane etc.), burning-off; sod-burning
brûler *v.t.* to burn, to scorch
brûleur *m.* burner; **b. antigelées** burner for frost protection; **b. à mazout** oil burner
brûlis *m.* burnt part of field, burn; *Fung.* zone where weeds have been killed by *Tuber* mycelium
brûlure *f.* burn, scald; **b. de feuilles** leaf scorch, damage caused by frost, wind or pesticide; **b. bactérienne du pommier** (*Canada*) fireblight (*Erwinia amylovora*) (= **feu bactérien**)
brume *f.* mist, fog
brumisateur *m.* = **brumiseur** (*q.v.*)
brumisation *f.* misting, fogging (= **nébulisation**)
brumiseur *m.* mister, fogger
brun,-e *a.* brown; **b. clair** *a. inv.* light brown; **b. rouille** *a. inv.* rust brown
brunâtre *a.* brownish
brunie *f.* *Brunia* sp.
brunir *v.t.* to make dark, to brown; *v.i.* to become brown, to turn brown
brunissement *m.* browning, darkening; **b. interne** internal browning; **b. marginal des feuilles (de laitue)** tipburn (of lettuce) (= **rand**)
brunissure *f.* browning; **b. interne** *PD* internal browning; *Vit.* red brown leaf disorder; **b. du cerisier** cherry leaf scorch (*Gnomonia erythrostoma*)
brut,-e *a.* raw, unrefined, crude
bruyère *f.* heather, heath (*Calluna, Erica* spp.); **b. arborescent** tree heath (*E. arborea*); **b. à balai** besom heath (*E. scoparia*); **b. cendrée** Scotch, bell heather (*E. cinerea*); **b. commune** heather, ling (*C. vulgaris); **b. du Cap** *Phylica ericoides* and other spp.; **b. des neiges** winter-flowering heath (*E. carnea*); **terre de bruyère** heath mould
bryobe *m.* *Ent.* *Bryobia* sp.; **b. des arbres fruitiers** apple and pear bryobia (*B. rubrioculus*)
bryone (*f.*) **dioïque** white bryony (*Bryonia dioica*)
bryophytes *f. pl.* Bryophyta
buchu *m.* buchu (*Barosma* sp.); **b. court, b. rond** *B. (Agathosma) betulina*; **b. long** *B. serratifolia*; **b. ovale** *B. crenulata*
buddleia *m.* buddleia, butterfly bush (*Buddleia* sp.) (= **lilas d'été**); **b. variable** *B. variabilis* (= *B. davidii*)
buée *f.* steam, vapour (on panes of glass etc.)
bugle *f.* bugle (*Ajuga* sp.); **b. rampante** common bugle (*A. reptans*) (see also **ajuga**)
buglosse *f.* bugloss, alkanet (*Anchusa* sp.); **b. azurée** large blue alkanet (*A. azurea*); **b. officinale** true alkanet (*A. officinalis*)
bugrane *f.* rest-harrow (*Ononis* sp.)
buis *m.* **(commun)** box (*Buxus sempervirens*)
buissaie, buissière *f.* area planted with box trees
buisson *m.* bush; **buissons de cassis** blackcurrant bushes; **b. d'ornement** ornamental bush; **b. ardent (d'Europe)** firethorn (*Pyracantha coccinea*); **b. ardent de Chine** Nepalese white thorn (*P. crenulata*); **b. ardent de Rogers** *P. rogersiana*; *Hort.* bush tree; **en buissons** as bush trees
buissonnant,-e *a.* bushy, growing in the shape of a bush
buissonner *v.i.* to grow like a bush
buissonneux,-euse *a.* bushy
bukku = **buchu** (*q.v.*)
bulbaison *f.* bulbing (of onions etc.)
bulbe *m.* or *f.* 1. *Bot.* bulb; **b. (plein), b. (solide)** corm; **b. défleuri** bulb after flowering; **b. écailleux** imbricated bulb; **b.-fils** (*pl.* **bulbes-fils**) daughter bulb; **b. mère** mother bulb; **b. préparé** heat-treated bulb; **b. tuniqué** tunicated bulb. 2. bulbous base (of palm)
bulbeux,-euse *a.* bulbous; *n.f.* bulbous or tuberous plant
bulbicole *a.* **région bulbicole** bulb-growing area
bulbiculteur *m.* bulb-grower
bulbiculture *f.* bulb-growing
bulbifère *a.* bulbiferous
bulbification *f.* bulbing
bulbiforme *a.* bulbiform
bulbille *f.* bulbil, bulblet
bulbillon *m.* cormlet, cormel
bulbocode *m.* spring meadow saffron (*Bulbocodium vernum*)
bullé,-e *a.* bullate, blistered, puckered
bulleux,-euse *a.* = **bullé** (*q.v.*)
bunchosie (*f.*) **des Andes** *Bunchosia armeniaca*, a small tree with edible fruits
buplèvre *m.* *Bupleurum* sp., hare's ear; **b. à feuilles rondes** thorow-wax (*B. rotundifolium*)
bupreste *m.* buprestid beetle
buprestidés *m. pl.* Buprestidae
buse *f.* 1. irrigation canal or water channel. 2. tube, nozzle, pipe; **b. à fente** fan nozzle; **b.**

à miroir deflector nozzle; **b. à turbulence** cone nozzle

busséole *f.* stem borer (*Busseola* sp.)

busserole (*f.*) **(raisin d'ours)** bearberry (*Arctostaphylos uva-ursi*) (= **raisin d'ours, uva-ursi**)

butéa *f. Butea* sp., including the Pulas tree (*B. frondosa*)

butiner *v.* (of bees etc.) to gather pollen and nectar from flowers, to forage

butineur,-euse *a.* (of bees etc.) nectar-gathering, honey-gathering; *n.f. Ap.* foraging bee

butôme (*m.*) **(à ombelles)** flowering rush (*Butomus umbellatus*)

buttage *m.* ridging, earthing-up; **b. mécanique** mechanical ridging; **sujet de buttage** earthed-up rootstock, rootstock in the stool- or layer-bed

butte *f.* rising ground, hillock; *Hort.* hill (for cucurbits, yams, fruit trees etc.), banked soil (at base of grapevines)

butter *v.t.* to ridge up, to earth up

butteur, buttoir *m.*, **butteuse** *f.* hand ridger: ridging plough, ridger (= **charrue billonneuse**)

buvable *a.* drinkable

buxacées *f. pl.* Buxaceae

byture *m. Byturus* sp., including raspberry beetle (*B. tomentosus*)

C

cabaret *m.* asarabacca (*Asarum europaeum*); **c. des murailles** pennywort (*Umbilicus rupestris*); **c. des oiseaux** teasel (= **cardère** *q.v.*)
cabine *f.* cabin; **c. de conservation** storage chamber
cabinet (*m.*) **de croissance** growth chamber; **c. de culture** growth chamber, cabinet (see also **chambre**)
câble (*m.*) **chauffant, c. de chauffage** heating cable
cabosse *f.* 1. *Bot.* pod e.g. cacao, kola; **c. momifiée** withered, mummified pod; **c. saine** healthy pod. 2. dried globe artichoke axillary shoot, used in propagation
cabuchage *m. Vit.* = **court-noué** (*q.v.*)
cabus *a.* **chou cabus** garden (headed) cabbage, oxheart cabbage
cacalie *f. Emilia* sp.; **c. écarlate** tassel flower (*E. flammea*) **c. à feuilles de laiteron** (*E. sonchifolia*)
cacao *m.* cacao, cocoa; **c. marchand** commercial, marketable cocoa; **beurre de cacao** cocoa butter
cacaoculture *f.* cocoa culture, cocoa growing
cacaoyer, cacaotier *m.* cacao tree (*Theobroma cacao*)
cacaoyère, cacaotière *f.* cacao plantation
cache-pot *m. inv.* flower pot case, cover, holder
cachexie *f. PD* cachexia (= **xyloporose**)
cachiman, cachiment *m. Annona* and *Rollinia* spp. (fruit or tree); **c. (cœur-de-bœuf)** bullock's heart (*A. reticulata*); **c. crème** wild sweetsop (*Rollinia mucosa*); **c. épineux** soursop (*A. muricata*); **c. sauvage** *Rollinia sibieri*; (see also **anone, corossol**)
cachimentier *m. Annona and Rollinia* spp. tree (see also **corossolier**)
cachou *m.* catechu (*Acacia catechu*)and its products; **c. pâle** gambir (*Uncaria gambir*)
cactacées *f. pl.* Cactaceae
cactées *f. pl.* = **cactacées**
cactier *m.* = **cactus; c. à fruits feuillés** West Indian gooseberry (*Pereskia aculeata*)
cactiforme, cactoïde *a.* cactus-like
cactus *m.* cactus
cadang-cadang *m.* (*Coconut*) cadang-cadang disease
cade *m.* 1. prickly, sharp juniper (*Juniperus oxycedrus*) (se also **huile**). 2. **cade, cadde** *m. Afr.* winterthorn, gawo (*Acacia albida*)
cadenelle *f.* berry of **cade** (*Juniperus oxycedrus*)
cadmium *m.* cadmium
cadran solaire *m.* sundial
cadre *m.* frame
caduc,-uque *a. Bot.* deciduous
caducifolié,-e *a.* deciduous (-leaved)
caducité *f.* deciduousness
cafamarine *f.* cafamarine
café *m.* coffee; **grain de café** coffee bean; **c. marchand** clean coffee; **c. (-)parche** parchment; **c. torréfié** roasted coffee; **c. vert** unroasted, raw, green coffee; **c. vert commercial, c. vert marchand** commercial raw coffee; **c. du Sénégal** coffee senna (*Cassia occidentalis*)
caféicole *a.* coffee-growing; **région caféicole** coffee-growing region
caféiculteur *m.* coffee grower
caféiculture *f.* coffee growing
caféier *m.* coffee tree, coffee bush (*Coffea* sp.); **c. d'Arabie** arabica coffee (*C. arabica*); **c. d'Inhambane** Inhambane coffee plant (*C. racemosa*)
caféière *f.* coffee plantation
caféine *f.* caffeine
cafetière *f. Vit.* pourer used for hot water treatment of vines
cage *f.* cage; **c. d'isolement** isolation cage; **c. vitrée** hand-light
cageot *m.* hamper, (fruit-) crate
cagette *f.* 1. small cage. 2. crate
caïeu, cayeu *m.*, *pl.* **-eux** offset bulb, bulbil, clove, sucker, slip
caïlcédrat *m. Afr.* African, Senegal, mahogany (*Khaya senegalensis*) (= **acajou du Sénégal**) and related trees
caillasse *f.* 1. marl, chalky bed under tilth 2. road-metal
caillebot *m.* guelder rose (*Viburnum opulus*) (see also **boule-de-neige, viorne**)
caille-lait *m. Galium* sp.; **c. (accrochant)** goose grass, cleavers (*G. aparine*) (= **gaillet**)
cailletier *m.* type of olive, also known as **olive de Nice**
caillou *m. pl.* **-oux** pebble
caillouteux,-euse *a.* stony, pebbly
cailloutis *m.* gravel, broken stones

caïmite *f.* star apple fruit; **petite caïmite** fruit of **petit caïmitier** (*q.v.*)
caïmitier *m.* star apple (*Chrysophyllum cainito*) (= **pomme étoilée**); **c. macoucou** *C. macoucou;* **petit caïmitier** *C. microcarpum*
caisse *f.* 1. case, packing case, chest, tank, frame (= **coffre**); **c. à plantes** plant tub; **c. de récolte** container for harvesting; **c. de stratification** stratifying box; **c. à vendange** *Vit.* box for carrying harvested grapes. 2. cash box, cash desk; **c. de stabilisation** stabilization fund, marketing board
caisse-palette *f., pl.* **caisses-palettes** bulk bin
caissette *f.* small box, tray (for fruit), flat (for punnets); **c. à semis** seed tray; **c. de fenêtre, c. de balcon** window box
cajeput *m.* punk-tree (*Melaleuca leucadendron*) and its oil
cajeputier *m.* = **cajeput** (*q.v.*)
cajou *m.* cashew (see **anacardier, noix d'acajou, pomme-cajou**)
cal *m., pl.* **cals** callus
calaba *m.* calaba (*Calophyllum tacamahaca*) (see also **baume**)
caladénie *f. Caladenia* sp.
caladion, caladium *m. Caladium*, sp. caladium
calalou *m.* dish which includes okra (in West Indies)
calament *m.* calamint (*Calamintha* sp.)
calamiforme *a.* calamiform, reed-like
calamondin *m.* calamondin (*Citrus madurensis*)
calandrinia, calandrinie *f.* rock-purslane (*Calandrinia* sp.)
calcaire *a.* calcareous, chalky; *n. m.* limestone; **"craint le calcaire"** (of plant) lime-hating, calcifugous
calcarifère *a.* calciferous, lime-bearing
calcéolaire *f. Calceolaria* sp., slipperwort; **c.ligneuse, c. à feuilles ligneuses** *C. rugosa* (= *C. integrifolia*)
calcicole *a.* calcicole, calcicolous, lime-loving
calcification *f. Mush.* chalking caused by *Fusarium* sp.
calcifuge *a.* calcifuge, calcifugous, lime-hating (see also **calcaire**)
calcimètre *m.* calcimeter
calcination *f.* calcination; **méthode de calcination** ashing method
calcique *a.* calcic; **nutrition calcique** calcium nutrition
calcium *m.* calcium
cale *f.* hold of a ship; **c. frigorifique c. frigorifique** refrigerated hold
calebasse *f.* calabash; calabash gourd, white-flowered gourd (*Lagenaria siceraria*) (= **gourde**)
calebassé,-e *a.* like a calabash; **poires calebassées** pears attacked by pear midge (*Contarinia pyrivora*)
calebassier *m.* calabash tree (*Crescentia cujete*)
calendule *f.* calendula (*Calendula officinalis*)
calfeutrage *m.* stopping-up chinks and cracks, making draught-proof
calibrage *m.* calibration, grading by size; **c. par le poids** grading by weigh
calibre *m.* calibre, bore: size, diameter; **de calibre moyen** of medium diameter
calibrer *v.t.* to calibrate, to measure
calibreur *m.* **calibreuse** *f.* calibrator, sorter (see also **arracheuse**)
calice *m. Bot* calyx; **stade du calice** fruitlet stage (of apple)
caliciflore *a. Bot.* calycifloral, calyciflorous
caliciforme *a.* calyciform
calicin,-ine *a.* calycinal
calicinaire *a.* calycinal
calicinal,-e *a., m. pl.* **-aux** calycinal, calycine
calicule *m.* calicle
caligni blanc *m. Licania incana* (= *L. crassifolia*), a neotropical tree with edible fruits
calla *f.* calla lily (*Zantedeschia aethiopica* = *Richardia africana*)
calleux,-euse *a.* callous
callicarpe *m.* French, purple mulberry (*Callicarpa* sp.)
callistémon *m. Callistemon* sp.
callistèphe de Chine *m.* China aster (*Callistephus chinensis*) (= **aster, reine-marguerite**)
callogène *a.* callus-forming; **pouvoir callogène** callus-forming capacity
callogenèse *f.* callus formation
callose *f. Bot.* callose
callosique *a.* relating to callose
callosité *f.* callus
callune *f.* heather, ling (*Calluna vulgaris*) (see also **bruyère**)
callus *m.* callus; *Hort.* union
calocoris *m. PC Calocoris* sp. capsid
calopogonium *m. Calopogonium* sp.
calorie *f.* calorie
calorifuge *a.* non-conducting, heat insulating; *n. m.* heat insulator
calorifugeage *m.* heat insulation
calorifuger *v.t.* to insulate, to lag (pipe)
calotte *f. Cit.* whole peel of citrus fruit
caltha, calthe *m. Caltha* sp.; **c. des marais** marsh marigold (*C. palustris*) (= **populage, souci d'eau**)
calus *m.* callus
calville *m.* or *f.* calville apple, queening apple
calycanthe (*m.*) **de la Caroline** Carolina allspice (*Calycanthus floridus*); **c. de la Californie** *C. occidentalis*
calyptre *f. Bot.* calyptra (= **coiffe**); *Vit.* calyptra
calyptré,-e *a. Bot.* calyptrate

camarine *f.* crowberry (*Empetrum* sp.)
camassia *f. Camassia* sp., wild hyacinth
cambium *m.* cambium; **c. liégeux** corky cambium
camélia, camellia *m.* camellia (*Cameilia* sp.)
caméline *f.* gold of pleasure (*Camelina sativa*)
camerounais,-aise *a.* of Cameroon
caminet *m.* cross-leaved heath (*Erica tetralix*) (see also **bruyère**)
camomille (romaine), c. d'Anjou *f.* chamomile (*Anthemis nobilis*); **c. à grandes fleurs** *Pyrethrum* sp.; **c. panachée** *A. mixta*; **c. puante** stinking chamomile (*A. cotula*) (= **maroute**); **petite camomille, c. allemande** wild chamomile (*Matricaria recutita*) (see also **anthrisque, maroute, matricaire**)
campagne *f.* 1. countryside; *Vit.* land and buildings of vineyard, 2. campaign *Ag.*, *Hort.* season
campagnol *m.* vole (*Arvicola* sp., *Microtus* sp. and other spp.); **grand campagnol, c. terrestre** ground vole (*A. terrestris*); **c. méditerranéen, c. provençal** Mediterranean pine vole (*Pitymys duodecimcostatus*)
campanacé,-e *a.* campanulate
campanelle *f.* bindweed (*Convolvulus arvensis*) (= **campanette, liseron**)
campanette *f.* = **campanelle** (*q.v.*)
campaniflore *a.* with bell-shaped flowers, campanulate
campaniforme *m.* campaniform
campanulacées *f. pl.* Campanulaceae
campanule *f. Campanula* sp., bellflower, campanula; **c. carillon** Canterbury bell (*C. medium*) (= **carillon**); **c. à feuille de lierre** ivy campanula (*Wahlenbergia hederacea*); **c. à grandes fleurs** Chinese bellflower (*Platycodon grandiflorus*); **c. à grosse fleur** = **c. carillon** (*q.v.*)
campanulé,-e *a.* campanulate
campanuliflore *a. Bot.* with bell-shaped flowers, campanulate
campernelle *f.* campernelle (*Narcissus odorus*) (= **grande jonquille**)
camphora (*m.*) **(vrai)** (= **camphrier** *q.v.*) (see also **cannellier**)
camphre *m.* camphor
camphrier *m.* camphor tree (*Cinnamomum camphora*); **c. de Bornéo** *Dryobalanops aromatica*
campylotrope *a.* campylotropous
canadienne *f.* spring-tine cultivator (= **cultivateur canadien**)
canaigre *f.* canaigre (*Rumex hymenosepalus*)
canal *m.*, *pl.* **-aux** 1. canal; **c. d'écoulement** drain; **c. d'évacuation** *Irrig.* overflow channel; **c. résinifère** resin duct; **c. sécréteur** *Physiol.* secretory canal. 2. groove e.g. in petiole of leaf (vine)
canaliculé,-e *a.* canaliculate
canalisation *f.* canalization, draining, piping; **c. en plastique** plastic piping
canaliser *v.t.* to canalize, to lay pipes, to pipe
canari *m. Fung.* yellow knight fungus (*Tricholoma flavovirens*) (= **chevalier**)
canarien,-ienne *a.* Canarian
cancerigène *a.* carcinogenic; *n.m.* carcinogen
candélabre *m. Hort.* pyramidal or espalier fruit tree
candidat,-e *n.* candidate; *a.* **arbre candidat** candidate tree
canéficier *m.* = **casse** *q.v.*
canillée *f.* duckweed (*Lemna* sp.) (= **lentille d'eau**)
canistel *m.* canistel, egg-fruit (*Lucuma rivicoa*) (= **jaune d'œuf**)
canna *m.* canna lily (*Canna* sp.); **c. florifère, c. de Crozy** (*C.* × *hortensis*) (see also **balisier**)
cannabin,-e *a.* cannabic
cannabiné,-e *a.* cannabinaceous
cannabis *m. Cannabis* sp., cannabis, hemp (see also **chanvre**)
cannaie *f.* area planted with sugar cane or reeds, canebrake
canne *f.* cane, reed, stick; **c. de bambou** bamboo cane; **c. d'Egypte, c. sauvage** (*Mauritius*) wild sugar cane (*Saccharum (aegyptiacum) spontaneum*); **c. faucheuse** slasher (for weeds etc.); **c. d'Inde** Indian shot plant (*Canna indica*) (= **balisier**); **c. à sucre** sugar cane (*S. officinarum*); **c. de bouche** *Sug.* sugar cane for chewing; **c. de Provence** giant reed (*Arundo donax*); **c. brulée** *Sug.* burned cane; **c. brute** *Sug.* harvested cane as delivered to the factory; **c. de graine** *Sug.* seedling cane; **c. de grande saison, de grande culture** *Sug.* full cane; **c. nette, propre** *Sug.* harvested cane after cleaning; **c. de petite saison, de petite culture** *Sug.* spring cane; **c. plantée, c. vierge** *Sug.* plant cane (see also **repousse**)
canneberge *f.* small European cranberry (*Vaccinium oxycoccus*) and other spp. (see also **airelle**)
cannelé,-e *a.* fluted, channelled, striated
cannelier *m.* see **cannellier**
cannelle *f.* cinnamon (bark); **c. blanche** wild cinnamon (*Canella alba*); **c. de Chine, c. de Cochinchine** (Chinese) cassia; **c. giroflée** clove bark (*Dicypellium caryophyllatum*); **c. de Magellan** Winter's bark (*Drimys winteri*) (= **écorce de Winter**) (see also **cannellier**)
cannellé,-e *a.* cinnamon-coloured
cannellier (*m.*) **(de Ceylan)** (Ceylon) cinnamon (*Cinnamomum zeylanicum*); **c. d'Annam** = **c. de Saïgon** (*q.v.*); **c. de Batavia** = **c de**

Malaisie (*q.v.*); **c. de Chine** Chinese cassia tree (*C. cassia*); **c. d'Indonésie** = **c. de Malaisie** (*q.v.*); **c. de Macassar** = **c. de Malaisie** (*q.v.*); **c. de Malaisie** Indonesian cassia (*C. burmanni*); **c. de Padang** = **c. de Malaisie** (*q.v.*); **c. de Saïgon** Saigon cassia (*C. loureirii*); **c. sauvage** wild cassia (*C. obtusifolium, C. pauciflorum, C. tamala*); black cinnamon (*Pimenta acris) (*= **quatre épices, piment âcre**); **c. du Tonkin** = **c. de Saïgon** (*q.v.*) (see also **cannelle**)

cannelure *f.* fluting, channelling, groove; *Bot.* deep striations; *PD* stem pitting, stem grooving (= **bois strié**)

cannisse *f.* screen made of canes (used for drying figs etc.)

canon *m.* cannon, gun, cylinder, pipe; **c. d'arrosage** boom sprinkler, sprayer; **c. effaroucheur** detonator (for bird scaring); **c. paragrêle** anti-hail gun

canopée *f.* canopy

cantaloup *m.* cantaloup (*Cucumis melo*)

caoutchouc *m.* 1. rubber, caoutchouc; **c. cru** raw rubber; **c. naturel** natural rubber 2. india rubber plant (*Ficus elastica*) (= **plante caoutchouc**); **c. de Ceara**, see **céara**; **c. de Pernambouc**, see **mangaba**; see also **hévéa**

caoutchouté,-e *a.* of rubber, rubberized

caoutchouteux,-euse *a.* rubbery; **coprah caoutchouteux** (*Coconut*) rubbery copra

caoutchoutier,-ière *a.* pertaining to rubber; *n. m.* india rubber plant (*Ficus elastica*)

capable *a.* capable; **c. de germer** (of seed) viable

capacité *f.* capacity; **c. germinative** viability; **c. d'échange** *Ped.* exchange capacity; **c. (moyenne) (de rétention) au champ** *Ped.* field capacity; **c. maximale** *Ped.* maximum capacity; **c. utile d'un sol** *Irrig.* moisture capacity

capendu *m.* *Hort.* type of apple with short peduncle (= **court-pendu**)

capillaire *a.* capillary

capillaire *m.* *Adiantum* sp., maidenhair fern (= **adiante**); **c. cheveux de Vénus** *A. capillus-veneris*; **c. en forme de croissant** *A. lunulatum*; **c. à lobes courbes** *A. curvatum*; **c. peu velue** *A. hispidulum;* **c. tetraphylle** *A. tetraphyllum*

capillarité *f.* capillarity

capité,-e *a.* capitate

capitule *m.* capitulum

capnodage *m.* hand picking of the pest, *Capnodis tenebrionis*

capoquier *m.* kapok tree (= **kapokier** *q.v.*)

capot *m.* *Hort.* (small) hotbed

capparidacées *f. pl.* Capparidaceae

câpre *f.* caper; **c. capucine** nasturtium seed, English caper (see also **capucine**)

capricorne *m.* *Ent.* longhorn beetle; **c. du chêne** oak longhorn beetle (*Cerambyx cerdo*)

câprier *m.* **(commun)** caper bush (*Capparis spinosa*)

câprière *f.* caper plantation

caprification *f.* caprification

caprifier *v.t.* to caprify

caprifiguier *m.* caprifig

caprifoliacées *f. pl.* Caprifoliaceae

capron *m.* hautboy strawberry (fruit)

capronier, capronnier *m.* hautboy strawberry (plant) (*Fragaria moschata*)

capselle *f.* small capsule; *Bot.* shepherd's purse (*Capsella bursa-pastoris*)

capsulaire *a.* capsular

capsule *f.* capsule

capsulifère *a.* *Bot.* capsule-bearing

capteur *m.* captor

capture *f.* capture; **c. d'électrons** electron capture

capua *f.* summer fruit tortrix moth (*Adoxophyes orana*) (= **tordeuse de la pelure**)

capuchon *m.* cap, cover, hood; **c. plastique** plastic hood

capuchon-de-moine *m., pl.* **capuchons-de-moine** monkshood (*Aconitum napellus*) (= **casque de Jupiter**)

capucine *f.* nasturtium (*Tropaeolum* sp.); **c. des Canaris** Canary creeper (*T. peregrinum*); **grande capucine** (ordinary) nasturtium (*T. majus*); **c. de Lobb** *T. peltophorum*; **petite capucine** dwarf nasturtium (*T. minus*) (see also **câpre**); **c. tubéreuse** tuberous nasturtium (*T. tuberosum*)

capuli *m.* Cape gooseberry (*Physalis peruviana*)

capulin *m.* capulin (*Prunus capollin*) (= **cerisier capulin**)

carabe *m.* (carabid) ground beetle

caracoli *m.* & *a. m.* *Coff.* peaberry, coffee berry with only one seed

caractère *m.* character; **c. dominateur** dominant character; **c. héréditaire** hereditary character; **c. récessif** recessive character

caractéristique *a.* characteristic; *n. f.* characteristic, feature; **c. culturale** cultural characteristic

carafe *f.* carafe

carafée *f.* = **giroflée jaune** (*q.v.*)

caragan, caragana *m.* *Caragana* sp.; **c. arborescent** pea tree (*C. arborescens*)

carambole *f.* carambola, Coromandel gooseberry

carambolier *m.* carambola tree (*Averrhoa carambola*) (= **pommier de Goa**)

caramélisé,-e *a.* caramelized

carapace *f.* carapace, shell; **c. (calcaire)** *Ped.* hard (calcareous) pan; **c. ferrugineuse** *Ped.* ironstone cap (see also **cuirasse**)

carbétamide *f.* carbetamide

carbo *m. Vit.* bacterial blight (= **maladie d'Oléron** *q.v.*)

carbonatation *f. Ped.* soil enrichment in calcium or magnesium carbonate

carbonate *m.* carbonate; **c. de chaux** calcium carbonate

carbone *m.* carbon

carboné,-e *a.* carbonaceous; **alimentation carbonée, nutrition carbonée** carbon nutrition; **enrichissement carboné, fertilisation carbonée, fumure carbonée (en serre)** CO_2 enrichment (in greenhouse)

carbonique *a.* carbonic; **bilan carbonique** carbon balance

carboxylant,-e *a.* carboxylating

carboxylation *f.* carboxylation

cardamine *f. Cardamine* sp.; **c. hérissée** hairy bittercress (*C. hirsuta*); **c. des prés** lady's smock (*C. pratensis*)

cardamome *m.* cardamom (*Elettaria cardamomum*); **graine de cardamome** cardamom (seed); **c. du Bengale** Bengal cardamom (*Amomum aromaticum*); **c. du Cameroun** Cameroons cardamom (*Aframomum hanburyi*); **c. d'Ethiopie** Korarima cardamom (*Aframomum korarima*); **c. d'Indochine, c. kravanh, c. krervanh** krervanh (*Amomum krervanh*); **c. de Madagascar** Madagascar cardamom (*Aframomum angustifolium*); **c. de Malabar** Malabar cardamom (*E. cardamomum* var. *minuscula*); **c. du Népal** Nepal cardamom (*Amomum subulatum*); **c. rond** round cardamom (*Amomum kepulaga*); **c. sauvage de Ceylan** Ceylon cardamom (*E. (cardamomum* var.*) major*)

carde *f.* 1. chard. 2. **c. (poirée)** Swiss chard (*B. vulgaris* var. *cicla*) (= **bette à carde, poirée**). 3. teasel head

cardère *f.* teasel; **c. commune, c. sylvestre** common teasel (*Dipsacus fullonum* ssp. *sylvestris*); **c. à foulon** fuller's teasel (*D. fullonum* spp. *fullonum*)

cardinale *f.* cardinal flower (*Lobelia cardinalis*); **c. bleue** blue cardinal flower (*L. syphilitica*)

cardon *m.* cardoon (*Cynara cardunculus*)

cardouille *f.* Spanish oyster plant (*Scolymus hispanicus*) (= **scolyme**)

carence *f.* deficiency; **c. d'eau** water shortage; **c. en bore** boron deficiency; **c. en cuivre** copper deficiency; **c. ferrique** iron deficiency

carencé,-e *a.* deficient; **plante carencée** deficient plant

carène *f. Bot.* keel

caréné,-e *a.* carinate, keeled

carex *m.* sedge (*Carex* sp.) (see also **choufa, laîche, souchet**)

carillon *m.* Canterbury bell (*Campanula medium*) (= **campanule carillon**)

carisse *f.* Natal plum (*Carissa grandiflora*)

carludovice *f. Carludovica* sp. including Panama hat plant (*C. palmata*)

carmantine *f. Adhatoda (Justicia)* sp.; **c. en arbre** *A. vasica* (= *J. adhatoda*)

carmin *m.* carmine (colour)

carminatif,-ive *a.* & *n. m.* carminative

carminé,-e *a.* carmine-coloured, ruby

carnauba *f.* carnauba wax (from *Copernicia cerifera*)

carné,-e *a.* flesh-coloured

caroncule *f. Bot.* caruncle

carotène *m.* carotene

caroténoïde *a.* & *n. m.* carotenoid

carottage *m.* taking cores (of soil etc.)

carotte *f.* carrot (*Daucus carota*); **carottes bottelées** bunched carrots; **c. grelot** small globular carrot; **c. à racine longue** long-rooted carrot; **c. sauvage** wild carrot (= **daucus carotte**); *Ped.* soil core, core sample

caroube *f.* carob bean

caroubier *m.* carob tree (*Ceratonia siliqua*); **c. de la Guyane** courbaril (*Hymenaea courbaril*) (= **courbaril, copalier d'Amérique**)

carpellaire *a.* carpellary

carpelle *m.* carpel

carpocapse *f.* **(des pommes et des poires)** codling moth (*Cydia pomonella* = *Laspeyresia (Carpocapsa) pomonella*); **c. des châtaignes** chestnut codling moth (*Laspeyresia splendana*); **c. des prunes** plum fruit moth (*Cydia funebrana*) (see also **ver**)

carpogenèse *f. Mush.* sporophore formation

carpologie *f.* carpology

carpologique *a.* carpological

carpomètre *m.* carpometer

carpomorphe *a.* fructiform

carpophage *a.* carpophagous, fruit-eating

carpophore *m. Bot.* carpophore; *Mush.* sporophore

carpophylle *m.* carpophyll

carré,-e *a.* & *n. m.* square; *Hort.* patch (of crop); **c. de choux** cabbage patch; **c. d'élevage (d'agrumes)** (citrus) nursery; **c. latin** *Stat.* Latin square; **c. moyen** *Stat.* mean square; **c. de semis** seedbed

carrière (*f.*) **souterraine** underground quarry

carrossable *a.* suitable for motor vehicles

carte *f.* 1. card; **c. perforée** punched card 2. map; **c. des sols** soil map

carthame *m.* safflower (*Carthamus* sp.) (= **safran bâtard**)

cartilagineux,-euse *a. Bot.* hard, tough

carton *m.* cardboard; **c. ondulé** corrugated cardboard, corrugated paper

cartouche *f. Mush.* portion of spawn (= **galette**)

carvi *m.* caraway (*Carum carvi*) (= **anis des Vosges, cumin des prés**); **c. noix de terre** pignut (*Conopodium majus*)
caryocinèse, caryokinèse *f.* karyokinesis, mitosis
caryogamie *f.* karyogamy
caryologie *f.* karyology, nuclear cytology
caryologique *a.* karyological
caryophyllacées *f. pl.* Caryophyllaceae
caryopse *m.* caryopsis
caryote *m.* *Caryota* palm; **c. brûlant** fish-tail palm (*C. urens*)
caryotype *m.* karyotype
casamangue *f. Afr.* golden apple (*Spondias cytherea*) (= **pommier cythère**)
cascade *f.* cascade
cascara *m.* cascara tree (*Rhamnus purshiana*); **cascara sagrada** *f.* cascara bark, cascara sagrada
cascarille *f.* cascarille bark (from *Croton eluteria*)
case *f.* small dwelling, hut; compartment, bin; **c. lysimétrique** lysimeter
caséolaire *a.* caseous
casier *m. Irrig.* 1. irrigated plot 2. sector (of irrigation scheme)
casque (*m.*) **de Jupiter** monkshood (*Aconitum napellus*) (= **capuchon-de-moine**)
casqué,-e *a. Bot.* galeate, galeiform
cassage *m.* breaking (see also **casse** 2)
cassant,-e *a.* brittle; **bois cassant** brittle wood
cassave *f.* cassava (= **manioc** *q.v.*)
casse *f.*, sometimes *m.* 1. cassia (*Cassia* sp.); **c. fistuleuse** purging cassia (*C. fistula*); **c. du Maryland** American senna (*C. marilandica*); **c. puante** coffee senna (*C. occidentalis*). 2. breaking, damage; *Rubb.* etc. **c. au vent** wind damage
cassement *m.* breaking; **c. partiel** *Hort.* brutting
casserie *f.* factory for breaking-up sugar, splitting peas etc.
cassie *f.*, **cassier** *m.* **c. (de Farnèse), c. (du Levant)** sweet acacia (*Acacia farnesiana*)
cassis *m.* black currant fruit (see also **groseille**)
cassissier *m.* black currant bush (*Ribes nigrum*) (see also **groseillier**)
cassonade *f. Sug.* brown sugar, moist sugar
cassure *f.* break, fracture
castanéicole *a.* relating to chestnuts (*Castanea* sp.); **production castanéicole** chestnut production; **zone castanéicole** chestnut-growing area (see also **châtaigne, châtaignier**)
castanéiculture *f.* chestnut growing
castos *m.* swamp bay (*Magnolia glauca*)
castration *f.* castration; *OP* etc. removal of male and female inflorescences
castreur *m.* *OP* etc. castrator
casuarine *f.* casuarina (*Casuarina* sp.)
catabolisme *m.* catabolism
cataire *f.* catmint (*Nepeta cataria*)
catalase *f.* catalase
catalpa *m.* catalpa (*Catalpa* sp.); **c. commun** common catalpa (*C. bignonioides*); **c. à fleurs roses** *C. fargesii*
catalpa-boule *m.* dwarf catalpa (*C. bignonioides* cv. Nana)
catalyse *f.* catalysis
catalyseur *m.* catalyst
catasète *m.* *Catasetum* sp.
caténulaire *a. Bot.* catenular
catillac, catillard *m.* warden pear
cation *m.* cation; **cations échangeables** exchangeable cations
cattleya *m.* cattleya (*Cattleya* sp.)
caucalide *f.* *Caucalis* sp., bur-parsley
caule *f.* caulis
caulescent,-e *a.* caulescent
caulicole *a.* caulicolous
caulifère *a.* cauliferous
cauliflore *a.* coliflorous
cauliflorie *f.* cauliflory
caulinaire *a. Bot.* cauline, caulinary, relating to the stem; **apex caulinaire** stem apex; **tissu caulinaire** stem tissue
caulogène *a.* stem forming; **effet caulogène** stem (forming) effect
caulogenèse *f.* stem formation, stem growth; bud formation (in stems and in *in vitro* culture)
cavage *m.* 1. excavation 2. storage (in a cellar)
cavaillon *m. Vit.* interrow strip not reached by the plough
cave *f.* cellar; *Vit.* **c. de vinification** winery
cavité *f.* cavity; **c. calicinaire (d'une pomme)** calyx cavity, calyx end, eye (of apple)
cayeu *m., pl.* **-eux** = **caïeu** (*q.v.*)
céanothe *m.* *Ceanothus* sp., Californian lilac
céara *m.* ceara rubber tree (*Manihot glazovii*)
cécidie *f.* gall (= **galle**)
cécidogène *a.* cecidogenous, gall-forming
cécidomyie, cécidomie *f.* gall midge (Cecidomyiidae); **c. du chou** swede midge (*Contarinia nasturtii*); **c. des feuilles du poirier** pear leaf midge (*Dasyneura pyri*); **c. des lavandes** lavender leaf midge (*Thomasiniana lavandulae*); **c. des lentilles** lentil Cecidomyiid (*Contarinia lentis*); **c. des poirettes** pear midge (*Contarinia pyrivora*)
cédrat *m.* citron fruit; **c. Etrog** = **citron-cédrat** (*q.v.*); **c. verruqueux** Karna lemon fruit
cédraterie *f.* citron grove or plantation
cédratier *m.* citron tree (*Citrus medica*); **c. Etrog** = **cédrat Etrog** (*q.v.*); **c. verruqueux** Karna lemon tree (*C. kharna*)
cèdre *m.* cedar; **c. de l'Atlas** Atlas cedar (*Cedrus atlantica*); **c. de Chypre** Cyprus cedar (*C. brevifolia*); **c. à encens** incense cedar

(*Libocedrus decurrens*); **c. de l'Himalaya** deodar, Indian cedar (*C. deodara*); **c. du Liban** cedar of Lebanon (*C. libani*); **c. de Virginie** pencil cedar (*Juniperus virginiana*); **c. nain du Liban** dwarf cedar of Lebanon (*C. libani* var. *nana*)

cédrel, cédrela *m.* *Cedrela* sp.; **c. odorant** cigarbox cedrela (*C. odorata*)

cédron *m.* cedron (*Simaba cedron*); **noix de cédron** *S. cedron* cotyledons

CEE *f.* **Communauté Economique Européenne** EEC European Economic Community

ceinturage *m. Arb.* ringing

ceinture *f.* girdle, belt; *PC* **c. gluante, c. de glu, c.-piège** sticky band (around tree)

céleri *m.* celery (*Apium graveolens*); **c. branche, c. à côtes, c.-côte** blanching celery; **pied de céleri** head of celery

céleri-pomme *m.* = **céleri-rave** (*q.v.*)

céleri-rave *m. pl.* **céleris-raves** celeriac, turnip-rooted celery (*Apium graveolens* var. *rapaceum*)

cellier *m.* storeroom

cellulaire *a.* cellular, relating to cells; **croissance cellulaire** cell growth

cellulase *f.* cellulase

cellule *f.* cell; **c. épidermique** epidermal cell; **c. foliaire** leaf cell; **c. de garde** guard cell; **c.-mère** mother cell; **c. photoélectrique** photo-electric cell; **c. végétale** plant cell; **c. vivante** living cell

cellulose *f.* cellulose

cellulosique *a.* cellulose

célosie *f.* *Celosia* sp., *Afr.* etc. *C. argentea* (used as vegetable); **c. crête-de-coq, c. à crête** cockscomb (*C. cristata*) (= **amarante crête-de-coq, passe-velours**)

cenchrus *m.* buffel grass (*Cenchrus ciliaris*)

cendre *f.* ash; **c. de bois** wood ash

cenelle *f.* haw, red berry

cenellier *m.* (*Canada*) hawthorn (*Crataegus* sp.)

centaine *f.* about a hundred

centaurée *f.* *Centaurea* sp., cornflower; **c. des Alpes** perennial cornflower (*C. montana*) (= **bleuet vivace**); **c. ambrette, c. de Malte** *C. melitensis*; **c. bleuet** cornflower (*C. cyanus*) (= **bleuet**); **c. musquée** sweet sultan (*C. moschata*); **c. jacée** brown-rayed knapweed (*C. jacea*); **c. noire** knapweed (*C. nigra*); **petite centaurée** centaury (*Centaurium erythraea*); **c. du solstice** St. Barnaby's thistle (*C. solstitialis*) (= **chardon doré**)

centranthe *m.* *Centranthus* sp., red, spur valerian

centrifugation *f.* centrifuging

centripète *a.* centripetal

centroséma *m.* centro, centrosema (*Centrosema pubescens*)

centrosome *m.* centrosome

cep *m.* **(de vigne)** *Hort.* vine plant

cépacé,-e *a.* cepaceous

cépage *m.* vine variety; vine cultivar; **c. de cuve** wine grape cultivar; **c. d'élite, c. noble** noble, high quality cultivar

cèpe *m.* **(de Bordeaux)** cep (*Boletus edulis*) (see also **bolet, indigotier, nonette**)

cépée *f. Arb.* stool shoots, coppice

cèphe (*m.*) **du poirier** pear sawfly (*Janus compressus*)

céraiste *m.* *Cerastium* sp., mouse-ear; **c. argenté** snow-in-summer (*C. tomentosum*) (see also **corbeille**); **c. aggloméré** clustered mouse-ear (*C. glomeratum*); **c. commun** common mouse-ear (*C. fontanum*); **c. laineux** = **c. argenté** (*q.v.*)

cérambycide *m.* cerambycid, longhorn beetle (Cerambycidae)

cératite *f.* Mediterranean fruit fly (*Ceratitis capitata*) (= **mouche méditerranéenne**)

cératonia *m.* *Ceratonia* sp., carob (see **caroube, caroubier**)

cercis *m.* *Cercis* sp., Judas tree (= **arbre de Judée**); **c. à fleurs blanches** *C. siliquastrum* var. *alba*

cercle *m.* circle; **c. de sorcière(s)** fairy ring (on lawn etc.); *Vit.* fruiting cane bent over and tied down

cercopide *m. Ent.* cercopid; **c. sanguin** ruddy cercopid (*Cercopis vulnerata* = *C. sanguinea*)

cercosporiose *f.* *Cercospora* and related diseases; **c. des agrumes** citrus leaf spot (*C. angolensis*); **c. du bananier** sigatoka disease (*Mycosphaerella musicola*); **c. noire du bananier** banana black leaf streak (*M. fijiensis*); **c. de la laitue** lettuce leaf spot (*C. longissima*); **c. du palmier à huile** *OP* freckle (*C. elaeidis*)

cérébriforme *a.* cerebriform

cereus *m.* *Cereus* sp.

cerfeuil *m.* chervil (*Chaerophyllum cerefolium*); **c. odorant, c. musqué** sweet cicely (*Myrrhis odorata*); **c. tubéreux** bulbous-rooted chervil (*C. bulbosum*)

cérifère *a.* wax-producing

cerisaie *f.* cherry orchard

cerise *f.* cherry; **c. acide, c. aigre** morello, sour cherry; **c. douce** sweet cherry; *Coff.* **c. (de caféier)** coffee berry, cherry, 'bean' (see also **cerisier**)

cerisette *f.* 1. dried cherry. 2. winter-cherry (*Physalis alkekengi*), nightshade (*Solanum* sp.)

cerisier *m.* 1. cherry tree; **c. acide, aigre** morello, sour cherry (*Prunus cerasus*) (= **griottier**); **c. capulin** = **capulin** *q.v.*; **c. doux**

sweet cherry (*P. avium*) (see also **merisier**); **c. à fleurs, c. du Japon** flowering cherry (*P. serrulata*); **c. franc** = **c. acide** (*q.v.*); **c. à grappes** bird cherry (*P. padus*); **c. merisier** = **merisier** (*q.v.*); **c. de Nanking** downy cherry (*P. tomentosa*); **c. des oiseaux** = **c. doux** *q.v.*; **c. de St. Julien, c. (de) Sainte-Lucie** mahaleb cherry (*P. mahaleb*); **c. nain des sables** *P. besseyi*; **c. des steppes** Mongolian cherry (*P. fruticosa*); **c. tardif** American black cherry (*P. serotina*); **c. de Virginie** choke cherry, Virginian bird cherry (*P. virginiana*). 2. Other genera: **c. d'amour** Jerusalem cherry (*Solanum pseudo-capsicum*); **c. des Antilles** West Indian cherry tree (*Malpighia glabra* = *M. punicifolia*) (= **acérolier**); **c. du Brésil** Brazilian cherry tree (*Eugenia brasiliensis*); **c. capitaine** (*West Indies*) stinging cherry (*M. urens*); **c. de Cayenne** Pitanga, Surinam cherry tree (*E. uniflora*); **c. du Cayor** = **c. du Sénégal** (*q.v.*); **c. carré** = **c. des Antilles** (*q.v.*); **c. de Jérusalem** = **c. d'amour** (*q.v.*); **c. noir (des Antilles)** *E. ligustrina;* **c. du Rio Grande** cherry of the Rio Grande (*E. aggregata*); **c. du Sénégal** jegidi (*Aphania senegalensis*); **c. de Surinam** = **c. de Cayenne** (*q.v.*); **c. de Tahiti** Otaheite gooseberry (*Phyllanthus acidus*)

cernage *m.* = **cernement** (*q.v.*)

cerne *m. Bot.* annual ring (= **couche annuelle**)

cerneau *m., pl.* **-eaux** green walnut (kernel)

cernement *m. Arb. Hort.* 1. girdling, ringing (of tree). 2. digging round a tree, undercutting (before lifting)

cerner *v.t. Arb., Hort.* 1. to girdle, to ring (a tree). 2. to dig round a tree or plant, to undercut (before lifting)

céroplaste (*m.*) **du figuier** wax scale (*Ceroplastes rusci*)

cérosie *f.* = **cire de la canne** *Sug.* wax

céroxyle *m.* Columbia wax-palm (*Ceroxylon andicola*) and other spp.

certificat *m.* certificate; **Certificat d'Obtention Végétale, C.O.V.** *PB* a certificate which protects the rights of the breeder

certification *f.* certification; **c. virologique des arbres fruitiers** certification of virus-tested fruit trees

cervidés *n. m. pl.* Cervidae, the deer family; **dégâts de cervidés** deer damage

cespiteux,-euse *a.* cespitose, turf-like

cestreau *m. Cestrum* spp., bastard jasmine

cétoine *f.* rose beetle, rose chafer; **c. dorée** rose chafer (*Cetonia aurata*); **c. mouchetée** spotted chafer (*Oxythyrea funesta*)

cétone *f.* ketone

cétonique *a.* ketonic

ceutorrhynque *f. Ceutorhynchus* weevil (see also **charançon**)

chadec *m.* (*West Indies*) pummelo, shaddock (*Citrus maxima*) (= **pamplemousse**)

chadouf *m.* shadoof

chai *m.* 1. wine-making plant on a vineyard 2. store for wine and spirits

chaîne *f.* chain; **c. à godets** bucket chain, Persian wheel (= **noria**); **c. à planter** planting line; **c. respiratoire** respiratory chain; **c. sans fin** endless chain

chaintre *m.* or *f.* headland; **plantation en chaintre** *Vit.* vine-growing at margins of fields, marginal vine culture; **taille en chaintre** *Vit.* training to a creeping habit

chair *f.* flesh; **pêche à chair jaune** yellow-fleshed peach; *PD* **chair molle** softening due to internal browning (= **brunissement interne**)

chalaze *f.* chalaza

chalcidien *m. Ent.* chalcid

chalcis *m. Ent.* chalcid; **c. de la pomme, c. du pommier** apple seed chalcid (*Torymus varians*)

chalef *m. Elaeagnus* sp., cherry elaeagnus (*E. multiflora*) (= **goumi**); **c. argenté** silver berry (*E. argentea*); **c. comestible** *E. edulis*; **c. à feuilles étroites** oleaster (*E. angustifolia*) (= **olivier de Bohème**); **c. en ombelle** *E. umbellata*; **c. panaché** *E. pungens aureomaculata*

chaleur *f.* heat, warmth; **les grandes chaleurs** the hot season; **"craint la chaleur"** "keep in a cool place"; **c. de fond** *Hort.* bottom heat; **c. radiante** radiant heat

chamaecyparis *m. Chamaecyparis* sp. (= **faux cyprès**); **c. du Japon** hinoki (*C. obtusa*)

chamaerops, chamérops *m.* European palm (*Chamaerops humilis*)

chambre *f.* room, chamber; **c. à atmosphère contrôlée** gas storage room; **c. à brouillard** mist chamber (= **miste**); **c. chaude** warm room; **c. climatisée, c. conditionnée** growth chamber (see also **cabinet**); **c. de culture** growing room; **c. de floraison** forcing room (for flowers); **c. de multiplication** propagation room

chaméphyte *f.* chamephyte

chamois *a. inv.* buff-coloured

champ *m.* field, plot; **au champ** in the field; **en plein champ** outside, out of doors, as a field crop; **c. d'action** field of action; **c. de comportement** (plot for) performance trial; **c. d'expériences, c. d'expérimentation, c. d'essai** field, plot for trials; **c. semencier** *Plant.* etc. seed garden, seed plot

champenois,-e *a.* of Champagne

champêtre *a.* rustic, rural; **arbre champêtre** isolated tree in field or hedge

champignon *m.* fungus, mould, mushroom,

toadstool; **c. à chapeau** agaric; **c. de couche** cultivated mushroom (*Agaricus bisporus*); **c. entomopathogène** entomopathogenic fungus; **c. de Ferney** (*Mauritius*) (*Tricholoma spectabilis* (see also **tricholome**); **c. de la paille du riz, c. des pailles** paddy straw mushroom (*Volvariella volvacea, V. diplasia*) (= **volvaire**); **c. de Paris** = **c. de couche** (*q.v.*); **c. parasite** parasitic fungus; **c. phytopathogène** pathogenic fungus; **c. saprophyte** saprophytic fungus

champignonnière *f.* mushroom bed, mushroom farm

champignonniste *m.* mushroom grower

chancre *m. Bot.* canker; **c. bactérien** bacterial canker; **c. bactérien des agrumes, c. citrique** bacterial canker of citrus (*Xanthomonas citri*); **c. bactérien de la tomate** bacterial canker of tomato (*Corynebacterium michiganense*); **c. du châtaignier** chestnut canker (*Cryphonectria (Endothia) parasitica*); **c. du collet** crown gall (of apple) (*Agrobacterium radiobacter*); **c. du houblon** hop canker (*Gibberella pulicaris*); **c. du pommier et du poirier** apple and pear canker and eye rot (*Nectria galligena*); **c. du rosier** rose stem canker (*Leptosphaeria coniothyrium*); **c. à tache** patch canker (of rubber) (*Pythium complectens*)

chancreux,-euse *a.* cankerous, cankered; **arbre chancreux** cankered tree

chandelle *f. Vit.* = **pissevin** (*q.v.*)

changement (*m.*) **de couleur** colour change

chanterelle *f.* chanterelle (*Cantharellus cibarius*) (= **girolle**)

chanvre *m.* hemp (*Cannabis sativa*) (see also **chènevis**); **c. d'Afrique** bowstring hemp (*Sansevieria zeylanica*); **c. bâtard** hemp nettle (*Galeopsis tetrahit*); **c. d'eau** (1) hemp agrimony (*Eupatorium cannabinum*), (2) bur-marigold (*Bidens tripartita*), (3) gypsywort (*Lycopus europaeus*); **c. de Guiné** kenaf (*Hibiscus cannabinus*); **c. de Manille** Manila hemp (= **abaca**); **c. de la Nouvelle Zélande** New Zealand flax (*Phormium tenax*)

chanvre-ortie *m.* wood nettle (*Laportea canadensis*)

chanvriculture *f.* hemp growing

chanvrine *f.* hemp agrimony (*Eupatorium cannabinum* (= **chanvre d'eau**)

chapeau *m.* hat; **c. d'évêque** bishop's hat (*Epimedium alpinum*); *Mush.* cap; **c. incliné** tilted cap

chapelet *m.* string (of objects); series (of objects); **c. d'oignons** string of onions

chapelle (*f.*) **de serre** bay, span (of greenhouse)

chaperon *m.* coping (on wall)

chapon *m. Vit.* vine which has not yet fruited; vine cutting

chaptalisation *f.* chaptalization, adding sugar to the must (in winemaking)

charançon *m.* weevil; **c. (noir) du bananier** banana weevil (*Cosmopolites sordidus*); **c. de la carotte** (*Canada*) carrot weevil (*Listronotus oregonensis*); **c. gallicole du chou, c. de la racine des crucifères** turnip gall weevil (*Ceutorhynchus pleurostigma*); **c. des pois** pea beetle (*Bruchus pisorum*); **c. de la prune** (*Canada*) plum curculio (*Conotrachelus nenuphar*); **c. des siliques** cabbage seed weevil (*Ceutorhynchus assimilis*); **c. de la tige du chou** cabbage stem weevil (*Ceutorhynchus quadridens*)

charançonné,-e *a.* weevilled, infested with weevils

charbon *m.* charcoal; **c. actif** activated charcoal; **c. de bois, c. végétal** charcoal; *PD* smut and other diseases; **c. de la canne à sucre** sugarcane smut (*Ustilago scitaminea*); **c. foliaire (du dahlia), c. du dahlia** leaf spot (of dahlia) (*Entyloma dahliae*); **c. (du poireau)** leek smut (*Urocystis cepulae*); **c. de la vigne** black spot of vine (*Elsinoë ampelina*) (= **anthracnose**)

charbonné,-e *a. PD* smutted

charbonnier *m. Fung. Russula cyanoxantha* (see also **palomet, russule**)

chardon *m.* 1. thistle; (*Mauritius*: Mexican prickly poppy, *Argemone mexicana*); **c. acaule** ground thistle (*Cirsium acaule*); **c. argenté** = **c.(-)Marie** (*q.v.*); **c. aux ânes** = **c. d'Écosse** (*q.v.*); **c. béni** blessed thistle (*Cnicus benedictus*); **c. bleu** sea holly (*Eryngium maritimum*) (see also **panicaut**), globe thistle (*Echinops* sp.); **c. bleu des Alpes** alpine eryngo (*Eryngium alpinum*); **c. des champs** creeping thistle (*Cirsium arvense*); **c. doré** St. Barnaby's thistle (*Centaurea solsticialis*); **c. d'Écosse** Scottish thistle (*Onopordon acanthium*); **c. à foulon** fuller's teasel (= **cardère** *q.v.*); **c.(-)Marie** blessed thistle, holy thistle (*Silybum marianum*); **c. (-) Marie à épines blanc d'ivoire** *S. eburneum*; **c. ordinaire** = **c. des champs** (*q.v.*); **c. à petits capitules** slender-flowered thistle (*Carduus tenuiflorus*); **c. Roland** field eryngo (*Eryngium campestre*) (see also **panicaut**); **c. à tête épaisse** Plymouth thistle (*Carduus pycnocephalus*). 2. **c. d'Inde** prickly pear (*Opuntia (ficus-indica)*)

chardonnière *f.* field of fuller's teasel

charge *f.* load, burden, bud load (vine etc.), carrier (in agrochemicals); **c. (en fruits)** fruit load; **c. en terre érodée** quantity of soil eroded; **charge (utile)** *AH* carrying capacity; **c. excessive** *AH* over-stocking

chargement *m.* loading, loading-up; *Hort.* retaining fruiting buds; **c. mécanique de cannes** *Sug.* mechanical loading of cane

charger *v.t.* to load, to fill; *Hort.* to retain fruit buds; **c. une couche** to cover hotbed with topsoil or mould

chargeur *m.*, **chargeuse** *f.* loader, charger (see also **arracheuse**)

chariot *m.* truck, trolley; **c. dévidoir, c. enrouleur (de tuyau d'arrosage)** unwinding/winding trolley (for garden hose), hose trolley

charme *m. Bot.* hornbeam (*Carpinus betulus*); **c. houblon, c. d'Italie** hop hornbeam (*Ostrya carpinifolia*)

charmille *f.* hedge, hedgerow, arbour of hornbeams **(charme)** and other trees

charnu,-e *a.* fleshy, plump; **fruits charnus** fleshy, pulpy fruit; **réceptacle charnu** fleshy receptacle

charpente *f.* framework; **c. d'un arbre fruitier** main branches of a fruit tree; **c. de base** basal framework

charpenté,-e *a.* built, constructed; **plante bien charpentée** plant with a good framework

charpentière *a.* & *n.f.* **(branche) charpentière** main branch. 2. *n. f.* link, tie

charretée *f.* cartload, cartful

charrette *f.* cart; **c. à bras** hand cart, barrow

charrue *f.* plough; **c. à bras** hand plough; **c. billonneuse** ridging plough, ridger; **c. décavaillonneuse** vineyard plough for the **cavaillon** (*q.v.*); **c. défonceuse** subsoiling plough, trenching plough; **c. à disques** disc plough; **c. fouilleuse** subsoiling plough; **c. vigneronne** vineyard plough

charruer *v.t., v.i.* to plough

chasmanthère *f. Chasmanthera* sp.

chasselas *m.* Chasselas grape variety

chasselatier *m.* grower of Chasselas grapes

chasse-taupe *m. pl.* **chasse-taupes** thorn-apple (*Datura stramonium*) (= **stramoine**)

châssis *m.* frame, case; *AM* under-frame (of trolley etc.); **c. (mobile)** sash (of frame); *Hort.* forcing frame, Dutch light; **c.-cloche** *pl. inv.* walled cloche; **c. froid** cold frame; **c. de multiplication** propagation case; **c.-vitré** frame with wooden walls and glass lights; **culture sous châssis** forcing

châtaigne *f.* 1. sweet chestnut, from septate fruit containing 2–5 seeds (see also **marron**) 2. (in West Indies) breadfruit seed; **c. d'eau** water chestnut (*Trapa natans*); **c. de la Guyane** fruit of provision tree (= **pachire aquatique**); **c. de l'Inhambane** Zanzibar oil vine (*Telfairia pedata*); **c. de terre** (1) earthnut pea (*Lathyrus tuberosus*) (= **gesse tubéreuse**) (2) earthnut, pignut (*Conopodium majus*) (= **carvi noix de terre**) (3) earthnut (*Bunium bulbocastanum*) (see also **châtaignier**)

châtaigneraie *f.* chestnut plantation, grove

châtaignier *m.* 1. (sweet) chestnut tree (*Castanea sativa*); **c. d'Amérique** American chestnut (*C. dentata*); **c. chincapin** chinquapin (*C. pumila*); **c. du Japon** Japanese chestnut (*C. crenata*); **c. de Chine** Chinese chestnut (*C. mollissima*) 2. (in West Indies) (seeded form of) breadfruit, breadnut (*Artocarpus altilis*) (= **arbre à châtaigne**); **c. du Brésil** Brazil nut tree (*Bertholletia excelsa*) (= **noyer du Brésil**); **c. de la Guyane** provision tree (*Pachira aquatica*); **c. de Tahiti** Polynesian chestnut (*Inocarpus edulis*); (see also **castanéicole, châtaigne**)

chataire *f.* = **cataire** (*q.v.*)

château *m., pl.* **-eaux** castle; **c. d'eau** water tower

chaton *m. Arb., Hort.* catkin

châtrer *v.t. Hort.* to cut back, to castrate (flower) (see also **castration**)

chaudière *f.* boiler

chauffage *m.* heating; **c. d'appoint** peak heating; **c. de fond** basal heat; **c. du sol** (supplying) bottom heat, soil heating; **c. solaire** solar heating

chauffé,-e *a.* heated

chauffer *v.t.* to heat

chaufferette *f.* heater; **c. anti-gel** heater for frost protection

chaufferie *f.* boiler room, boiler house

chaulage *m.* liming

chauler *v.t.* to lime (soil), to whitewash trees

chaulmoogra *m.* chaulmoogra oil (from *Hydnocarpus* sp. and *Taraktogenos* sp.)

chaume *m.* grass stem, stubble; thatch (on lawn)

chaussage *m. Hort.* earthing up

chausser *v.t. Hort.* to earth up; **c. un arbre** to earth up a tree

chauve-souris *f.* bat (Chiroptera)

chaux *f.* lime; **c. éteinte** slaked lime; **c. fusée** air-slaked lime; **c. grasse** fat lime; **c. hydratée** = **c. éteinte**; **c. magnésienne** magnesian lime; **c. maigre** quiet lime; **c. vive** quicklime; **eau de chaux** lime water; **lait, blanc de chaux** lime-wash, whitewash; **pierre à chaux** limestone

chayote, chayotte *f.* chayote (*Sechium edule*) (= **christophine**)

chéimatobie *f.* winter moth (*Operophtera brumata*) (= **phalène hiémale**)

chéiranthus *m. Cheiranthus* sp., wallflower (see also **giroflée**)

chéiroptérogame *a.* pollinated by bats

chéiroptérogamie *f.* pollination by bats, chiropterogamy

chélate *m.* chelate
chélidoine *f. Chelidonium* sp.; **grande chélidoine** greater celandine (*C. majus*)
chélone *f.* turtlehead (*Penstemon barbatus*)
chemin *m.* way, path, track; **c. d'accès** = **c. de service**; **c. de roulement** rollers in packing shed; **c. de service** access road (in plantation, vineyard)
cheminée *f.* chimney
chênaie *f.* oak grove, plantation
chêne *m.* oak; **c. de Banister** scrub oak (*Quercus ilicifolia*); **c. blanc (de Provence)** white oak (*Q. pubescens*); **c. blanc d'Amérique** white oak (*Q. alba*); **c. chevelu** Turkey oak (*Q. cerris*); **c. écarlat** scarlet oak (*Q. coccinea*); **c. à kermès** kermes oak (*Q. coccifera*); **c. liège** cork oak (*Q. suber*); **c. des marais** pin oak (*Q. palustris*); **c. pédonculé** common oak (*Q. robur*); **c. pubescent** = **c. blanc (de Provence)**; **c. pyramide** cypress oak (*Q. robur fastigiata*); **c. quercitron** black oak (*Q. velutina*); **c. rouge d'Amérique** red oak (*Q. borealis*); **c. rouvre** Durmast oak (*Q. petraea*); **c. saule** willow oak (*Q. phellos*); **c. sessile** = **c. rouvre** (*q.v.*); **c. tauzin** *Q. toza*; **c. vélani** Valonia oak (*Q. aegylops* and allied species) (= **vélani**); **c. vert, c. yeuse** holm oak (*Q. ilex*) (= **yeuse**)
chéneau *m., pl.* **-eaux** gutter (of roof), channel
chêneau *m., pl.* **-eaux** 1. young oak. 2. wall germander (*Teucrium chamaedrys*), mountain avens (*Dryas octopetala*) (see also **véronique**)
chêne-liège *m., pl.* **chênes-lièges** cork-oak (*Quercus suber*)
chènevière *f.* hemp field
chénevis m. hemp seed (see also **chanvre**)
chènevotter *v.i. Vit.* etc. to grow weakly (like hemp fibres)
chenille *f.* caterpillar; **c. arpenteuse** looper caterpillar; **c. bourrue** arctid moth caterpillar (see also **écaille**); **c. défeuillante** defoliating caterpillar; **c. fileuse** = **hyponomeute** (*q.v.*); **c. à fourreau de l'asperge** asparagus sheath caterpillar (*Hypopta caestrum*); **c. légionnaire** "army worm"; **c. limace** (*Cacao*), *Coff.* stinging caterpillar (*Parasa vivida*); **c. marteau, c. queue de rat** *Coff.* caterpillar of *Epicampoptera strandi* and others; **c. mineuse** borer (see **mineur**); **c. processionnaire** processionary caterpillar; **c. à tête noire** (*Coconut*) black-headed caterpillar (*Nephantis serinopa*); **c. tordeuse des feuilles** leaf-roller; **c. verte et noire (des agrumes)** caterpillar of orange dog butterfly (*Papilio demodocus*)
chénopode *m. Chenopodium* sp.; **c. blanc** fat hen (*C. album*); **c. à feuille de viorne** grey goosefoot (*C. opulifolium*); **c. des murs** wall goosefoot (*C. murale*)
chercheur *m.* **(scientifique)** research worker, researcher
cherelle *f.* (*Cacao*) cherelle, pod at an early stage of development
chergui *m.* hot, desiccating wind (in North Africa)
chérimbélier *m.* Otaheite gooseberry (*Phyllanthus acidus*) (= **cerisier de Tahiti, surette**)
chérimole *f.* cherimoya fruit
chérimolier *m.* cherimoya (*Annona cherimolia*); **c. des terres basses** ilama (*A. diversifolia*) (see also **anone, corossol**)
chérimoyer = **chérimolier** (*q.v.*)
chervi(s) *m.* skirret (*Sium sisarum*) (= **berle des potagers**); caraway (see **carvi**)
chétif,-ive *a.* weak
chevalier *m. Nung.* yellow knight fungus (= **canari**) (see also **tricholome**)
chevauchant,-e *a.* overlapping; *Bot.* equitant
chevelu,-e *a. Hort.* comose; *n. m.* **c. (racinaire)** root hairs; (fibrous) root system
chevelure *f.* 1. *Bot.* coma, 2. **chevelure de Vénus** = **cheveux de Vénus** (*q.v.*)
cheveux (*m.*) **de Vénus** maidenhair (*Adiantum capillis-veneris*) (= **capillaire**); love-in-a-mist (*Nigella damascena*) (= **nigelle de Damas**)
chevillage *m.* pegging; *Hort.* **c. (de l'artichaut)** increasing globe artichoke bract volume by inserting pegs in the capitulum
cheville *f.* peg, pin
chèvrefeuille *m.* 1. honeysuckle (*Lonicera* sp.); **c. des Alpes**, *L. alpigena*; **c. des bois** woodbine, honeysuckle (*L. periclymenum*); **c. d'Italie** *L. etrusca*; **c. du Japon** Japanese honeysuckle (*L. japonica*); **c. des jardins** perfoliate honeysuckle (*L. caprifolium*); **c. en arbuste, c. des haies** fly honeysuckle (*L. xylosteum*) 2. Other genera. **c. blanc des marais** *Azalea viscosa*
chevrier *m. Hort.* small green kidney bean
chevron *m.* rafter
chia *m.* chia (*Salvia chia*)
chiche *m.* **(pois) chiche** chick pea (*Cicer arietinum*)
chicon *m.* cos lettuce: chicory (head), chicon
chicorée *f.* chicory; **c. amère, c. sauvage** chicory (*Cichorium intybus*); **c. barbe-de-capucin** variety of *C. intybus*; **c. de Bruxelles, c. witloof** witloof chicory (*C. intybus*) (= **endive**); **c à café** chicory grown for its root, used as a coffee additive; **c. frisée** (curled-leaved), **c. scarole** (broad-leaved) endive (*C. endivia*); **c. rouge de Vérone** radicchio-type chicory (*C. intybus*)
chicot *m.* 1. stump; *Hort.* stub, snag. 2. young chicory sprout (after leaf cutting). 3. **chicot**

du Canada Kentucky coffee tree (*Gymnocladus dioica*)
chiendent *m.* **(rampant)**, couch grass (*Agropyron repens*) (= **petit chiendent**); **c. (pied de poule)** Bermuda grass (*Cynodon dactylon*) (= **gros chiendent**); (in *Mauritius*, **chiendent** = *C. dactylon*; **c. patte de poule, gros chiendent** = goosegrass, *Eleusine indica*)
chiffonne *f.* slender brindle with wood and fruit buds
chignon *m. Hort.* top of Brussels sprout plant
chilli *m.* red pepper (*Capsicum frutescens*) (= **piment enragé**)
chimère *f.* chimera, graft hybrid
chimérique *a.* chimeral
chimioluminescence *f.* chemiluminescence
chimiostérilant,-e *a.* & *n.m.* chemosterilant
chimiostérilisation *f. Ent.* chemosterilization
chimiotaxonomie *f.* chemotaxonomy
chimiotaxonomique *a.* chemotaxonomic
chimiothérapie *f.* chemotherapy
chimiotropisme *m.* chemotropism
chimiotype *m.* chemotype
chimonanthe (odorant) *m.* winter sweet (*Chimonanthus fragrans*)
china-grass *m.* ramie, China grass (*Boehmeria nivea*)
chincapin *m.* chinquapin, dwarf chestnut (*Castanea pumila*)
chinois *m.* 1. *Hort.* see **bigaradier chinois**. 2. small candied green orange
chiocoque *m.* snowberry (*Chiococca racemosa*)
chionodoxa *m.* glory-of-the-snow (*Chionodoxa* sp.)
chionophile *a.* (of plant) which can live in the snow
chiquage *m.* chewing (of tobacco, betel etc.)
chique *f.* quid (of tobacco, betel etc.)
chiquer *v.t.* to chew (tobacco, betel etc.)
chiqueur,-euse *n.* chewer (of tobacco, betel etc.)
chirimoyer *m.* = **chérimolier** (*q.v.*)
chironome *m.* chironomid midge
chitinase *f.* chitinase
chitine *f.* chitin
chitoun *m.* (*Breton*) *PD* cauliflower bract proliferation (= **bractation**)
chlorambène *m.* chloramben
chloramphénicol *m.* chloramphenicol
chloranthie *f.* chloranthy (= **virescence**)
chlorate *m.* chlorate; **c. de soude** sodium chlorate
chlore *m.* chlorine
chloré,-e *a.* chlorinated
chlorhydrate *m.* hydrochlorate
chlorophyllase *f.* chlorophyllase
chlorophylle *f.* chlorophyll
chlorophyllien,-enne *a.* chlorophyllous, chlorophyllian
chlorophyllogenèse *f.* chlorophyll production
chloropicrine *f.* chloropicrin
chloroplaste *m.* chloroplast; **c. d'épinard** spinach chloroplast; **c. isolé** isolated chloroplast
chloroplastique *a.* relating to chloroplasts; **membrane chloroplastique** chloroplast membrane; **résistance chloroplastique** chloroplast resistance
chlorosant,-e *a.* chlorosis-inducing
chlorose *f. Bot.* chlorosis; **c. calcaire** lime-induced chlorosis; **c. ferrique** iron chlorosis; **c. infectieuse** *Cit.* infectious variegation, infectious chlorosis virus; **c. et nécrose foliaire des géraniacées** *PD* pelargonium leaf curl virus; **c. des nervures du groseillier épineux** *PD* gooseberry vein-banding virus
chlorosé,-e *a.* chlorotic
chloroser *v.i.* to become chlorotic
chloroxuron *m.* chloroxuron
chlorthiamide *f.* chlorthiamid
chlorure *m.* chloride; **c. d'ammonium** ammonium chloride; **c. de polyvinyle** polyvinyl chloride (PVC); **c. de potassium** potassium chloride; **c. de sodium** sodium chloride
choc *m.* shock, impact; **c. physiologique** physiological shock; *Physiol.* pulse; **chocs CO_2** CO_2 pulses
cholate *m.* cholate; **c. de sodium** sodium cholate
choline *f.* choline
chomé,-e *a.* (*Tunisia*) overripe (of olives)
chondrille (*f.*) **à tige de jonc** skeleton weed (*Chondrilla juncea*)
chondriome *m.* chondriome
chorologie *f.* chorology, science of geographical distribution
chou *m.* 1. cabbage (*Brassica oleracea*); **c. blanc** white cabbage; **c. borgne** cabbage which fails to heart; **c. brocoli** broccoli; **c. brocoli à jets** calabrese, sprouting broccoli; **c. de Bruxelles** Brussels sprouts; **c. cabus** headed cabbage; **c. caraïbe** cocoyams (*Colocasia antiquorum* and *Xanthosoma sagittifolium*); **c. de Chine, c. chinois** Chinese cabbage (*B. chinensis, B. pekinensis*) (see also **pak-choï, pe-tsai**); **c. à choucroute** sauerkraut cabbage; **c. frisé** borecole, curly kale; **c. frisé d'Afrique** Texel greens (*B. carinata*); **c. d'hiver** winter cabbage; **c. hiverné** overwintered cabbage; **c. marin** seakale (*Crambe maritima*); **c. de Milan, c. frisé-pommé** Savoy cabbage; **c. noir** black mustard (*B. nigra*) (= **moutarde noire**); **c. d'ornement** ornamental cabbage; **c. pommé** cabbage with well-rounded heart; **c. de printemps** spring cabbage; **c. rouge, c. roquette** red cabbage; **c. vert** kale (see also

chou-fleur, chou-navet, chou-palmiste, chou-rave) 2. Other species. **c. de Kerguélen** Kerguelen cabbage (*Pringlea antiscorbutica*)
choucas *m.* jackdaw (*Corvus monedula*)
choucroute *f.* sauerkraut
choufa *m.* chufa (*Cyperus esculentus*) (= **souchet comestible**)
chou-fleur *m., pl.* **choux-fleurs** cauliflower (*Brassica oleracea* var. *botrytis*); **c.-fleur d'hiver** winter cauliflower (= **(c.) brocoli**)
chou-gras *m., pl.* **choux-gras** (*Canada*) fat hen (*Chenopodium album*)
chou-moutarde *m., pl.* **choux-moutardes** Indian mustard (*Brassica juncea*) (= **moutarde brune**)
chou-navet *m., pl.* **choux-navets** swedish turnip, swede (*Brassica napus* var. *napo-brassica*) (= **rutabaga**)
chou-palmiste *m., pl.* **choux-palmistes** palm cabbage (growing point of various palm species eaten as a vegetable)
chou-rave *m., pl.* **choux-raves** kohl-rabi (*Brassica oleracea* var. *caulorapa*)
cho-yo *m.* China paeony (*Paeonia albiflora*)
christe-marine see **criste-marine**
christophine *f.* = **chayote** (*q.v.*)
chromatine *f.* chromatin
chromatographie *f.* chromatography; **c. sur couche(s) mince(s)** thin-layer chromatography; **c. d'échange d'ions** ion exchange chromatography; **c. (en phase) gazeuse (CPG)** gas chromatography; **c. gaz-liquide** gas-liquid chromatography; **c. sur papier** paper chromatography; **c. liquide haute performance** high performance liquid chromatography; **c. liquide sous haute pression** high pressure liquid chromatography
chromatographique *a.* chromatographic
chromatropisme *m.* chromatropism
chrome *m.* chromium
chromoplaste *m.* chromoplast
chromoprotéine *f.* chromoprotein
chromosome *m.* chromosome; **c. annulaire** ring chromosome; **c. linéaire** rod chromosome
chromosomique *a.* chromosome, chromosomal; **aberration chromosomique** chromosome aberration; **nombre chromosomique** chromosome number
chronique *a.* chronic
chronoséquence *f.* chronosequence; **c. de sols** soil chronosequence
chrysalidation *f.* pupation
chrysalide *f* chrysalis, pupa
chrysalider(se) *v. p.* to turn into a chrysalis (of caterpillar)
chrysanthème *m.* chrysanthemum (*Chrysanthemum* sp.); **c. d'automne** autumn-flowering chrysanthemum; **c. à carène** tricolor chrysanthemum (*C. carinatum*); **c. chinois** *C. sinense*; **c. frutescent** marguerite, Paris daisy (*C. frutescens*); **c. de l'Inde** Japanese chrysanthemum (*C. indicum*); **c. insecticide** *Pyrethrum* sp.; **c. japonais** *C. japonicum*; **c. des jardins, c. à couronne** garland chrysanthemum, crown daisy (*C. coronarium*); **c. des moissons** corn marigold (*C. segetum*); **c. à petites fleurs** Michaelmas daisy (*Aster* sp.) = **marguerite d'automne**)
chrysanthémiste *m.* chrysanthemum grower
chrysobalanier *m.* cocoplum, icaco plum tree (*Chrysobalanus icaco* = *C. purpureus*) (= **icaquier**)
chrysorrhée *f.* brown tail moth (*Euproctis chrysorrhoea*)
chufa *m.* = **choufa** (*q.v.*)
chute *f.* drop, fall; **c. de grêle** hailstorm; **c. des aiguilles (d'un conifère)** needle cast (of conifer); **c. des bourgeons du poirier** *PD* pear bud drop virus; **c. des boutons, c. des bourgeons** bud drop; **c. d'été (de l'olivier)** fruit drop (of olive); **c. des feuilles** leaf fall; **c. de juin, c. (physiologique) des fruits** June drop; **c. des pétales** petal fall; **c. printanière** = **c. de juin; c. des fleurs femelles** (*Coconut*) button shedding; **c. des noix immatures** (*Coconut*) nut fall
cible *f.* target; **faune non cible** *PC* non-target fauna
ciboule *f.* Welsh onion (*Allium fistulosum*): spring onion (*Allium cepa*); **c. vivace** perennial Welsh onion (*A. lusitanicum*)
ciboulette *f.* chives (*Allium schoenoprasum*)
cicadelle *f.* leafhopper (Cicadellidae); **c. du rosier** rose leafhopper (*Typhlocyba rosae*); **c. flavescente, c. verte de la vigne** green vine leafhopper (*Empoasca flavescens*)
cicatrice *f. Bot.* scar; **c. foliaire** leaf scar
cicatriciel,-elle *a.* scar, cicatricial; **bourrelet cicatriciel** scar tissue
cicatrisant,-e *a.* (wound) healing; **mastic cicatrisant** grafting compound
cicatrisation *f.* cicatrization, wound healing, scarring over
cicatriser *v.t.* to heal; *v.i.* to heal up, to scar over
cidre *m.* cider
cidricole *a.* cider-producing, relating to cider; **verger cidricole** cider-apple orchard
cierge *m. Bot.* (large) *Cereus* sp. and other spp.; **c. barbe de vieillard** old man cactus (*Cephalocereus senilis*); **c. candélabre** *Lemaireocereus weberi* (= *Cereus candelabrum*); **c. lézard** common night-blooming cereus (*Hylocereus undatus*); **c. liane, c. rampant** *Hylocereus triangularis*; **c. monstreux du Pérou, c. rocher** rock cactus

(*Cereus peruvianus* var. *monstrosus*); **c. rouge, c. rampant à grandes fleurs** night-blooming cereus (*Selenicereus grandiflorus*); **c. à tête blanche** *Cephalocereus leucocephalus*
cigale *f.* cicada
cigare *m.* cigar; *Ban.* leaf cylinder, top shoot of growing plant; *Hort.* dried globe artichoke axillary shoot for propagation
cigarier, cigareur *m. Ent.* leaf roller; (hazel) leaf roller weevil (*Byctiscus betulae*) (= **urbec**)
ciguë *f.* hemlock (*Conium maculatum*); **petite ciguë** fool's parsley (*Aethusa cynapium*)
cil *m. Bot.* stiff hair on leaf margins
cilié,-e *a.* ciliate
cime *f.* top (of tree), summit (of hill)
cimentation *f. Ped.* cementation
cimicifuge *f.* bugbane (*Cimicifuga foetida*)
cinéraire *f. Cineraria* sp., cineraria; **c. des fleuristes, c. des horticulteurs** (*Cineraria cruenta* = *Senecio cruentus*); **c. hybride** hybrid cineraria; **c. maritime** *Senecio cineraria* (= *Cineraria maritima*)
cinétine *f.* kinetin
cinétique *a.* kinetic; *n. f.* kinetics
cinnamome *m.* 1. cinnamon tree (*Cinnamomum zeylanicum*) and *Cinnamomum* sp. (see **cannellier**). 2. cinnamon
cintré,-e *a.* arched; bent, curved; **toit cintré** barrel roof
cirage *m.* waxing, polishing; **c. des pommes** waxing of apples (see also **cire**)
circadien,-enne *a.* circadian
circoncision *m. Hort.* ringing of fruit trees
circulaire *a.* circular
circulation *f.* circulation; **c. hydrique** water flow, water transfer (in plant); **c. de la sève** flow of sap
circumnutation *f.* circumnutation
cire *f.* wax; **c. d'abeilles** beeswax; **c. cuticulaire** cuticular wax; **c. à greffer** grafting wax (= **mastic à greffer**); **c. de pomme** apple wax; **c. de la canne** *Sug.* wax
ciré,-e *a.* waxed, polished
cireux,-euse *a.* waxy (see also **bouclier**); **structures cireuses** wax structures
cirier,-ière *a.* wax-producing; *n. m.* wax myrtle (*Myrica cerifera*)
cirral,-e *a., m. pl.* **aux** cirrous, having cirri
cirre *m. Bot.* cirrus, tendril
cirreux,-euse *a. Bot.* cirrous, having cirri
cirrifère *a.* cirriferous
cirriforme *a.* cirriform
cisaille *f. sing.* or *pl.* shears, cutter; **c. à bordures, c. à ébarber, c. à gazon** edging shears; **c. électrique** electric shears; **c. à haies** hedge-clippers, hedge shears
cisailleuse *f.* mechanical pruner, cutter
ciseau *m.* chisel; **c. à rejet** (*Date*) offshoot cutter (= **pince à rejet**); *pl.* scissors, shears; **ciseaux à gazon** edging shears; *Vit.* **c. à vendange** grape shears
ciselage *m. Hort., Vit.* fruit thinning; (*Date*) etc. bunch thinning
ciseler *v.t. Hort.* to thin (out) fruit
cisse, cissus *m. Cissus* sp.
ciste *m. Cistus* sp., rock rose; **c. cotonneux** hoary cistus (*C. albidus*)
citerne *a.* cistern; *Hort.* pitcher (of bromeliad)
citragon *m.* balm (*Melissa officinalis*)
citrange *m.* citrange (*Citrus sinensis* × *Poncirus trifoliata*); **c. Troyer** Troyer citrange
citrate *m.* citrate
citrique *a.* citric
citron *m.* 1. lemon (see also **citronnier**); **c. verruqueux** rough lemon (*Citrus jambhiri*); **c. vert** lime (see also **lime**) 2. **c. de mer** *Afr.* seaside plum, wild lime, wild olive (*Ximenia americana*) (= **prunier de mer, ximénie**)
citron-cédrat *m., pl.* **citrons-cédrats** Etrog citron (*Citrus limonomedica*)
citronnade *f.* lemon squash
citronnelle *f. Bot.* citronella (*Andropogon nardus*); southernwood (*Artemisia abrotanum*) (= **aurone**), and also other plants which smell of lemon e.g. *Melissa officinalis* (see also **lemon-grass**)
citronnelle-verveine *f.* lemon verbena (*Lippia citriodora*)
citronnier *m.* lemon tree (*Citrus limon*) (see also **citron**); **c. bigaradier** sour orange tree (*C. aurantium*) (= **bigaradier**); **c. à trois feuilles** trifoliate orange tree (*Poncirus trifoliata*) (see also **limettier**)
citrouille *f.* pumpkin, summer squash (*Cucurbita peop*)
civ = **culture *in vitro*** (*q.v.*)
cive, civette *f.* = **ciboulette** *q.v.*
civière *f. Hort.* hand barrow
cladiaie *f.* bog with fen sedge (*Cladium mariscus*) (see also **marisque**)
cladode *m.* cladode
cladonie *f.* cladonia
cladosporiose *f.* disease caused by *Cladosporium* sp.; **c. des cucurbitacées** cucumber gummosis (*C. cucumerinum*) (= **nuile grise**); **c. de la tomate** tomato leaf mould (*Fulvia fulva* = *Cladosporium fulvum*)
claie *f.* 1. wattle, hurdle, screen; **claies d'ombrage, claies à ombrer** shading screens; **c. de roseaux** reed hurdle. 2. screen, riddle
clair,-e *a.* clear, bright, light; **vert clair** *a. inv.* light green (see **sombre**)
clairière *f.* clearing, glade
clairsemé,-e *a.* sparse, thinly-sown, scattered
clarkia *m.*, **clarkie** *f. Clarkia* sp., clarkia; **clarkia**

élégant *C. elegans*; **clarkie gentille** *C. pulchella*
classe *f.* class, division, grade
classification *f.* classification
classique *a.* standard, approved; **fumure classique** standard dressing; **méthode classique** standard method; **traitement classique** standard treatment
clasterosporium *m. Clasterosporium* sp., peach shot hole (*C. carpophilum* = *Stigmina carpophila*) (= **maladie criblée**)
clavaire *f. Fung. Clavaria* sp.
clavalier *m. Zanthoxylum* (*Xanthoxylum*) sp. (= **xanthoxylum** *q.v.*); **c. à feuilles de frêne** *X. fraxineum*; **c. poivrier** Japan pepper (*Z. piperitum*) (= **poivrier du Japon**)
clavifolié,-e *a.* bearing clavate leaves
claviforme *a. Bot.* clavate, club-shaped
clayette *f.* flat, hurdle, tray; *Mush.* quantity of mushrooms equal to 24 **maniveaux** (1 **maniveau** = 2–3 litres); *Vit.* drying rack for grapes
clayon *m. Vit.* = **clayette**
claytone de Cuba *f.*, **c. perfoliée** winter purslane (*Claytonia perfoliata*) (= **pourpier d'hiver**)
claytonie *f.* = **claytone**
clé ficheuse *f.* metal attachment for pushing in stakes with the foot
cléistogame *a.* cleistogamic
cléistogamie *f.* cleistogamy
clématite *f.* clematis (*Clematis* sp.); **c. azurée** *C. patens;* **c. blanche odorante** *C. flammula* (= **flammule**); **c. bleue** *C. viticella*; **c. à grandes fleurs** *C. campaniflora*; **c. des haies** traveller's joy (*C. vitalba*) (= **herbe aux gueux**); **c. laineuse** *C. lanuginosa*; **c. toujours verte** virgins' bower (*C. cirrhosa*)
clémentine *f.* clementine
clémentinier *m.* clementine tree (*Citrus clementina*)
clémentinieraie *f.* clementine grove or plantation
cléome *m. Cleome* sp., spider flower
clèthre *m. Clethra* sp., including white alderbush (*C. acuminata*)
climacique *a.* relating to a climax; **végétation climacique** climax vegetation
climat *m.* climate; **c. méridional** southern (French) climate; *Vit.* locality in which vineyards are established
climatérique *a.* climacteric; *n. f.* climacteric
climatique *a.* climatic
climatisation *f.* air conditioning; **c. avec nébulisation** conditions for mist propagation
climatisé,-e *a.* air-conditioned
climatiseur *m.* air-conditioner
climatologie *f.* climatology; **c. agricole** agricultural climatology, agroclimatology
climatologique *a.* climatological
clinandre *m.* clinandrium (in orchids)
clitocybe *m. Fung. Clitocybe* sp.; **c. nébuleux** cloudy clitocybe (*C. nebularis*)
clivia *m.*, **clivie** *f.* clivia (*Clivia* sp.)
clochage *m. Hort.* growing plants under cloches; *Vit.* vine fumigation under cover
cloche *f. Hort.* bell-glass, cloche; **c. plastique** plastic cloche, hot cap; **cloches d'Irlande** = **clochettes d'Irlande** (*q.v.*)
clochette *f.* small bell; *Hort.* small bell-shaped flower; **c. d'hiver** snowdrop (*Galanthus nivalis*) (= **perce-neige**); **clochettes d'Irlande** bells of Ireland (*Moluccella laevis*) (= **fleur-coquillage**)
cloison *f.* partition, division; **c. cellulaire** cell wall
clonage *m.* cloning
clonalisation *f.* cloning
clone *m.* clone; **c. de greffe** *Rubb.* etc. set of trees budded with budwood from the same tree; **c. sanitaire** healthy clone (see also **jeune clone**)
cloner *v.t.* to clone
cloporte *m.* woodlouse
cloque *f.* lump, swelling; *Hort.* blight; **c. de l'azalée** azalea leaf gall (*Exobasidium japonicum*); **c. du pêcher** peach leaf curl (*Taphrina deformans*); **c. du poirier** pear leaf blister (*T. bullata*)
cloqué,-e *a. Hort.* blighted, diseased, curled (leaf)
clos *m.* enclosure; *Vit.* **c. (de vigne)** walled vineyard
clôture *f.* enclosure, fence; **c. infranchissable** impassable fence; **c. en lattis** paling fence
clou (*m.*) **de girofle** clove (see also **giroflier**)
cloucourde *f.* cherry pie (*Heliotropium peruvianum*) (= **héliotrope du Pérou**); **coquelourde** (*q.v.*); hepatica (*Hepatica triloba* = *Anemone hepatica*)
coagulation *f.* coagulation
coagulum *m.* 1. coagulum. 2. coagulant
cobalt *m.* cobalt
cobéa *m.*, **cobée** *f. Cobaea* sp., cup and saucer plant, Mexican ivy (*C. scandens*)
coca *m.* coca
cocaïer *m.* coca shrub (*Erythroxylum coca, E. novogranatense*)
coccidés *m. pl.* Coccidae (see also **cochenille**)
coccidiphage *a. Ent.* scale-eating
coccinelle *f.* coccinellid, ladybird; **c. prédatrice** predatory ladybird
coccolobe *f.* sea grape (*Coccolobis uvifera*) (= **raisinier**)
cocculus *m. Cocculus* sp., including *C. pendulus*, with medicinal properties, and *C. laurifolius*, an ornamental shrub
cochenille *f.* scale insect, scale, mealybug; **c. à bouclier** shield scale; **c. australienne** *Cit.*

cottony cushion scale (*Icerya purchasi*); **c. blanche** = **c. farineuse** (*q.v.*); **c. blanche du palmier dattier** white scale of date palm (*Parlatoria blanchardi*); **c. cannelée** = **c. australienne** (*q.v.*); **c. chinoise** *Cit.* (Chinese) wax scale (*Ceroplastes sinensis*); **c. diaspine** diaspid scale (Diaspididae); **c. farineuse** mealybug (Pseudococcidae); **c. farineuse de l'ananas** pineapple mealybug (*Dysmicoccus brevipes*); **c. flutée** = **c. australienne** (*q.v.*); **c. du mûrier** mulberry scale (*Pseudaulacaspis pentagona*); **c. noire de l'olivier** black scale (*Saissetia oleae*); **c. ostréiforme** European fruit scale (*Quadraspidiotus ostreaeformis*); **c. plate (de l'oranger)** *Cit.* soft brown scale (*Coccus hesperidum*) (= **pou des Hespérides**); **c. rouge du poirier** pear scale (*Epidiaspis leperii*) (= **diaspis**); **c. du rosier** scurfy scale, rose scale (*Aulacaspis rosae*); **c. serpette (des agrumes)** *Cit.* glover scale (*Lepidosaphes gloverii*); **c. tortue** = **c. noire de l'olivier** (*q.v.*); **c. virgule** mussel scale (*Lepidosaphes ulmi*) (see also **pou**)

cochenillier *m.* nopal (*Nopalea cochenillifera*)

cochléaire *a.* cochlear

cochléaria *m.* scurvy grass (*Cochlearia* sp. including *C. officinalis*)

cochylis *f.* grape moth (*Clysiana (Clysia) ambiguella*) (= **teigne de la vigne**)

coco *a.* applied to a **demi-sec** (*q.v.*) variety of bean (*Phaseolus*)

coco *m.*, **noix de coco** coconut; **eau de coco** coconut water, nut water (aqueous liquid within the cavity of the nut); **fibre de coco** coir, coconut fibre; **huile de coco** coconut oil; **lait de coco** coconut milk (milky extract from the fresh ripe kernel); **coco de mer** coco-de-mer, double coconut (*Lodoicea maldivica*); **c. rapé** grated coconut

cocon *m.* cocoon

cocoteraie *f.* coconut plantation, grove

cocotier *m.* coconut palm (*Cocos nucifera*); **c. semi-grand** semi-tall coconut; **grand cocotier** tall coconut; **c. nain** dwarf coconut; **c. du Chili** Chile wine palm (*Jubaea spectabilis* = *J. chilensis*) (= **jubée remarquable**); **c. des Maldives, c. des Séchelles, c. de l'île Praslin** = **coco de mer** (*q.v.*)

coefficient *m.* coefficient; **c. d'héritabilité** coefficient of heritability; **c. thermique** temperature coefficient; **c. de variation** variation coefficient

cœur *m.* heart; **en (forme de) cœur** heart-shaped; *Bot.* heart (of plant, fruit, tree etc.); *Arb.* heartwood; **c.-de-bœuf** bullock's heart (*Annona reticulata*); **c. brun (des pommes)** internal breakdown (of apples); **c. brun du rutabaga** brown heart of swede; **c. creux** hollow heart (of plant); **(maladie du) cœur qui penche** (*Date*) bending head disorder

cœur-de-Jeannette *m.*, *pl.* **cœurs-de-Jeannette, cœur-de-Marie** *m.*, *pl.* **cœurs-de-Marie** bleeding heart (*Dicentra spectabilis*)

cofacteur *m.* cofactor

coffre *m.* coffer, chest, frame, enclosure (for propagation etc.); case (of frame); **c. de bouturage** *Plant.* propagating bed; **c. de germination** germinator

cognasse *f.* wild quince (see also **coing**)

cognassier *m.* quince tree (*Cydonia vulgaris*) (see also **coing**); **c. du Japon** japonica (*Chaenomeles japonica*)

cognée *f.* axe, hatchet

cohésion *f.* cohesion, cohesiveness

coiffe *f.* cover; **c. protectrice** protective cover; *Bot.* root cap, calyptra

coiffé,-e *a. Hort.* overlapping (of leaves forming the heart of lettuce, cabbage etc.), tied (of cos lettuce)

coin *m.* 1. corner. 2. wedge

coing *m.* quince fruit (see also **cognassier**)

coir *m.* coir, coconut fibre (= **fibre de coco**)

coître (de la vigne) *m. Vit.* = **rot blanc** (*q.v.*)

cokrigeage *m. Stat.* co-kriging

col de cygne *m. PD* swan-neck (of tulip)

cola *m.* cola, kola; **noix de cola** kola nut

cola-demi *m. Afr. Cola acuminata* tree

colatier *m.* cola, kola tree (*Cola* sp. including *C. nitida*, gbanja kola) (= **kolatier**); **c. sauvage** abata kola (*C. acuminata*)

colature *f. Irrig.* outlet

colchicine *f.* colchicine

colchique *m. Colchicum* sp.; **c. d'automne** autumn crocus, meadow saffron (*C. autumnale*)

coléophore *m. Coleophora* sp. (Lepidoptera)

coléoptère *m.* coleopterous insect, beetle; *pl.* Coleoptera

coléoptile *f.* coleoptile

coléus *m. Coleus* sp. (see also **pomme de terre de Madagascar**)

colladés *m. pl.* see **collengal**

collant,-e *a.* sticky, tacky; *n. f. OP* seedling in which leaf remains folded

collecte *f.* collecting, gathering; **c. des données** data gathering

collecteur *m. Irrig.* main channel, drain

collembole *m.* springtail (Collembola)

collenchyme *m.* collenchyma

collengal *m.*, *pl.* **colladés** *Afr.* topographical depressions subject to flooding

collerette *f.* collar, ruff; *Bot.* involucre; *Mush.* annulus

collet *m. Bot.* hypocotyl; *Hort.* collar, crown; **c. vert (de la tomate)** *PD* greenback (of tomato)

collétie *f. Colletia* sp., especially *C. ferox*

collier *m.* collar, tie (for trees etc.); *Sug.* collar; **les deux triangles de jonction du collier** dewlap (= **ocréa**)
colloïdal,-e *a., m. pl.* **-aux** colloidal
colloïde *a.* & *n. m.* colloid
colloque *m.* symposium
colluvial,-e *a., m. pl.* **-aux** colluvial
colluvion *f.* colluvium, colluvial deposits; **colluvions sableuses** sandy colluvial deposits
collybie *f. Fung. Collybia* sp.
colmatage *m. Hort.* warping, silting (of land): filling in, clogging up
colmate *f.* warped, silted land
colmater *v.t. Hort.* to warp (land): to fill in, to clog up, to silt; **se colmater** to clog up, to become choked, to silt up
colocase, colocasie *f.* cocoyam, taro (*Colocasia antiquorum*)
colombette *f. Fung.* dove-coloured tricholoma (*Tricholoma columbetta*) (see also **champignon, tricholome**)
colombine *f.* 1. pigeon manure, bird, chicken manure. 2. columbine (*Aquilegia* sp.)
colombo (*m.*) **d'Amérique** American columbo (*Frasera carolinensis*)
colonisation *f.* colonization, invasion (of weeds, pathogens etc.)
coloniser *v.i.* (of plant etc.) to spread; **colonisant facilement** spreading easily
colonnade *f.* colonnade
colonne *f.* column; **c. d'eau** water column, waterspout; *Bot.* column (in Orchidaceae) (= **gynostème**); *Hort.* pillar
coloquinte *f.* colocynth (*Citrullus colocynthis*)
colorant,-e *a.* colouring; *n. m.* colouring matter, dye
coloration *f.* colouring, colour; *Hort.* **c. de fond** ground colour (of fruit etc.)
coloré,-e *a.* coloured
colorimétrie *f.* colorimetry
colorimétrique *a.* colorimetric; **mesure colorimétrique** colorimetric measurement
coloris *m.* colour, colouring
colrave *m.* kohl-rabi (= **chou-rave**)
columelle *f.* columella
comaret *m.* marsh cinquefoil (*Comarum palustre* = *Potentilla comarum*)
combara, combava *m.* Mauritius papeda (*Citrus hystrix*)
combler *v.t.* to fill, to fill in, to fill up; **c. les lacunes, c. les vides** to fill (up) the gaps, to gap up
combustible *m.* fuel; **c. gazeux** gas (fuel); **c. liquide** liquid fuel; **c. solide** solid fuel
comifère *a. Bot.* criniferous
commerce *m.* commerce, trade; **c. du café** coffee trade; **c. de la fleur coupée** cut flower trade
commercialisable *a.* marketable
commercialisation *f.* marketing
communauté *f.* community; **c. végétale** plant community; **Communauté** (*f.*) **Economique Européenne (CEE)** European Economic Community (EEC)
compacité *f.* compactness, density
compact,-e *a.* compact
compactage *m.* compaction
compagnon *m. Bot.* **c. blanc** white campion (*Silene alba*); **c. rouge** red campion (*S. dioica*)
compartimentage *m.* partitioning
compatibilité *f.* compatibility; *Hort.* **c. de greffe, c. au greffage** graft compatibility; **c. hybride** hybrid compatibility
compatible *a.* compatible
compensation *f.* compensation; **point de compensation du CO_2** CO_2 compensation point
complaisant,-e *a.* (of plant) adaptable
complant *m.* collective term for plantations or vineyards on more than one piece of land
complanter *v.t.* to plant with trees, to plant under trees
compléter *v.t.* to complete; *Hort.* = **combler** (*q.v.*)
complexe *a.* & *n. m.* complex; **c. de virus** virus complex; **c. absorbant** *Ped.* absorbing complex
comporte *f. Vit.* etc. wooden tub (used for grape harvest)
comportement *m.* behaviour, habit, performance; **c. au champ** field performance; **c. au froid** reaction to cold; **c. à la torréfaction** *Coff.* roasting quality; **carré de comportement** *Plant.* etc. plot for habit or behaviour observation; **verger de comportement** experimental orchard
composacées *f. pl.* Compositae
composant,-e *a.* & *n.* component, constituent; **composantes du rendement** yield components
composé,-e *a.* & *n. f. Bot.* composite; **composées** *n. f., pl.* Compositae; *n. m. Chem.* compound; **c. odorant** flavour compound; **c. organique** organic compound; **c. phénolique** phenolic compound
compost *m.* compost; **c. d'ordures (ménagères, des villes)** domestic waste, town refuse compost; **c. tourbeux** peat compost; **c. urbain** town compost; **c. usé** spent compost
compostage *m. Hort.* composting
composter *v.t. Hort.* to treat land with compost
compote *f.* compote (of fruits), stewed fruit; **en compote** stewed
compressibilité *f.* compressibility
compression *f.* compression, compaction
comprimé,-e *a.* compressed
comptage *m.* count, enumeration, numbering
compte-gouttes *m. inv.* pipette, eye-dropper

cônaison *f.* cone development (in hops)
concassage *m.* breaking, crushing, pounding
concasser *v.t.* to break, to crush, to pound
concasseur *m.* breaker, crusher, crushing mill; *OP* **c. à palmiste** palm-kernel mill
concave *a.* concave; **c. gum** *Cit.* concave gum (psorosis)
concentration *f.* concentration; **c. létale minimum** minimum lethal concentration
concentré,-e *a.* concentrated; *n. m.* concentrate; **c. émulsifiable** *PC* emulsifiable concentrate
conceptacle *m.* conceptacle
concolore *a.* concolorous, of uniform colour
concombre *m.* cucumber (*Cucumis sativus*); **c. amer** (i) bitter cucumber (*Momordica charantia*) (= **margose**) (ii) *Afr.* egusi (see **goussi**); **c. des Antilles, c. à épines** West Indian gherkin (*Cucumis anguria*); **c. grimpant** climbing cucumber (*Cyclanthera pedata*); **c. porte-bornes** horned cucumber (*Cucumis metuliferus*); **c. porte-soies** *Cucumis dipsaceus*; **c. des prophètes** globe cucumber (*Cucumis prophetarum*); **c. de serre** greenhouse cucumber; **c. de table** table cucumber (see also **cornichon, courge**)
concrescence *f.* concrescence
concrescent,-e *a.* concrescent
concrète *f.* concrete (in perfumery)
concrétion *f.* *Ped.* concretion
concurrence *f.* *Bot.* competition; **c. (des) adventice(s)** weed competition; **c. radiculaire** root competition
concurrencer *v.t.* to compete with
condensation *f.* condensation
condiment *m.* condiment, seasoning
condimentaire *a.* relating to condiments; **plantes condimentaires** condiment plants
condition (*f*) **de croissance** growing conditions
conditionnement *m.* conditioning; packaging; **c. de l'air** air conditioning (= **climatisation**); **centre de conditionnement** grading shed, packing shed
conditionner *v.t.* to condition; to put into good condition; to package
conditionneur *m.* conditioner; **c. de sols** soil conditioner
conductance *f.* conductance, conductivity; **c. électrique** electrical conductivity; **c. foliaire** leaf conductance; **c. stomatique** stomatal conductance
conductibilité *f.* conductivity
conductimétrie *f.* conductimetry
conductivité *f.* conductivity
conduite *f.* 1. management; **c. des cultures** crop management; *Hort.* training; **c. haute** high training. 2. *Irrig.* **c. d'eau** water channel, pipe
condurango *m.* eagle-vine (*Marsdenia (Gonolobus) condurango*)
cône *m.* cone; *Hort.* = **fuseau** (*q.v.*); **c. de houblon** hop cone; **c. de pin** pine cone (= **pomme de pin**)
confire *v.t.* to preserve, to candy
confiserie *f.* confectionery
confiture *f.* jam, marmalade; **c. de singe** genipap (*Genipa americana*) (= **génipa**)
confluent,-e *a.* *Bot.* confluent
conformé,-e *a.* **bien conformé** well-formed; **mal conformé** misshapen
confusion *f.* confusion; **c. sexuelle** *PC* mating disruption
congélation *f.* freezing, deep-freezing, solidification; **c. rapide** quick freezing
congeler *v.t.* to freeze, to deep-freeze, to solidify
conidie *f.* conidium; *pl.* conidia
conifère *a.* coniferous; *n. m.* conifer
conjugaison *f.* conjugation
conné,-e *a.* connate
connectif,-ive *a.* connective; *n. m.* *Bot.* connective (of anther)
connivent,-e *a.* connivent
consanguinité *f.* *PB* inbreeding
conscrit *m.* *Vit.* small grape bunch on lateral shoot
conseiller (*m.*) **horticole** horticultural adviser
conservabilité *f.* storability
conservation *f.* conservation, preservation, storage; **c. en atmosphère contrôlée** controlled atmosphere storage; **c. frigorifique** cold storage; **c. du sol** soil conservation
conserve *f.* preserve, preserved food; **en conserve** preserved, canned; **de conserve, pour la conserve** for canning, for preserving; **haricots en conserve** tinned beans
conserverie *f.* 1. canning industry. 2. cannery
consistance *f.* consistence, consistency; *Ped.* cohesion
console *f.* bracket, wall-bracket
consommable *a.* eatable, edible
consommateur *m.* consumer
consommation *f.* consumption; **c. d'eau** water consumption; **c. en oxygène** oxygen consumption
consommé,-e *a.* = **décomposé** (of manure) (*q.v.*)
consoude *f.* comfrey (*Symphytum* sp.); **c. commune, grande consoude** common comfrey (*S. officinale*)
constante *f.* constant; **c. de Michaelis** Michaelis constant
constituant,-e *a.* constituent; *n. m.* constituent; **constituants volatils** volatile constituents
contagiosité *f.* contagiousness
container *m.* container (see also **conteneur**)
containerisation *f.* = **conteneurisation** (*q.v.*)
contamination *f.* contamination

contaminé,-e *a.* contaminated, infected
contaminer *v.t.* to contaminate, to infect
contenant,-e *a.* containing; *n. m.* (plant) container, jardinière; **c. cache-pot**, jardinière, decorative urn; **c. fixe** fixed (plant) container; **c. mobile** moveable (plant) container
conteneur *m.* container; **c. frigorifique** refrigerated container
conteneurisation *f.* containerization
contenu *m.* content
contingentement *m.* quota system of distribution: apportioning quotas
contour *m.* 1. outline; **c. du limbe** leaf margin. 2. *SC* contour. 3. circuit
contrainte *f.* constraint; **c. hydrique** water stress
contrayerva *f.* *Dorstenia contrajerva*, a perennial herbaceous plant with medicinal properties
contre-bourgeon *m. Vit.* basal bud, dormant bud at base of cane, secondary dormant bud
contre-espalier *m.* row of fruit trees, trained as for **espalier**, but not against a wall
contre-fiche *f.* strut
contrefort *m.* buttress, abutment
contre-greffe *f., pl.* **contre-greffes** double-graft
contre-greffer *v.t.* to double-graft
contre-parasite *m.* predator (of pest)
contreplantation *f.* interplanting, transplanting, lining-out
contreplanter *v.t.* to interplant, to transplant, to line out
contre-poison *m.* antidote
contre-saison *f., pl.* **contre-saisons** out-of-season flowers; **à contre-saison** *adv. phr.* out-of-season
contreventement *m.* wind bracing
contrôle *m.* 1. roll, list, register. 2. inspection, supervision, checking, monitoring; *Hort.* **c. entomologique** entomological check; **c. phytosanitaire** plant health inspection, plant quarantine; **c. de qualité** quality inspection; **c. sanitaire (en pépinière)** roguing (in nursery); **c. des semences** seed inspection, seed testing (see also **lutte**)
contrôlé,-e *a.* inspected, supervised, checked, controlled; **entreposage en atmosphère contrôlée** controlled atmosphere (CA) storage
contrôler *v.t.* to inspect, to supervise, to check, to control, to monitor
contrôleur *m.* inspector, examiner
convection *f.* convection
convexe *a.* convex
convoluté,-e *a.* convolute, convoluted
convolvulacées *f. pl.* Convolvulaceae
convolvulus *m. Convolvulus* sp., convolvulus, morning glory, bindweed (see also **volubilis**)
convoyeur *m.* conveyor
conyse *f.* ploughman's spikenard (*Inula conyza*)
coopératif,-ive *a.* cooperative; *n. f.* cooperative society; **c. bananière** banana (farmers') cooperative; **c. fruitière** fruit (growers') cooperative; **c. horticole** horticultural cooperative
copahier *m.* = **copaïer** (*q.v.*)
copahu *m.* copaiba
copaïer, copayer *m.* copaiba tree (*Copaifera reticulata*) and related species
copal *a.* & *n.* copal resin
copalier *m.* copal tree (*Copaifera* sp.); **c. d'Amérique** courbaril (*Hymenaea courbaril*) (= **caroubier de la Guyane, courbaril**); **c. de Madagascar** *H. verrucosa*
copalme *m.* copalm; **c. d'Amérique** red gum, sweet gum (*Liquidambar styraciflua*)
copeau *m., pl.* **-eaux** shaving, chip of wood or metal
copra, coprah *m.* copra; **c. frais** fresh coconut; **huile de copra** coconut oil (= **huile de coco**)
coprin *m. Coprinus* sp., ink-cap fungus; **c. chevelu** shaggy ink cap (*C. comatus*); **c. noir d'encre** ink cap (*C. atramentarius*)
coque *f.* shell, husk; *Bot.* coccus; **coque du Levant** Indian berry (*Anamirta paniculata*); **c. de cabosse** (*Cacao*) cocoa pod shell; **c. (de la noix)** (*Coconut*) coconut shell; **à c. mince** *OP* var. *pisifera*; **à c. moyenne** *OP* var. *tenera*; **à c. épaisse** *OP* var. *dura*; **à c. très épaisse** *OP* var. *macrocarya*
coquecigrue *f. Hort.* smoke tree (*Cotinus coggygria*) (= **arbre à perruque**)
coquelicot *m.* field, corn poppy (*Papaver rhoeas* and other spp.) (see also **pavot**)
coquelourde (*f.*) **des jardins** rose campion (*Lychnis coronaria*); **c. rose du ciel** rose of heaven (*L. coeli-rosa*) (see also **croix de Malte**)
coqueluchon *m.* aconite (*Aconitum* sp.)
coqueret *m. Physalis* sp., winter cherry (*P. alkekengi, P. franchetii*); **c. anguleux** cowpops (*P. angulata*); **c. comestible, c. pubescent** downy groundcherry, dwarf Cape gooseberry (*P. pubescens*); **c. du Mexique** = **c. anguleux**; **c. du Pérou** Cape gooseberry (*P. peruviana*) (see also **alkékenge**)
coquette *f.* = **zeuzère du poirier** leopard moth (*Zeuzera pyrina*)
coquille *f.* shell, nut-shell
corallien,-ienne *a.* coralline; **sol corallien** coral soil
corbeau *m.* crow (*Corvus* sp.) (= **corneille**); **corbeau freux** rook (= **freux**)
corbeautière *f.* rookery
corbeille *f.* 1. open basket; **c. à fleurs** flower basket; **c. d'argent** garden arabis (*Arabis caucasica*) (= **arabette**); sweet alyssum

(*Alyssum maritimum*); candytuft (*Iberis sempervirens*); snow-in-summer (*Cerastium tomentosum*); **c. d'or** yellow alyssum (*Alyssum saxatile*). 2. flower bed

cordé,-e *a.* cordate; *Vit.* applied to a rounded leaf; **c.-pentagonale** applied to a pentagonal leaf

cordeau *m., pl.* **-eaux** garden line, tracing line, planting line

cordiforme *a.* cordiform, heart-shaped

cordon *m.* strand (of rope etc.), cord, string; *Bot.* funicle; **c. pétiolaire** *Vit.* etc. petiolar vascular traces; **c. vasculaire** vascular strand; *Hort.* (grass) (tree) border; **c. de gazon** turf border; *Arb.* cordon; **c. horizontal** horizontal cordon; **c. oblique** oblique cordon; **c. vertical** upright cordon; **c. mycélien** *Mush.* mycelial strand

cordonnet *m. Mush.* small mycelial strand

cordyla *m.* bush mango (*Cordyla africana*)

coréopsis *m.*, **coréopside** *f. Coreopsis* sp., coreopsis; **coréopside élégante** *C. tinctoria*

corète (*f.*) **du Japon** jew's mallow (*Kerria japonica*); **c. potagère** jute (*Corchorus olitorius*), grown as a vegetable

coriace *a.* tough, leathery

coriandre *f.* coriander (*Coriandrum sativum*)

corme *f.* service berry

cormier *m.* service tree (*Sorbus domestica*) (see also **alisier, sorbier**)

cormus *m.* corm (= **bulbe solide**)

cornaret *m. Martynia fragrans*; also applied to *M. lutea* (= **bicorne**)

corne *f.* horn; **c. rapée** horn chips; **c. torréfiée** horn meal; *Vit.* primary trunk ramification

corné,-e *a.* 1. corneous. 2. horn-like

corne de bélier *f.* = **corne de cerf** 2 (*q.v.*)

corne de cerf *f.* 1. buck's-horn plantain (*Plantago coronopus*). 2. *Aloe arborescens*. 3. **laitue vivace** (*q.v.*)

corne d'élan *f.* staghorn fern (*Platycerium alcicorne*)

corneille *f.* corvid (*Corvus* sp.); **c. chauve** rook (*C. frugilegus*) (= **freux**); **c. d'église, c. des clochers** jackdaw (*C. monedula*) (= **choucas**); **c. mantelée, c. cendrée, c. grise** hooded crow (*C. corone cornix*); **c. noire** carrion crow (*C. corone corone*)

cornet (*m.*) **(de l'enveloppe florale)** cornet

cornichon *m.* gherkin (*Cucumis sativus, C. anguria*) (see also **concombre, courge**)

cornichonier *m.* bilimbi (*Averrhoa bilimbi*)

cornicule *f. Ent.* cornicle

corniculé,-e *a.* corniculate

cornière *f.* angle-iron, angle-bar (in greenhouses etc.)

corniolle *f.* 1. fruit of water chestnut (*Trapa natans*) (= **châtaigne d'eau**). 2. fruit of scorpion senna (= **coronille**)

cornouille *f.* cornel-berry

cornouiller *m.* **(mâle)** cornelian cherry (*Cornus mas*); **c. de la Floride** flowering dogwood (*C. florida*); **c. panaché** *C. alba variegata*; **c. sanguin** dogwood (*C. sanguinea*)

cornu,-e *a. Bot.* cornute, spurred

corollacé,-e *a.* corollaceous

corollaire *a.* corolline; *n. m.* corollary tendril

corolle *f. Bot.* corolla

corollé,-e *a.* corollate

corollifère *a.* corolliferous

corolliflore *a.* corolliflorous

corolliforme *a.* corolliform

corollin,-e *a.* corolline

coronille *f. Coronilla* sp.; **c. glauque** *C. glauca*; **c. des jardins** scorpion senna (*C. emerus*)

coronule *f.* corona, cup (see also **couronne**)

corossol (*m.*) **(épineux)** soursop; **c. bâtard** mountain soursop; **c. cœur-de-bœuf** bullock's heart, soncoya; **c. écailleux** custard apple, sweetsop (= **pomme cannelle**); **c. des marais** pond apple; **c. du Pérou** cherimoya; **c. sauvage** biriba

corossolier (*m.*) **(épineux)** soursop tree (*Annona muricata*) (see also **anone**); **c. bâtard** mountain soursop tree (*A. montana*); **c. cœur-de-bœuf** bullock's heart tree (*A. reticulata*), soncoya (*A. purpurea*); **c. écailleux** custard apple tree, sweetsop (*A. squamosa*) (= **pommier cannelle**); **c. des marais** pond apple tree (*A. glabra*); **c. du Pérou** cherimoya tree (*A. cherimola*) (= **chérimolier**); **c. sauvage** biriba tree (*Rollinia pulchrinervia*)

corozo *m.* corozo nut, ivory nut, vegetable ivory (*Phytelephas macrocarpa*)

correctif,-ive *a.* corrective; **fertilisation corrective** corrective fertilisation

corrélation *f.* correlation; **coefficient de corrélation** correlation coefficient

corrigiole *f. Corrigiola* sp., strapwort

corsé,-e *a.* full-bodied (of wine)

corset *m. Hort.* tree guard

cortex *m.* cortex

cortical,-e *a., m. pl.* **-aux** cortical

corticifère *a.* corticiferous

corticole *a. Ent.* bark-dwelling

cortinaire *m. Cortinarius* sp.

cortiqueux,-euse *a.* corticous, corticose

corvicide *m.* product for corvid (bird) control

corvifuge *m.* corvid repellent

corydale, corydalis *f. Corydalis* sp., corydalis

corymbe *m.* corymb

corymbé,-e, corymbeux,-euse *a.* corymbose

corymbifère *a.* corymbiferous

corymbiforme *a.* corymbiform

corynéum *m.* shot hole disease (*Stigmina carpophila* = *Coryneum beijerinckii*)
cosmétologique *a.* cosmetological
cosmopolite *a.* cosmopolitan
cosmos *m.* *Cosmos* sp., Mexican aster, cosmos
cosse *f.* pod, shell, hull, husk, rind; **à c. jaune** (of bean) wax-podded; **à c. plate** flat-podded; **à c. ronde** round-podded; **à c. violette** purple-podded (see also **gousse**)
cossette *f.* dried sugar-beet or chicory root
cosson *m.* pea beetle (*Bruchus pisorum*)
cossus *m.* *Cossus* sp.; **c. gâte-bois, c. ronge-bois** goat moth (*C. cossus*)
costière *f.* = **côtière** *q.v.*
costus *m.* *Costus* sp., ginger lily
cot *m.* *Vit.* renewal spur
côte *f.* 1. rib, midrib, ridge, striation; **c. d'une feuille** midrib of leaf. 2. *Vit.* hillside, slope
côté *m.* side; **c. du soleil** sunny side
coteau *m., pl.* **-aux** slope, hillside, sometimes used for vineyard
côtelé,-e *a.* ribbed, ridged, striated
côtière *f.* *Hort.* bed placed against a wall, hedge or fence
cotinus *m.* *Cotinus* sp.
cotonéastre, cotonéaster *m.* cotoneaster (*Cotoneaster* sp.); **c. divariqué** *C. divaricata*; **c. à feuille de saule** *C. salicifolia*; **c. de Franchet** *C. franchetii*; **c. d'Henry** *C. henryana*; **c. horizontal** herringbone cotoneaster (*C. horizontalis*); **c. à petites feuilles** *C. microphylla*
cotonneux,-euse *a.* cottony, downy, woolly (of fruit)
cottis *m.* *Vit.* chlorosis
cotylédon *m.* cotyledon
cotylédonaire *a.* cotyledonary
cotylédoné,-e *a.* cotyledonous
couchage *m.* layering (of plants) (see also **marcottage**)
couche *f.* bed layer, stratum (see also **champignon**); *Hort.* frame; hotbed; **c. d'abscission** abscission layer; **c. annuelle** *Bot.* annual ring (= **cerne**); **c. arable** plough layer, topsoil, tilth; **c. chaude, c. chauffée** hotbed; **c. de couverture** *Mush.* casing soil; **c. drainante (d'un pot)** *Hort.* roughage (to drain pot); **c. de forçage** forcing bed; **c. froide** cold frame; **c. de fumier** (1) hotbed, (2) dressing of FYM; **c. génératrice libéro-ligneuse** cambium (= **cambium**); **c. de séparation, c. séparatrice** = **c. d'abscission** (*q.v.*); **c. de sol** soil layer; **c. de terre** layer of earth
couche-couche *f.* (*West Indies*) cush-cush yam (*Dioscorea trifida*) (see also **igname**)
coucher *v.t.* to lay (something) down; **se coucher** (of plants) to lodge
couchis *m.* = **couchage** *q.v.*
coucou *m.* *Bot.* cowslip (see **primevère**); daffodil (see **narcisse**); ragged robin (*Lychnis flos-cuculi*) (= **fleur de coucou**); barren strawberry (*Potentilla sterilis*)
coucourzelle *f.* cocozelle (*Cucurbita pepo*) (= **courge d'Italie**)
coude *m.* *Hort.* (sharp) bend, elbow
coudé,-e *a.* *Hort.* (sharply) bent, elbowed
couder *v.t.* to bend (into an elbow)
coudraie *f.* hazel grove, hazel wood
coudrier *m.* hazel tree, cobnut tree (*Corylus avellana*) (see also **avelinier, noisette, noisetier**); **c. charmeur** witch hazel (*Hamamelis* sp.); **c. du Levant** Turkish hazel (*C. colurna*)
couffe *f.* basket
couffin *m.* = **couffe** (*q.v.*)
cougourde *f.*, **cougourdon** *m.* pumpkin, marrow (*Cucurbita pepo*) (in south of France)
coulant *m.* runner of plant; **c. de fraisier** strawberry runner
coulard,-e *a.* (of plant) producing runners; *Vit.* aborting, failing to set fruit; **cépage coulard** *Vit.* cultivar subject to flower abortion
coulé,-e *a.* *Hort.*, *Vit.* aborted; **grappe coulée** aborted bunch
coulemelle *f.* **(grande) coulemelle** parasol mushroom (*Lepiota procera*) (see also **lépiote, golmotte** (1))
couler *v.i.* *Hort.*, *Vit.* to abort, to fail to set fruit, to run off (of blackcurrants etc.); (*USA*) to shell
couleur *f.* colour; **c. de fond** ground colour
couleuvrée *f.* 1. traveller's joy (*Clematis vitalba*) 2. white bryony (*Bryonia dioica*)
coulinage *m.* *Hort.* singeing (of fruit trees)
couliner *v.t.* *Hort.* to singe (fruit trees)
coulure *f.* *Hort.*, *Vit.* abortion, coulure, failure to set fruit, running off (of blackburrants etc.); (*USA*) shelling
coumarine *f.* coumarin
coumarou *m.* tonka bean (*Dipteryx odorata*) (= *Coumarouna odorata*) (= **fève tonka**)
coup *m.* knock, blow, stroke; **c. de bêche** spit (of soil); **c. de feu** *Hort.* rapid heating (in compost heap etc.); **c. de fouet** boost, stimulus (to vegetation from nitrogen fertilizers etc.); **c. de pouce** *Vit.* sun-scald, sunburn; **c. de soleil** sun scorch
coupe *f.* cutting (off), harvesting by cutting, lopping, pruning: incision: pruning cut; section; **c. en biseau** oblique, slanting cut; **coupes répétées** repeated cuts or mowings (of lawn etc.); **c. transversale** cross-section
coupe-asperges *m. inv.* asparagus knife
coupe-bordures *m.* edge trimmer (for lawns)
coupe-bourgeon *m., pl.* **coupe-bourgeons** *Hort.*,

Vit. general name for leaf-eating beetles; **c.-bourgeon (de l'abricotier)** *Rhynchites* sp.

coupe-coupe *m. Afr.* cutlass, machete

coupe-gazon *m. inv.* lawn mower (see also **tondeuse**)

coupe-paille *m. inv.* straw cutter

coupe porte-greffes *m. inv. Vit.* cutter for rootstock cuttings

couper *v.t.* to cut, to cut off; **c. à l'ongle** to cut with the thumbnail; **c. l'onglet** to snag

coupe-sarments *m. inv. Vit.* brush cutter, shredder (= **hache-sarments**)

coupe-sève *m. inv.* ringing shears

coupeuse-écimeuse-andaineuse *f. Sug.* cutter-topper windrower

coupure *f.* cut, incision, slit

courant *m.* current, flow; **c. d'air** draught (of air); **c. de sève** flow of sap

courbaril *m.* courbaril (*Hymenaea courbaril*) (= **caroubier de la Guyane, copalier d'Amérique**)

courbe *f.* curve; **c. de niveau** *SC* contour

courbé,-e *a.* curved, bent

courbure *f.* curve, bend

coureuse *a.* & *n. f.* running plant, plant which scion roots

courge *f.* cultivated cucurbitaceous plant, gourd; **c. aubergine, c. à la moelle** vegetable marrow (*Cucurbita pepo*); **c. baleine** winter squash (*C. maxima*) of Mammoth group; **c. cannelée** *Afr.* fluted pumpkin (*Telfairia occidentalis*); **c. cireuse** wax gourd (*Benincasa hispida*); **c. citrouille** pumpkin (*C. pepo*) (= **citrouille**); **c. éponge** loofah (*Luffa cylindrica*); **c. d'Italie, c. d'Orient** squash (*C. pepo*); **c. musquée** pumpkin (*C. moschata*); **c. potiron** winter squash (*C. maxima*) (= **potiron**); **c. de Siam** Malabar gourd (*C. ficifolia*); **c. torchon** = **c. éponge** (*q.v.*)

courgette *f.* small **courge**, courgette, zucchini (*Cucurbita pepo*)

couronne *f.* crown; **c. de fleurs** wreath; **c. de cérosie** *Sug.* wax band, wax ring; **c. impériale** crown imperial (*Fritillaria imperialis*); *Bot.* corona, cup (of narcissus etc.); **à petite couronne** small cupped; *Hort.* head (of tree), crown

couronné,-e *a. Bot.* coronate, coronated

courson *m.*, **coursonne** *f. Hort.* spur, fruit-bearing shoot; **coursonne artificielle** artificial, induced spur; **coursonne naturelle** natural spur (see also **branche coursonne, b. fruitière**)

coursonnage *m.* = **coursonnement** (*q.v.*)

coursonnement *m.* spur formation

court-bois *m. Vit.* = **courson** *q.v.*

courte-queue *n., pl.* **courtes-queues** short-peduncled cherry

courtilière *f.* mole-cricket (*Gryllotalpa gryllotalpa*)

court-noué,-e *a.* **bouture court-nouée** cutting affected by court-noué; *n. m. Vit. PD* fanleaf virus, infectious degeneration

court-pendu *m., pl.* **court-pendus** *Hort.* court-pendu, type of apple with short peduncle

coussin *m.* cushion, pad; **c. d'argent** sea wormwood (*Artemisia maritima*)

coussinant,-e *a.* cushion-forming (of plant)

coussinet *m.* small cushion, pad; *Bot.* small European cranberry (*Vaccinium oxycoccus*); **c. des marais** bilberry (*V. myrtillus*); **c. (floral)** (*Cacao*) floral cushion; **c. (d'une main de bananes)** *Ban.* crown (of hand of bananas)

coût *m.* cost; **coûts de fonctionnement** running costs; **coûts de production** costs of production; **coûts et recettes** costs and returns

couteau *m., pl.* **-eaux** knife; *Hort.* **c. à découper** edging knife; **c. à vendange** *Vit.* scissors or secateurs for cutting or thinning bunches

couteau-scie *m., pl.* **couteaux-scies** knife with serrated edge

coutelassage *m.* cutlassing

coûteux,-euse *a.* costly, expensive

coutre *m.* coulter; **c. circulaire, c. à disque** disc coulter

couvert *m.* cover, shelter; **c. végétal** plant cover, plant canopy, cover of vegetation; **c. gazonnant, c. gazonné** turf cover

couverture *f.* covering, cover; **c. (du sol)** soil cover; *Hort.* cover of vegetation, mulch; **c. morte** leaf litter, cut grass mulch (see also **paillage**); **c. thermale** thermal screen; **engrais en couverture** surface dressing, top dressing; **plante de couverture** cover crop

couvrant,-e *a. Hort.* forming a canopy, forming a cover; **plante couvrante** plant which forms a cover

couvre-sol *a. inv.* soil-cover(ing); **espèce couvre-sol** soil-cover species; *n.m., pl.* **couvre-sols** soil cover (plant)

couvrir *v.t.* to cover

C.O.V. = **Certificat d'Obtention Végétale** (*q.v.*)

coyol (*m.*) **du Mexique** coyal (*Acrocomia mexicana*)

CPG = **chromatographie** (*q.v.*) **en phase gazeuse**

crabe *m.* crab; (*Coconut*) **c. voleur** "robber" crab (*Birgus latro*)

crachin *m.* fine drizzle

craie *f.* chalk; **c. pulverisée** powdered chalk

craindre *v.t.* to fear; **"craint l'excès d'humidité"** (of plant) "do not overwater"

crambe, crambé (maritime) *m.* seakale (*Crambe maritima*) (= **chou-marin**); **c. à feuilles en cœur** flowering seakale (*C. cordifolia*)

cram-cram *m. Afr.* prickly burr grass, karengiya (*Cenchrus biflorus*)
cramoisi,-e *a.* crimson; **c. clair** *a. inv.* light crimson
crampon *m. Bot.* aerial root, clinging root tendril
cramponnant,-e *a.* shaped like a **crampon**
cran *m. Hort.* 1. (strong) notch (see also **incision**) 2. horseradish (= **raifort** *q.v.*)
cranson *m.* horseradish (= **raifort** *q.v.*)
crapaud *m.* toad (*Bufo* and related genera)
craquant,-e *a.* cracking, crackling
craquelure *f.* crack; **c. longitudinale de l'écorce** bark longitudinal crack; **c. en étoile du pommier** *PD* apple star crack(ing) virus
crassicaule *a. Bot.* thick-stemmed
crassifolié,-e *a. Bot.* thick-leaved
crassulacées *f. pl.* Crassulaceae
crassule *f.* 1. crassula. 2. *Crassula* sp.
craterelle *f. Fung.* horn of plenty (*Craterellus cornucopioides*) (= **trompette des morts**)
crédit *m.* credit
crémaillère *f.* toothed rack; *Hort.* cloche-peg (for tilting cloches etc.); *Vit.* gauge for measuring the diameter of stock and scion wood
crémant *m. Vit.* creaming wine
crème *a. inv.* cream, cream coloured
crémeux,-euse *a.* creamy
crénelé,-e *a.* crenate, with scalloped margin
crénulé,-e *a.* crenulate, with margins minutely crenated
crepis *m.* crepis; **c. élégant** *C. pulchra*
crépu,-e *a.* crinkled, curly (of leaf) (= **crispé**)
crépuscule *m.* twilight, dusk
cressiculteur *m.* watercress grower
cressiculture *f.* watercress growing
cresson *m.* 1. cress; **c. alénois** garden cress (*Lepidium sativum*) (= **nasitor**); **c. de fontaine** watercress (*Nasturtium officinale*); **c. de jardin** = **c. de terre** (*q.v.*); **c. du Para** Para cress (*Spilanthes oleracea*); **c. des prés** = **cressonnette** *q.v.*; **c. de terre** American cress (*Barbarea praecox*) (= *B. verna*) (= **barbarée précoce**) 2. **c. de cheval** brooklime (*Veronica beccabunga*)
cressonnette *f.* lady's smock (*Cardamine pratensis*)
cressonnière *f.* watercress bed or pond
crétacé,-e *a.* cretaceous, chalky
crête *f.* crest, ridge; **c. (mitochondriale)** crista
crête-de-coq *f., pl.* **crêtes-de-coq** 1. cockscomb (*Celosia cristata*). 2. yellow rattle (*Rhinanthus* sp.)
crételle (*f.*) **(des prés)** crested dogstail (*Cynosurus cistatus*)
creuser *v.t.* to dig, to excavate, to scoop out; **c. des fossés** to ditch; **radis "ne creusant pas"** radish which does not become hollow
creux,-euse *a.* hollow, sunken; *n. m.* hollow, hole; **c. de plantation** planting hole
crevard,-e *a. PD* affected by, or likely to be affected by calyx-splitting (of carnation); *n. m.* carnation with split calyx
crevasse *f.* crack, split, crevice; **crevasses longitudinales du bois** *Cit.* stem pitting virus
crevassé,-e *a.* cracked
crevasser *v.t.* to crack, to split; **se crevasser** to crack, to split
criblage *m.* sieving, riddling; screening (of cultivars etc.)
crible *m.* sieve, riddle
criblé,-e *a.* riddled, pitted (see also **vaisseau**)
cribler *v.t.* to sift, to riddle, to screen
criblure *f.* siftings, riddlings, screenings; *PD* shot-hole or similar disease; **c. des amygdalées, c. du cerisier** peach shot-hole, cherry shot-hole (*Stigmina carpophila*); **c. du melon** melon shot-hole disease (virus)
crin *m.* horsehair; **c. végétal** plant fibre; **c. de Tampico** *Agave heteracantha*
crinkle-scurf *m. Cit.* crinkle scurf
criocère *m. Ent. Crioceris* sp.; **c. de l'asperge** asparagus beetle (*C. asparagi*); **c. du lis** lily beetle (*Lilioceris lilii*)
criollo *m.* (*Cacao*) criollo
crique *f.* creek, cove, bay
criquet *m.* locust, (short-horned) grasshopper; **c. migrateur tropical** African migratory locust (*Locusta migratoria migratorioides*); **c. nomade** red locust (*Nomadacris septemfasciata*); **c. pèlerin** desert locust (*Schistocerca gregaria*); **c. puant** *Afr.* variegated grasshopper (*Zonocerus variegatus*) (see also **sauterelle**)
crispation *f.* crispation, shrivelling up; **c. des pousses** shrivelling of shoots
crispé,-e *a.* crinkled, curly (of leaf) (= **crépu**)
crispiflore *a.* crispifloral, with curled or crinkled petals
crispifolié,-e *a.* crispifoliar
cristacortis *m. Cit. PD* cristacortis (virus)
cristalliser *v.t. & i.* to crystallize
cristalloïde *a.* crystalloid; *n. m.* crystalloid
cristé,-e *a.* cristate, crested
criste-marine, christe-marine *f.* samphire (*Crithmum maritimum*)
crithme *m. Crithmum* sp., samphire (*C. maritimum*) (= **c(h)riste-marine**)
croc *m.* hook; **c. à fumier** manure fork; **c. à pommes de terre** potato fork; **c. à sarcler** weed lifter, hoe
crochet *m.* hook; **taille en crochet** fan-training (e.g. for peach); **c. de provignage** *Vit.* fork for holding layers in the soil; *Hort.* (1) stub, (2) tie
crocus *m.* crocus (*Crocus* sp.) (see also **safran**);

c. printanier spring crocus (*C. vernus*) (= **safran printanier**); **c. rouge** spring meadow saffron (*Bulbocodium vernum*)
croisement *m. PB* 1. crossing, breeding, hybridization. 2. cross, hybrid; **c. diallèle** diallel cross; **c. en retour** back-cross; **c. multiple** multiple cross; **c. réciproque** reciprocal cross; **c. trois-voies** three-way cross
croisillon *m.* cross-bar
croissance *f.* growth; **c. cellulaire** cell growth; **c. en longueur** longitudinal growth; **à c. déterminée** of determinate growth; **à c. indéterminée** of indeterminate growth; **à c. rapide** quick-growing; **c. par poussées, c. saccadée** growth in flushes; **c. végétative** vegetative growth; **type de croissance** habit of growth; *Mush.* **c. du blanc** spawn run
croissant *m.* **(à élaguer)** billhook
croître *v.i.* to grow
croix de Malte, c. de Jérusalem *f.* Maltese cross, Jerusalem cross (*Lychnis chalcedonica*) (see also **coquelourde**)
croskill *m.* croskill roller, toothed roller
croskillage *m.* rolling with a croskill roller
croskiller *v.t.* to roll with a croskill roller
croskillette *f.* small crosskill roller
crosne (*m.*) **(du Japon)** Chinese artichoke (*Stachys affinis*) (= **stachys tubéreux**)
crossette *f. Hort.* (mallet) heel cutting (= **bouture à crossette**)
crotalaria *m.*, **crotalaire** *f. Crotalaria* sp., sunn-hemp
croton *m. Croton* sp.; croton (*Codiaeum variegatum*)
crottin *m.* dung, droppings
croupir *v.i.* 1. (of water) to stagnate, to grow foul. 2. to rot in stagnant water
croût'.ge *m. Ped.* crusting, hardening (of soil)
croûte *f.* crust; *Ped.* **c. du sol** soil crust; **c. désertique** duricrust; **c. superficielle** superficial crust
cru,-e *a.* raw; **fruit cru** raw fruit; **légumes crus** raw vegetables
cru *m.* high quality vineyard: high quality wine
crucianelle *f. Crucianella* sp., cross-wort
cruciféracées *f. pl.* Cruciferae
crucifère *a.* cruciferous; *n. f.* crucifer, cruciferous plant
crue *f.* rising, swelling of river, flood; **crue du Niger** flood of the Niger; **rivière en crue** river in spate
crustacé,-e *a.* crustaceous
crustiforme *a. Ped.* crusty
cryoconservation *f.* cryoconservation
cryodéshydraté,-e *a.* freeze-dried
cryodéshydration *f.* freeze-drying
cryodessication *f.* = **cryodéshydration** (*q.v.*)
cryogénie *f.* cryogenics
cryptogame *a.* cryptogamous, cryptogamic; *n. f.* cryptogam
cryptogamique *a.* cryptogamic; **maladies cryptogamiques** fungal diseases
cryptomérie *f.* cryptomerism
cryptoméria *m.*, **cryptomérie** *f. Cryptomeria* sp.; **c. du Japon** Japanese cedar (*C. japonica*)
crystalline *f.* ice plant (*Mesembryanthemum (Cryophytum) crystallinum*)
cubèbe *m.* cubeb (*Piper cubeba*) and its fruit (= **poivre à queue**)
cucullé,-e *a.* cucullate, hooded
cuculle *f.* hood, cowl; *Bot.* cucullus
cuculliforme,-e *a.* cuculliform, cowl-shaped
cucumifome *a.* cucumiform, formed like a cucumber
cucurbitacé,-e *a.* cucurbitaceous: gourd-shaped; *n. f. pl.* Cucurbitaceae
cueillage *m.* = **cueille** (*q.v.*)
cueillaison *m.* = **cueille** (*q.v.*)
cueille *f.* 1. picking, gathering; **de c. facile** easily picked; **première cueille** first picking; **c. unique** once-over harvest. 2. season of picking
cueille-fleurs *m. inv.* flower picker (implement)
cueille-fruits *m. inv.* long-arm (fruit) picker
cueilleteuse *f.* fruit picking machine
cueillette *f.* gathering, picking, harvesting of fruit and other plants; **c. aisée** easy harvesting; **c. libre-service** pick-your-own; **c. mécanique** mechanical harvest(ing); **(époque de) cueillette** picking season
cueilleur,-euse *n.* picker
cueillir *v.t.* to pick, to harvest (fruit etc.)
cueilloir *m.* 1. fruit picker (implement), 2. basket for fruit
cuiller, cuillère *f.* spoon; **en cuiller** spoon-shaped (see also **feuille**)
cuirasse *f.* armour; **c. (ferrugineuse)** *Ped.* ironstone cap, crust
cuisse-de-nymphe *f.*, *pl.* **cuisses-de-nymphe** white rose with pink tints; *a. inv.* its colour
cuivre *m.* copper
cuivré,-e *a.* copper-coloured; **rouge cuivré** coppery red
cuivreux,-euse *a.* coppery, cuprous
cuivrique *a.* cupric; **nutrition cuivrique** copper nutrition
cul-brun *m.* brown-tail moth (*Euproctis chrysorrhea*)
culée *f.* anchorage for trellis wires
culmifère *a.* culmiferous
culot *m.* residue (of chemical process etc.)
-culteur *m. suffix* grower, e.g. **viticulteur** vinegrower
cultigène *m.* cultigen
cultivable *a.* cultivable, arable
cultivar *m.* cultivar; **c.-clone** clone cultivar; **c.**

lignée pure (pure) line cultivar; **c. population** assemblage cultivar; **c. synthétique** synthetic cultivar (F_1 hybrids etc.)

cultivateur,-trice *a. & n.* 1. cultivator, commercial grower, farmer. 2. grubber, cultivator; **c. canadien** (= **canadienne**); **c. à dents flexibles** spring-tine cultivator; **c. à dents rigides** rigid-tine cultivator; **c. à disques** disc cultivator, disc harrow; **c. fouilleur** chisel cultivator; **c. porté** mounted cultivator

cultivé,-e *a.* cultivated, under cultivation, cultured

cultiver *v.t.* to cultivate, to grow, to farm; **c. des choux** to grow cabbages

cultriforme *a.* cultriform, cultrate

cultural,-e *a., m. pl.* **-aux** pertaining to culture (agriculture, horticulture etc.); **méthodes culturales** growing methods, cultural methods

culture *f.* 1. cultivation, culture, growing, tillage; **la grande culture** large-scale growing; **la petite culture** small-scale growing; **c. abritée** protected cultivation; **c. en ados** bed culture; **c. de plantes aromatiques et médicinales** herb growing; **c. en bandes** strip cropping; **c. sur billons** growing crops on ridges; **c. de(s) bulbes** bulb growing; **c. sous châssis** forcing; **c. en courbes de niveau** contour cultivation; **c. de décrue** *Afr.* etc. growing crops on receding flood, on falling water table; **c. sur film nutritif** nutrient film technique; **c. fruitière** fruit growing; **c. sur gravier** gravel culture; **c. hâtée** forcing; **c. horizontale** contour cropping; **c. hors-sol** soilless culture; **c. hydroponique** hydroponics; **c. intensive** intensive cultivation; **c. itinérante** shifting cultivation; **c. en larges planches** broadland cultivation; **c. maraîchère** market gardening; **c. mécanique** mechanized cultivation; **c. minimale** minimum tillage; **c. mixte** mixed cropping; **c. en mottes** growing in blocks; **c. nue** bare soil cultivation, bare soil treatment; **c. à plat** growing crops on the flat; **c. en pots** pot culture; **c. protégée** protected cultivation; **c. retardée** retarded culture; **c. sur sable** sand culture; **c. sans sol** soilless culture, hydroponics; **c. de serre, c. sous verre** glasshouse cultivation; **c. en terrasses** terraced culture, cultivation; **c. unie** unterraced cultivation. 2. land under cultivation, crop; **culture de** – crop of –; **c.-abri** nurse crop; **c. améliorante** soil-improving crop; **c. annuelle** annual crop; **c. arborescentre** tree, perennial woody crop; **c. arbustive** tree crop; **c. associée** associated, companion crop, intercrop(ping); **c. basse** low-growing crop; **c. sur billons** crop grown on ridges; **c. de couverture** cover crop; **c. couvrante** crop which forms a closed cover or canopy; **c. dérobée** catch crop; **c. à engrais vert** green manure crop; **c. épuisante** (soil-) exhausting crop; **c. familiale** cottage crop, *Afr.* compound crop; **c. florale** flower crop; **c. forcée** forced crop; **c. fruitière** fruit crop; **c. industrielle** industrial crop; **c. installée** established crop; **c. intercalaire** interplanted crop, intercrop; **c. légumière** vegetable crop; **c. manquée** failed crop; **c. maraîchère** market garden crop; **c. marchande** cash crop; **c. en mélange, c. mixte** mixed crop; **c. nettoyante** cleaning crop; **c. pérenne** perennial crop; **c. à plat** crop grown on the flat; **c. en plein air, c. en pleine terre** outdoor crop; **c. de plein champ** field-grown crop; **c. pluviale, c. sous pluie** rain-fed crop; **c. porte-graines** crop grown for seed; **c. potagère** vegetable crop, garden crop; **c. propre** clean crop; **c. protégée** protected crop; **c. pure** straight, pure crop; **c. de rapport** cash crop; **c. de remplacement** replacement crop; **c. rémunératrice** paying crop, economic crop; **c. retardée** late, delayed crop; **c. sarclée** hoed, root crop; **c. sèche** upland, non-irrigated crop; **c. de serre** glasshouse crop; **c. serrée** close-planted crop; **c. de sidération** green manure crop; **c. vivrière** food crop. 3. culture; **c. d'apex** apical culture; **c. de méristème** (meristem) tissue culture; **c. de tissus, c.-tissu** (*pl.* **cultures-tissus**) tissue culture

cumin *m.* cumin (*Cuminum cyminum*); **c. des prés** = **carvi** (*q.v.*)

cunéifolié,-e *a.* with cuneate, wedge-shaped leaves

cunéiforme *a.* cuneate, wedge-shaped

cunifuge *m.* rabbit and hare repellent

cuphaea *m. Cuphea* sp.; **c. du Mont Jorullo** (*C. jorullensis = C. micropetala*); **c. pourpre** *C. lanceolata*

cupidone *f.* Cupid's dart (*Catananche caerulea*)

cuprique *a.* cupric

cupulaire *a.* cupular, cup-shaped

cupule *f. Bot.* cupule; husk (of hazel nut etc.); cup (of acorn)

cupulé,-e *a.* cupulate, cup-shaped

cupulifère *a.* cupuliferous; *n. f. pl.* Cupuliferae

cupuliforme *a.* cupuliform, cup-shaped

curage *m.* 1. clearing, cleaning (of drain, channel etc.); **c. des fossés** ditch clearing. 2. *pl.* spoil (from clearing)

curare *m.* curare

curatif,-ive *a. & n. m.* curative, remedial (agent); **méthode curative** remedial method

curculionidés *m. pl.* Curculionidae, weevils

curcuma, cucurmin *m.* turmeric (*Curcuma longa*) (= **safran des Indes**)

curer *v.t.* to clean out, to dredge

cureuse *f.* ditch-cleaning machine
curuba *f.* banana passion fruit (*Passiflora mollissima*)
curure *f.* spoil, muck (from drain, ditch etc.)
cuscute *f.* dodder (*Cuscuta* sp.)
cuspidé,-e *a.* cuspidate
cuspidifolié,-e *a.* with cuspidate leaves
cusson *m.* vine beetle (*Sinoxylon sexdentatum, Schistocerus bimaculatus*)
cuticulaire *a.* cuticular
cuticule *f.* cuticle
cutine *f.* cutin
cutinisé,-e *a.* cutinized
cutiniser (se) *v. p.* to become cutinized
cuve *f.* vessel for fermentation, tank, tub, vat (see also **raisin**)
cuvée *f. Vit.* vatful (of wine); **vin de première cuvée** wine of the first pressing
cuvette *f.* basin, depression; **la c. congolaise** the Congo basin; **c. jaune** *Ent.* yellow-pan trap; *Hort.* **c. calicinale (de la pomme)** calyx end (of apple); **c. de l'œil, c. oculaire** depression of eye (e.g. of apple); **c. de plantation** *Plant.* etc. depression caused by planting; **irrigation par cuvette** *Irrig.* basin irrigation
C.V. = **coefficient de variation** (*q.v.*)
cyamopse (*f.*) **(à quatre ailes)** *Afr.* cluster bean (*Cyamopsis tetragonoloba*)
cyanamide *f.* cyanamide; **c. de chaux, c. calcique** calcium cyanamide
cyanhydrique *a.*, **acide cyanhydrique** prussic acid
cyanure *m.* cyanide; **c. de potassium** potassium cyanide
cyathe *m.* cyathium
cyathea *m.*, **cyathée** *f.* tree fern (*Cyathea* sp.); **c. en arbre** *C. arborea*; **c. blanchâtre** *C. dealbata*; **c. épineux** *C. spinulosa*; **c. à moelle** *C. medullaris*; **c. remarquable** *C. insignis*
cycadacées *f. pl.* Cycadaceae, cycads
cycadé,-e *a.* cycadaceous
cycas *m.* cycad
cyclamen *m.* cyclamen (*Cyclamen* sp.); **c. d'Afrique** *C. africanum*; **c. de Cilicie** *C. cilicium*; **c. d'Europe** *C. europaeum*; **c. de l'île de Cos** *C. coum*; **c. du Liban** *C. libanoticum*; **c. de Naples** *C. neapolitanum*; **c. de Perse** florist's cyclamen (*C. persicum*)
cycle *m.* cycle, crop cycle; **c. annuel** annual cycle; **c. cultural** cultural, growth cycle; **c. évolutif** life cycle; **c. reproducteur** reproductive cycle; **c. végétatif** growth cycle (of plant); **c. vital, c. de vie** life cycle
cycloheximide *f.* cycloheximide
cyclopentadiène *m.* cyclopentadiene
cydonia *m.* *Cydonia* sp., quince (see **coing**)
cylindre *m.* cylinder; **c. central** *Bot.* stele; **c. de coupe** cutting cylinder (of mower)
cylindrique *a.* cylindrical
cymbalaire *f.* ivy-leaved toadflax (*Linaria (Cymbalaria) muralis*) (= **ruine de Rome**)
cymbiforme *a.* cymbiform
cyme *f. Bot.* cyme; **c. hélicoïde** helicoid cyme; **c. scorpioïde** scorpioid cyme
cynanque, cynanche *f.* *Cynanchum* sp., including *C. lineare*
cynips *m.* gall wasp (Cynipidae); **c. du rosier** bedeguar gall wasp (*Diplolepis (Rhodites) rosae*)
cynodon *m.* *Cynodon* sp., bermuda grass
cynoglosse *f.*, **cynoglossum** *m.* *Cynoglossum* sp., hound's tongue
cynorrhodon *m. Bot.* hip
cyperacées *f. pl.* Cyperaceae
cypérus *m.* *Cyperus* sp., especially nutgrass (*C. rotundus*)
cyprès *m.* cypress; **c. de l'Arizona** Arizona cypress (*Cupressus arizonica*); **c. de Busaco** = **c. du Portugal** (*q.v.*); **c. chauve, c. de (la) Louisiane** swamp cypress (*Taxodium distichum*); **c. d'été** summer cypress (*Kochia trichophylla*); **c. faux-thuya** white cypress (*Chamaecyparis thyoides*); **c. de Goa** = **c. de Busaco** (*q.v.*); **c. horizontal** spreading form of Mediterranean cypress (*Cupressus sempervirens horizontalis*); **c. de Lambert, c. de Monterey** Monterey cypress (*Cupressus macrocarpa*); **c. de Lawson** Lawson's cypress (*Chamaecyparis lawsoniana*); **c. méditerranéen** Mediterranean cypress (*Cupressus sempervirens*); **c. de Nootka** Nootka cypress, yellow cypress (*Chamaecyparis nootkatensis*); **c. du Portugal** Mexican cypress (*Cupressus lusitanica*); **c. pyramidal** *Cupressus sempervirens pyramidalis*; **c. toruleux** Bhutan cypress (*Cupressus torulosa*)
cypripède *m.* *Cypripedium* sp., lady's slipper (= **sabot de Vénus**)
cystolithe *m. Bot.* cystolith
cytise *m.* 1. *Cytisus (Sarothammus)* sp.; **c. blanc (du Portugal)** white Spanish broom (*C. albus*) (= **genêt blanc**). 2. **c. (faux ébénier)** common laburnum (*Laburnum anagyroides*); **c. des Alpes** Scotch laburnum (*L. alpinum*) (= **ébénier des Alpes**)
cytochimie *f.* cytochemistry
cytochimique *a.* cytochemical
cytochrome *m.* cytochrome
cytocinèse *f.* cytokinesis
cytodème *m.* cytodeme, a unit in a taxon which differs cytologically (usually in chromosome number)
cytodiérèse *f.* cytokinesis
cytokinine *f.* cytokinin
cytologie *f.* cytology
cytologique *a.* cytological

cytoplasme *m.* cytoplasm
cytoplasmique *a.* cytoplasmic; **stérilité cytoplasmique** cytoplasmic sterility

cytotoxicité *f.* cytotoxicity
cytotoxique *a.* cytotoxic

D

daba *f. Afr.* short-handled hoe

dactyle (pelotonné) *m.* cocksfoot, orchard grass (*Dactylis glomerata*)

dahlia *m.* dahlia (*Dahlia* sp.); **d. à fleurs d'anémone** anemone-flowered dahlia; **d. cactus** cactus dahlia; **d. décoratif** decorative dahlia; **d. à fleurs simples** single dahlia; **d. à fleurs simples à collerette** collarette dahlia; **d. à fleurs de pivoine** paeony-flowered dahlia; **d. pompon** pompom dahlia; **d. semi-cactus** semi-cactus dahlia

daïkon *m.* daikon, Japanese radish (*Raphanus sativus longipinnatus*)

dalbergia *m.*, **dalbergie** *f.* *Dalbergia* sp. dalbergia

dallage *m.* paving, flagging, pavement; **plantes pour dallages fleuris** plants for planting between paving stones or in cracks in paving stones

dalle *f.* flagstone, paving-stone; slab (of stone etc.)

damas *m.* damson (*Prunus insititia* = *P. damascena*)

dame *f. Hort.* earth rammer, earth tamper

dame-d'onze-heures *f., pl.* **dames-d'onze-heures** star of Bethlehem (*Ornithogalum umbellatum*) (= **belle-d'onze-heures, étoile de Bethléem**)

damer *v.t.* to ram, to compact (earth)

daphné *m.* bay (*Daphne* sp.); **d. bois gentil, d. bois-joli** mezereon (*D. mezereum*)

dard *m. Hort.* dard; **d. couronné** dard with a fruit bud; **d. simple** dard without a fruit bud

date (*f.*) **de semis** sowing date

datte *f.* date; **d. de Trébizonde** oleaster (*Elaeagnus angustifolia*) (= **chalef à feuilles étroites, olivier de Bohème**)

dattier *m.* date-palm (*Phoenix dactylifera*); **d. des Canaries** Canary date-palm (*P. canariensis*); **d. du désert** desert date (*Balanites aegyptiaca*) (= **balanites, soump**); **d. de Roebelin** pygmy date palm (*P. roebellini*) (see also **faux dattier**)

datura *m. inv.* *Datura* sp., datura, thorn-apple (*Datura stramonium*) (= **stramoine**) and other spp.; **d. cornu** *D. ceratocaula;* **d. d'Egypte, d. fastueux** *D. fastuosa*

dauciforme *a.* carrot-shaped

daucus *m.* *Daucus* sp.; **d. carotte** wild carrot (*D. carota*) (= **carotte sauvage**)

dauphinelle *f.* larkspur (*Delphinium* sp.); **d. staphysaigre** *D. staphysagria* (see also **pied-d'alouette**)

dé *m. Vit.* footing for stake

débâchage *m.* removing (plastic) cover from crop

débâcher *v.t.* to remove (plastic) cover from crop

débéquiller *v.t. Hort.* to remove props from (branches etc.) (see also **béquiller**)

débile *a.* sickly (of plant)

débit *m.* output, delivery rate, sale; **d. d'arrosage** discharge of water (when watering)

déblaiement *m.* clearing land, removing earth from land

déblayer *v.t.* to clear land, to remove earth from land

déboisement *m.* deforestation, clearing trees

déboiser *v.t.* to deforest, to clear trees

débouché *m.* outlet, market

débourber *v.t.* to clear mud (from ditch etc.), to sluice (drain etc.); (in winemaking) to draw off clear must after allowing solid matter to settle

débourrage *m.* (*Coconut*) = **défibrage** (*q.v.*)

débourrement *m.* bud burst, bud break, bud flushing

débourrer *v.i.* (of buds) to burst, to break; *v.t.* to remove the husk from, to (de)husk (e.g. coconuts)

débourreur *m.* (*Coconut*) one who husks, husker

débridage *m.* = **débridement** (*q.v.*)

débridement *m. Hort.* cutting (bark etc.) to relieve tension

débrider *v.t. Hort.* to make a cut (in bark etc.) to relieve tension

débris *m. pl.* remains, left-overs, debris

débroussaillant,-e *a.* scrub killing; *n. m.* scrub killer, brushwood killer (herbicide)

débroussaillement *m.* clearing undergrowth, slash control

débroussailler *v.t.* to clear undergrowth

débroussailleuse *f.* brush cutter, tree-dozer

débroussement *m.* clearing bush, brush

débrousser *v.t.* to clear bush, brush

débuttage *m.* ploughing away from base of plant exposing base or roots; removing earth from base of tree, vine; splitting ridges

débutté,-e *a.* with earth removed from the base (of plant); with split ridges
débutter *v.t.* to plough away from base of plant to expose base or roots; to remove earth from base of tree, vine (see also **déchausser**); to split ridge
décaféiné,-e *a.* decaffeinated; **café décaféiné** decaffeinated coffee
décalage *m.* staggering (of planting dates etc.)
décalcarisation *f.* soil impoverishment in limestone
décalcifié,-e *a.* decalcified
décalcifier *v.t.* to decalcify; **se décalcifier** *v.i.* to become decalcified
décamètre (*m.*) **(à ruban)** 10-metre measuring tape
décapage *m.* cleaning the surface, scraping (off)
décapé,-e *a.* (surface-) cleaned
décaper *v.t.* to clean the surface, to scrape (off)
décapitation *f. Hort.* heading back
décapiter *v.t. Hort.* to head back
décarboxylant,-e *a.* decarboxylating; **enzyme décarboxylante** decarboxylating enzyme
décarboxylation *f.* decarboxylation
décavaillonnage *m. Vit.* ploughing between rows; removal of soil after **débuttage**
décavaillonner *v.t. Vit.* to plough between rows; to remove soil after **débuttage**
décavaillonneur *m.*, **décavaillonneuse** *f.* plough (including vineyard plough) or cultivator for **décavaillonnage**
décelable *a.* discernible, detectable; **non décel-able** not detectable
décharge *f.* discharge, outlet; **d. (publique)** rubbish tip, dump
déchargeur *m.* dumper, unloader
déchaumage *m.* 1. skim ploughing, stubble ploughing. 2. breaking ground
déchaumer *v.t.* 1. to plough stubble. 2. to break ground
déchaussage, déchaussement *m.* laying bare roots of trees, unearthing, frost-lifting; *Vit.* (= **débuttage**); **d. en cuvette** making hollows around vines to collect rainwater
déchausser *v.t. Hort.* to bare, to expose (roots etc.), to unearth (see also **débutter**)
déchausseuse *f.* vineyard plough for **déchaussage**
déchet *m.* waste, refuse; **déchets de case** *Afr.* household refuse; **déchets urbains** town refuse
déchicotage *m. Sug.* = **dessouchage**
déchiqueté,-e *a.* jagged, indented; *Bot.* laciniate
déchiqueter *v.t.* to cut, tear to shreds, to shred
déchirer *v.t.* to tear
déchirure *f.* tear, rent: cleft
décidu,-e *a.* deciduous
déclencheur *m.* release, trip (of machine etc.)
déclin *m. PD* decline
décliné,-e *a.* declinate, hanging downwards
décollement *m.* unsticking, separation (of stock and scion), breaking of graft union
décoller *v.t.* to unstick, to loosen; **se décoller** to come unstuck, to separate (of stock and scion); **l'écorce se décolle** the bark is coming off
décolletage *m. Hort.* cutting off the tops, topping
décolleter *v.t. Hort.* to cut off the tops, to top
décoloration *f.* discolouration, fading; *Cit.* **d. (discontinue) des nervures** vein clearing
décoloré,-e *a.* discoloured, faded, bleached
décolorer *v.t.* to discolour, to fade, to bleach; **se décolorer** to lose colour, to fade
décombant,-e *a.* decumbent
décompacter *v.t.* to loosen (soil)
décomposé,-e *a.* decomposed, rotten
décomposer *v.t.* to decompose; **se décomposer** *v.i.* to decompose, to putrefy
décomposition *f.* decomposition, breakdown
déconseillé,-e *a.* not recommended
déconseiller *v.t.* to advise against (doing) something
décorticage *m.*, **décortication** *f.* decortication, shelling, hulling, peeling; **d. annulaire** ring barking
décorticose *f. Cit., PD* scaly bark
décortiquer *v.t.* to decorticate, to shell, to hull, to peel
décortiqueur,-euse *n.* decorticator; *n. m. Vit.* brush or sacking to remove old bark from vines
découlement *m.* drainage
découpé,-e *a. Bot.* denticulate
découpe-bordure *m.* edge-cutter (for lawns)
découpure *f. Bot.* denticulation
décroissance *f.* decrease, decline (of radio-activity etc.), decay
décroûtage *m.* **décroûtement** *m.* breaking the (soil) crust
décrue *f.* fall, subsidence (of river); **zone de décrue** *Afr.* land uncovered by the receding (annual) flood
décurrent,-e *a.* decurrent
décuscutage *m.* removing dodder
décuscuter *v.t.* to remove dodder
décussé,-e *a.* decussate
dédifférenciation *f.* dedifferentiation
dédifférencié,-e *a.* dedifferentiated
dédossement *m. Hort.* splitting, dividing plants
dédosser *v.t.* to split, to divide plants
dédoublement *m. Hort.* dividing, cutting up, splitting
dédoubler *v.t. Hort.* to divide, to cut up, to split
défanage (*m.*) **(de la pomme de terre)** desiccation (of potato tops)
défanant *m.* desiccating herbicide, product
défavorable *a.* unfavourable

déferrification *f. Ped.* soil impoverishment in iron
défeuillage *m.* leaf removal, defoliation; *Vit.* removal of surplus or unwanted leaves to enhance fruit ripening
défeuillaison, défoliation *f.* defoliation, leaf fall; **d. artificielle** artificial defoliation; *Coff.* **d. en mannequin d'osier** "wicker-basket" defoliation
défeuillé,-e *a.* stripped of leaves, defoliated, without petals (of flower)
défeuiller *v.t.* to strip the leaves from, to defoliate, to top (onions); **se défeuiller** to shed leaves (of plant), to shed petals (of flower)
défibrage *m.* (*Coconut*) husking; **pieu de défibrage** husking blade
déficience *f.* deficiency; **d. magnésienne** magnesium deficiency
déficit *m.* deficit, deficiency, shortage; **d. en eau, d. hydrique** water shortage, moisture deficit; **d. en eau des feuilles** leaf water deficit; **d. thermique** heat deficit
déflecteur, -trice *a.* deflecting; *n.m. AM* deflector
défleuraison *f.* fall of blossom, flower shedding
défleurir *v.i.* (of plant) to lose blossom, to cease flowering; *v.t.* to remove flowers, blossom
défloré,-e *a.* deflorate, stripped of flowers
déflorer *v.t. Hort.* to remove, to strip, flowers from plant
défoliaison *f.* defoliating
défoliant,-e *a.* defoliating; *n. m.* defoliant
défoliateur,-trice *a.* defoliating; **lépidoptère défoliateur** leaf-eating lepidopteran; *n.* defoliator
défoliation *f.* defoliation (= **défeuillaison**)
défonçage, défoncement *m. Hort.* deep ploughing, breaking up land, trenching; Vit. **d. au large, d. en plein** deep ploughing over the whole (vineyard) area; **d. en lignes** deep ploughing in rows
défoncer *v.t. Hort.* to plough deeply, to break up land, to trench
défonceuse *f.* heavy plough
déforestation, déforestage *m.* deforestation
déformation *f.* deformation; **déformations foliaires** leaf deformation; **d. liégeuse des racines** corky root condition; **d. mosaïque (du prunier)** prune dwarf virus
déformé,-e *a.* deformed
déformer *v.t.* to deform; **se déformer** to become deformed
défourcheur *m.* long-handled pruner (smaller than **ébrancheur** *q.v.*)
défraîchi,-e *a.* shrivelled, withered, faded (of flower, stalk etc.)
défrichement *m.* clearing, grubbing
défricher *v.t.* to clear, to grub, to reclaim land, to bring into cultivation
dégarni,-e *a.* depleted, stripped; **arbre dégarni** tree bare of leaves
dégât *m.* damage; **d. d'insectes** insect damage; **d. de pulvérisation** spray damage
dégazage *m.* release of gas
dégel *m.* thaw
dégeler *v.i.* & *v.t.* to thaw
dégénérescence *f.* degeneration, die-back; *Vit.* **d. infectieuse** infectious degeneration (= **court-noué**)
dégermage *m.* removal of (young) shoots; **d. des tubercules (de pommes de terre)** removal of shoots from potatoes
dégivrage *m.* defrosting
déglaçage *m.*, **déglacement** *m.* thawing, removing ice, de-icing
dégradation *f.* weathering (of rocks); degradation, deterioration; **produit de dégradation** degradation product
dégrain *m. Ban.* fruit drop at maturation time
dégraisser *v.t. Hort.* to leach out nutrients
dégrapper *v.t.* = **égrapper** (*q.v.*)
degré *m.* degree (of heat etc.); **d. de liberté** *Stat.* degree of freedom; **d. de ploïdie** *PB* degree of ploidy; **d. réfractométrique** *Sug.* refractometric index
degré-jour *m.* degree-day
dégustateur,-trice *n.* taster (of coffee, tea, wine etc.)
déhiscence *f.* dehiscence
déhiscent,-e *a.* dehiscent
déjection *f.* droppings, frass (of insects), cast (of worms)
délabrement *m.* decline
délicat,-e *a.* delicate; *Hort.* frost-tender
déligaturage *m. Hort.* removal of tie from bud union
déligaturé,-e *a. Hort.* with tie removed (of bud union)
déligaturer *v.t. Hort.* to remove tie from bud union
délou *m. Afr.* water-lifting system using a self-emptying leather bottle with folding neck
delphinelle, delphinette *f.* = **dauphinelle** (*q.v.*)
delphinium *m. Delphinium* sp., larkspur (see also **dauphinelle, pied-d'alouette**); **d. de (la) Chine** bouquet larkspur (*D. grandiflorum*)
deltoïde *a.* deltoid
démaquisage *m.* removing scrub, clearing land of scrub
démariage *m. Hort.* thinning, singling
démarier *v.t. Hort.* to thin, to single
démarieuse *f.* type of harrow
démarrage *m.* start; *Hort.* starting (plants)
démarrer *v.t.* to start; *Hort.* to start (plants); *v.i.* to grow, to sprout (of plant, bud etc).

dématophore *m. Fung. Dematophora necatrix* (see also **pourridié**)
demeure *f.*, **à demeure** *Hort.* "at stake", in place
demi-double *a.* semi-double (of flower)
demi-fleuron *m., pl.* **demi-fleurons** ligulate floret
demi-hâtif,-ive *a.* semi-early
déminéralisation *f.* demineralization
demi-noix *f.* (*Coconut*) split coconut
demi-rustique *a.* half hardy
demi-sec *a.* 1. applied to bean (*Phaseolus vulgaris*) cultivars in which the seeds are eaten green (see also **coco**) 2. (of wine) medium dry
demi-spirale *f. Rubb.* half spiral
demi-tige *a.* & *n. m. Hort.* half standard; **arbre demi-tige** half-standard (tree) (not above 1.50 m high)
demi-valeur *f., pl.* **demi-valeurs** half-value
demi-variogramme *m.* semi-variogram
demi-vie *f.* half-life
demoiselle *f. Ban.* sword sucker
démontable *a.* that can be taken to pieces, dismountable
démucilagination *f. Coff.* mucilage removal
dendrobie *f. Dendrobium* sp. (orchid)
dendroclimatologie *f.* dendroclimatology
dendrographe *m.* one who studies trees
dendrographie *f.* study of trees (= **dendrologie**)
dendroïde *a. Bot.* dendroid, branching
dendrologie *f.* dendrology, study of trees
dendromètre *m.* dendrometer
déneigé,-e *a.* (*Canada*) free of snow
déneigement *m.* snow removal
dénitrification *f.* denitrification
dénitrifier *v.t.* to denitrify
dénombrement *m.* enumeration, counting
dénoyauter *v.t.* to stone fruit
dénoyauteur *m.* fruit stoning machine
denrée *f.* produce, foodstuff; **denrées alimentaires** food products
densité *f.* density; **d. d'aspersion** *Irrig.* precipitation per hour; **d. de plantation** *Hort.* planting rate, stand; **d. de semis** *Hort.* seed rate, sowing rate; **d. de peuplement** *Hort.* stand, planting density; **d. du sol** *Ped.* soil density
densitomètre *m.* densitometer
dent *f.* tooth; *Hort.* tine; **d. de fourche** prong of a fork; **d. de herse** harrow tine
dent-de-chien *f. pl.* **dents-de-chien** dog's tooth violet (*Erythronium dens-canis*)
dent-de-lion *m., pl.* **dents-de-lion** = **pissenlit** (*q.v.*)
denté,-e *a.* dentate, toothed; **d. en scie** serrate
dentelaire *f. Plumbago* sp., plumbago; **d. du Cap** Cape leadwort (*P. capensis*); **d. de Lady Larpent** *Ceratostigma plumbaginoides* = *P. larpentae*
dentelé,- *a.* indented, serrate
dentelure *f.* serration (of leaf)
denticulation *f.* denticulation
denticulé,-e *a.* denticulate
denture *f.* serrated edge; **d. d'une feuille** serrated edge of a leaf
dénudé,- *a.* denuded, bare, stripped
dénuder *v.t.* to denude, to bare; **se dénuder** to become bare
dépaisselage *m. Vit.* vine prop removal
dépaisseler *v.t. Vit.* to remove vine props
dépalissage *m.* loosening or detaching branches of an espalier fruit tree
dépalisser *v.t.* to loosen or detach branches of an espalier fruit tree (see also **palisser**)
dépanneauter *v.t.* to remove glazed tops of frames etc.
déparchage *m. Coff.* hulling
départ (*m.*) **de la végétation, d. végétatif** bud burst (= **débourrement**), sprouting
dépérir *v.i.* to wither, to decay
dépérissement *m.* decay, decline, withering, dieback; **d. brutal** *OP* etc. sudden dieback; **d. du champignon** mushroom dieback; **d. du concombre** black root rot (*Phomopsis sclerotioides*); **d. de l'orme** Dutch elm disease (*Ceratocystis ulmi*); **d. bactérien du pêcher** peach bacterial canker (*Pseudomonas morsprunorum*); **d. de Molières (du cerisier doux)** Molières decline (of sweet cherry)
déphasage *m.* dephasing, difference in phase (of growth etc.)
dépigmentation *f.* loss of pigment, chlorosis
dépiquer *v.t. Hort.* to transplant, to prick out
dépistage *m.* detection, monitoring
dépistillage *m. Ban.* removal of dried floral parts
déplacage *m.* **(de gazon)** lifting, cutting (of turf) (see also **déplaquer**)
déplacement *m.* displacement, shift, drift, movement (e.g. of water); *Phys.* translocation
déplantation *f.*, **déplantage** *m.* lifting (plant), uprooting, transplanting; **d. automnale** autumn lifting
déplanter *v.t.* to lift (plant), to uproot, to transplant
déplanteuse *f.* plant lifting machine or tool
déplantoir *m.* (garden) trowel, hand fork
déplaquer *v.t.* **(le gazon)** to lift, to cut (turf) (see also **déplacage**)
déplié,-e *a.* unfolded
déplier *v.t.* to unfold
dépôt *m.* 1. depository, store. 2. deposit, deposition, settling; **d. fluvial** alluvial deposit
dépotage *m.* removal of plant from pot, potting out, planting out
dépoter *v.t.* to remove plant from pot, to potout, to plant out
dépréciation *f.* depreciation

déprédateur *m.* pest
dépresseur *m. Hort.* growth retardant
dépressif,-ive *a.* depressive, depressing; **effet dépressif** depressive effect
dépression *f.* 1. depression, hollow. 2. (physiological) depression, set-back (of plant etc.)
déprimé,-e *a.* depressed, flattened
déprotéiné,-e *a.* deproteinized; **extrait déprotéinisé** deproteinized extract
dépulpage *m. Coff.* pulping
dépulper *v.t.* to remove the pulp, to pulp
dépulpeur *m.* pulping machine
dépuratif,-ive *a.* depurative, blood-cleansing; *n.m.* depurative, blood cleanser
déracinable *a.* capable of being uprooted
déracinage *m.* uprooting
déraciné,-e *a.* uprooted
déracinement *m.* uprooting
déraciner *v.t.* to uproot, to eradicate
déracineur,-euse *n.* uprooter
déraflage *m.* removal of stalks from grapes etc.
dérâpage *m. Vit.* removal of grapes from stalks (= **égrappage**)
dérâper *v.t.* to remove grapes from stalks (= **égrapper**)
dératisation *f.* rodent control
dérayure *f.* furrow
dérivé,-e *a.* derived; *n. m.* derivative
dérobé,-e *a.*, **culture dérobée** catch crop
dérochage, dérochement *m.* removal of stones
déroulement *m.* 1. unwinding, uncoiling. 2. development, progress
dérouleur *m.*, **dérouleuse** *f.* **(de film plastique)** plastic mulch-laying machine
dérouleuse-planteuse de film plastique *f.* mulcher-transplanter
déruellage *m. Vit.* = **débuttage** (*q.v.*)
désacidification *f.* deacidification
désagréger *v.t.* to disintegrate; *v.i.* to weather
désaisonné,-e *a.* out-of-season; **culture désaisonnée** out-of-season crop
désalinisation *f.* desalinization (see also **dessaler**)
désamérisation *f.* removing bitterness (from olives etc.)
désaturé,-e *a.* desaturated; **sol désaturé** desaturated soil
descendance *f. PB* progeny, descent; **d. homozygote** homozygous progeny
descendant,-e *a.* descending; *n.* progeny
désembourber *v.t.* to extricate (tractor etc.) from mud
désensachage *m.* bag removal
déséquilibre *m.* lack of balance, imbalance; **d. alimentaire** nutritional imbalance; **d. hydrique** water imbalance; **d. physiologique** physiological disorder
désertification *f.* desertification
désespoir (*m.*) **des peintres** London pride (*Saxifraga (spathularis ×) umbrosa*) (*S. umbrosa*) (= **saxifrage ombreuse**)
désherbage *m.* weed control, weed killing, herbicide application, hoeing (to remove weeds)
désherbant *m.* herbicide, weed killer; **d. sélectif** selective herbicide; **d. total** non-selective herbicide
désherbé,-e *a.* weeded, treated with herbicide; without grass
désherber *v.t.* to control weeds, by mechanical or chemical methods
désherbeuse *f.* weeding machine
déshydratation *f.* dehydration; *PD* dry tip-burn of lettuce
déshydrogenase *f.* dehydrogenase
désinfectant,-e *a.* disinfectant; *n. m.* disinfectant
désinfecter *v.t.* to disinfect
désinfection *f.* disinfection; *Hort.* **d. des semences** seed dressing; **d. du sol** soil disinfection; **d. du sol à la vapeur** soil sterilization, soil steaming
désinsectisation *f.* insect control, insect removal, disinfestation
désintégreur *m.* "splitter" (in systematics) (see **rassembleur**)
désonglettage *m. Hort.* snagging
désongletter *v.t. Hort.* to snag (= **couper l'onglet**)
désordre *m. PD* disorder
désorption *f.* desorption
désoxyribonucléique *a.*, **acide désoxyribonucléique** desoxyribonucleic acid
désquamation *f. Arb.* bark shedding, peeling
dessalage *m.* salt removal
dessaler *v.t.* to remove salt from, to de-salt
desséchant,-e *a.* drying; **vent desséchant** drying, desiccating wind
desséché,-e *a.* dry, withered
dessèchement *m.* drying up, desiccation; **d. floral (de la tulipe)** tulip flower bud blast; **d. de la rafle des raisins** *Vit.* grape stalk necrosis; **d. des tiges** (du framboisier) (raspberry) cane blight (*Leptosphaeria coniothyrium*); **d. des chérelles, d. physiologique (des jeunes cabosses)** (*Cacao*) cherelle wilt
dessécher *v.t.* to dry up, to desiccate; **se dessécher** to dry up, to become dry
desserte *f.*, **allée de desserte** *Plant.* etc. access track, service road
dessiccation *f.* desiccation
dessouchage, dessouchement *m.* removal of stumps, uprooting, grubbing of stumps, stumping; *Sug.* removing stubble (= **relevage, déchicotage**)
dessoucher *v.t.* to remove stumps, to stump
dessoucheur *m.* stumping machine, stumper
déstolonnage *m.* stolon, runner removal

déstolonner *v.t.* to remove stolons or runners
détaillant,-e *n.* retailer (see also **grossiste**)
détar *m. Afr. Detarium* sp., including tallow tree (*D. senegalensis*)
détérioration *f.* deterioration, decay; **d. de la structure du sol** soil structure deterioration
détergent,-e *a.* detergent; *n.m.* detergent
déterminé,-e *a.* determinate (of plant); **à croissance déterminée** of determinate growth
déterminisme (*m.*) **génétique** genetical control; **d. du sexe** sex determination
déterrage, déterrement *m.* removal from the soil, unearthing, digging up
déterrer *v.t.* to remove from the soil, to unearth, to dig up
détonateur (*m.*) **épouvantail** detonator for bird scaring
détour *m. Vit.* rejected grapes
détoxication *f.* detoxication
détrempé,-e *a.* waterlogged, soaked
détritoir *m.* olive crusher
détritus *m.* refuse, rubbish; **d. de jardin** garden refuse
deutérium *m.* deuterium, heavy hydrogen
deutzie *f. Deutzia* sp., including Japanese snowflower (*D. gracilis*)
dévasement *m.* dredging, removal of silt
dévaser *v.t.* to dredge, to remove silt
développement *m.* development
déverdir *v.t. Cit.* etc. to degreen; *v.i.* to degreen
déverdissage *m. Cit.* etc. degreening
déverser *v.t.* to divert (canal, channel); to pour (water), to discharge (water), to dump (earth etc.); **se déverser** to empty, to flow, to fall
déversoir *m.* draining channel, outfall
dévidoir *m.* reel, winder, hose-reel
dévitalisation *f.* devitalization, weakening (e.g. of weed by herbicide)
dextrogyre *a.* dextrorotatory
DF = **diagnostic foliaire** (*q.v.*)
diable *m.* two-wheeled trolley, barrow or truck
diadelphe *a.* diadelphous
diagnose *f.* diagnosis
diagnostic (*m.*) **foliaire** foliar, leaf analysis, diagnosis; **d. ligneux** *Vit.* vine cane analysis, diagnosis
diagramme *m.* diagram; **d. floral** floral diagram
diakhatou *m. Afr.* (*Wolof*) African eggplant (*Solanum aethiopicum, S. macrocarpum* and related species)
diallèle *a.* diallel
dialypétale *a.* dialypetalous
dialyse *f.* dialysis
dialysépale *a.* dialysepalous
dianthiculteur *m.* carnation grower
diapausant,-e *a.* diapausing; **larves diapausantes** diapausing larvae
diapause *f.* diapause
diaphragme *m. Bot.* diaphragm
diaspis *m. Diaspis* sp. and related spp; **d. du poirier** pear scale (*Epidiaspis leperii* = *Diaspis ostreaeformis*) (= **cochenille rouge du poirier**)
diastase *f.* diastase
dibasique *a.* dibasic
dibble *m.* (multiple) dibbler
DBE = **dibromoéthane** (*q.v.*)
dibromoéthane *m.* ethylene dibromide, EDB
dicaryon *m.* dikaryon, dicaryon
dicentre *m. Dicentra* sp., including bleeding heart (*D. spectabilis*) (= **cœur-de-Jeannette, diélytre**)
dichogame *a.* dichogamous
dichogamie *f.* dichogamy, anther and ovule maturation at different times in the same flower
dichotome *a.* dichotomous
dichotomie *f.* dichotomy
dichroanthe *a. Bot.* dichromatic
dicksonia *m. Dicksonia* sp. tree fern
dicline *a.* diclinous
dicotyle, dicotylé,-e *a.* dicotyledonous
dicotylédone, dicotylédoné,-e *a.* dicotyledonous; *n. f.* dicotyledon
dicotylédonicide *a.* **efficacité dicotylédonicide (d'un herbicide)** anti-dicotyledon efficiency (of a herbicide), anti-broad-leaved weed efficiency (see also **antidicotylé, antidicotylédones**)
dictame (blanc) *m.* dittany, fraxinella (*Dictamnus albus* = *D. fraxinella*); **d. (de Crète)** dittany of Crete (*Origanum dictamnus*) (see also **origan**)
dictyosome *m.* dictyosome
didyme *a.* didymous, growing in pairs
didyname *a.* didynamous, with four stamens, two long and two short
dieffenbachie *f.* **dieffenbachia** *m.* dumb cane (*Dieffenbachia* sp.)
diélytre *f. Dicentra (Dielytra)* sp. (= **dicentre**)
diervilla *m.* weigela, bush honeysuckle (*Diervilla* sp.)
diététique *a.* dietetic; *n. f.* dietetics
différence *f.* difference; **d. de température** difference in temperature
différenciation *f.* differentiation; **d. des bourgeons** bud differentiation; **d. florale** floral differentiation; **d. tissulaire** tissue differentiation
difforme *a.* deformed, misshapen
diffus,-e *a.* diffused, diffuse; **lumière diffuse** diffused light
diffusé,-e *a.* diffused; **lumière diffusée** diffused light, indirect lighing
diffuseur *m.* diffuser; *Irrig.* microjet

diffusion *f.* diffusion; **d. atmosphérique** atmospheric diffusion; **d. gazeuse** gaseous diffusion
digitaire *f. Digitaria* sp. (= **digitaria**); **d. couchée** smooth finger grass (*D. ischaemum*)
digitale *f. Digitalis* sp. foxglove; **d. à grandes fleurs** (*Digitalis ambigua* = *D. grandiflora*); **d. laineuse** *D. lanata*; **d. pourpre, d. pourprée** foxglove (*D. purpurea*) (= **gant de Notre-Dame**)
digitalé,-e *a.* resembling foxglove
digitaline *f.* digitalin
digitaria *m. Digitaria* sp. (= **digitaire** *q.v.*)
digité,-e *a.* digitate
digue *f.* dyke, dam, causeway, embankment
diguette *f. Irrig.* bund
dihybridisme *m.* dihybridism
dika (du Gabon) *m.* dika, wild mango (*Irvingia gabonensis*) (= **manguier sauvage**)
dilacéré,-e *a.* torn to pieces, dilacerated
dilatation *f.* expansion
dillénie *f. Dillenia* sp.; **d. élégante** *D. indica*, a small tree with edible fruits
dilobé,-e *a.* bilobed (= **bilobé**)
diluant *m.* diluent
dilué,-e *a.* diluted, dilute
diluer *v.t.* to dilute
diluvium *m. Ped.* diluvium, river drift
dimère *a.* dimerous
dimorphisme *m.* dimorphism; **d. sexuel** sexual dimorphism
dimorphothéca *m.* or *f. Dimorphotheca* sp., star of the veldt, African daisy
diœcie *f.* dioecism
dioïque *a.* dioecious
dionée *f.* dionaea, Venus' fly-trap (*Dionaea muscipula*)
dioscoréacées *f. pl.* Dioscoreaceae
diosgénine *f.* diosgenin
diospyros *m. Diospyros* sp. (see **ébénier**)
dioxyde *m.* dioxide; **d. de carbone** carbon dioxide; **d. de soufre** sulphur dioxide
dipétale *a.* dipetalous
diploïde *a.* diploid
diploïdisé,-e *a.* diploidized
diplotaxis (*m.*) **des murs** sand rocket (*Diplotaxis muralis*)
dipsacacées, dipsacées *f. pl.* Dipsacaceae
diptère *a. Ent.* dipterous; *n. m.* dipteran, *pl.* Diptera
discage *m. Hort.* discing, disking
discoïde *a.* discoid, disc-shaped
discolore *a.* 1. bicoloured. 2. of different colours
disépale *a.* with two sepals
disette *f.* scarcity, want (of food); **d. d'eau** drought
disperme *a.* two-seeded
dispersion *f.* dispersion
disponibilité *f.* availability; **d. en nitrate** nitrate availability; **disponibilités hydriques** water supply, water resources
dispostif *m.* 1. disposition, arrangement, layout; *Stat.* **d. expérimental** experimental design. 2. apparatus, device; **d. d'attelage** hitch, linkage; **d. de protection** protective device
disposition *f.* disposition, arrangement, layout; **d. des feuilles** phyllotaxis
disque *m.* disc, disk; **d. foliaire** foliar disc; **fleur du disque** disk floret (in Compositae)
diss *m. Ampelodesma tenax*, a densely-tufted perennial grass
dissémination *f.* dissemination, dispersal, spreading, scattering; **d. naturelle** natural dissemination
dissocié,-e *a.* dissociated
dissoudre *v.t.* to dissolve, to melt
dissous,-oute *a.* dissolved
dissuadant *m. Ent.* deterrent
distachyé,-e *a.* distichous
distal,-e *a., pl.* **-aux** distal, terminal
distançage *m. Hort.* spacing; **d. en travers** cross-blocking
distance *f.* distance; **d. de plantation** *Hort.* planting distance, plant spacing; **d. sur le rang** *Hort.* spacing within the row; **d. d'isolement** *PB* isolation distance
distancer *v.t. Hort.* to space out
distanceuse *f. Hort.* gapper, thinner; **d. électronique** selective thinner
distillation *f.* distillation
distique *a. Bot.* distichous
distributeur *m.* distributor; *AM* feed (of sower etc.); **d. d'engrais** fertilizer spreader
distribution *f.* distribution
diurétique *a.* & *n.m.* diuretic
diuron *m.* diuron
divariqué,-e *a.* divaricate
divergence *f.* divergence
divergent,-e *a.* divergent
diversicolore *a.* variegated
diversiflore *a.* diversiflorous
diversifolié,-e *a.* with leaves of different shapes
diversiforme *a.* diversiform
divinatoire *a.* divinatory; **baguette divinatoire** divining rod
divisé,-e *a.* divided, parted
diviser *v.t.* to divide
division *f.* division; **d. cellulaire** cell division; **d. de touffe** dividing a clump, tuft (of plant)
djakattou *m. Afr.* = **diakhatou** (*q.v.*)
djebar *m.* (*North Africa*) date offshoot
DL = **dose létale** *PC* lethal dose
doe-koe *m.* fruit of langsat (*Lansium domesticum*)
doigt *m.* finger; *Ban.* finger
dolique, dolic *m. Dolichos* sp., *Vigna* sp.; **d.**

asperge yard-long bean (*V. sesquipedalis*); **d. bulbeux** yam bean (*Pachyrrhizus erosus*); **d. d'Egypte, d. lablab** hyacinth bean (*D. lablab*) (= **fève d'Egypte**)
dolomie *f.* dolomite, magnesian limestone
domatie *f.* domatium, cavity in a plant in which live insects or mites in symbiosis
dôme *m.* dome
dominance *f.* dominance; **d. apicale** apical dominance
dominant,-e *a.* dominant; **espèce dominante** dominant species
dommage *m.* damage, injury
dommageable *a.* detrimental
dompte-venin *m. inv.* white swallowwort, vincetoxicum (*Cynanchum vincetoxicum*)
donnée *f.* datum, *pl.* data; **données de base** basic data
donneur *m.* donor, giver (see also **receveur**)
doradille (*f.*) **nid d'oiseau** bird's nest fern (*Asplenium nidus*) (= **fougère nid d'oiseau**); **d. des murailles** wall rue (*A. ruta-muraria*) (see also **asplenium**)
doré,-e *a.* gilded, gilt
dorine *f.* alternate-leaved golden saxifrage (*Chrysorplenium alternifolium*), opposite-leaved golden saxifrage (*C. oppositifolium*)
dormance *f.* dormancy; **d. embryonnaire** embryo dormancy
dormant,-e *a.* dormant; **graine dormante** dormant seed; **bourgeon, œil dormant** dormant bud
doronic, doronicum *m.* *Doronicum* sp., including leopard's bane (*D. pardalianches*)
dorsal,-e *a., m.pl.* **-aux** dorsal
dortoir *m.* *PC* roost (of birds)
doryphore *m.* Colorado beetle (*Leptinotarsa decemlineata*)
dosage 1. quantitative analysis, determination; **d. de l'azote** nitrogen determination; **d. des éléments assimilables** determination of assimilable elements. 2. dosage, rate; **d. en CO_2** CO_2 application
dose *f.* 1. proportion, amount, rate. 2. dose (of compound etc.); **dose létale** *PC* lethal dose
doser *v.t.* 1. to determine the quantity of a substance, to proportion. 2. to divide into doses
doseur (*m.*) **d'engrais chimique** fertilizer applicator
doublage *m.* lining, with plastic etc.
double *a.* double (of flower) (see also **vitrage**)
double-greffage *m.* double-working, double grafting
douçain *m.* = **doucin** *q.v.*
douce-amère *f.* bittersweet, woody nightshade (*Solanum dulcamara*)
doucette *f.* lamb's lettuce (= **mâche**)
doucin *m.* type of wild apple used as rootstock (*Malus pumila* var. (*acerba* =) *sylvestris*)
douglas *m.* *Coll.* Douglas fir (*Pseudotsuga menziesii*)
doum *m.* dum palm (*Hyphaene thebaica*) (= **palmier fourchu**)
doumier *m.* = **doum** (*q.v.*)
dourian *m.* = **durian** *q.v.*
douve *f.* 1. *Hort.* trench, (drainage) ditch. 2. spearwort; **grande douve** greater spearwort (*Ranunculus lingua*)
dracéna *m.* dragon tree (*Dracaena draco*) (= **dragonnier**); **d. (des horticulteurs)** *Cordyline* sp.
dracontium *m.* *Dracontium* sp. (Araceae)
dragée *f.* sugared almond
drageon *m.* root sucker
drageonnage, drageonnement *m.* 1. suckering, producing suckers, stooling. 2. dividing, using suckers for propagation
drageonnant,-e *a.* producing suckers, suckering, stooling
drageonner *v.i.* to sucker, to produce suckers, to stool (of plant); **plante qui drageonne très peu** plant which does not produce many suckers
drageonnicide *m.* sucker control chemical (for tobacco etc.)
dragonnier *m.* dragon tree (*Dracaena draco*) (= **dracéna**)
drague *f.* dredger; **d. à bras** hand dredger; **d. à godets** bucket dredger
draguer *v.t.* to dredge
dragueur *m.* dredger
drain *m.* drain, drain-pipe
drainage *m.* drainage; **d. par fossés** open cut drainage; **d. taupe** mole drainage; **d. par tuyaux** pipe drainage
draine *f.* missel thrush (*Turdus viscivorus*) (see also **grive**)
drainer *v.t.* to drain
draineur *m.* drainer; **équipement draineur** draining equipment
draineuse *f.* trencher with drain-laying attachment
drave *f.* whitlow grass (*Draba* sp.)
drèche, drêche *f.* brewer's grains
dressage *m.* straightening by tying (shoots etc.)
dressé,-e *a.* (of plant) upright, erect
dresse-bordure *m.* edging tool (for lawns)
drogue *f.* drug
droséra *m.*, **drosère** *f.* *Drosera* sp. (= **rossolis**)
drosophile *f.* *Ent.* drosophila, fruit fly
dru,-e *a.* dense, thick-set; **semer dru** to sow thickly
drupacé,-e *a.* drupaceous
drupe *m.* or *f.* drupe
drupéole *m.* small drupe, drupel(et)

dryopteris *m. Dryopteris* sp. fern
duboisie *f. Duboisia*, a genus which contains species with medicinal properties
dune *f.* dune
dura *m. OP dura;* **d. Deli** Deli *dura*
durabilité *f.* durability, keeping quality
durable *a.* lasting, durable
duramen *m.* duramen, heart-wood
duraminisation *f.* heartwood formation
durcir *v.t.* to harden (off)
durcissement *m.* hardening
durée *f.* 1. keeping quality, life (of cut flower). 2. duration; **d. d'action** duration of action; **d. de l'éclairage, d. d'éclairement** duration, period, of illumination; **d. de floraison** length of flowering (period); **d. germinative** seed life; **de longue durée** long-term
dureté *f.* 1. hardness, firmness. 2. difficulty
durian, durion *m.* durian tree (*Durio zibethinus*)
durio *m. Durio* sp., including durian
durione *f.* durian fruit
duveté,-e *a.* downy
duveteux,-euse *a.* downy, fluffy
dystrophie *f.* dystrophy

E

eau *f.* water; **d. d'arrosage** water for watering; **e. bouillante** boiling water; **e. capillaire** *Ped.* capillary water; **e. de coco** (*Coconut*) coconut water; **e. de condensation** condensation water; **e. courante** running water; **e. distillée** distilled water; **e. douce** fresh water, soft water; **e. dure** hard water; **e. d'égout** sewage; **e. de gravité, e. de gravitation** *Ped.* gravitational water, infiltration water; **e. industrielle** water from an industrial process; **e. d'irrigation** irrigation water; **e. liée** bound water; **e. de mare** pond water; **e. oxygénée** hydrogen peroxide; **e. de pluie** rainwater; **e. potable** drinking water; **e. de puits** well water; **e. salée** salt, saline water; **e. saumâtre** brackish water; **e. du sol** ground water; **e. de source** spring water; **e. tritiée** tritiated water

eau-de-vie *f., pl.* **eaux-de-vie** brandy, spirits

ébarbage *m.* trimming, clipping

ébarber *v.t.* to trim, to clip

ébauche *f. Bot.* primordium; **ébauches florales** floral primordia

ébénier *m.* ebony tree; **é des Alpes** Scotch laburnum (*Laburnum alpinum*) (= **cytise des Alpes**); **é. faux-anagyre** laburnum (*Laburnum anagyroides*) (= **faux ébénier**); **é. de l'ouest africain** West African ebony (*Diospyros mespiliformis*) (= **néflier africain**)

ébénisterie *f.* cabinet making

éborgnage *m.* bud removal, disbudding; *PD* bud fall

éborgné,-e *a.* disbudded

éborgner *v.t. Hort.* to remove buds, to disbud

ébouillantage *m.* scalding

ébouillanter *v.t.* to scald

ébouillanteur *n. m.*, **ébouillanteuse** *n. f.* machine for scalding

ébourgeonnage *m.* deshooting, bud removal; *Vit.* water soot removal

ébourgeonné,-e *a.* deshooted, disbudded, with buds removed

ébourgeonnement *m.* = **ébourgeonnage** (*q.v.*)

ébourgeonner *v.t.* to deshoot, to remove buds; **é. l'onglet** to snag

ébourgeonneur *m.* bullfinch and other bud-eating bird spp. such as hawfinch

ébourgeonnoir *m.* disbudding tool, disbudder

éboutonnage, éboutonnement *m. Hort.* bud rubbing, disbudding

éboutonner *v.t.* to disbud

ébouturer *v.t.* to remove buds, shoots or suckers

ébracté,-e *a. Bot.* without bracts, ebracteate

ébranchage, ébranchement *m.* cutting, lopping (branches)

ébrancher *v.t.* to cut, to lop (branches)

ébrancheur *m.*, **ébrancheuse** *f.* pruner, long-handled cutter, implement for cutting off branches, lopping shears

ébranchoir *m.* lopping bill

ébullition *f.* boiling

écabossage *m.* (*Cacao*) breaking, cutting open pods

écabosser *v.t.* (*Cacao*) to break or cut open pods

écaille *f.* 1. *Bot.* scale; **é. de bourgeon** bud scale; **é de bulbe** bulb scale. 2. arctid caterpillar (= **chenille bourrue**); **é. fermière** cream-spot tiger (moth) (*Arctia villica*); **é. fileuse** fall webworm (*Hyphantria cunea*); **é. martre** garden tiger (moth) (*A. caja*)

écaillement *m.* scaling, peeling off

écailleux,-euse *a.* scaly, squamous; **bourgeons écailleux du poirier** scaly buds of pear

écale *f.* shell, pod, husk, hull, testa

écaler *v.t.* to shell, to pod, to husk, to hull; **s'écaler** to fall out of the pod (peas), to burst the husk (walnut) etc.

écalot *m.* nut without a shell

écalure *f.* tough skin of certain seeds, husk

écarlate *a.* scarlet

écart *m.* deviation, divergence; **écarts journaliers de température** daily variations in temperature

écartement *m.* spacing; **é. des lignes** row-spacing; **à grand écartement** at a wide spacing. **un é. plus serré** a closer spacing

écarter *v.t.* to space, to space out

ecballium *m.* squirting cucumber (*Ecballium elaterium*)

écéper, écepper *v.t.* to uproot grapevines

échalas *m.* stake, prop, vine prop, hop pole

échalassage, échalassement *m.* propping, staking

échalasser *v.t.* to prop, to stake

échalier, échallier *m.* fence of stakes; stile

échalote *f.*, **é. grise** shallot (*Allium ascalonicum*); **é. d'Espagne** = **rocambole** (*A. scorodoprasum*); **é. de Jersey, é. rose** Jersey shallot; **é. nouvelle** spring onion, scallion

échamp *m.* interrow in vineyard

échampelé,-e *a.* frost-damaged; **vigne écham-**

pelée vine whose buds have not formed before the hot weather
échamplure *f. Vit.* delayed bud burst
échancré,-e *a.* notched, cut, nicked
échancrure *f.* notch, cut, nick
échange *m.* exchange; **é. de bases** base exchange; **é. cationique** cation exchange; **é. gazeux** gas exchange
échangeable *a.* exchangeable
échantillon *m.* sample; **é. de graines** seed sample; **é. de terre** soil sample; **é. végétal** plant sample
échantillonnage *m.* sampling; **é. foliaire** leaf sampling; **méthode d'échantillonnage** sampling method
échantillonner *v.t.* to sample
échantillonneur *m.* sampler
échanvrer *v.t.* to hackle
échanvroir *m.* hackle
échardonnage *m.* removing thistles
échardonner *v.t.* to remove thistles
échardonnet *m.*, **échardonnette** *f.* thistle hook, weed hook
échardonnoir *m.* thistle hook, weed hook
échau *m., pl.* **-aux** drainage channel, drain
échaudage *m.* 1. scalding. 2. sun scorch
échaudé,-e *a.* scalded, scorched
échauder *v.t.* to scald
échaudure *f.* scald; *PD* apple scald; **é. superficielle** superficial scald
échauffé,-e *a.* fermented, heated
échauffement *m.* heating, overheating
échauffer *v.t.* to cause to ferment, to heat
échelle *f.* 1. ladder; **é. de Jacob** Jacob's ladder (*Polemonium coeruleum*). 2. scale; **à grande échelle** on a large scale; **é. de notation** scale of marking
échelonné,-e *a.* spaced out, spread out, staggered (of trees, sowing dates etc.)
échelonnement *m.* spacing out, spreading out, staggering (of trees, sowing dates etc.)
échelonner *v.t.* to space out, to spread out, to stagger (trees, sowing dates etc.)
échenillage *m. Hort.* removing (and destroying) caterpillars from fruit trees etc.
écheniller *v.t. Hort.* to remove (and destroy) caterpillars from fruit trees etc.
échenilleur *m.* person who removes (and destroys) caterpillars
échenilloir *m.* tree pruner, long-arm pruner, branch cutter
écheveria *m.*, **échevérie** *f. Echeveria* sp.; **é. écarlate** scarlet echeveria (*E. coccinea*)
échicotage *m.* stump, stub removal
échicoter *v.t.* to remove the stumps, stubs (from branches, trees etc.)
échiné,-e *a.* echinate, furnished with spines or prickles
échinocactus *m. Echinocactus* sp., hedgehog cactus
échinocarpe *a.* with spiny fruit, echinocarpous
échinocéréus *m. Echinocereus* sp.
échinochloa *f. Echinochloa* sp.
échinope, échinops *m.* globe thistle (*Echinops* sp.)
échiquier *m.* chess-board; **en échiquier** chequered, in chequer pattern
échitès *m. Echites* sp., including *E. rubro-venosa*
écidie *f. Fung.* aecidium
écimage *m. Hort.* topping, pollarding, tip pruning
écimer *v.t. Hort.* to top, to pollard, to tip prune
écimeuse *f. Hort.* topping machine
éclaboussure *f.* splash, spatter
éclairage *m.* lighting, illumination; **é. artificiel** artificial lighting; **é. supplémentaire** supplementary lighting
éclaircie *f. Hort.* thinning
éclaircir *v.t. Hort.* to thin; **é. sur la ligne** to chop out, to gap, to single
éclaircissage, éclaircissement *m.* 1. *Arb., Hort.* thinning; **é. des branches** branch thinning; **é. chimique** chemical thinning; **é. des coursonnes** spur thinning; **é. des fleurs** *Vit.* thinning of flower clusters; **é. des fruits** fruit thinning; **é. des grappes** *Vit.* cluster thinning; **é. mécanique** mechanical thinning; **é. sur le rang** chopping out, gapping, singling. 2. clearing; **é. des nervures** yellow net vein virus
éclaircir *v.t. Arb., Hort.* to thin
éclaircisseur *m.*, **éclaircisseuse** *f.* hoe for thinning
éclaire *f.* **(grande) éclaire** greater celandine (*Chelidonium majus*) (= **chélidoine**); **petite éclaire** lesser celandine (*Ficaria ranunculoides*) (= **ficaire**)
éclairé,-e *a.* lit, illuminated, in the light
éclairement *m.* illumination, lighting; **é. artificiel** artificial light(ing); **é. continu** continuous illumination; **é. naturel** natural light(ing)
éclairer *v.t.* to light, to illuminate
éclat *m.* 1. splinter, chip; *Hort.* fragment of plant for propagation, cutting, slip, sett; **propagation par éclat de souche** propagation by division. 2. brilliance, bloom; **chrysanthèmes dans tout leur éclat** chrysanthemums at peak flowering, in all their brilliance
éclatage *m. Hort.* division of plant into fragments for propagation
éclatant,-e *a.* (of colour) bright, vivid
éclaté,-e *a.* burst, split; *Hort.* divided into fragments
éclatement *m.* bursting, splitting; *Hort.* bud burst; division of plant for propagation; *PD* cracking (e.g. of tomato, citrus fruit etc.); fanging (of roots)

éclater *v.t.* to burst, to split; *Hort.* to divide plants into fragments for propagation; *v.i.* 1. to burst. 2. to sparkle, to glitter
éclisse *f.* wooden wedge
éclore *v.i.* to hatch (of eggs); *Bot.* to open (of flowers), to blossom, to bloom, to burst (of buds)
éclosion *f.* hatching (of eggs); *Bot.* opening (of flowers), blossoming, blooming, bursting (of buds)
écoclimatologie *f.* ecoclimatology
écœurage *m. Ban.* removal of terminal bud to stimulate shooting of side buds
écœurer *v.t. Ban.* to remove the terminal bud to stimulate shooting of side buds
écologie *f.* ecology
économie *f.* economy, management; saving; **é. d'énergie en serre** energy saving in the greenhouse
écope *f.* ladle, scoop
écophysiologie *f.* ecophysiology
écophysiologique *a.* ecophysiological
écorçage *m.* stripping, peeling of bark, barking (= **écorcement**)
écorce *f.* bark (of plant, tree etc.), peel, rind; **é. d'angusture** angostura bark (see also **galipée**); **é. écailleuse** *PD* citrus psorosis (scaly bark) virus; **é. liégeuse** *PD, Vit.* corky bark virus; **é. de résineux** conifer bark
écorcement *m.* stripping, peeling of bark, barking (= **écorçage**)
écorcer *v.t.* to bark (tree), to peel, to husk
écosser *v.t.* to husk, to hull, to shell (beans etc.)
écosseur,-euse *n.* person who husks or shells; *n. f.* shelling, husking machine
écosystème *m.* ecosystem
écot *m. Hort.* lopped tree or branch, loppings
écoté *a.* lopped (of tree or branch)
écotype *m.* ecotype
écoulement *m.* outflow, discharge; **é. (du latex)** *Rubb.* latex flow; *Comm.* sale, (rapid) disposal
écouler *v.t. Comm.* to sell, to dispose of; **s'écouler** to flow out, to pour out
écourcelage *m. Vit.* removing water shoots
écran *m.* screen; **é. thermique** thermal screen
écraser *v.t.* to crush, to bruise, to squash
écrivain (*m.*) **(de la vigne)** *Vit.* vine chrysomelid beetle, western grape root worm (*USA*) (*Adoxus obscurus*) (= **eumolpe, gribouri**)
ectomycorhize *f.* ectomycorrhiza, ectotrophic mycorrhiza
ectomycorhizien,-enne *a.* ectomycorrhizal, relating to ectomycorrhizas; **inoculation ectomycorhizienne** ectomycorrhizal inoculation; **champignons ectomycorhiziens** ectomycorrhizal fungi
ectomycorhizogène *a.* ectomycorrhizal
ectoparasite *a.* ectoparasitic; *n. m.* ectoparasite
ectophyte *m.* ectophyte
ectotrophe *a.* ectotrophic
écume *f.* froth, foam, scum; *Sug.* **écumes de défécation** scums, filter mud
écusson *m. Hort.* bud for grafting, bark shield, bud shield
écussonnable *a.* capable of being budded, buddable
écussonnage *m. Hort.* (shield) budding, grafting; **é. en anneau** ring-budding; **é. avec un éclat d'écorce** chip-budding; **é. en flute** flute budding; **é. à œil dormant** budding with a dormant bud; **é. en pied** budding, grafting about 10 cm from the ground, low working; **é. en tête** budding, grafting at about 1.80 m from the ground, high working
écussonné,-e *a.* (shield) budded, grafted
écussonner *v.t.* to (shield) bud, to graft
écussonnoir *m.* budding knife, grafting knife (= **entoir, greffoir**)
édaphique *a.* edaphic; **facteur édaphique** soil factor
edelweiss *m.* edelweiss (*Leontopodium alpinum = Gnaphalium leontopodium*)
édulcorant,-e *a.* sweetening; *n.m.* sweetener
édulcorer *v.t.* to sweeten, to edulcorate
édule *a. Bot.* edible, esculent
effaneuse *f. AM* potato haulm remover
effarouchement *m.* startling, frightening; **moyen d'effarouchement** means of repelling, scaring (birds etc.)
effaroucher *v.t.* to startle, to frighten
effaroucheur *m.* scarer, banger (for scaring birds)
effet *m.* effect, result; **e. bénéfigue** beneficial effect; **e. de bordure** border effect; **e. dépressif** depressive effect; **e. morphogène** formative effect; **e. résiduel** residual effect; **e. de serre** greenhouse effect
effeuillage *m.* leaf-thinning, defoliation
effeuillaison *f.* leaf fall
effeuillé,-e *a.* defoliated, leafless
effeuillement *m.* 1. = **effeuillaison** (*q.v.*) 2. leaflessness
effeuiller *v.t.* to thin out leaves or petals, to defoliate
effeuilleur,-euse *n.* stripper (of leaves)
effeuillure *f.* leaves detached from tree, fallen leaves
efficacité *f.* efficacy, efficiency; **e. herbicide** herbicidal efficiency
effilé,-e *a.* slender, tapering
effleurage *m.* flower removal
effloraison *f.* flowering
efflorescence *f. Bot.* 1. flowering, efflorescence. 2. bloom (on grapes etc.) (= **fleur**) 3. *Ped.* efflorescence

efflorescent,-e *a. Bot.* efflorescent
égaliser *v.t.* 1. to equalize. 2. to level, to make even
égauler *v.t.* to prune (new) shoots
égayer *v.t. Hort.* to prune, to lop
églantier *m.* wild rose bush; **é. (commun)** dog rose (*Rosa canina*) (= **rosier des chiens**); **é. des champs, é. rampant** trailing rose (*R. arvensis*); **é. odorant** sweet briar (*R. rubiginosa*)
églantine *f.* wild rose flower; **é. odorante, é. d'Elisabeth d'Angleterre** sweet briar
égoïne, égohine *f.* pruning saw, small handsaw (= **scie égoïne**); **é. fermante** folding pruning saw; **é. à lame fixe** pruning saw with fixed blade
égopode *m.* ground elder (*Aegopodium podagraria*) (see also **ægopodium**)
égourmandage *m.* sucker removal
égourmander *v.t.* to remove suckers
égouttage *m.* drainage, drying
égoutté,-e *a.* drained, dried
égouttement *m.* draining, drying; **é. de surface** surface draining, drying
égoutter *v.t.* to drain, to dry
égouttoir *m. Irrig.* trickler
égrain, égrin *m.* young apple or pear tree seedling; young crab apple or wild pear tree (= **aigrin**)
égrainage *m.* = **égrenage** *q.v.*
égrainer *v.t.* = **égrener** *q.v.*
égrappage *m.* picking off grapes, currants, dates etc. from bunch; removal of grape stalks, stalking of grapes, destemming
égrapper *v.t.* to pick off grapes or currants from bunch; to remove grape stalks, to stalk grapes, to destem
égrappeur,-euse *n.* grape, currant etc. picker
égrappoir *m.* machine for removing grapes from bunches or stalks from grapes
égrenage *m.* shelling peas etc., picking off grapes, currants, gooseberries etc.
égrené,-e *a.* shelled (of peas etc.) picked off (of grapes, currants, gooseberries etc.)
égrener *v.t.* to shell peas etc., to pick off grapes, brussels sprouts, currants, gooseberries etc.
égreneur,-euse *n.* person who shells peas etc., picks off grapes, currants, gooseberries etc.; **égreneuse** *n. f.* machine for shelling peas etc., or picking off grapes, currants, gooseberries etc.
égrin = **égrain** (*q.v.*)
éhoupage *m.* topping, pollarding (tree)
éhouper *v.t.* to top, to pollard (tree)
éjecteur *m. AM* ejector; **é. de fanes** haulm ejector (on potato harvester etc.)
élæicole *a.* relating to oil palm; **zone élæicole** oil palm growing zone
élæiculture *f.* oil palm growing
élæis *m.* oil palm (*Elaeis guineensis*) (= **palmier à huile**)
élagage *m. Hort.* pruning, lopping, trimming; *pl.* prunings, loppings; **é. hivernal** dormant pruning; **palmes d'élagage** pruned fronds of palm
élagué,-e *a. Hort.* pruned, lopped, trimmed
élaguer *v.t. Hort.* to prune, to lop, to trim
élagueur *m. Hort.* 1. pruner (person). 2. pruning hook or shears; long-arm pruner (= **échenilloir**)
élancé,-e *a.* tall and slim, slender
élargir *v.t.* to widen, to enlarge
élarvement *m.* removal of insect larvae (caterpillars, sawfly larvae etc.) from plants
élastique *a.* elastic; *n. m.* rubber, indiarubber
elastomère *a.* elastomeric; *n. m.* elastomer
élatérion *m.* squirting cucumber (*Ecballium elaterium*) (= **ecballium**)
électrification *f.* electrification (of greenhouse etc.)
électrisation *f.* electrification (of substance, particle)
électrofocalisation *f.* electrofocusing
électrolyte *m. Ped.* electrolyte
électrophorèse *f.* electrophoresis
électropompe *f.* electric pump
éléis *m.* = **élæis** (*q.v.*)
élément *m.* 1. element; **é. (minéral) majeur** major element; **é. marqué** labelled element; **é. mineur** minor element; **é. nutritif** nutrient, nutritive element. 2. component, constituent; **é. du milieu** environmental factor
élémi, élémide *m.* resin from West Indian birch (*Bursera gummifera*) and others (see also **iciquier**)
éléophage *a.* & *n.* olive-eating, olive-eater
élévateur *m.* elevator, lift, hoist; **é. à fourche** fork lift
éliciteur *m.* elicitor
élimination *f.* elimination; **é. des types aberrants, é. des plants malades, é. sélective** culling, roguing
ellébore *m.* hellebore (*Helleborus* sp.); **e. blanc** false hellebore (*Veratrum album*); **e. fétide** stinking hellebore (*H. foetidus*); **e. d'hiver** winter aconite (*Eranthis hyemalis*) (= **éranthe**); **e. livide** *H. lividus*; **e. noir** Christmas rose (*H. niger*) (= **rose de Noël**); **e. vert** green hellebore (*H. viridis*)
ellipsoïde *a.* ellipsoidal; *n.m.* ellipsoid
elliptique *a.* elliptical
élongation *f.* elongation
éluvial,-e *pl.* **-aux** *Ped.* eluvial
éluviation *f. Ped.* eluviation

élyme *m.* **(des sables)** lyme grass (*Elymus arenarius*)
émarginé,-e *a. Bot.* emarginate
émasculation *f.* emasculation
emballage *m.* packing, packing material; **e. consigné** returnable pack(ing); **e. perdu** non-returnable, expendable pack(ing)
emballer *v.t.* to pack, to crate, to bale, to wrap
emballeur *n. m.* packer; **emballeuse** *n. f.* packing, wrapping machine
embase *f.* base (of post etc.), seating (of machine)
embêchage *m.* depth of digging with spade, spit
embellissement *m.* embellishing, improving
emblavure *f.* land under crop
embout *m.* nozzle (of hose, pipe etc.)
embranchement *m.* 1. branching. 2. phylum
embrassant,-e *a. Bot.* stem-clasping (= **amplexicaule**)
embroussaillé,-e *a.* covered with bushes
embrun *m.* often in *pl.* spray; **embruns de pulvérisation** *PC* spray drift
embryogène *a.* embryogenic
embryogenèse *f.* embryogenesis
embryoïde *m.* embryoid
embryon *m.* embryo
embryonnaire *a.* embryonic, embryonary
émergé,-e *a.* emerged
émergence *f.* emergence
émétique *a.* emetic
émettre *v.t.* to emit; **é. des pousses** to grow, to produce shoots
émietté,-e *a.* crumbled
émietter *v.t.* to crumble
émietteur *m.* = **brise mottes** (*q.v.*)
émissaire *m.* **(de drainage)** *Irrig.* drainage channel, outlet; **é. bouché** blocked up outlet
émission *f.* emission
émissivité *f.* emissivity
emmagasinage *m.* storage, storing
emmagasiné,-e *a.* stored
emmagasiner *v.t.* to store
emmagasineur *m.* person who stores, store-keeper, warehouseman
emmaillotage *m.* binding up, swathing
emménagogue *a.* & *n. m.* emmenagogue
emmotté,-e *a.* with a ball of earth around the roots, balled
emmotter *v.t.* to place soil around plant roots, to ball up a plant
émondage, émondement *m. Arb., Hort.* 1. pruning, trimming. 2. cleaning of seed etc., blanching of almonds
émonder *v.t. Arb., Hort.* 1. to prune, to trim. 2. to clean seed etc., to blanch almonds
émondes *f. pl. Arb., Hort.* prunings
émondeur,-euse *n.* pruner, trimmer
émondoir *m.* pruning or trimming tool
émottage, émottement *m.* breaking clods
émotter *v.t.* to break clods
émotteur,-euse *n.* (crosskill) roller; harrow for breaking clods
émottoir *m.* clod-breaking implement
émoussage *m. Hort.* removing moss
émousser *v.t. Hort.* to remove moss
émoussoir *m.* moss knife, wire brush (to remove moss from trees); **é.-grattoir** combined moss knife and scraper
empaillage, empaillement *m.* packing in straw; *Hort.* covering plants with straw, matting up (plants)
empailler *v.t.* to pack in straw; *Hort.* to cover plants with straw, to mat up (plants)
empaisselage *m.* = **échalassage** *q.v.*
empapillotage *m.* wrapping (fruit) in paper
empapilloter *v.t.* to wrap (fruit) in paper
empaquetage *m.* 1. packing. 2. packing materials
empaqueter *v.t.* to pack
empattement *m.* foundation, base of erection, (thickened) starting point of branch, base of plant, branch, base of bud; **e. des charpentes** *Hort.* crotch
empêtre, empetrum *m. Empetrum* sp.; **e. à fruits noirs** crowberry (*E. nigrum*)
empilage *m.* stacking; **e. en quinconce** *Mush.* checkerboard stacking
empilement *m.* stacking; **e. des thylakoïdes** thylakoid stacking
empiler *v.t.* to stack
empileur *m.* stacker, stacking machine
emplacement *m.* site, place, spot
emplanter *v.t.* to plant land
emploi *m.* use, employment; **"d'un emploi facile"** "easy to use"
empois *m.* starch paste
empoisonnement *m.* poisoning; **e. par le sélénium** selenium poisoning
empoisonner *v.t.* to poison
emporte-pièce *m. inv.* punch (tool)
empotage *m.* potting (up) of plants
empoté,-e *a.* potted; **plante empotée** potted plant
empotement *m.* = **empotage** *q.v.*
empoter *v.t.* to pot (plants)
empoteuse *f.* potting machine
empoussiéré,-e *a.* dusty
empoussièrement *m.* covering with dust, dusting
empoussiérer *v.t.* to dust
empreinte *f.* impression, imprint; **e. pétiolaire** *Vit.* etc. petiolar scar; **e. sporale** *Fung.* spore print
empyreumatique *a.* empyreumatic(al)
émulsifiable *a.* emulsifiable; **concentré émulsifiable** emulsifiable concentrate
émulsion *f.* emulsion
émulsionnable *a.* emulsible, emulsifiable
en *prep., pron. & adv.;* **en U** U-shaped; **en V** V-

shaped; **en cuiller** spoon-shaped; **en entonnoir** funnel-shaped
énation *f.* enation
encadrement *m.* 1. **personnel d'encadrement** managerial, supervisory staff; **un e. important** a large supervisory staff. 2. framework, frame; **e. de gazon** turf border
encaissement *m. Hort.* planting in tubs, boxes
encaisser *v.t. Hort.* to plant in tubs, boxes
enceinte *f.* enclosed space, enclosure, chamber; **e. climatisée** growth chamber, controlled environment chamber
encens *m.* incense
encépagement *m.* the range of the varieties in a vineyard or region
enchausser *v.t. Hort.* to earth up, to cover with straw
enchevêtré,-e *a.* tangled, intergrown
encoche *f.* small cut; **e. (de saignée)** *Rubb.* tapping cut; **e. sèche** *Rubb.* dry cut
encocher *v.t.* to make a small cut; *Rubb.* to tap
encombrant,-e *a.* encumbering, cumbersome, in the way
encre *f.* ink; **maladie de l'encre (du châtaignier)** chestnut ink disease (*Phytophthora cambivora, P. cinnamomi, P. syringae*); **maladie de l'encre (de l'iris)** iris ink disease (*Mystrosporium adustum*)
encroûtement *m. Ped.* crusting
endémique *a.* endemic
endémisme *m.* endemism
endiguer *v.t.* to dam up, to embank, to dike
endive *f.* 1. endive (*Cichorium endivia*). 2. witloof chicory (*C. intybus*) (see also **chicorée**)
endivette *f.* low grade of witloof chicory
endivier,-ière *a.* relating to witloof chicory growing; *n. m.* witloof chicory grower
endocarpe *m.* endocarp
endoderme *m.* endoderm
endogame *a.* endogamous
endogamie *f.* endogamy, inbreeding
endogé,-e *a.* found in the soil
endogène *a.* endogenous
endogenèse *f.* endogeny
endommagé,-e *a.* damaged, injured
endommagement *m.* damage, injury; **e. mécanique** mechanical damage
endomycorhization *f.* endomycorrhizal infection, endomycorrhizal development
endomycorhize *f.* endomycorrhiza, endotrophic mycorrhiza
endomycorhizien,-enne *a.* relating to endomycorrhizas
endomycorhizogène *a.* endomycorrhizogenous
endoparasite *a.* endoparasitic; *m.* endoparasite
endophyte *a.* & *n. m.* endophyte
endoplèvre *f.* endopleura
endosmose *f.* endosmosis
endosperme *m. Bot.* endosperm; (*Coconut*) endosperm, meat
endospermé,-e *a. Bot.* endospermic
endospore *m. Bot.* endospore
endosymbiote *m.* endosymbiont
endothérapique *a.* systemic (= **systémique** *q.v.*)
endotrophe *a.* endotrophic
enduction *f.* coating
enduire *v.t.* to coat with, to cover with, to plaster
enduit *m.* paste or liquid for spreading; **e. cireux** waxy compound; *Ped.* coating
endurci,-e *a.* hardened; **plante endurcie** hardened plant
endurcir *v.t.* to harden, to harden off
endurcissement *m.* hardening off
endymion *m. Endymion* sp., bluebell (*E. non-scriptus*) (= *Scilla non-scripta*)
énergide *m.* energid
énergie *f.* energy; **é. de bourgeonnement** shooting vigour; **é. germinative** germination vigour; **é. lumineuse** light energy; **e. rayonnate** radiant energy; **é. solaire** solar energy
énervé,-e *a. Bot.* enervate
éneyer *v.t. Sug.* to remove nodes
enfascié,-e *a. Bot.* fasciated
enfeuiller *v.t.* to cover with leaves; **s'enfeuiller** to break into leaf
enfonce-pieux *m.* post-driver
enfouir *v.t.* to bury, to plough in, to hoe in
enfouissable *a.* capable of being buried in the ground
enfouissage *m.* = **enfouissement** (*q.v.*)
enfouissement *m.* burying in the ground
enfouisseur *m.* implement which buries something in the ground
enfourchure *f.* fork, crotch of tree
engainant,-e *a.* sheathing (leaf)
engazonné,-e *a.* turfed, grassed; **surface engazonnée** grassed surface
engazonnement *m.* turfing, sowing with grass seed
engazonner *v.t.* to turf, to sow with grass seed
engluage *m. Arb., Hort.* 1. banding, covering with wax, smearing with (grafting) compound. 2. grafting wax
englué,-e *a. Arb., Hort.* banded, smeared with compound, waxed (of graft)
engluement *m.* = **engluage** (*q.v.*)
engluer *v.t. Arb., Hort.* to band, to smear with compound; **e. (un greffage)** to wax a graft
engorgé,-e *a.* blocked (up), clogged (up), waterlogged (of soil); **sol engorgé** waterlogged soil
engorgement *m.* blocking, clogging; **e. (en eau)** waterlogging
engorger *v.t.* to block (up), to clog (up), to waterlog (soil); **s'engorger** to become

blocked (up), to become clogged (up), to become waterlogged (of soil)
engrais *m.* manure, fertilizer; **e. artificiel** artificial manure; **e. azoté** nitrogenous fertilizer; **e. chimique, e. minéral** chemical fertilizer; **e. combiné, e. composé** compound fertilizer; **e. complet, e. ternaire** complete fertilizer; **e. complexe** compound fertilizer; **e. en couverture** top dressing; **e. à action lente, e. à libération lente** slow-release fertilizer; **e. liquide** liquid manure (= **purin**); **e. organique** organic manure; **e. phosphaté** phosphatic fertilizer; **e. potassique** potash fertilizer; **e. retard** slow-release fertilizer; **e. soufré** sulphur fertilizer; **e. vert** green manure
enguirlander *v.t.* to garland, to engarland
enherbé,-e *a.* 1. turfed, grassed down. 2. invaded by (grass) weeds
enherbement *m.* 1. grassing down; **e. d'un verger** grassing down an orchard. 2. (grass) weed infestation
enherber *v.t.* to turf, to grass down
enjambement *m.* crossing over (in cell division)
enjambeur *m.* straddling or high-clearance implement for work in high crops (see also **tracteur**)
enjaugé,-e *a.* heeled in
enjaugeage *m.* heeling in
enjauger *v.t.* to heel in
enjonçage *m.* furnishing with rushes
enjoncer *v.t.* to furnish with rushes; *Vit.* to bury rushes in the loose soil of vineyards to stabilize it
enlèvement *m.* removal, removing
enlisement *m.* sinking into a swamp, bog etc.
enliser *v.t.* to suck in, engulf (of swamp etc.); **s'enliser** to sink, to become bogged
ennéagyne *a.* enneagynous
ennéandre *a.* enneandrous
enneigé,-e *a.* snow-covered
ennemi *m. Hort.* pest; **e. naturel** natural enemy, predator (of pest)
ennoyage *m.* submergence, waterlogging
ennoyé,-e *a.* submerged, waterlogged
ennoyer *v.t.* to submerge, to waterlog
enothère *m.* = **œnothère** (*q.v.*)
énoyautage *m.* stoning of fruit
énoyauter *v.t.* to stone fruit
énoyauteur *m.* machine for stoning fruit
enquête *f.* study, survey, enquiry; **e. scientifique** scientific investigation, research
enracinable *a.* capable of being rooted
enraciné,-e *a.* rooted, deep-rooted
enracinement *m.* 1. planting. 2. rooting, taking root; **e. (d'une bouture)** striking, rooting (of cutting); **à e. peu profond** shallow-rooted; **à e. profond** deep-rooted; **aptitude à l'enracinement** rooting ability
enraciner *v.t.* 1. to dig in. 2. to root; **s'enraciner** to take root, to root, to become established
enrayer (*v.t.*) **un champ** to plough the first furrow in a field; **e. les sillons** to lay out the furrows
enrayure *f.* the first furrow ploughed
enrésinement *m. Arb.* underplanting with conifers
enrichi,-e *a.* enriched
enrichissement *m.* enrichment
enrobage *m.* coating, covering, waxing (of fruit); *Hort.* **e. (des graines)** seed dressing, pelleting
enrobé,-e *a.* coated, covered; **graines enrobées** dressed seed, pelleted seed
enrober *v.t.* to coat, to cover, to dress (seed), to pellet (seed)
enroulage, enroulement *m.* rolling up; *PD* leaf roll, crinkle; **e. (des feuilles)** leaf curl, roll; **e. chlorotique** chlorotic leaf curl (virus); **e. des feuilles du cerisier** cherry leaf roll virus; **e. des feuilles de la pomme de terre** potato leaf roll virus; **e. de la vigne** *Vit.* grapevine leaf roll
enrouler *v.t.* to roll up
enrouleur *m.* winder (for garden hose etc.) (see also **chariot**)
enrouleuse *f. Ent.* leaf roller; **e. pâle** (*Canada*) pale apple leaf roller (*Pseudexentera mali*)
ensachage *m.* bagging, putting into bags; **e. des régimes de bananes** bagging banana bunches
ensaché,-e *a.* bagged, put into bags
ensachement *m.* = **ensachage** (*q.v.*)
ensacher *v.t.* to bag, to put into bags
ensacheur,-euse *n.* 1. bagger. 2. *n.* bag-filling machine
ensachoir *m.* bag-filling machine
ensemencement *m.* sowing
ensemencer *v.t.* to sow, to culture, to spawn (mushroom)
enserrer *v.t. Hort.* to put into a greenhouse
ensifolié,-e *a. Bot.* with ensiform leaves
ensiforme *a.* ensiform, sword-shaped
ensilage *m.* silage
ensiler *v.t.* to ensile
ensileur *m.* person who ensiles
ensimage *m.* oiling, greasing (of textile)
ensoleillé,-e *a.* sunny (position etc.)
ensoleillement *m.* insolation
entada *m. Entada* sp., including nicker bean (*E. scandens*)
entaillage *m.* notching. (*Coconut*) etc. notching the husk to facilitate germination
entaille *f.* incision, notch, nick: gash, cut
entailler *v.t.* to make an incision, to nick, to notch: to gash, to cut; **e. en dessous de l'empattement (du bourgeon)** bud nicking; **e. au dessus de l'empattement** bud notching
entamer *v.t.* to cut into, to penetrate

entassement *m.* heaping
entasser *v.t.* to heap, to put in heaps
ente *f. A, Hort.* scion, graft (union), grafted shoot
entement *A, m. Hort.* grafting
enter *v.t. A, Hort.* to graft
enterrable *a.* capable of being buried
enterré,-e *a.* buried, sunken (of drain etc.); **e. au ¾** three-quarters buried
enterrement *m.* burial
enterrer *v.t.* to put in earth, to bury
entêtage *m.* topping, pollarding
entier,-ère *a.* entire, whole
entoir *m.* grafting knife, budding knife (= **écussonnoir, greffoir**)
entolome (*m.*) **livide** *Fung.* leaden entoloma (*Entoloma (Rhodophyllus) lividus*)
entomique *a.* entomic, relating to insects
entomocénose *f. Ent.* insect community, insect balance
entomofaune *f.* insect fauna
entomogame *a.* entomophilous
entomogamie *f.* entomophily
entomogène *a.* entomogenous
entomologie *f.* entomology
entomologique *a.* entomological
entomologiste *m.* entomologist
entomopathogène *a.* entomopathogenic, pathogenic to insects; **champignon entomopathogène** entomopathogenic fungus
entomophage *a.* entomophagous
entomophile *a. Bot.* entomophilous
entomophthore *m. Fung. Entomophthora* sp., (parasitic on insects)
entomosporiose *f.* **(du poirier et du cognassier)** quince leaf blight (*Entomosporium maculatum*)
entonnoir *m.* funnel (for pouring liquids); **en entonnoir** funnel-shaped
entophyte *m.* entophyte
entraînement *m. Ped., PC* etc. leaching, drift
entre-cœur *m. Vit.* (small) lateral (shoot) arising from bud close to dormant bud (= **prompt-bourgeon**)
entrecueillir *v.t.* to pick fruits as they ripen; to pick over; to thin fruits
entre-deux *m. inv., Vit.* interrow strip not reached by the plough (= **cavaillon**)
entrée *f. Ag., Hort.* input
entre-écorce *m.* natural graft between two branches or between a branch and the trunk
entre(-)feuille *f.* 1. secondary leaf. 2. interval between leaves (on grapevine etc.)
entreligne *f.* interrow
entre-nœud *m., pl.* **entre-nœuds** *Bot.* internode (= **mérithalle**)
entre-plant *m., pl.* **entre-plants** interplants for the rejuvenation of vineyard etc.
entreplantation *f.* interplanting
entreposage *m.* warehousing, storage; **e. en atmosphère contrôlée** CA storage; **e. frigorifique** cold storage
entreposer *v.t.* to warehouse, to store
entrepôt *m.* warehouse, store; **en entrepôt** in store; **e. frigorifique** cold store
entre-rang *m., pl.* **entre-rangs** space between rows, interrow
entretien *m.* upkeep, maintenance
entretoise *f.* strut, brace: tie
énucléation *f. Hort.* stoning of fruit
énucléer *v.t. Hort.* to stone fruit
envahissant,-e *a.* spreading, infesting, invasive (of plant); **mauvaise herbe envahissante** (invasive) weed
envahissement *m.* invasion (by weeds etc.); **e. (du mycélium)** *Mush.* spawn-running
envasement *m.* silting
envaser *v.t.* to silt, to choke up; **s'envaser** *v.i.* to silt up
enveloppe *f.* envelope, cover, covering; **e. séminale** seed coat
environnement *m.* environment
environnemental,-e *a., m. pl.* **-aux** environmental
enzymatique *a.* enzymatic
enzyme *m.* or *f.* enzyme
éolien,-enne *a.* aeolian; **érosion éolienne** wind erosion; *n.f.* air-motor, wind-engine; windmill for pumping
épacride *f. Epacris* sp., Australian heath
épaillage *m. Sug.* defoliation, trashing
épailler *v.t. Sug.* to defoliate sugar cane
épais,-aisse *a.* thick; **bouillie épaisse** thick (spray) mixture; **feuillage épais** dense foliage; **haie épaisse** close-set hedge
épaississement *m.* thickening
épamprage, épamprement *m. Vit.* removing water shoots or leaves, pruning, thinning (leaves)
épamprer *v.t. Vit.* to remove water shoots, to remove leaves, to prune, to thin (leaves)
épandage *m.* spreading, application (of fertilizer); **e. d'herbicide** herbicide application
épandeur *m.* spreader; **é. d'engrais** fertilizer spreader; **é. de fumier** manure spreader; **é. de microgranulés** microgranule applicator; **é. à la volée** (fertilizer) broadcaster
épandre *v.t.* to spread, to scatter; **s'épandre** to spread
épanoui,-e *a. Hort.* in full bloom; **rose épanouie** full-blown rose
épanouir *v.t. Hort.* to cause (flowers) to open; **s'épanouir** to open, to bloom (of flowers)
épanouissement *m. Hort.* opening (of flowers); **solution d'épanouissement** opening solution
épars,-e *a.* scattered

épaulé,-e *a.* stepped (of scion etc.)
épaulement *m.* step (of scion etc.); **greffe en fente à épaulements** shoulder grafting
épépinage *m.* removal of seeds or pips from fruit
épépiné,-e *a.* with seeds or pips removed
épépiner *v.t.* to remove seeds or pips from fruit
éperon *m.* spur
éperonnière *f.* larkspur (*Delphinium* sp.); columbine (*Aquilegia* sp.); toadflax (*Linaria* sp.)
épervière *f.* hawkweed (*Hieracium* sp.)
éphémère *a.* ephemeral, short-lived *n. f. Hort. Tradescantia* sp.; **é. de Virginie** common spiderwort, flower of a day (*T. virginiana*) (= **misère**)
éphippigère *m.* long-horned grasshopper, bush cricket (*Ephippiger* sp.)
épi *m.* ear, head of cereal, spike of flower; **é. de maïs** maize cob; **é. de lait, é. de la Vierge** French asparagus (*Ornithogalum pyrenaicum*)
épiaire *m.* woundwort (*Stachys* sp.)
épiaison *f.* ear emergence, heading (of cereals, grasses); **pleine épiaison** full heading
épibiote *m.* = **greffon** (*q.v.*)
épicalice *m. Bot.* epicalyx
épicarpe *m. Bot.* epicarp; **à l'é. mince** thin-walled
épicarpique *a. Bot.* epicarpic
épice *f.* spice
épicéa *m.* spruce (*Picea* sp.); **é. commun** common, Norway spruce (*P. abies*); **é. d'Engelmann** *P. engelmannii*; **é. glauque** white spruce (*P. glauca*); **é. de Glehn** Saghalien fir (*P. glehnii*); **é. de Hondo** Hondo spruce (*P. koyamai*); **é. d'Orient** Oriental spruce (*P. orientalis*); **é. piquant** Colorado spruce (*P. pungens*); **é. à queue de tigre** tiger-tail spruce (*P. polita*); **é. de Serbie** Serbian spruce (*P. omorika*); **é. de Sitka** Sitka spruce (*P. sitchensis*); **é. de Yéso** Yezo spruce (*P. jezoensis*) (see also **sapinette**)
épicotyle *m.* epicotyl
épicuticulaire *a.* epicuticular
épidémiologie *f.* epidemiology
épidendrum, épidendre *m. Epidendrum* sp., dragon's mouth orchid
épiderme *m.* epidermis; **é. dorsal** upper epidermis; **é. foliaire** leaf epidermis; **é. ventral** lower epidermis
épidermé,-e *a.* covered by epidermis
épidermique *a.* epidermal, epidermic
épidermoïde *a.* epidermoid
épié,-e *a.* spicate; headed (of cereals, grasses)
épierrage, épierrement *m.* removal of stones
épierrer *v.t.* to remove stones
épigé,-e *a.* epigeal
épigyne *a.* epigynous
épigynie *f.* epigyny
épillet *m.* spikelet
épilobe *m. Epilobium* sp., willow herb
épinage *m.* protecting bases of trees with thorns
épinaie *f.* spiny thicket
épinard *m.* spinach (*Spinacia oleracea*); **é. de Cayenne** *Phytolacca icosandra*; **é. de Ceylan** waterleaf (see **grassé**); **é. de Chine** = **é. indien** (*q.v.*); **é. cochon** = **é. du Soudan** (*q.v.*); **é. fraise** strawberry blite (*Chenopodium capitatum*); **é. géant** garden orache (*Atriplex hortensis*) (= **arroche**); **é. à graines rondes** non-prickly spinach; **é. à graines piquantes** prickly spinach; **é. indien, é. de Malabar** Indian spinach (*Basella rubra*) (= **baselle, brède d'Angola**); **é. de la Nouvelle Télande, é. d'été** New Zealand spinach (*Tetragonia expansa*) (= **tétragone**); **é. perpétuel** = **patience** (*q.v.*); **é. piquant** = **é. du Soudan** (*q.v.*); **é. sauvage** Good King Henry (*Chenopodium bonus-henricus*); **é. du Soudan** African spinach (*Amaranthus* spp., including *A. hybridus*) (see also **amarante**)
épinaste *a.* epinastic
épinastie *f.* epinasty
épinçage *m.* pinching off buds, disbudding
épincer *v.t.* to pinch off buds, to disbud
épine *f.* 1. thornbush; **é. blanche** hawthorn (*Crataegus monogyna, C. oxyacantha*); **é. du Christ** Christ's thorn (*Paliurus spina-Christi*) (= **porte-chapeau**), *Ziziphus spina-Christi*, crown of thorns (*Euphorbia splendens*); **é. ergot de coq** = **ergot de coq** (*q.v.*); **é. fleurie** scorpion-broom (*Genista scorpius*); **é. à fruit noir** Hungarian thorn (*C. nigra*); **é. noire** blackthorn, sloe (*Prunus spinosa*) (= **prunellier**); **é. petit corail** Washington thorn (*C. phoenopyrum*); **é. de rat** butcher's broom (*Ruscus aculeatus*) (= **fragon**). 2. thorn, prickle. 3. spine; **é. foliaire** leaf spine
épinette *f.* = **épicéa, sapinette** (*q.v.*)
épineux,-euse *a.* thorny, prickly, spiky; *n.m.* thornbush
épine-vinette *f., pl.* **épines-vinettes** barberry (*Berberis vulgaris*) (= **vinettier**) and *Berberis* sp.; **é.-v. pourpre** *B. thunbergii atropurpurea*; **é.-v. de Thunberg** *B. thunbergii*
épiphléose *m.* plant epidermis
épiphylle *a.* epiphyllous
épiphylle *f. Epiphyllum* sp., leaf-flowering cactus
épiphyte *a.* epiphytic; *n. m.* epiphyte
épiphytie *f.* epiphytic disease
épiphytique *a.* epiphytic
épirhize, épirrhize *a.* epirrhizous
épisperme *m.* episperm
épistaminé,-e *a.* gynandrous
épistasie *f. PB* epistasis, inhibiting effect of a

gene on a character controlled by another non allelic gene
épistillage *m. Ban.* removal of dried flowers
éplanage *m. Vit.* disbudding (= **éborgnage**)
épluchage, épluchement *m.* 1. picking over. 2. peeling, paring (fruit, vegetables etc.). 3. thinning out fruit tree
éplucher *v.t.* 1. to pick over. 2. to peel, pare (fruit, vegetables etc.). 3. to thin out fruit trees
éplucheur,-euse *n.* person who picks over, peels or pares (fruit, plants, vegetables etc.), person who thins fruit trees; *n.f.* peeling tool, machine
épluchoir *m.* paring knife
épluchure *f.* peeling, paring
épointage *m. Hort.* stopping; *Vit.* removing tips of grape bunches
épointer *v.t. Hort.* to stop; *Vit.* to remove tips of grape bunches
éponge *f.* 1. sponge; **é. végétale** vegetable sponge, loofah (*Luffa* sp.). 2. bedeguar, rose-gall (*Rhodites rosae*)
époque *f.* time, period; **é. du semis** sowing season, sowing time
épouvantail *m.* scarecrow
épreuve *f.* test, trial; **é. variétale, é. de variétés** variety trial; **é. à la coupe (des fèves)** (*Cacao*) cut test
éprouvette *f.* test-tube
épuisant,-e *a.* exhausting; **culture épuisante** (soil-) exhausting crop
épuisé,-e *a.* exhausted
épuisement (*m.*) **du sol, é. des terres** soil exhaustion
épulpeur *m.*, **épulpeuse** *f.* depulper
épuration (*f.*) **des eaux** water purification; **é. (sanitaire)** roguing (of crops)
épurer *v.t.* to purify, to filter, to rogue (crops)
épurge *f.* spurge (*Euphorbia* sp.); caper spurge (*E. lathyrus*)
équatorial,-e *a., m. pl.* **-aux** equatorial
équeutage *m.* removing stalks, stems, stalking (of fruit), topping (onions etc.)
équeuter *v.t.* to remove stalks, stems, to stalk (fruit), to top (onions etc.)
équeuteuse *f.* fruit stalking machine; stem-removing machine; **é. d'oignons** onion topper
équilibre *m.* equilibrium, balance; **e. cationique** cation equilibrium; **é. hormonal** hormonal balance; **é. K-Mg** K-Mg balance
équilibré,-e *a.* in equilibrium, balanced
équipe *f.* gang (of labourers etc.)
équipement *m.* equipment, outfit; **é. draineur** drainage equipment; **é. rigoleur** trenching equipment; **é. sillonneur** furrowing equipment
équivalent,-e *a.* equivalent
érable *m.* 1. maple (*Acer* sp.); **é. argenté** silver maple (*A. saccharinum*); **é. champêtre** field maple (*A. campestre*); **é. à feuilles de frêne = é. négundo** (*q.v.*); **é. du Japon, é. japonais (rouge)** (red) Japanese maple (*A. palmatum*) (*atropurpureum*); **é. jaspé** moosewood (*A. pennsylvanicum*); **é. de montagne = é. sycomore** (*q.v.*); **é. de Montpellier** Montpellier maple (*A. monspessulanum*); **é. négundo** box elder (*A. negundo*) (= **négundo**); **é. plane** Norway maple (*A. platanoides*); **é. rouge** red maple (*A. rubrum*); **é. à sucre** sugar maple (*A. saccharum*); **é. sycomore** sycamore (*A. pseudoplatanus*) (= **sycomore**). 2. **é. d'appartement** 'flowering maple' (*Abutilon hybridum*)
érablière *f.* (*Canada*) sugar maple plantation
éraillé,-e *a.* frayed, scratched
éranthe *f.* **eranthis** *m.* (*Eranthis* sp.); **é. d'hiver** winter aconite (*E. hyemalis*) (= **ellébore d'hiver, renoncule d'hiver**)
éranthème *m. Eranthemum* sp. (Acanthaceae), a genus containing ornamental tropical plants
ercissement *m.* sun scorch
érémostachys *m.* desert rod (*Eremostachys laciniata*)
érémure, érémurus *m. Eremurus*, a genus containing ornamental Liliaceae
ergot *m.* spur; **en ergot** spur-shaped; *Hort.* snag, stub; *PD* ergot (*Claviceps* sp.)
ergot (*m.*) **de coq** cockspur thorn (*Crataegus crus-galli*) (= **épine ergot de coq**)
érianthe *m.* ravenna grass (*Erianthus ravennae*) and other spp.
éricoïde *a.* ericoid
érigé,-e *a.* erect
érigéron *m. Erigeron* sp., fleabane; **é. âcre** blue fleabane (*E. acer*); **é. du Canada** Canadian fleabane (*E. canadensis*) (see also **vergerette**)
érine *f. Erinus* sp., including summer starwort (*E. alpinus*)
érinée *f.* = **érinose** (*q.v.*)
érineum *m.* = **érinose** (*q.v.*)
érinose *f. PC* erinose; **é. bronzée** tomato russet mite (*Vasates destructor*); **é. de la vigne** vine leaf blister mite (*Eriophyes (Phytoptus) vitis*) (= **phytopte de la vigne**)
ériodendron *m.* 1. *Eriodendron* sp. 2. silk cotton tree (see **fromager**)
ériophyide *m.* blister mite; **é. du pommier** apple leaf and bud mite (*Aculus schlechtendali*)
érodé,-e *a.* eroded; *Bot.* eroded, erose
éroder *v.t.* to erode; **s'éroder** to become eroded
érodibilité *f.* erodibility; **indice d'érodibilité** erodibility index

érodium *m. Erodium* sp., storksbill; **é. fausse mauve** soft storksbill (*E. malacoides*)
érosion *f.* erosion; **é. éolienne** wind erosion; **é. hydrique** water erosion; **é. en nappe, é. par couches** sheet erosion; **é. ravinante, é. en ravins** gully erosion
erreur (*f.*) **moyenne** *Stat.* standard error
érubescent,-e *a.* erubescent, rubescent, becoming red
éruca *f. Eruca* sp., rocket (see also **roquette**)
érucastrum (*m.*) **à feuilles de cresson** *Erucastrum nasturtiifolium*
érusser *v.t.* to remove leaves from tree shoots, fibre plants etc.
éryngium *m.*, **érynge** *f. Eryngium* sp., sea holly
érysimum, érysimon *m. Erysimum* sp., treacle mustard (= **vélar**); **e. nain compact jaune d'or** *E. murale*
érysiphe *m. Erysiphe* sp., powdery mildew
érythrée *f.* **(centaurée)** lesser centaury (*Centaurium pulchellum* = *Erythraea pulchella*)
érythrine *f. Erythrina* spp.; **é. chauve-souris** *E. vespertilis*; **é. crète de coq** common coral tree (*E. crista-galli*); **é. du Sénégal** Senegal coral tree (*E. senegalensis*)
esca *f. Vit.* esca, vine apoplexy (*Uncinula necator* and *Fomes (Phellinus) igniarius*)
escamotable *a.* retractable; **rampe escamotable** retractable boom
escarbille *f.* clinker, ash, cinder
escarbot *m.* chafer or other beetle; **e. doré** rose chafer (*Cetonia aurata*) (= **cétoine**); *Ban. (Martinique) Lysergus* sp.
escargot 1. snail. 2. *Hort.* snail medic (*Medicago scutellata*)
escarole *f.* endive (= **scarole**)
escarpé,-e *a.* steep, precipitous; **vignobles escarpés** vineyards on steep slopes
eschscholtzie *f.* Californian poppy (*Eschscholtzia* sp.)
escionnement *m.* disbudding
escionner *v.t.* to disbud
escourgée *f. Vit.* fruit cane
esherber *v.t.* = **désherber** (*q.v.*)
espace *m.* space; **e. vert** open space (with lawns, trees etc.)
espacement *m.* spacing, planting distance
espacer *v.t.* to space, to space out
espalier *m.* espalier, row of fruit trees growing against a wall (see also **contre-espalier**); *Vit.* espalier training
espargoute *f.* spurrey (*Spergula* sp.)
espèce *f.* species; *Hort.* **e. fruitière** fruit species; **e. de jardin** garden species; **e. vectrice** *Ent.* vector species
espèce-hôte *f.*, *pl.* **espèces-hôtes** *PD* etc. host species
espoudassage *m. Vit.* preliminary pruning
espoudasser *v.t. Vit.* to give a preliminary pruning
esquille *f.* splinter
essai *m.* trial, test, experiment; **e. biologique** bioassay test; **e. à blanc** blank test; **e. de comportement** performance trial; **e. d'écartement** spacing trial; **e. éliminatoire** screening test; **e. factoriel** factorial trial; **e. de fumure** manurial trial, experiment; **e. de germination** germination test; **e. de rendement** yield trial; **e. d'uniformité** uniformity trial
essaim *m.* swarm (of bees)
essaimage *m.* swarming (of bees)
essarmenter *v.t.* to remove vine stems, climbing stems or bines
essart *m.* freshly cleared land
essartage, essartement *m.* grubbing, clearing (of land)
essarter *v.t.* to grub up (trees, roots), to clear (land); **machine à essarter** grubber
essence *f.* essential oil; *Arb.* species; **e. d'alignement** avenue tree; **e. feuillue** broad-leaved tree; **e. d'ornement** ornamental tree or shrub; **é. résineuse** coniferous species
essorage *m.* drying
essorer *v.t.* to put out to dry, to dry
essouchage, essouchement *m.* removal of stumps, stumping; grubbing
essoucher *v.t.* to remove stumps, to stump
essoucheur *m.* implement or machine for stump removal
ester *m.* ester
estérifié,-e *a.* esterified
estérifier *v.t.* to esterify
estimation *f.* estimation, assessment; **méthode d'estimation des pertes de récolte** crop loss assessment method
estival,-e *a.*, *m. pl.* **-aux** (a)estival, summer (flower, plant etc.)
estivation *f. Bot.* aestivation (= **préfloraison, vernation**)
estragon *m.* tarragon (*Artemisia dracunculus*)
ésule *f.* spurge (*Euphorbia* sp.); **é. ronde** petty spurge (*E. peplus*) (= **euphorbe des jardins**)
établissement *m.* establishment; **é. horticole** horticultural establishment, market garden
étage *m.* ecological zone; *Hort.* tier; *Arb.* storey
étagé,-e *a.* terraced; arranged in tiers
étagement *m. Hort.* horizontal spacing out of branches; *Arb.*, *Vit.* etc. terracing, horizontal spacing of rows
étager *v.t.* to terrace; to arrange in tiers
étagère *f.* shelf, rack
étain *m.* tin
étalé,-e *a.* (of plant) spreading; **étalée** *n. f. Vit.* leaf with horizontally-placed blade
étalement *m. PC* spreadability (of spray etc.)

étalon *m.* standard; **arbre étalon** control, check tree
étalonnage *m.* standardization, gauging, calibration
étalonner *v.t.* to standardize, to gauge, to calibrate
étamine *f.* stamen
étanche *a.* impervious; **é. à l'eau** waterproof; **é. aux insectes** insect-proof
étanchéité *f.* imperviousness; **é. à l'eau** watertightness; **é. à l'air** air-tightness
étang *m.* pond
état *m.* state; **é. sanitaire** state of health; **à l'é. spontané** (of plant) self-sown
étayage *m.* propping up, buttressing, supporting
étayer *v.t.* to support, to buttress, to prop up
été *m.* summer
étendard *m. Bot.* standard (petal)
étendeuse (*f.*) **de film plastique** plastic film spreader (see also **dérouleuse**)
étêtage, étêtement *m.* heading, topping, pollarding
étêter *v.t.* to head, to top, to pollard
éthane *m.* ethane
éthanol *m.* ethanol
éthylène *m.* ethylene
étiage *m.* low water, lowest water level of river
étiochloroplaste *m.* etiochloroplast
étiolé,-e *a.* etiolated
étiolement *m.* etiolation
étioler *v.t.* to etiolate; **s'étioler** to become etiolated
étiologie *f.* aetiology, etiology
étioplaste *m.* etioplast
étiquetage *m.* labelling
étiqueté,-e *a.* labelled, ticketed
étiqueter *v.t.* to label
étiquette *f.* label
étirer *v.t.* to stretch, to draw out
étoc *m.* stock, trunk (of tree)
étoile *f.* star; **é. d'argent** edelweiss (*Leontopodium alpinum*); **é. de Bethléem** star of Bethlehem (*Ornithogalum umbellatum*) (= **dame-d'onze-heures**); also applied to *Campanula isophylla* and other spp.; **é. de Noël** = **poinsettia** (*q.v.*)
étoilé,-e *a.* star-shaped, stellate
étouffée *f. Hort.* **à l'étouffée** under high relative humidity, under glass, under a cloche
étouffement *m.* smothering, choking; *Hort.* placing under high relative humidity, placing under glass
étouffer *v.t.* to smother, to choke; *Hort.* to place under high relative humidity, to place under glass
étoupe *f.*, **é. (blanche)** tow; **é. de coton** cotton waste
étourneau *m.*, *pl.* **-eaux** starling (*Sturnus* sp.)
étranglement *m. Hort.* constriction (made by ringing with a metal ring, label-tie etc.)
étronçonner *v.t.* to head, to top, to pollard (a tree)
étuve *f.* stove; **séché à l'étuve** oven-dried
eucalyptus *m.* eucalyptus
eucharis *m. Eucharis* sp., eucharis lily
eudémis *m.* **(de la vigne)** grape moth (*Lobesia (Polychrosis) botrana*)
eugénie *m. Eugenia* sp., fruiting myrtle
eugénol *m.* eugenol
eulalie *f.* zebra-striped rush (*Miscanthus sinensis* = *Eulalia japonica*)
eumolpe *m. Vit.* vine chrysomelid beetle, western grape root worm (*USA*) (*Adoxus obscurus*) (= **écrivain, gribouri**)
eupatoire *f. Eupatorium* sp., including **e. à feuilles de chanvre** hemp agrimony (*E. cannabinum*); **e. pourprée** joe-pye weed (*E. purpureum*) (see also **herbe du Laos**)
euphorbe *f. Euphorbia* sp., spurge; **e. très belle** poinsettia (*E. pulcherrima*); **e. du Cayor** balsam spurge (*E. balsamifera*); **e. épurge** caper spurge (*E. lathyrus*); **e. des jardins** petty spurge (*E. peplus*) (= **ésule ronde**); **e. panachée** snow-on-the-mountain (*E. marginata*)
euphorbiacées *f. pl.* Euphorbiaceae
euryale *f.* gorgon waterlily (*Euryale ferox*), a plant with edible seeds and shoots used as vegetables
euryhalin,-e *a.* euryhaline
eutrophe *a.* **eutrophic**
eutrophisation *f.* eutrophization
eutypiose *f. PD Eutypa* canker and dieback of fruit trees and grapevines; *Vit.* dieback caused by *E. armeniacae*
évacuation (*f.*) **d'eau** drainage
évaporateur *m.* drying oven, evaporator
évaporation *f.* evaporation
évaporimètre *m.* evaporimeter
évapotranspiration *f.* evapotranspiration; **é. potentielle, ETP** potential evapotranspiration
évapotranspiromètre *m.* evapotranspirometer
évasé,-e *a.* bell-mouthed, funnel-shaped
éventable *a.* capable of being exposed to the air or wind
éventage *m.* exposure to the air or wind
éventail *m.* fan; **en éventail** fan-shaped; **un éventail de variétés** a range of varieties; **é. de vigueur** range of vigour; *Hort.* fan training (for stone fruit trees etc.) (= **queue de paon**)
éventement *m.* 1. exposure to the air or wind, ventilation. 2. going stale or flat (of food or drink)
éventer *v.t.* to expose to the air or wind

éverdumer *v.t.* to remove the green colour from, to degreen
évidé,-e *a.* hollow, cut away
évidement *m.* hollow, groove, cavity
évider *v.t.* to hollow out, to cut away; **é. une touffe** to remove shoots from the centre of a clump
évolage *m.* wet period of pond, water-meadow or flood plain dried periodically for cultivation (see **assec**); arrangement of such a pond, water-meadow or flood plain
évolutif,-ive *a.* evolutionary; evolutive; **phases évolutives** developmental phases
évolution *f.* evolution, course of development
évonyme *m. Euonymus* sp., spindle tree (see also **fusain**)
évrillage *m. Vit.* removal of tendrils
exalbuminé,-e *a.* exalbuminous
exanthéma, exanthème *m. Cit.* exanthema (Cu deficiency)
excécaire, excécarie *f. Excoecaria* sp. (see **arbre aveuglant**)
excès *m.* excess
excisé,-e *a.* excised; **embryon excisé** excised embryo
exciser *v.t.* to excise, to cut out
excitation *f.* excitation, stimulation
excorié,-e *a. PD* affected by dead-arm disease, excoriosis
excoriose (*f.*) **(de la vigne)** *Vit.* dead-arm disease, excoriosis (*Phomopsis viticola*)
excrétion *f.* excretion
excroissance *f.* excrescence, tumour, growth
exercice *m. Adm.* financial year
exfoliatif,-ive *a.* exfoliative
exfoliation *f.* exfoliation
exfolier *v.t.* to exfoliate
exhaure *f.* pumping (out)
exigeant,-e *Hort. a.* demanding (of plant), difficult to grow
exigence *f.* requirement (of plant etc.); **e. en froid** chilling requirements; **exigences culturales** cultural requirements; **aux exigences variables** with variable requirements
exine *f.* extine, exine
exocortis *m. Cit., PD* exocortis
exogène *a.* exogenous
exogyne *a.* exogynous
exorhize, exorrhize *a.* exorrhizal
exosmose *f.* exosmosis
exospore *m.* exospore
exosporé,-e *a.* exosporous
exostème *m. Exostema* sp.; some species have medicinal properties
exostome *m.* exostome
exostose *f.* exostosis, knot-growth on trunk etc.
exotique *a.* exotic
expérience *f.* experience, experiment
expérimentateur,-trice *n.* research worker
expérimentation *f.* experimentation, experimenting; **e. au champ** field experimentation
expérimenter *v.t.* to test, to try, to experiment
explant, explantat *m.* explant
exploitant *m.* grower, cultivator (of land); **e. horticole** *Hort.* commercial grower
exploitation *f.* exploitation, working; *Hort.* holding; **e. familiale** family holding; **e. horticole** market garden, nursery
explosif,-ive *a.* explosive; *n. m.* explosive; **explosifs agricoles** agricultural explosives (for stump removal etc.)
expoliation *f.* removal of dead wood, parts
expolier *v.t.* to remove dead wood, parts
exportation *f.* export, exportation
exposition *f.* aspect, exposure
expression *f.* expression; **e. du sexe (des palmiers)** sex expression (in palm trees)
expurgation *f.* thinning of thickets etc.
exsert,-e *a.* exserted
exsudat *m.* exudate; **e. caulinaire** stem exudate; **e. racinaire** root exudate
exsudation *f.* exudation
extensif,-ive *a.* extensive
extirpable *a.* eradicable
extirpage *m.* cultivation to destroy weeds, grubbing
extirpateur,-trice *n.* 1. extirpator, uprooter. 2. (implement) cultivator, scarifier, weeder
extirpation *f.* eradication, uprooting
extirper *v.t.* to extirpate, to eradicate
extra-axillaire *a.* extra-axillary
extractibilité *f.* extractability
extractible *a.* extractable; **e. par l'eau** water-soluble
extraction *f.* extraction
extrafoliacé,-e, extrafolié,-e *a.* extrafoliaceous
extraire *v.t.* to extract
extrait *m.* extract; **e. de feuilles** leaf extract; **e. sec** dry matter
extrémité *f.* extremity, end, tip; **e. de la racine, e. radiculaire** root tip
extrorse *a.* extrorse
exudat *m.* exudate
exuvie *f. Ent.* exuvia

F

F. à B. = **franco à bord** free on board
fabagelle *f.*, **fabago** *m.* bean caper (*Zygophyllum fabago*)
face *f.* face; **f. inférieure, supérieure (d'une feuille)** lower, upper surface (of leaf)
facies, faciès *m. Bot.* facies, appearance
facilité *f.* ease; **f. de cueillette** ease of picking (of fruit etc.)
façon (culturale) *f. Hort.* (soil) cultivation; **petites façons** final harrowing etc. for fine tilth
façonnage *m.* **(du sol)** cultivation
facteur *m.* factor; **f. climatique** climatic factor; **f. de croissance** growth factor; **f. édaphique** soil factor; **f. limitant** limiting factor; **le f. lumière** the light factor; **f. de rendement** yield factor
faculté *f.* faculty, power; **f. germinative** germination capacity; **f. végétative** viability, vitality, growth vigour
fade *a.* insipid, flavourless, tasteless, flat
fagara, fagare *m.* 1. **fagarier** fruit. 2. **fagara** *Afr. Fagara* sp.; **f. bouche ouverte** *F. heitzii*; **f. à grandes feuilles** *F. macrophylla*
fagarier *m.* prickly ash (*Zanthoxylum fraxineum*) (= **frêne épineux**)
fagonie *f. Fagonia* sp., including *F. cretica*
fagot *m.* faggot, bundle of (fire-) wood
faham *m.* faham (*Angraecum fragrans*) (see also **thé de Bourbon**)
faible *a.* feeble, weak
faim *f.* hunger; **f. d'azote** nitrogen deficiency
faîne *f.* beechnut, *pl.* beechmast
faire (*v.t.*) **valoir** to develop (land, estates etc.)
faisabilité *f.* feasability
faisceau *m.* bundle, cluster; **f. libéro-ligneux, f. vasculaire** vascular bundle
faîtage *m.* ridge board, ridge (of roof), crest (of roof)
faîte *m.* 1. ridge (of roof), crest (of hill), 2. top (of tree etc.)
faîtière *a.* & *f.* **tuile faîtière** ridge tile, crest tile
falciforme *a.* falciform, falcate, sickle-shaped
falun *m.* shell-marl
falunage *m.* application of shell-marl
faluner *v.t.* to apply shell-marl
famille *f.* family; **f. naturelle** natural family
fanage *m.* drying, wilting
fanaison *f.* wilting; **f. permanente** permanent wilting
fane *f.* 1. dead, dried leaf, haulm; **fanes de navet** turnip tops; **fanes d'oignons** onion tops. 2. *Bot.* involucre
faner *v.t.* to dry, to wilt plants, plant material etc.; **se faner** (of plant) to dry, to flag, to wilt, to fade; (of colour) to fade
faneur,-euse *n.* drier, wilter (of plants, plant material etc.); *n.f.* drying, wilting machine
fardeler *v.t.* to bundle up, to do up into a bundle
farigoule *f.* thyme (*Thymus* sp.) (in south of France)
farine *f.* flour, meal; **f. de coco** coconut meal; **f. de cuir** ground leather; **f. de moutarde** mustard flour; **f. de pois** pease meal
farineux,-euse *a.* mealy, farinaceous
fasciation *f.* fasciation; **f. des tiges (du fraisier)** *PD* cauliflower disease, leafy gall (of strawberry) (*Corynebacterium fascians*)
fasciculé,-e *a.* fasciculate, growing in bunches; **racines fasciculées** bunched (fibrous) roots
fascie *f.* fasciated shoot or stem
fascié,-e *a.* fasciated, fasciate
fascine *f.* faggot
faséole *f.* haricot bean (*Phaseolus vulgaris*) (see **haricot**)
fastigié,-e *a.* fastigiate, conical
fatigue (*f.*) **du sol** soil sickness
faucard *m.* long-handled scythe for cutting water weeds
faucardage *m.* use of **faucard**; **f. motorisé** mechanized water weed cutting
faucarder *v.t.* to clear water weeds
faucardeur *m.* boat with mechanical weed cutter
fauchable *a.* which can be mown, mowable
fauchage *m.* mowing, cutting
fauchard *m.* slasher, slash-hook
faucher *v.t.* to cut down, to mow
fauchet *m.* 1. wooden hay rake. 2. billhook
fauchette *f.* billhook
faucheur,-euse *n.* 1. mower; **faucheuse** *n. f.* mowing machine, mower; **f. auto-chargeuse** cutter loader; **f. rotative** rotary scythe; **f. de talus** bank mower. 2. = **faucheux** (*q.v.*)
faucheux, faucheur *m. PC* harvest-spider, harvester; **f. des agrumes** *Phyllocoptruta oleivora*
faucille *f.* sickle; **f. électrique** electric scythe
fausse arpenteuse du chou *f.* cabbage looper (*Trichoplusia ni*)

fausse-chenille *f.* sawfly larva (see also **tenthrède**)
fausse chérimole *f.* biriba (*Rollinia deliciosa*)
fausse cloison *f.* septum (in siliqua)
fausse cloque *f. PD* gall of azaleas (*Exobasidium azaleae*)
fausse épervière *f.* bristly ox-tongue (*Picris echioides*)
fausse golmotte *f. Fung.* false blusher (*Amanita pantherina*) (see also **golmotte**)
fausse noix muscade *f.* calabash nutmeg (*Monodora myristica*)
fausse roquette *f.* crested bunias (*Bunias erucago*)
fausse truffe *f. Mush.* false truffle (*Diehliomyces (Pseudobalsamia) microsporus*)
fauve *a.* fawn-coloured, buff; *n. m.* fawn colour
faux *f.* scythe
faux acacia *m.* = **robinier** (*Robinia pseudoacacia*)
faux alkékenge *m.* apple-of-Peru (*Nicandra physaloides*)
faux aloès *m.* red hot poker (*Kniphofia* sp.)
faux baume du Pérou *m.* blue fenugreek (*Trigonella caerulea*) (= **lotier odorant, mélilot bleu**)
faux bourgeon *m.* = **bourgeon anticipé** (*q.v.*)
faux cotonnier *m.* shrubby milkweed (*Asclepias fruticosa*); floss silk tree (*Chorisia monstruosa*)
faux cyprès *m.* false cypress (*Chamaecyparis* sp.) (= **chamaecyparis**)
faux dattier *m.* false date palm (*Phoenix atlantica*)
faux ébénier *m.* laburnum (*Laburnum anagyroides*) (= **aubour**)
faux fruit *m. Bot.* false fruit
faux indigo *m.* false indigo (*Amorpha fruticosa*)
faux karité *m.* false shea (*Lophira alata*)
faux liseron *m.* black bindweed (*Polygonum convolvulus*)
faux mangoustanier *m.* *Garcinia cochinchinensis*, a tree cultivated for its fruits, eaten fresh or preserved
faux mousseron *m.* fairy ring champignon (*Marasmius oreades*) (= **marasme d'oréade**)
faux muscadier *m.* calabash nutmeg tree (*Monodora myristica*) (= **muscadier de calebasse**)
faux narcisse *m.* daffodil (*Narcissus pseudonarcissus*) (see also **jonquille**)
faux nard *m.* *Allium victorialis*
faux nerprun *m.* see **argousier**
faux pistachier *m.* bladder nut (*Staphylea pinnata*) (= **nez coupé, staphylier**)
faux poivrier *m.* pepper tree (*Schinus molle*) (= **poivrier d'Amérique**)
faux puceron = **psylle** (*q.v.*)
faux quinquina = **exostème** (*q.v.*)
faux séné *m.* bladder senna (= **baguenaudier** *q.v.*)
faux souchet *m.* cyperus sedge (*Carex pseudocyperus*)
faux tronc (du bananier) *m.* pseudostem (of banana)
faux vernis du Japon tree of heaven (*Ailanthus altissima*) (= **vernis du Japon**)
faux ver rose *m. Ent.* false codling moth (*Cryptophlebia leucotreta*)
favisme *m.* favism
fébrifuge *a.* & *n.m.* febrifuge, antifebrile
fécondable *a.* capable of being fertilized
fécondateur,-trice *a.* fertilizing; *n.* fertilizing agent
fécondation *f.* fertilization; **f. artificielle** artificial fertilization, pollination; **f. contrôlée, f. dirigée** controlled fertilization, pollination; **f. croisée** cross-fertilization, outcrossing; **f. directe** self-fertilization, selfing; **fécondations illégitimes** *Rubb.* illegitimate pollination; **f. naturelle** natural fertilization, pollination
fécondé,-e *a.* fertilized
féconder *v.t.* 1. to fecundate. 2. to make land fertile (of stream etc.)
fécondité *f.* fecundity, fruitfulness, fertility
fécule *f.* starch; **f. de pomme de terre** potato starch
féculence *f.* starchiness
féculent,-e *a.* starchy; *n. m.* starchy substance
féijoa *m.* feijoa (*Feijoa sellowiana*)
félicie *f.* *Felicia* sp.
fêlure *f.* crack; **fêlures concentriques des baies** *Vit.* concentric cracking of berries
femelle *a. Bot.* female, pistillate (flower)
fendillement *m. Hort.* cracking, splitting
fendiller *v.t.* to crack; **se fendiller** to crack, to become cracked
fendoir *m. Hort.* grafting tool
fendu,-e *a.* split, cleft
fenouil *m.* **(commun), f. officinal** fennel (*Foeniculum vulgare*); **f. amer** bitter fennel (var. *piperitum*); **f. bâtard** dill (*Anethum graveolens*); **f. doux, f. de Florence** sweet fennel, Florence fennel (var. *dulce*); **f. d'ours, f. des Alpes** signel (*Meum athamanticum*)
fente *f.* crack, split; **f. de l'écorce** bark split (ting); **f. d'insolation** sun-crack; **f. du noyau (des pêches)** *PD* split-pit (of peaches)
fenugrec *m.* fenugreek (*Trigonella foenum-graecum*) (= **sainegrain**)
fer *m.* iron; *Hort.* spit (of spade)
ferme *f.* farm, holding; **f. expérimentale** experimental farm, holding
ferment *m.* ferment
ferment gum *m. Cit.* Florida gummosis
fermentation *f.* fermentation; **f. alcoolique**

alcoholic fermentation; **f. dirigée et contrôlée** *Mush.* peak heating
fermenter *v.i.* to ferment
fermentescible *a.* fermentable
fermeté *f.* firmness; **f. de fruit** firmness of fruit
fermeture *f.* closing, shutting; **f. de la grappe** *Vit.* bunch closure; **f. des stomates** closing of stomata
ferrarie *f. Ferraria* sp., including *F. undulata*, black iris
ferrallitique *a. Ped.* ferrallitic; **sol ferrallitique** ferrallitic soil (= **latosol**)
ferrallitisation *f. Ped.* ferrallitization (= **latéritisation**)
ferrique *a.* ferric, relating to iron; **nutrition ferrique** iron nutrition
ferrocyanhydrique *a.* ferrocyanic
ferrocyanure *m.* ferrocyanide
ferrugineux,-euse *a.* ferruginous; **argile ferrugineuse** iron clay; **carapace ferrugineuse, cuirasse ferrugineuse** *Afr.* ironstone cap; **grès ferrugineux** ironstone
fertile *a.* fertile
fertilement *adv.* fertilely, fruitfully
fertilisable *a.* fertilizable
fertilisant,-e *a.* fertilizing; **éléments fertilisants** fertilizer elements; *n. m.* fertilizer
fertilisation *f.* fertilization, application of fertilizer
fertiliser *v.t.* to fertilize, to apply fertilizer
fertiliseur *m.* fertilizer
fertilité *f.* fertility; **f. potentielle** *Vit.* potential fertility (number of flowers per sprouted bud); **f. pratique** *Vit.* practical fertility (number of flowers per bud left at pruning); **f. du sol** soil fertility
fertirrigation *f.* fertigation
férule *f. Ferula* sp., asafoetida
feston *m. Hort.* festoon of flowers etc.
festonné,-e *a.* festooned
fétide *a.* fetid
fétidité *f.* fetidness, foulness (of smell etc.)
fétu *m.* (bit of) straw
fétuque *f.* fescue (*Festuca* sp.); **f. crin d'ours** *F. crinum-ursi*; **f. élevée** tall fescue (*F. arundinacea*); **f. hétérophylle** various-leaved fescue (*F. heterophylla*); **f. ovine** sheep's fescue (*F. ovina*); **f. des prés** meadow fescue (*F. pratensis*); **f. rouge** red fescue (*F. rubra*); **f. rouge de Nouvelle Zélande** Chewings fescue (*F. rubra* spp. *commutata*); **f. rouge traçante** creeping red fescue (*F. rubra* ssp. *rubra*)
feu *m., pl.* **feux** fire; **f. bactérien** *PD* fire blight (*Erwinia amylovora*); **f. de brousse** *Afr.* bush fire; **f. précoce** *Afr.* early burn; **f. tardif** *Afr.* late burn
feuillage *m.* foliage; **f. d'asperge** asparagus "fern"; **à feuillage caduc** deciduous; **à feuillages larges** (*Canada*) broad-leaved (of weeds)
feuillagé,-e *a.* in leaf
feuillaison *f.* 1. foliation; vernation. 2. spring, spring-time
feuille *f.* 1. leaf, leaf-like organ; **f. morte** dead leaf; **f. primordiale** primary leaf; **f. sèche** dried leaf; **f. séminale** seed leaf, cotyledon; **f. d'artichaut** (globe) artichoke bract; **f. de rose** rose petal; **à feuilles caduques** deciduous; **à feuilles persistantes** evergreen; **à feuilles rugueuses** rough-leaved; **feuilles en baïonnettes** *OP, PD* little leaf (B deficiency); **feuilles en cuiller** leaf-cupping deformation caused by unsuitable herbicide; **feuilles entortillées du cerisier** *PD* cherry twisted leaf virus; **feuilles étroites (du cassis)** *PD* blackcurrant reversion virus; **stade "2 feuilles", "3 feuilles"** two-, three-leaf stage. 2. sheet (of paper, metal etc.); **f. d'emballage** sheet for wrapping; **f. isolante** insulating film; **f. de plastique** plastic film; **f. fumée** *Rubb.* smoked sheet
feuillé,-e *a.* leafy, in full leaf, foliate
feuiller *v.i.* to grow leaves
feuillu,-e *a.* leafy, foliate; *Arb.* broad-leaved, deciduous; *n. m.* broad-leaved tree; **f. d'ornement** ornamental tree
feutrage *m.* felting
feutre *m.* felt
feutré,-e *a.* felted
fève *f.* 1. **f. (des marais)** broad bean (*Vicia faba*); **f. de Calabar** Calabar, ordeal bean (*Physostigma venenosum*); **f. d'Egypte** = **dolique lablab** (*q.v.*); **f. de Saint Ignace** Ignatius bean (*Strychnos ignatii*) (= **ignatie**); **f. tonka** tonka bean (*Coumarouna odorata*) (= **coumarou**). 2. cocoa bean, coffee bean; **f. beurrée** (*Cacao*) bean with white spot; **f. monstrueuse** *Coff.* elephant bean; **f. en parche** *Coff.* coffee bean in parchment; **f. plate** (*Cacao*) flat bean; **f. puante, f. trop fermentée** (*Cacao*), *Coff.* stinker (bean)
févier *m. Gleditschia* sp.; **f. (d'Amérique), f. à trois épines** honey locust (*G. triacanthos*)
févillée *f. Fevillea* spp., neotropical cucurbitaceous climbers, some of which have medicinal properties
fibre *f.* fibre; *Sug.* fibre; **f. de verre** glass fibre
fibreux,-euse *a.* fibrous, stringy
fibrillaire *a.* fibrillar, fibrillary
fibrille *f.* fibril, fibrilla
filbrillum *m.* (*Date*) fibre (= **liff**)
ficaire *f.* lesser celandine (*Ficaria ranunculoides*)
ficelle *f.* string; **f. répulsive** treated string for repelling pests (rabbits etc.)
fichage *m. Vit.* staking
fiche *f.* 1. peg, stake. 2. memorandum, card,

label; **f. ampélographique** *Vit.* document with details of the characteristics of a vine; **f. technique** technical data sheet

ficoïde *f.* fig marigold, mesembryanthemum (*Mesembryanthemum* sp.) (= **mésembryanthème**); **f. comestible** Hottentot fig (*M. edule*); **f. cristalline, f. glaciaire, f. glaciale** ice plant (*M. crystallinum*); **f. épinard** *f. Cryophytum (Mesembryanthemum) angulatum*

ficoïdé,-e *a.* resembling **ficoïde**

fiente *f.* dung, droppings; **fientes de poules** chicken, poultry manure; **fientes d'oiseaux** bird droppings; **fientes de volailles** chicken, poultry manure

figue *f.* 1. fig; **f. blanche** green fig; **f. noire** purple fig; **f.-fleur** (*pl.* **figues-fleurs**), **f. d'été, première figue** early, first crop fig (May–June) (ripening in its second season); **f. d'automne, f. ordinaire** fig harvested at the normal time (August–September). 2. Other plants: **f. (-)banane** = **banane figue** (*q.v.*); **f. de Barbarie** prickly pear fruit; **f. des Hottentots** Hottentot fig (*Mesembryanthemum edule*) (= **ficoïde comestible**)

figue-caque *f., pl.* **figues-caques** persimmon (*Diospyros* sp.) (see also **kaki, plaqueminier**)

figuerie *f.* fig plantation or grove

figuier *m.* 1. fig tree (*Ficus carica*); **f. bifère** double cropping fig tree; **f. unifère** single cropping fig tree; **f. des banians** banyan tree (*Ficus benghalensis*); **f. sycomore** sycomore fig (*F. sycomorus*); 2. Other plants: **f. de Barbarie, f. d'Inde, f. à raquettes** prickly pear (*Opuntia ficus-indica* and other spp.) (see also **nopal, raquette**); **f. étrangleur, f. tueur d'arbres** strangling fig (*Ficus parasitica*)

fil *m.* thread; string (in bean pod); **f. du bois** grain (in wood); **f. à planter** planting line; **f. de fer** wire; **f. de fer barbelé** barbed wire

filage *m.* 1. cudweed (*Gnaphalium* sp.). 2. *PD, Vit.* loss of floral primordia, disorder in which bunches are partially or totally changed into tendrils. 3. see **filer**

filament *m.* filament, fibre, thread

filamenteux,-euse *a.* filamentous, fibrous

filandre *f.* fibre; strings (of vegetables etc.)

filandreux,-euse *a.* tough, stringy

filao *m.* casuarina (*Casuarina* sp.)

filasse *f.* tow, textile fibre

filé,-e *a.* (of plant) etiolated

filer *v.i.* (of plant) to become etiolated (see also **filage**)

filet *m.* net; **f. protecteur** net for bird protection; *Bot.* filament; *Hort.* 1. runner (of strawberry etc.), stolon 2. tender green pod (of bean) (see **haricot I.**)

filiation *f.* consanguinity, affiliation; **hybrides de filiation directe** *PB* direct line hybrids

filiforme *a.* filiform, thread-like

filipendula, filipendule *f. Filipendula* sp., meadowsweet

fille de l'air *f.* Spanish moss (*Tillandsia usneoides*) (= **barbe de vieillard**)

film *m.* film; **f. continu** unperforated film (mulch); **f. perforé** perforated film; **f. plastique refléchissant** reflecting plastic film; **f. P.V.C.** PVC film

filosité *f.* **(des tubercules)** *PD* potato spindle tuber virus

filtrage *m.* filtering, percolation

filtrant,-e *a.* (of soil) free draining; **sol filtrant** free-draining soil

filtrat *m.* filtrate

filtre *m.* filter

filtrer *v.t.* to filter

fimbrié,-e *a.* fimbriate, fringed

fin,-e *a.* fine, of high quality (the opposite of **grossier** *q.v.*); **vin fin** choice wine

finement *adv.* finely

fines herbes *f. pl.* culinary herbs

finesse *f.* fineness, high quality, choiceness (of fruit etc.)

finissage *m.* finishing off; **f. à la main** finishing off by hand

fiole *f.* flask, small bottle

fisanier *m.* akee (*Blighia sapida*)

fissuré,-e *a.* fissured, split

fistuleux,-euse *a.* fistulous

fistuline *f.* beefsteak fungus (*Fistulina hepatica* (= **langue-de-boeuf**)

fixation *f.* fixation; **f. de l'azote** nitrogen fixation; **f. bactérienne** bacterial fixation; **f. symbiotique** symbiotic fixation

flabelliforme *a.* flabelliform

flacon (*m.*) **poudreur** *OP* etc. pollen (dusting) flask

flagelliforme *a.* flagelliform

flageolet *m.* flageolet, small kidney bean (*Phaseolus vulgaris*)

flambage *m. Vit.* flaming, singeing

flamboyant *m.* 1. flamboyant, flame of the forest (*Poinciana regia*). 2. *Colvillea racemosa*

flammette *f.* 1. traveller's joy (*Clematis vitalba*). 2. lesser spearwort (*Ranunculus flammula*)

flammule *f. Clematis flammula* (see also **clématite**)

flavedo *m. Cit.* flavedo

flavescence *f.* **(de la vigne)** *PD* rougeot, reddening disorder (of grapevine); **f. dorée** flavescence dorée virus

flavine *f.* flavin

flavonique *a.* flavonoid

flavonoïde *m.* flavonoid
flavonol *m.* flavonol
fléau *m., pl.* **-aux** 1. *Ag.* flail 2. plague, pest
fléchage *m. Sug.* arrowing, tasselling
flèche *f. Hort.* leading shoot, main stem, terminal bud, spear (of palm); stigma (of saffron), top (of tree); *Sug.* arrow; **f. d'eau** common arrowhead (*Sagittaria sagittifolia*) (= **sagittaire**)
fléchière *f.* = **flèche d'eau** (*q.v.*)
fléchissement *m.* bending, flexing, toppling (of plant)
flein *m.* punnet
fléole (des prés) *f.* timothy (*Phleum pratense*)
flétrir *v.t.* to fade, to wilt, to wither; **se flétrir** to become faded, to fade, to flag, to wilt, to wither
flétrissage *m.* withering (of harvested tea leaves)
flétrissement *m.* wilt, wilting; **point, seuil de flétrissement** wilting point; **f. bactérien du haricot** bean bacterial wilt (*Corynebacterium flaccumfaciens*); **f. bactérien de la tomate** tomato bacterial wilt (*Pseudomonas solanacearum*); **f. du fraisier** leather rot (*Phytophthora cactorum*)
flétrissure *f.* fading, withering; *PD* blight, wilt; **f. sud-américaine des feuilles d'hévéa** South American leaf blight (*Dothidella ulei*); **f. de la tige (de la canne à sucre)** *PD* dead heart (of sugar cane)
fleur *f.* flower, bloom (including bloom (= **efflorescence**) on grapes, plums etc.), blossom; **à fleurs** flower-bearing, flowering; **à fleurs doubles** double-flowered; **en fleur, en fleurs** in flower, flowering; **bouton à fleur** flower bud; **f. d'amour** columbine (*Aquilegia* sp.), larkspur (*Delphinium* sp.), love-lies-bleeding (*Amaranthus* sp.), night-blooming cereus (*Selenicereus grandiflorus*); **f. d'Arménie** sweet william (*Dianthus barbatus*) (= **œillet des poètes**); **f. de Candie** resurrection plant (*Anastatica hierochuntia*); **fleurs de cassia** cassia buds (*Cinnamomum cassia*) (see also **cannelle**); **f. de coucou** ragged robin (*Lychnis flos-cuculi*) (= **lychnide des prés**); **f. coupée** cut flower; **fleur(s) à couper** flower(s) for cut flower production; **f. des dames** cherry pie (*Heliotropium peruvianum*); **fleurs encapuchonnées** *Vit.* abnormal flowers which retain their corollas; **f. femelle** female flower; **f. de Jupiter** flower of Jove (*Lychnis flos-jovis*); **f. de léopard** *Belamcanda chinensis* (= **iris tigré**); **f. de mai** May lily (*Maianthemum bifolium*); **f. mâle** male flower; **f. de muscade** = **macis** (*q.v.*); **f. de Pâques** pasque flower (*Pulsatilla vulgaris*) (= **herbe de Pâques**); **f. de la Passion** blue passion flower (*Passiflora caerulea* and other spp.); **f. du prophète** prophet flower (*Arnebia echioides*); **f. sèche** dried flower; **f. de soufre** flowers of sulphur; **f. terminale** terminal flower; **f. de la Toussaint** chrysanthemum; **f. de(s) veuve(s)** sweet scabious (*Scabiosa atropurpurea*)
fleuraison *f.* flowering (time)
fleur-coquillage *f., pl.* **fleurs-coquillages** bells of Ireland (= **clochettes d'Irlande** *q.v.*)
fleurette *f.* small flower
fleuri,-e *a.* in flower, decorated with flowers
fleurir *v.i.* to flower; *v.t.* to deck, decorate with flowers
fleurissant,-e *a.* flowering
fleuriste *n.* & *a.* floriculturist, flower grower, flower trader; florist; **jardin fleuriste** flower garden; **jardinier fleuriste** flower grower
fleuron *m.* floret
fleuronné,-e *a.* floreted
fleuronner *v.i.* to grow florets, to blossom
flexible *a.* flexible
flexion *f.* bending
flexueux,-euse *a.* flexuous; *Bot.* flexuose
floculation *f.* flocculation
floraison *f.* flowering, blooming, blossoming, flowering time; *Sug.* arrowing; **à f. tardive** late-flowering, serotinous; **f. terminée** post-blossom stage, fruitlet stage (in apple and pear etc.)
floral,-e *a., m. pl.* **-aux** floral; **bouton floral** flower bud
floralies *f. pl.* flower show
flore *f.* flora; **f. adventice** weed flora; **f. dominante** dominant (plant) species, dominant weeds
floribond,-e *a.* floriferous
floribondité *f.* floriferousness
floribunda *a.*, **rosier floribunda** *Hort.* floribunda rose
floricole *a.* flower-dwelling (of insect)
floriculture *f.* floriculture, flower growing; **f. commerciale** commercial floriculture
florifère *a.* bearing flowers, floriferous
floriforme *a.* floriform, flower-shaped
florigène *a.* relating to factors determining flowering
floripare *a.* floriparous, producing flowers
floriste *n.* 1. florist. 2. student of flora
floristique *a.* floristic; *n. f.* flora science
florule *f.* florule, florula, small flora
flosculeux,-euse *a.* flosculous
flottant,-e *a.* floating; **plante flottante** floating plant
flotteur *m.* float
flouve odorante *f.* sweet vernal grass (*Anthoxanthum odoratum*)
fluctuation *f.* fluctuation
fluor *m.* fluorine

fluoré,-e *a.* fluorinated; **pollution fluorée** fluorine pollution
fluorescence *f.* fluorescence
fluorescent,-e *a.* fluorescent
fluorodensitométrie *f.* fluorodensitometry
fluorure *m.* fluoride
flûte *f. Hort.* narrow vase
flûteau *m., pl.* **-eaux** water plantain (*Alisma plantago-aquatica*) (= **alisma, plantain d'eau**)
flux *m.* flow; **f. aqueux** water flow
focalisation *f.* focusing; **f. isoélectrique** isoelectric focusing
foin *m.* 1. hay. 2. choke (of globe artichoke)
foisonner *v.i.* to abound; (of soil, lime etc.) to swell, to expand
foliacé,-e *a.* foliaceous
foliaire *a.* foliar; **enroulement foliaire** leaf rolling; **traitement foliaire** leaf treatment
foliaison = **foliation** (*q.v.*)
foliation *f.* foliation, leafing (of plants)
folié,-e *a. Bot.* foliate, foliated
foliole *f.* leaflet
foliolé,-e *a.* foliolate; *n. f. Vit.* deeply-indented leaf
folle-avoine *f.* wild oat (*Avena fatua, A. ludoviciana*)
folletage *m. Vit.* vine apoplexy, sudden death of part or whole of vine
folleté,-e *a. Vit.* suffering from **folletage**
follette *f.* garden orache, mountain spinach (*Atriplex hortensis*) (= **arroche des jardins, belle-dame**)
follicule *m.* follicle
folliculeux,-euse *a.* follicular
fomès *m. Rubb., PD* white rot (*Leptoporus (Rigidoporus) lignosus*)
foncé,-e *a.* dark (of colour); **rouge foncé** *a. inv.* dark red
fondant,-e *a.* melting; **poire fondante** pear that melts in the mouth, juicy pear
fondatrice *f. Ent.* fundatrix
fongicide *a.* fungicidal; *n. m.* fungicide; **f. à base de cuivre** copper fungicide; **f. systémique** systemic fungicide
fongique *a.* fungal; **spore fongique** fungal spore
fongistatique *a.* & *n. m.* fungistatic (compound)
fongitoxicité *f.* fungitoxicity
fongitoxique *a.* fungitoxic; **action fongitoxique** fungitoxic action
fongueux,-euse *a.* fungous
fontaine *f.* 1. spring, pool of running water. 2. fountain; **f. d'arrosage** fountain-type sprinkler. 3. cistern
fonte (*f.*) **des semis** *Hort.* damping-off disease (see also **toile**); **f. (de la tomate)** tomato foot rot and damping-off (*Phytophthora cryptogea, P. parasitica*)
forage *m.* 1. drilling, boring. 2. bore-hole, drill hole
forçage *m.* forcing; **f. en appartement** indoor (home) forcing; **f. du bourgeonnement** forced bursting of buds; **f. hâtif** early forcing
force (*f.*) **à gazon** turf clippers
forcé,-e *a.* forced
forcer *v.t.* to force
forcerie *f.* forcing house, greenhouse; forcing bed
foresterie *f.* forestry
forêt-galerie *f., pl.* **forêts-galeries** gallery forest
foreur *m.*, **foreuse** *f.* borer, driller; *Ent.* borer; **f. des gousses (du haricot)** *Ent.* legume pod borer (*Maruca testutalis*)
forficule *f.* earwig (= **perce-oreille**)
formaldéhyde *m.*, sometimes *f.* formaldehyde
formamide *m.* formamide
formation *f.* formation; *Hort.* 1. **f. (d'un arbre fruitier)** training (of fruit tree); **f. basse** low training; **f. défectueuse** unsatisfactory training; **f. haute** high training; **f. de la table de cueillette** (*Tea*) tipping; 2. **f. végétale** plant community
forme *f.* form, shape; *Hort.* shape obtained by training; **en forme de** – shaped like a –; **en forme de bouteille** bottle-shaped; **f. juvénile** juvenile form; **f. spécialisée** *Fung.* forma specialis
formol *m.* formol
formulation *f.* formulation
formule *f.* formula; **f. d'un engrais** fertilizer formula; **f. d'encépagement** *Vit.* proportion of various varieties in a vineyard; **f. de fumure** manuring formula
forsythia *m.*, **forsythie** *f.* forsythia (*Forsythia* sp.)
fort,-e *a.* strong, (of soil) heavy; **terre forte** heavy soil
fosse *f. Hort.* pit, hole, sump; **f. à compost** compost pit; **f. à fumier** manure pit; **f. à purin** (liquid) manure pit, sump
fossé *m.* ditch, trench, drain; **f. collecteur** main drain; **f. de drainage, f. d'écoulement** drainage ditch
fosse-piège *f., pl.* **fosses-pièges** pit trap
fossette *f.* indentation, small cavity
fossoir *m.* vineyard plough or hoe
fossoyage *m. Vit.* spring hoeing, cultivation
fotètè *m. Afr.* African spinach (*Amaranthus hybridus*)
foudre *f.* lightning; **frappé par la foudre** struck by lightning
foudre *m.* large cask, tun
fouet *m.* whip; *Hort.* runner (of strawberry); **f. foliaire** *Sug.* spindle; *Vit.* = **long bois** (*q.v.*)
fouetteur *m. AM* beater
fougeraie *f.* area planted with ferns, fern grove

fougère *f.* fern; **f. aigle** bracken (*Pteridium aquilinum*); **f. d'Allemagne** ostrich fern (*Matteuccia struthiopteris*); **f. d'appartement** fern for indoor growing; **f. femelle** lady fern (*Athyrium filix-femina*); **grande fougère** = **f. aigle** *q.v.*; **f. mâle** male fern (*Dryopteris filix-mas*); **f. nid d'oiseau** bird's nest fern (*Asplenium nidus*) (= **doradille**); **f. plume d'autruche** = **f. d'Allemagne**; **f. de la résurrection** resurrection plant (*Selaginella lepidophylla*); **f. royale** royal fern (*Osmunda regalis*) (= **osmonde royale**)
fougerole *f.* small fern
fouillage *m. Hort.* grubbing
fouiller *v.t.* to dig, to excavate; *Hort.* to grub
fouisseur,-euse *a.* burrowing; *n.* burrower; **insecte fouisseur** burrowing insect
foulage *m.* pressing, crushing, treading (see also **piétinement**)
fouler *v.t.* to press, to crush, to tread
four *m.* oven, kiln; **f. artisanal** peasant kiln; **f. haute fréquence** microwave oven
fourche *f.* 1. fork (implement), pitchfork; **f. américaine, f. à fumier** manure fork; **f. à bêcher, fourche bêche** digging fork, garden fork; **f. crochue** rake fork (= **griffe à fumier**); **f. élévatrice** fork lift truck; **f. à fleurs** hand fork 2. fork (in tree, road etc.)
fourchette *f. Vit.* tendril
fourchine *f. Vit.* small forked prop
fourchon *m.* fork or branch
fourchu,-e *a.* forked; **racine fourchue** fanged root (of carrot etc.)
fourchure *f.* fork (of tree, road etc.)
fourmi *f.* ant; **f. blanche** white ant, termite; **f. coupe-feuilles** leaf-cutting ant (*Atta* sp. and other spp.); **f. fileuse** tailor ant (*Oecophylla* sp.)
fournisseur,-euse *n.* supplier, caterer; **f. de graines** seed merchant
fourré *m.* thicket
fourrière *f. Hort.* headland
fousseux *m. Vit.* hand-hoe
foutanien,-enne *a. Afr.* of the Fouta-Djallon
fovéolé,-e *a.* foveolate, pitted
foyer *m.* (of disorder or disease) focus, seat; **f. primaire** primary focus
fractionnement *m.* dividing into parts, splitting up (of fertilizer), split dressing
fragment *m.* fragment, chip
fragon *m.* butcher's broom (*Ruscus aculeatus*) (= **petit houx**); **f. à grappes** Alexandrian laurel (*Danaë racemosa*)
fragrance *f.* fragrance, perfume
fraîcheur *f.* 1. coolness; **maintenir à la fraîcheur** to keep in a cool place. 2. freshness
frais,-aîche *a.* fresh
frais *m. pl.* costs, expenses; **f. supplémentaires** additional costs
fraisage *m.* rotary tillage, rotavating
fraise *f.* 1. strawberry; **f. des bois** wild strawberry (see also **fraisier**); 2. rotary hoe, rotavator; **f. à bêcher** rotary digger
fraiser *v.t. Hort.* to rotavate, to cultivate with a rotary hoe
fraiseraie *f.* strawberry field, strawberry bed
fraiseuse *f.* rotary hoe, rotavator
fraisiculture *f.* strawberry growing
fraisier *m.* 1. strawberry plant (*Fragaria* sp.); **f. du Canada** = **f. de Virginie** (*q.v.*); **f. de Chiloé, f. du Chili** Chilean strawberry (*F. chiloensis*); **f. écarlate** = **f. de Virginie** (*q.v.*); **f. élevé** hautboy strawberry plant (*F. moschata*); **f. à gros fruits** large-fruited strawberry; **f. des quatre saisons** perpetual fruiting strawberry; **f. de Virginie** scarlet strawberry (*F. virginiana*). 2. **f. en arbre** = **arbousier** (*q.v.*); **f. du désert** strawberry cactus (*Echinocereus enneacanthus*). 3. strawberry grower
fraisière *f.* strawberry plantation, strawberry bed
fraisiériste *m.* strawberry grower
framboise *f.* raspberry
framboisé,-e *a.* with raspberry aroma
framboiseraie, framboisière *f.* area planted to raspberries
framboisier *m.* raspberry plant; **f. (commun)** common raspberry (*Rubus idaeus*); **f. fraise** Cape bramble (*R. rosaefolius*)
franc,-che *a.* free; **sol franc** (easily worked) loam; **f. de pied, de pied franc** seedling, sprung from seed, on its own roots; **plant de semis franc** free stock; *n. m.* seedling, free stock, tree on its own roots
franco-américain,-aine *a. Vit.* applied to *Vitis vinifera* × American vine hybrids
frange *f.* fringe; **f. capillaire** *Ped.* capillary fringe
frangé,-e *a.* fringed
frangipane *f.* frangipani fruit
frangipanier *m.* frangipani (*Plumeria rubra*)
frangule *f.* alder buckthorn (*Frangula alnus* = *Rhamnus frangula*
frankenia *f. Frankenia laevis*, sea-heath, and other spp.
frasère *f. Frasera* sp. (see **colombo**)
fraxinelle *f.* burning bush, dittany, fraxinella (*Dictamnus albus* = *D. fraxinella*) (= **dictame**)
freesia *m.*, **freesie** *f. Freesia* sp., freesia
freinage *m.* braking, check
freinant (*m.*) **de croissance** growth retardant
freiner *v.t.* to brake, to check; **f. le développement** to check growth
frelon *m.* hornet (*Vespa crabro*)

frênaie *f.* area planted to ash

frêne *m.* ash; **f. blanc d'Amérique** white ash (*Fraxinus americana*); **f. commun** common ash (*F. excelsior*); **f. à écorce dorée** golden ash (*F. excelsior* var. *aurea*); **f. épineux** prickly ash (*Zanthoxylum fraxineum*) (= **fagarier**); **f. à fleurs, f. à manne** flowering ash, manna ash (*F. ornus*)

fréquence *f.* frequency

freux *m.* rook (*Corvus frugilegus*) (see also **corbeautière**)

friable *a.* friable, crumbly

friche *f.* waste land, fallow: young secondary scrub; **en friche** fallow(ed)

frigo *m. Coll.* refrigerator, cold store; **en frigo** in cold storage; **plant (de) frigo** cold-stored runner (of strawberry)

frileux,-euse *a.* sensitive to cold; **plante frileuse** tender plant

fripé,-e *a.* crumpled

frisé,-e *a.* curly; **laitue frisée** curly lettuce

frisée, friselée *f. PD* crinkle virus (e.g. of potato); **f. du fraisier** *PD* strawberry crinkle virus

frisolée *f. PD* crinkle virus (e.g. of potato), crinkly leaf virus (of citrus)

frisure *f.* curling, curliness *PD* crinkle, (leaf)-curl; **f. de l'anémone** anemone crinkle

fritillaire *f.* fritillary; **f. damier, f. méléagride, f. pintade** snake's head fritillary (*Fritillaria meleagris*); **f. (couronne) impériale** crown imperial (*F. imperialis*) (= **couronne impériale**)

froid,-e *a.* cold; **f. printanier** cold spell in spring; *n. m.* cold; refrigeration; **industrie du froid** refrigeration industry

fromager *m. Ceiba* sp., especially *C. pentandra*, silk-cotton tree; *Bombax* sp. (see also **kapokier**)

fromental *m.* tall oat grass (*Arrhenatherum elatius*) (= **avoine élevée**)

froncé,-e *a.* wrinkled, puckered

frondaison *f.* 1. foliation, leafing. 2. foliage

fronde *f. Bot.* 1. frond. 2. (spring) foliage

frondifère *a.* frondiferous

frondiforme *a.* leaflike, frondose

fructiculteur *m.* fruit-grower, orchardist

fructiculture *f.* fruit growing, fruit culture

fructifère *a.* fructiferous, bearing fruit; **branche fructifère** fruiting branch

fructifiant,-e *a.* fruitful

fructification *f.* fructification, fruit formation, fruiting; **f. sur extrémité** *Hort.* tip bearing

fructifier *v.i.* to fructify, to fruit

fructiforme *a.* fruit-like, fructiform

fructigène *a.* living or growing on fruit; **champignon fructigène** fungus growing on fruit

fructose *m.* fructose

frugifère *a.* frugiferous, fruit-bearing

frugivore *a.* frugivorous, fruit-eating; **chauve-souris frugivore** fruit bat (= **roussette**)

fruit *m.* fruit; **petits fruits** small fruits, berried fruits; **f. ainé** older fruit, mature fruit (on tree); **f. atrophié** *PD* (apple) chat fruit virus; **f. bouilli vert** *Ban.* fruit which has begun to ripen; **f. bosselé** *PD* (apple) green crinkle virus; **f. cadet** young(er) fruit (on tree); **f. à chair jaune** *Ban.* fruit with dark pulp; **f. confit** glacé fruit, crystallized fruit; **f. à confiture** fruit for jam making; **f. à couteau** dessert fruit; **f. creux (de la tomate)** *PD* hollow fruit (of tomato); **f. frais** fresh fruit; **f. légumier** fruit from herbaceous or vegetable plant (tomato etc.); **f. de marché** = **f. à couteau** (*q.v.*); **f. miracle, f. miraculeux** sweetberry (*Synsepalum dulcificum*); **f. mûr** ripe fruit; **f. non pulpeux** = **f. sec** (*q.v.*); **f. à noyau** stone fruit; **f. de la passion** passion fruit (*Passiflora* sp.); **f. à pépins** pome fruit; **f. pommelé** *PD* (apple) dapple; **f. sec** dried fruit; fruit without pulp (nuts etc.) (= **f. non pulpeux**); **f. véreux** maggoty fruit; **f. vert** green, unripe fruit

fruité,-e *a.* fruity, tasting of fruit (of wines etc.); *n.m.* fruity taste

fruiter *v.i.* to fruit, to bear fruit

fruiterie *f.* 1. fruit store. 2. fruit shop

fruitier,-ère *a.* fruit bearing; **arbre fruitier** fruit tree; **station fruitière** fruit-packing station

fruitier,-ère 1. *n.* fruit (and vegetable) seller. 2. *n. m.* fruit tree. 3. *n. m.* fruit shed, store: shelves for fruit storage. 4. *n. m.* orchard

frutescent,-e *a.* frutescent, shrubby

fruticetum *m.* fruticetum, shrubbery

frutiqueux,-euse *a.* frutescent, shrubby

fuchsia *m.* fuchsia (*Fuchsia* sp.); **f. de Californie** Californian fuchsia (*Zauschneria californica*); **f. du Cap** Cape figwort (*Phygelius capensis*)

fuel *m.* fuel oil

fugace *a.* fleeting, transient

fugacité *f.* evanescence, transience

fuligineux,-euse *a.* fuliginous, sooty

fumade *f.*, **fumage** *m.* = **fumure** (*q.v.*)

fumagine *f. PD* sooty mould caused by *Fumago* sp., *Capnodium* sp.; **f. de l'oranger** *Cit.* sooty mould (*C. citri*)

fumago *m. Fumago* sp.

fumaison *f.* = **fumure** (*q.v.*)

fumé,-e *a.* smoked, smoke-cured; *Hort.* manured, fertilized

fumer *v.t. Hort.* to apply manure, to fertilize

fumeterre *f.* fumitory (*Fumaria officinalis* and other spp.); **f. grimpante** white fumitory (*F. capreolata*)

fumier *m.* manure, dung: dung or manure heap;

f. artificiel compost; **f. bien fait, f. consommé, f. décomposé** well-made, well-rotted manure; **f. de cheval** horse manure; **f. d'étable, f. de ferme** farmyard manure; **f. de mouton** sheep manure; **f. de porc, f. de porcherie** pig manure; **f. de poule** chicken manure; **f. de vache** cow manure

fumigant *m.* fumigant

fumigateur *m.* fumigator

fumigation *f.* fumigation

fumigé,-e *a.* fumigated; **sol fumigé** fumigated soil

fumiger *v.t.* to fumigate

fumure *f.* fertilizer, manure: fertilizing, manuring; **f. azotée** nitrogen(ous) manure, manuring; **f. carbonée, f. carbonique** CO_2 enrichment; **f. d'entretien** maintenance, routine dressing; **f. foliaire** foliar manuring; **f. de fond, f. d'investissement** basal dressing; **f. en ligne** row placement of fertilizer; **f. minérale** chemical fertilizer, fertilizing; **f. organique** organic manure, manuring

funicule *m. Bot.* funicle

funkia, funkie *f. Funkia* (*Hosta*) sp., plantain lily

funtumia *f. Funtumia* sp., including *F. elastica*, silkrubber tree

fusain *m. Euonymus* sp., spindle tree (*E. europaeus*) (= **bonnet carré**); **f. du Japon** Japanese spindle tree (*E. japonicus*); **f. panaché** *E. fortunei*; **f. vert** = **f. du Japon**

fusarien,-enne *a. PD* relating to *Fusarium*; **flétrissure fusarienne** *Fusarium* wilt

fusariose *f. PD Fusarium* infection or disease; **f. (vasculaire) du palmier dattier** bayoud disease of date (*F. oxysporum* f. sp. *albedinis*) (= **bayoud**); **f. du palmier à huile** vascular wilt disease of oil palm (*F. oxysporum* f. sp. *elaeidis*); **f. vasculaire de la tomate** tomato *Fusarium* wilt (*F. oxysporum* f. sp. *lycopersici*)

fusarium *m. PD Fusarium* sp.

fuseau *m.* spindle, spindle-shaped tree or bush; *Biol.* nuclear spindle; *Hort.* dwarf pyramid

fusée *f.* rocket; **f. paragrêle** anti-hail rocket

fusiforme *a.* fusiform, spindle-shaped

fustet *m.* fustet, smoke-tree, Venetian sumach (*Rhus cotinus*)

fût *m.* 1. bole, stem of tree; **arbre à fût décroissant** tapering tree. 2. cask, barrel

futaie *f.* wood, forest (of tall trees); **f. de châtaigniers** chestnut grove; **f. fleurie** grove with a carpet of flowers

G

gabarit *m. Hort.* gauge, planting frame
gabion *m. Hort.* large basket
gadelier *m.* (*Canada*) red currant bush (*Ribes rubrum*)
gadelle *f.* (*Canada*) red currant
gadoue *f.* night soil, sewage sludge; **g. noire** fermented, processed night soil; **g. verte** raw night soil
gagéa, gagée *f. Gagea* sp., yellow star of Bethlehem
gaïac *m. Guaiacum* sp.; **g. officinal** lignum vitae (*G. officinale*); **résine de gaïac** guaiacum (medicinal resin); **g. de Cayenne** tonka bean (*Coumarouna odorata*) (= **fève tonka**)
gaillarde *f. Gaillardia* sp., gaillardia, blanket flower; **g. peinte** *G. pulchella* var. *picta*; **g. peinte double variée** *G. pulchella* var. *lorenziana*; **g. vivace** blanket flower (*G. aristata*)
gaillet *m. Galium* sp., **gaillet (-gratteron), g. accrochant** goosegrass, cleavers (*Galium aparine*) (see also **caille-lait**)
gain (*m.*) **de rendement** increase in yield
gainage *m. Ban.* sheathing, covering, sleeving (of bunch)
gaine *f.* cover, wrapping, sheath; **g. chauffante** (plastic) tubing for heating; **g. foliaire** leaf sheath; **g. d'irrigation** (plastic) tubing for irrigation; **g. radiante** = **g. chauffante**
gainé,-e *a. Ban.* etc. sheathed, sleeved
gainer *v.t.* to sheath
gainier *m.* Judas tree (*Cercis siliquastrum*) (= **arbre de Judée**)
galactose *m.* or *f.* galactose
galane *f.* beard tongue (*Penstemon* sp.); **g. barbue** *P. barbatus*
galanga *m.* 1. small galangal (*Alpinia officinarum*) (= **petit galanga, g. vrai, g. de la Chine**); **grand galanga** greater galangal (*A. galanga*). 2. galanga (*Kaempferia galanga*)
galanthe, galanthus *m. Galanthus* sp., including snowdrop (*G. nivalis*) (see **perce-neige**)
gale *f. PD* scab; **arbre à la gale** (*Canada*) poison ivy (*Rhus radicans*), poison oak (*R. toxicodendron*); **g. argentée (de la pomme de terre)** silver scurf (of potato) (*Helminthosporium atrovirens* = *Spondylocladium atrovirens*); **g. commune, g. ordinaire (de la pomme de terre)** potato scab (*Streptomyces scabies*); **g. poudreuse (de la pomme de terre)** (potato) powdery scab (*Spongospora subterranea*); **g. rugueuse du céleri** celery root rot (*Phoma apiicola*); **g. verruqueuse, g. noire (de la pomme de terre)** (potato) wart disease (*Synchytrium endobioticum*)
galé *m.* sweet gale (*Myrica gale*) (see also **arbre à cire, myrica**)
galéga *m.* goat's rue (*Galega officinalis*)
galène *f. PD* silver leaf (*Stereum purpureum*) (= **plomb**)
galénique *a.* galenic(al)
galéopsis *m. Galeopsis* sp., hemp-nettle
galerie *f.* gallery; **g. forestière** gallery forest; **g. de taupe** mole tunnel; **g. de termites** termite gallery; **g. de ver de terre** worm hole, channel
galéruque *f. Galerucella* sp. and other chrysomelid beetles
galet *m.* pebble; **gros galet** boulder
galette *f. Mush.* portion of spawn (= **cartouche**)
galeux,-euse *a.* scabby
galinsoga *m.* **(à petites fleurs)** gallant soldier (*Galinsoga parviflora*)
galipée *f. Galipea* sp. (see also **écorce**)
galium *m. Galium* sp. (see **gaillet**)
galle *f. PD* gall; **g. du collet, g. en couronne (du pommier)** crown gall (of apple) (*Agrobacterium tumefaciens*); **galles foliaires (de l'azalée)** azalea gall (*Exobasidium vaccinii*); **g. verruqueuse (de la pomme de terre)** wart disease (of potato) (*Synchytrium endobioticum*)
gallicole *a.* gall-forming (of insect)
galline *f.* poultry manure
gambier *m.* gambier (plant) (*Uncaria gambir*)
gambir *m.* gambir, gambier
gamète *m.* gamete
gamétique *a.* gametic
gamétocide *m.* gametocide
gamétogénèse *f.* gametogenesis
gamétophyte *m.* gametophyte
gamme *f.* scale, range; **une g. d'herbicides** a range of herbicides; **la g. d'hôtes** the host range
gamopétale *a.* gamopetalous
gamophylle *a.* gamophyllous
gamosépale *a.* gamosepalous
gamostyle *a.* gamostelic
gangrène *f. Bot., Hort.* canker
gant *m.* glove; **g. de jardinage** garden glove; **g. à mailles d'acier** steel-mesh glove (for cleaning tree bark etc.); **g. de Notre-Dame** foxglove

(*Digitalis* sp.) (= **digitale**), columbine (*Aquilegia* sp.)
gantelée *f.* nettle-leaved bell flower (*Campanula trachelium*)
gao *m. Afr.* winterthorn, gawo (*Acacia albida*) (= **cadde**)
garance *f.* madder (*Rubia tinctorum*)
garancière *f.* field of madder: area where dyeing with madder is carried out
garanti (*a.*) **sans virus (connu)** virus tested
garcie *f. Garcia nutans*, a Mexican plant whose seeds yield a drying oil
garcinie *f. Garcinia* sp.
garde (*f.*) **au sol** *AM* ground clearance
gardénia *m.*, **gardénie** *f.* gardenia (*Gardenia* sp.)
gardoquie *f. Gardoquia* sp.
gargouille *f.* spout of gutter, culvert, drain (of embankment)
gari *m. Afr.* garri (processed cassava)
garigue, garrigue *f.* garigue, fallow or uncultivated land in the south of France
garou *m.* garou bush (*Daphne gnidium*) (= **sainbois**); **g. des bois** mezereon (*D. mezereum*) (= **bois gentil**)
garrottage *m. Hort.* ringing by constriction
gaspiller *v.t.* to waste, to squander
gâté,-e *a.* spoilt; **fruits gâtés** spoilt fruit
gattilier *m. Vitex* sp.; **g. commun** chaste tree, tree of chastity (*Vitex agnus-castus*) (= **agneau-chaste, arbre au poivre**); **g. en arbre** Chinese chaste tree (*V. negundo*)
gaude *f.* dyer's rocket (*Reseda luteola*) (see also **réséda**)
gaufrage *m.* crinkling, puckering; *PD* creasing (of citrus)
gaufré,-e *a.* crinkled, puckered; *PD* creased (of citrus)
gaulage *m.* beating fruit or nut trees with a **gaule**
gaule *f.* pole, long stick; *Vit.* = **long bois** (*q.v.*)
gaulée *f.* beating fruit or nut trees with a **gaule**; fruit or nuts brought down in this way
gauler *v.t.* to beat a fruit tree or nut tree with a **gaule** to bring down the fruits or nuts
gaulette *f.* small **gaule**: slat
gauleur *m.* person who brings down fruit or nuts with a **gaule**
gaulthérie *f. Gaultheria* sp. (see also **thé du Canada**)
gaura *m. Gaura* sp.; **g. bisannuel** *G. biennis*; **g. de Lindheimer** *G. lindheimeri*
gaz *m.* gas; **g. ammoniac** ammonia; **g. carbonique** carbon dioxide; **g. de fumier** biogas
gazanie *f. Gazania* sp., treasure flower
gazeux,-euse *a.* gaseous, aerated (of water), fizzy (of lemonade etc.)
gazon *m.* short grass, turf, sward, lawn, green: turf, sod; **g. blanc** Venus' navelwort (*Omphalodes linifolia*) (see also **petite bourrache**); **g. d'Espagne, g. d'Olympe** thrift (*Armeria maritima*); **g. japonais, g. russe** mixture of flower seeds and/or grass seed; **g. turc** Dovedale moss, Eve's cushion (*Saxifraga hypnoides*) (= **saxifrage mousseuse**)
gazonnage *m.* turfing, edging with turf
gazonnant,-e *a.* turf-forming, caepitose
gazonné,-e *a.* turfed, covered with short grass
gazonnée *n. f.* area covered with short grass
gazonnement *m.* = **gazonnage** *q.v.*
gazonner *v.t.* to turf, to cover with short grass; *v.i.* to become covered with turf
gazonneux-euse *a.* covered with turf or grass, grassy
gel *m.* frost; **g. printanier** spring frost, late frost
gélatine *f.* gelatin(e); **g. végétale** (agar-)agar (= **gélose**)
gelé,-e *a.* frosted, frost-damaged, frozen
gelée *f.* 1. frost; **g. blanche** white frost; **g. nocturne** night frost; **g. printanière** spring frost, late frost. 2. jelly; **g. de groseilles** red currant jelly; **g. royale** *Ap.* royal jelly
geler *v.t.* & *v.i.* to freeze
gélif,-ive *a.* frost-cleft, frost-riven: (of tree) susceptible to frost; **espèce gélive** frost-susceptible, tender species (of plant); **place gélive** frost pocket
gélifiant,-e *a.* gelling
gélification *f.* gelling, gelification
gélivité *f.* (of tree) liability to crack from frost
gélivure *f.* frost crack
gélose *f.* (agar-)agar (= **gélatine végétale**)
gélosé,-e *a.* **milieu gélosé** medium containing agar
gelsémine *f. Gelsemium* sp., including Carolina jasmine (*G. sempervirens*)
gemmacé,-e *a.* gemmaceous, bud-like
gemmal,-e *a., pl.* **-aux** gemmaceous, relating to buds
gemmation *f.* gemmation, bud formation, period of bud formation
gemme *f. Bot.* 1. (leaf) bud; **g. secondaire** secondary bud. 2. offset bulb. 3. *Biol.* gemma. 4. *Arb.* resin (from pines)
gemmer *v.i.* to grow buds, to bud
gemmifère *a.* gemmiferous, bearing buds
gemmiforme *a.* gemmiform, bud-shaped
gemmipare *a.* gemmiparous
gemmule *f.* gemmule
gendarme *m. Afr.* village weaver(bird) (*Ploceus cucullatus*)
gène *m.* gene; **g. dominant** dominant gene; **g. marqueur** marker gene; **g. récessif** recessive gene
génécologie *f.* genecology
génépi *m. Artemisia* spp. (including *A. umbelliformis, A. genipi*) and other spp. growing in

the Alps and Pyrenees, used to give aroma to liqueurs; **g. des glaciers** *A. glacialis*

générateur *m.* generator; **d. d'air chaud** hot air heater

génératif,-ive *a.* generative; *Hort.* **multiplication générative** propagation by seed

génération *f.* generation

générique *a.* generic

genestrole, genestrolle, genestrelle *f.* dyer's greenweed (*Genista tinctoria*) (= **genêt des teinturiers**)

genêt *m.* *Genista* sp., broom and other genera; **g. à balai, g. commun** common broom (*Sarothamnus (Cytisus) scoparius*); **g. blanc (du Portugal)** white Spanish broom (*Cytisus albus (= C. multiflorus)*); **g. épineux** gorse (*Ulex europaeus*) (= **ajonc**); **g. d'Espagne** Spanish broom (*Spartium junceum*); **g. à fibre** = **g. d'Espagne** (*q.v.*); **g. à fleurs** = **g. à balai** (*q.v.*); **g. d'or** = **g. à balai** (*q.v.*); **g. du Portugal** = **g. blanc** (*q.v.*); **g. des teinturiers** dyer's greenweed (*Genista tinctoria*)

généticien *m.* geneticist (= **génétiste**)

génétique *a.* genetic; *n. f.* genetics

génétiste *m.* geneticist

genévrier *m.* (common) juniper (*Juniperus communis*); **g. à balai, g. des Montagnes Rocheuses** rocky mountain juniper (*J. scopulorum*); **g. cade, g. oxycèdre** prickly juniper (*J. oxycedrus*) (= **cade**); **g. de Chine** Chinese juniper (*J. chinensis*); **g. drupacé** Syrian juniper (*J. drupacea*); **g. à l'encens** Spanish juniper (*J. thurifera*); **g. de Phénicie** *J. phoenicea*; **g. sabine** savin (*J. sabina*); **g. (cèdre) de Virginie** pencil cedar (*J. virginiana*)

genévrière *f.* juniper plantation

géniculé,-e *a.* geniculate

génie (*m.*) **rural** agricultural engineering

genièvre *m.* juniper, juniper seed

génipa, génipayer *m.* genipap, marmalade box (*G. americana*) (= **confiture de singe**)

génipape *m.* genipap fruit

génique *a.* genic; **mutation génique** genic mutation

géniteur *m.* *PB* parent; **g. femelle** female parent; **g. de palmier à huile** oil-palm parent

génôme *m.* genome

génotype *m.* genotype

genouillé,-e *a.* = **géniculé** (*q.v.*)

genre *m.* genus

gentianacées *f. pl.* Gentianaceae

gentiane *f.* gentian; **g. de l'Abbé Coste** *Gentiana costei*; **g. acaule** gentianella (*G. (acaulis) excisa*); **g. des Alpes** (*G. (acaulis* var.*) alpina*); **g. des Alpes Dinariques** *G. (acaulis* var.*) dinarica*; **g. croisette** *G. cruciata*; **g. de Dahurie** *G. dahurica*; **g. de Fetisovi** *G. fetisovii*; **g. à feuilles étroites** *G. angustifolia*; **g. à feuilles de phlox** *G. phlogifolia* (= *G. cruciata phlogifolia*); **g. gracile** *G. gracilipes*; **g. à grandes feuilles** *G. macrophylla*; **g. jaune** = **grande gentiane** (*q.v.*); **g. de Kauffmann** *G. kauffmanniana*; **g. de Koch** *G. kochiana*; **g. de Lhassa** *G. lhassica*; **g. d'Olivier** *G. olivieri*; **g. pourpre** *G. purpurea*; **g. printanière** vernal gentian (*G. verna*); **g. retombante** *G. decumbens*; **grande gentiane** *G. lutea*

géocarpique *a.* geocarpic

géologique *a.* geological

géomètre *f.* geometrid moth, looper caterpillar (= **arpenteuse**); **g. des fleurs (d'agrumes)** *Cit. Gymnoscelis pumilata*

géonoma, géonome *m.* *Geonoma*, a genus of palms

géoréaction *f.* geotropism

géothermal,-e *a., m. pl.* **-aux** geothermal

géothermie *f.* heat from the earth, geothermics

géothermique *a.* geothermic

géothermomètre *m.* soil thermometer

géotropisme *m.* geotropism

géranium *m.* geranium, cranesbill (*Geranium* and *Pelargonium* spp.); **g. élégant** = **g. à grandes fleurs** (*q.v.*); **g. (à feuille(s) de) lierre** trailing pelargonium, ivy-leaved geranium (*P.* × *hederaefolium* = *P. peltatum*) (= **pélargonium à feuilles de lierre**); **g. des fleuristes** show, fancy geranium (*P.* × *domesticum*); **g. à feuilles rondes** round-leaved cranesbill (*Geranium rotundifolium*); **g. à grandes fleurs** *P. grandiflorum*; **g. des horticulteurs** zonal pelargonium (*P. zonale*); **g. des jardins** garden geranium (*P.* × *hortorum*); **g. rosat, g. à la rose** rose geranium (*P. capitatum, P. graveolens* etc.) (see also **pélargonium**); **g. à tiges grêles** small-flowered cranesbill (*Geranium pusillum*)

gerbable *a.* stackable

gerbage *m.* 1. binding, of sheaves etc. 2. stacking, piling, of bales etc.

gerbe *f.* sheaf; **g. de fleurs** (funeral) bouquet

gerber *v.t.* 1. to bind, to sheave. 2. to stack, to pile

gerbera *m.* gerbera, Barberton daisy (*Gerbera* sp.)

gerce *f.* crack, fissure (e.g. in wood)

gercer *v.t.* to crack; **se gercer** to become cracked, to crack

gerçure *f.* crack, cleft, fissure

gerçuré,-e *a.* cracked, fissured

germandrée *f.* germander (*Teucrium* sp.); **g. botryde** cut-leaved germander (*T. botrys*); **g. petit chêne** wall germander (*T. chamaedrys*) (= **chêneau**)

germe *m.* germ, sprout, eye (of tuber); **pousser des germes** to sprout
germé,-e *a.* sprouted
germer *v.i.* to germinate, to shoot, to sprout
germinabilité *f.* germinability
germinateur,-trice *a.* germinative
germinatif,-ive *a.* germinative; **capacité germinative** viability (of seed); **énergie germinative** germination energy; **essai germinatif** germination trial; **faculté germinative, pouvoir germinatif** power of germination
germination *f.* germination; **en germination** germinating
germoir *m. Hort.* seedbed, pre-nursery; germinator, hot bed
gesnérie *f. Gesneria* sp.
gesse *f.* vetch, everlasting pea (*Lathyrus* sp.); **g. annuelle** annual vetch (*L. annuus*); **g. chiche** dwarf chickling pea (*L. cicera*); **g. commune, g. cultivée** chickling pea (*L. sativus*); **g. à grandes fleurs** large-flowered vetch (*L. grandiflorus*); **g. à larges feuilles** everlasting pea (*L. latifolius*) (= **pois vivace**); **g. odorante** sweet pea (*L. odoratus*) (= **pois de senteur**); **g. des prés** yellow meadow vetchling (*L. pratensis*); **g. tubéreuse** earth-nut pea (*L. tuberosus*) (= **châtaigne de terre, gland de terre**)
gestion *f.* management
gibbérelline *f.* gibberellin
gibbérellique *a.* gibberellic
gibier *m.* game (wild animals)
giboulée *f.* sudden shower (often with snow or hail); **giboulées de mars** "April showers"
gicleur *m.* sprayer, spray nozzle, jet
gigantisme *m.* gigantism
gilie *f. Gilia* sp., including prickly phlox (*G. californica*)
gingembre *m.* ginger (*Zingiber officinale*)
ginkgo *m.* ginkgo (*Ginkgo biloba*) (= **arbre aux quarante écus**)
ginseng *m.* ginseng (*Panax* sp.)
girasol *m.* sunflower (*Helianthus* sp.)
giraumon(t) *m.* type of turban or winter squash (*Cucurbita maxima*) (see also **bonnet turc, courge**)
girofle *m.* clove; **clou de girofle** clove (of commerce); **huile de girofle** oil of cloves; **g.-mère** mother of cloves (see also **giroflier**)
giroflée (*f.*) **des jardins, grande giroflée, g. rouge** Brompton stock (*Matthiola incana*); **g. quarantaine** annual stock (*M. incana* var. *annua*) (= **quarantaine**); **g. quarantaine kiris** *M. incana var. graeca*; **g. jaune, g. des murailles, g. ravenelle** wallflower (*Cheiranthus cheiri*) (= **rameau d'or, violier**); **g. de Mahon** Virginian stock (*Malcomia maritima*) (= **julienne de Mahon, mahonille**); **cannelle giroflée** clove tree bark
giroflier *m.* clove tree (*Eugenia cayophyllus*)
girolle, girole *f.* chanterelle (*Cantharellus cibarius*) (= **chanterelle**)
girouette *f.* weathercock, vane
gisement *m.* layer, bed, stratum
givre *m.* hoar-frost, rime
glabre *a.* glabrous
glabrescent,-e *a.* glabrescent
glaçage (*m.*) **du sol** soil capping
glaçant,-e *a.* capping (of soil); **terre glaçante** capping soil
glace *f.* ice
glacé,-e *a.* 1. frozen, chilled, icy. 2. glazed, glossy
glaciale *f.* ice plant (*Mesembryanthemum crystallinum*) (= **ficoïde glaciale**)
glacis *m.* slope, bank
glaçogène *a.* ice-nucleating, ice-forming; **bactéries glaçogènes** ice-nucleating bacteria
gladié,-e *a. Bot.* gladiate, sword-shaped
gladiolé,-e *a.* resembling gladiolus
glaïeul *m.* gladiolus (*Gladiolus* sp.); **g. cardinal** *G. cardinalis*; **g. à grandes fleurs, g. de Gand** (*G.* × *gandavensis*); **g. à grandes macules, g. de Lemoine** *G.* × *lemoinei*; **g. de Nancy** *G.* × *nanceianus*; **g. perroquet** *G. psittacinus*; **"g. puant"** stinking iris (*Iris foetidissima*) (= **iris gigot**); **g. triste** *G. tristis*
glaise *f.* clay, loam; **terre glaise** clay
glaiser *v.t.* to clay, to dress with clay; to line with clay
glaiseux,-euse *a.* clayey, loamy
glaisière *f.* clay pit
glanage *m.* gleaning
gland *m.* acorn; **g. de terre** = **gesse tubéreuse** (*q.v.*)
glande *f.* gland; **g. nectarifère** nectar gland; **g. oléifère** oil gland
glandifère *a.* glandiferous
glandiforme *a.* glandiform
glandulaire *a.* glandular
glanduleux,-euse *a.* glandulous
glandulifère *a.* glanduliferous
glane *f.* gleaning; *Hort.* **g. d'oignons** rope of onions; **g. de poires** cluster of pears
glaner *v.t.* to glean, to gather fruit etc.
glaucescence *f.* glaucescence
glaucescent,-e *a.* glaucescent
glaucienne = **glaucier** (*q.v.*)
glaucier *m.*, **glaucière** *f. Glaucium* sp.; **g. jaune** yellow horned poppy (*G. flavum*) (= **pavot cornu**); **g. (rouge)** red horned poppy (*G. corniculatum*)
glauque *a.* glaucous
glaux *m.* sea milkwort (*Glaux maritima*)
glèbe *f.* clod of earth, sod

glèchome, glécome *m.* ground ivy (*Glechoma hederacea*) (= **lierre terrestre**)
gley *m. Ped.* gley
glissement *m.* 1. sliding, slipping. 2. landslide, landslip
globulaire *f. Globularia* sp.; **g. vulgaire** common globularia (*G. vulgaris*)
globule *m.* globule
globuleux,-euse *a.* globular, globulose
gloéosporiose *f.* perennial canker and fruit rot (of apples) (*Pezicula malicorticis* = *Gloeosporium perennans*)
gloire (*f.*) **des neiges** glory-of-the-snow (*Chionodoxa* sp.)
glomérule *m.* glomerule, cluster of short-stalked flowers
glomérulé,-e *a.* glomerulate
gloriette *f.* arbour, summer house
glorieuse *f.* gloriosa lily (*Gloriosa* sp.)
glouteron *m.* 1. burdock (*Arctium* sp.) (= **bardane**). 2. bedstraw (*Galium* sp.)
gloxinia *m. inv.*, **gloxinie** *f.* gloxinia (*Sinningia speciosa* and hybrids)
gluant,-e *a.* sticky, gummy
glucane *f.* glucan
glucide *m.* glucid(e), soluble carbohydrate
glucose *m.* & *f.* glucose
glucoside *m.* glucoside
glumacé,-e *a.* glumaceous
glume *f.* glume
glumelle *f.* lemma
glutamine *f.* glutamine
glutineux,-euse *a.* glutinous
glycérie *f. Glyceria* sp.; **g. aquatique** reed-grass (*G. maxima*)
glycérine *f.* glycerine
glycine *f.* 1. wistaria; **g. d'Amérique** *Wistaria frutescens*; **g. de (la) Chine** *W. sinensis*; **g. frutescente** = **g. d'Amérique** (*q.v.*); **g. du Japon** *W. floribunda*; **g. tubéreuse** potato bean (*Apios tuberosa*) (= **apios tubéreux**). 2. *Chem.* glycine
glycogène *m.* glycogen
glycolate *m.* glycolate
glycolyse *f.* glycolysis
glycolytique *a.* glycolytic
glycopeptide *m.* glycopeptide
glycoprotéine *f.* glycoprotein
glycorégulation *f.* glycoregulation
glycoside *m.* glycoside
glycyrrhiza *m. Glycyrrhiza sp.*, liquorice
gnaphale, gnaphalium *m. Gnaphalium* sp.; **g. des marais** marsh cudweed (*G. uliginosum*)
gnète *f.*, **gnetum** *m. Gnetum* sp., including *G. gnemon*, a small Malaysian tree with edible fruit
gnidie *f. Gnidia*, a genus of southern African evergreen flowering shrubs
gnomonia *m.* cherry leaf scorch (*Gnomonia erythrostoma*)
gobelet *m. Hort.* goblet, open-centre tree; **g. différé** delayed open-centre tree
gobe-mouche(s) *m. inv. Bot.* spreading dogbane (*Apocynum androsaemifolium*) and other insectivorous plants (see also **attrape-mouche**); *PC* insect trap
gobetage *m. Mush.* casing
gobeter *v.t. Mush.* to case
godet *m.* bowl, cup, mug, bucket (of Persian wheel); *Hort.* small pot; **g. en papier** paper pot; **g. de sphaigne, g. de tourbe** peat pot
godétia *m.*, **godétie** *f. Godetia* sp., godetia
goémon *m.* seaweed
goethée *f. Goethea* sp.; some spp. are tropical evergreen ornamental shrubs
golden *f. Coll.* Golden Delicious apple
golmette *f.* = **golmotte** (*q.v.*)
golmot(t)e *f. Fung.* 1. parasol mushroom (*Lepiota procera*) (= **coulemelle**). 2. blusher (*Amanita rubescens*) (see also **fausse golmotte**)
gomart *m.* West Indian birch (*Bursera gummifera* = *B. simaruba*)
gombo *m.* okra, ladies' fingers (*Hibiscus esculentus*)
gomme *f.* gum; **g. arabique** gum arabic; **g. élastique** rubber, eraser; **g. laque** shellac; *PD* gummosis; **g. alvéolaire** *Cit., PD* concave gum
gomme-gutte *f.*, *pl.* **gommes-guttes** gamboge (*Garcinia cambogia* and other spp.)
gommeux,-euse *a.* gummy; **plante gommeuse** gum-yielding plant
gommier *m.* gum tree, including *Eucalyptus, Acacia* and other species; **g. bleu** blue gum (*E. globulus*); **g. noir** black gum (*Nyssa sylvatica*)
gommose *f. PD* gummosis; *Sug.* gumming disease (*Xanthomonas vasculorum*); **g. bacillaire de la vigne** vine gummosis (= **nécrose bactérienne** *q.v.*); **g. du concombre** cucumber gummosis (*Cladosporium cucumerinum*)
gomphide *m. Fung. Gomphidius* sp.
gomphrène *f. Gomphrena* sp., including *G. globosa*, globe amaranth, globe everlasting
gonflable *a.* inflatable; **serre gonflable** inflatable, "bubble" greenhouse
gonflant,-e *a.* swelling; **sol argileux gonflant** swelling clay soil
gonflement *m.* inflation, swelling; bud swell (of fruit trees)
gonfler *v.i.* to swell (of buds etc.)
gonolobe *f. Gonolobus*, an American genus with some medicinal species

goodyère *f. Goodyera* sp., (terrestrial orchid), adder's violet
goptage *m.* = **gobetage** (*q.v.*)
gopter *v.t.* = **gobeter** (*q.v.*)
gorge *f. Bot.* throat; *Sug.* internal surface of collar
goudron *m.* tar
goudronné,-e *a.* tarred; **toile goudronnée** tarpaulin
goudronner *v.t.* to tar
gouet *m.* 1. billhook. 2. cuckoo-pint (*Arum maculatum*) (see also **attrape mouche**)
gouge *f.* gouge; **g. à asperge** asparagus knife; *Rubb.* tapping knife
gougeage *m.* gouging; (*Pineapple*) gouging of heart to reduce crown size at harvest
goulet *m.* narrow part, neck; **g. d'étranglement** bottleneck
goulotte, goulette *f.* 1. (small) channel. 2. spout (of hopper etc.)
goumi *m.* **(du Japon)** cherry elaeagnus (*Elaeagnus multiflora*) (see also **chalef**)
gourde *f.* gourd; **g. (commune)** calabash gourd (*Lagenaria siceraria*) (= **calebasse**); **g. buffalo** buffalo gourd (*Cucurbita foetidissima*)
gourmand,-e *a.*, **branche gourmande** sucker; *n.m. Hort.* sucker, sap shoot, water shoot, water sprout; **g. de tige** *Sug.* late tiller or sucker
gousse *f.* pod, shell, husk (of legume etc.): clove; **g. d'ail** garlic clove; **g. de vanille** vanilla pod
goussi *m. Afr.* egusi (*Cucumeropsis edulis, C. manii*)
goût *m.* taste; **g. de bois** woody taste; **g. de moisi** musty, mouldy taste; **g. piquant** pungent taste; **g. terreux** earthy taste; **g. de vert** grassy taste
goutte *f.* drop of liquid; **goutte à goutte** drop by drop, drip, trickle (of irrigation); **g. pendante** hanging drop
goutte-de-sang *f.*, *pl.* **gouttes-de-sang** pheasant's eye (*Adonis annua* and other spp.)
gouttelette *f.* small drop
goutteur *m. Irrig.* dripper, trickler
gouttière *f.* gutter; channel, gully (in Nutrient Film Technique); **g. en V** V-shaped gutter; *Bot.* etc. groove; **en gouttière** (of leaf) folded; *Vit.* leaf folded along the central vein
goyave *f.* guava fruit; **g. fraise** strawberry guava; **g. noire** *Alibertia edulis*; **g. poire** pear-shaped guava fruit; **g. pomme** apple-shaped guava fruit
goyavier *m.* guava tree (*Psidium guajava*); **g. acide, g. du Brésil** Brazilian guava (*P. araca*); **g. de Cattley, g. de Chine** strawberry guava (*P. cattleianum*); **g. du Chili** Chilean guava, ugni shrub (*Myrtus ugni*) (= **lucet musqué, myrte musqué**); **g. citronnelle** *Campomanesia aromatica* (*P. aromaticum*); **g. cochon** *P. cattleianum* var. *coriaceum*; **g. de Costa Rica** Costa Rican guava (*P. friedrichsthalianum*); **g.-fraise** = **g. de Cattley** (*q.v.*); **g. à grandes fleurs, g. sauvage** *Campomanesia grandiflora* (*P. grandiflorum*)
gradient *m.* gradient; **g. de fertilité** fertility gradient; **g. de sensibilité** sensitivity gradient; **g. thermique** temperature gradient
grain *m.* 1. grain, particle, speck; *Mush.* pinhead; **g. de café** coffee bean; **g. de haricot** bean seed; **g. de moutarde** mustard seed; **g. de poivre** peppercorn; **g. de pollen** pollen grain; **g. de raisin** grape. 2. squall
graine *f.* 1. seed (see also **enrobé**); **g. potagère** vegetable seed; **graines germées** germinated seed(s); **g. de capucin** stavisacre seed (*Delphinium staphisagria*); **g. de girofle** cardamom; **g. de moutarde** mustard seed; **g. oléagineuse** oilseed; **monter en graine** to (run to) seed, to bolt. 2. **graines d'Avignon** fruit of Avignon berry (*Rhamnus infectorius*); **graines de paradis** grains of paradise (*Aframomum melegueta*) (= **maniguette**)
graineterie *f.* seed trade; seed-shop; seed store
grainetier,-ère *a.* relating to the seed trade; relating to seeds; *n.* seed merchant
grainier,-ère *a.* relating to seeds; **champ grainier** field for seed production; **production grainière** seed production; *n.* seed merchant; **grainier-fleuriste** seedsman; *n. m.* seed collection, seed store; **jardin grainier cocotier** coconut seed garden
graisse *f.* grease, fat; **g. alimentaire** edible fat; **g. extraite de graines** seed fat; **g. commune (du haricot)** *PD* common blight, bacterial blotch of bean (*Xanthomonas phaseoli*); **g. à halo (du haricot)** *PD* halo blight of bean (*Pseudomonas phaseolicola*)
graisser *v.i.* to become oily
gramen *m.* lawn grass
graminacées *n. f., pl.* Gramineae, grasses
graminé,-e *a.* graminaceous; *n. f.* grass (species); **les graminées** Gramineae, grasses; **graminée à gazon** lawn grass, turf grass
graminicide *a.* graminicidal; *n. m.* graminicide, grass herbicide
graminifolié,-e *a.* graminifolious
graminiforme *a.* graminiform
granaire *a.* relating to **granum**
grand,-e *a.* tall, large, big; **grand cru** high quality vineyard or wine; **à grandes feuilles** large-leaved; **grand liseron** bellbine (*Calystegia sepium*); **grande pervenche** greater periwinkle (*Catharanthus (Vinca) major*);

grand piment allspice (*Pimenta officinalis*) (= **toute épice, poivre de la Jamaïque**); **grand plantain** great plantain (*Plantago major*); **grand pourpier** Ceylon spinach (*Talinum triangulare*); **grand soleil** sunflower (*Helianthus annuus*) (= **tournesol, hélianthe**); **grande tayove** giant alocasia (*Alocasia macrorrhiza*)

grande-épurge *f.* caper spurge (*Euphorbia lathyrus*)

grandissement *m.* growth, increase

granifère *a.* graniferous

graniforme *a.* graniform

granivore *a.* granivorous

granofibre *f.* granulated wood fibre

granulaire *a.* granular

granulation *f.* granulation

granule *m.* granule

granulé,-e *a.* granulated, pelleted; *n. m.* granulated product, granule, pellet; **g. anti-limace** slug pellet

granuler *v.t.* to granulate

granuleux,-euse *a.* granular

granulométrie *f. Ped.* grain size distribution

granulose *f.* granulosis; **virus de la granulose** granulosis virus

granum *m.* granum

grapefruit, grape-fruit *m.* grapefruit (*Citrus paradisi*) (see also **pamplemousse**)

graphiose *f.* **(de l'orme)** Dutch elm disease (*Ceratocystis ulmi*) (= **maladie hollandaise de l'orme**)

graphique *m.* graph

grappe *f.* cluster, bunch of grapes, currants etc.; **en grappes** in bunches; **g. de fleurs** flower cluster; **g. de fruits** fruit cluster; **g. lâche** loose bunch; **g. d'oignons** string of onions; **grappe ailée** *Vit.* winged bunch (bunch with a secondary small bunch near the base of the main stem); **g. épaulée** *Vit.* "shouldered" bunch, with large branches at the base; **g. lâche** *Vit.* loose bunch; **g. secondaire** *Vit.* sprig (of grapes); *Bot.* raceme

grapperie *f.* 1. grape growing, grape production. 2. greenhouse for grape production, vinery

grappier,-ère *a.* grape-producing

grappillage *m.* (*Date*) picking dates from the bunch (as they mature); *Vit.* gleaning

grappiller *v.t.* (*Date*) to pick dates as they mature; *Vit.* to glean

grappilleur,-euse *n.* person doing **grappillage**

grappillon *m. Vit.* small bunch; second crop; **g. vert** cluster of unripe grapes

grappu,-e *a.* laden with grapes

gras,-asse *a.* fat, fatty; **matières grasses** fats: greasy, oily, thick; **plante grasse** succulent plant

grassé *m. Afr.* water leaf, Ceylon spinach (*Talinum triangulare*), a leaf vegetable

grassette *f.* butterwort (*Pinguicula* sp.)

gratiole *f.* gratiole (*Gratiola* sp., including *G. officinalis*)

grattage *m.* scraping, scraping off; **g. des écorces (d'arbres fruitiers)** scraping the bark (of fruit trees); **g. du sol** soil scratching, superficial cultivation

gratte-cul *m. inv. Coll.* (rose) hip

gratteron *m.* cleavers, goosegrass (*Galium aparine* (= **gaillet (-gratteron)**)

grattoir *m.* scraper; **grattoir-émoussoir à écorce** tree scraper

graveleux,-euse *a.* gravelly; **terre graveleuse** gravelly soil; gritty (of pear)

gravelle *f.* **(du poirier)** *PD* (pear) stony pit virus

graves *f. pl. Vit., Ped.* tertiary soils in the Gironde department formed from a mixture of siliceous rocks, sand and other elements, with a subsoil which may be clayey, calcareous or sandy

gravier *m.* gravel, grit

greening *m. Cit.* greening

greffable *a.* capable of being grafted

greffage *m.* (see also **greffe**) grafting, budding; **g. en anneau** ring budding; **g. de bourgeon** terminal tip grafting; **g. d'écusson boisé** chip budding; **g. en flûte** flute budding; **g. en pied** budding near the base; **g. sur place** field grafting; **g. sur table** bench grafting; **g. en tête** top-working; **double-greffage** double working

greffe *f.* (see also **écussonnage, greffage**) graft, scion, grafted plant; **g. d'un an** maiden tree, one-year whip (= **scion**); **g. (à l') anglaise simple, g. (en fente) (à l') anglaise** splice graft, whip graft; **g. (à l') anglaise (compliquée)** whip-and-tongue graft; **g. anglaise double** whip-and-tongue graft; **g. (anglaise) (en fente) à cheval** saddle graft; **g. anglaise à cheval compliquée** strap graft; **g. par approche** approach graft, bridge grafting by approach, inarching; **g. par approche (à la bouteille)** bottle graft; **g. par approche en arc-boutant** inarch (graft); **g. de côté** side graft; **g. de côté sous l'écorce** side graft under bark; **g. en coulée** side rind graft; **g. en couronne** crown graft; **g. en couronne avec incrustation** kerf graft; **g. d'écorce** ring graft, ring budding; **g. en écusson** shield budding; **g. d'écusson boisé** chip budding; **g. en fente (simple)** cleft graft; **g. en fente double** double cleft graft; **g. en fente terminale** terminal cleft graft; **g. herbacée** herbaceous graft; **g. en incrustation** coin graft; **g. en incrustation latérale** side graft; **g. ligneuse** *Vit.* ligneous graft; **g. sur moignon** graft on a stub; **g.**

d'œil, **g. par œil détaché** budding; **g. à œil dormant** graft with a dormant bud; **g. à œil poussant** graft with a pushing bud; **g. en placage, g. en plaque, g. "à la plancha"** (*Morocco*) veneer graft; **g. sur place** graft in the field or nursery; **g. en pont** inarching; **g. de racine** root graft; **g. par rameau détaché** (ordinary) graft; **g. sur table** bench graft; **g. à tenon et mortaise** tenon and mortise graft; **g. en tête** top graft; **g. en vert** green graft; **g. sur vieux pied(s)** *Vit.* graft on old vine(s)

greffé,-e *a.* grafted; **g. en pied, g. au ras du sol** low worked; **g. en tête** topworked

greffer *v.t.* to graft, to bud, to work (tree)

greffé-soudé *m.* (successfully) grafted plant or cutting (see also **soudé**)

greffeur *m.* grafter, budder

greffoir *m.* grafting knife, budding knife (= **écussonnoir, entoir**); **g. à fendre** cleft grafting knife; **g. à spatule fermante** clasp grafting knife; **g. à spatule fixe** grafting knife with fixed blade

greffon *m. Hort.* scion

grêle *a.* slender, thin; **rameau grêle** thin branch

grêle *f.* hail

grêlé,-e *a. Hort.* damaged by hail; **culture grêlée** hail-damaged crop

grêlimètre *m.* a device to measure hail intensity

grêlon *m.* hailstone

grelot *m.* small round bell, sleigh-bell; **g. blanc** snowdrop (= **perce-neige**); **en grelot** shaped like a small bell

grémil *m.* gromwell (*Lithospermum* sp.); **g. des champs** corn gromwell (*L. arvense*)

grenade *f.* pomegranate fruit

grenadier *m.* pomegranate plant (*Punica granatum*)

grenadille *f.* passion flower (*Passiflora* sp. esp. granadilla, *P. edulis*) (see also **barbadine, passiflore, pomme-liane**); **g. douce** sweet grenadilla (*P. ligularis*); **g. à fleurs rouges** scarlet passion flower (*P. coccinea*)

grenadin *m. Hort.* grenadin, type of carnation

grenaille *f.* refuse grains, tailings; **g. (du framboisier)** *PD* (raspberry) crumbly berry

grenat *a. inv.* garnet red

grener *v.i.* (of cereals etc.) to seed, to set seed

grèneterie *f.* 1. seed trade. 2. seedsman's shop

grènetier,-ière *n.* seedsman, seedswoman

grenier *m.* 1. granary, storehouse. 2. attic

grenu,-e *a.* granular, grained

grewie *f. Grewia* sp., including *G. asiatica*, a tree with edible fruits

gribouri *m. Vit.* chrysomelid beetle, western grape root worm (*USA*) (*Adoxus obscurus*) (= **écrivain, eumolpe**)

griffage *m.* forking with **griffe**

griffe *f.* 1. (a) *Hort.* bunch of short thick roots or rhizomes; **g. d'asperge** crown, root bunch of asparagus; **g. de muguet** lily-of-the-valley crown, (b) **griffes de girofle** *Comm.* clove stems. 2. tendril (of vine). 3. **g. (à fleurs)** (small) garden fork; **g. à fumier** rake fork (= **fourche crochue**); **g. sarcleuse** multi-tine hand hoe

griffe-bineuses *f., pl.* **griffes-bineuses** Dutch hoe

griffer *v.t. Hort.* to fork with a **griffe**

grignon *m.* olive cake

grillage *m.* 1. roasting; *Hort.* scorch(ing), leaf scorch, leaf damage, sunburn. 2. metal grating, lattice work; **g. en fil de fer** wire-netting; **g. fin** fine netting

grillagé,-e *a.* surrounded by wire netting

grille *f.* grating, screen

grillé,-e *a.* roasted; *Hort.* scorched

grille-midi *m.* rock-rose (*Helianthemum* sp.)

grillure *f.* = **grillage** 1. *q.v.*

grimpage *m.* climbing

grimpant,-e *a.* climbing, creeping; **plante grimpante** climber, creeper; **rose grimpante** climbing rose

grindélie *f. Grindelia* sp., including Californian gum plant (*G. robusta*)

griotte *f.* sour cherry

griottier *m.* sour cherry tree (*Prunus cerasus*) (= **cerisier acide**)

gris,-e *a.* grey; **g. ardoise** *a. inv.* slate grey; **g. fumée** *a. inv.* smoky grey; *n. m.* **g. du groseillier** American mildew of gooseberry and currant (*Sphaerotheca mors-uvae*)

grisard *m. Hort.* grey poplar (*Populus canescens*)

grisâtre *a.* greyish

grise *f.* applied to plant damage caused by various pests (including *Tetranychus* sp.) and diseases

griset *m.* sea buckthorn (*Hippophaë rhamnoides*) (= **argousier**)

grisette *f. Fung.* grisette (*Amanita vaginata*) (= **amanite engainée**); *Vit.* capsid bug (*Lopus sulcatus*); wine weevil (*Peritelus griseus* = *P. sphaeroides*) (= **péritèle gris**)

grive *f.* thrush (*Turdus* sp.)

gros-bec (casse-noyaux) *m., pl.* **gros-becs** hawfinch (*Coccothraustes coccothraustes*)

gros chiendent *m.* (*Mauritius*) goosegrass (*Eleusine indica*)

groseille *f.*, **g. à maquereau, g. verte** gooseberry; **g. à grappes** currant (white or red); **g. rouge** red currant; **g. noire** black currant (= **cassis**); **g. de Chine** Chinese gooseberry

groseillier *m.* 1. currant, gooseberry bush; **g. commun, g. rouge** red currant bush (*Ribes rubrum*); **g. doré** golden currant bush (*R. aureum*); **g. épineux, g. à maquereau** gooseberry bush (*R. grossularia*); **g. à grappes**

currant bush (white or red) (*R. sativum, R. rubrum*); **g. noir** black currant bush (*R. nigrum*) (= **cassissier**); **g. sanguin** flowering currant bush (*R. sanguineum*); **g. d'ornement** ornamental currant. 2. Other genera: **g. de(s) Barbade(s)** West Indian gooseberry (*Pereskia aculeata*); **g. du Cap** Cape gooseberry (*Physalis peruviana*); **g. de Ceylan** Ceylon gooseberry (*Dovyalis hebecarpa*), emblic myrobalan (*Phyllanthus emblica*); **g. de Chine** Chinese gooseberry plant (*Actinidia chinensis*) (= **actinidia, yang-tao**)

gros minet *m.* hare's tail grass (*Lagurus ovatus*)

gros-patte *m. Fung. (Mauritius) Tricholoma mauritiana (see also* **tricholome**)

grosseur *f.* size, volume; **de la grosseur d'une noix** nut-sized

grossier,-ère *a.* coarse (the opposite of **fin**)

grossissement *m.* increase in size, swelling; **g. des fruits** fruit swelling

grossiste *n.* wholesaler (see also **détaillant**)

gru-gru *m.* Paraguay palm (*Acrocomia sclerocarpa*) (= **noix de Coyol**)

grume *f.* log; grape (in Burgundy)

grumeleux,-euse *a.* gritty (of soil, pear etc.)

guaco *m.* guaco (*Mikania guaco*)

guano *m.* guano

guarana *m.* guarana (*Paullinia cupana*)

guayule *m.* guayule (*Parthenium argentatum*)

guazuma *m. Guazuma* sp., including *G. grandiflora*

guède *f.* woad (*Isatis tinctoria*) (= **pastel des teinturiers**)

guêpe *f.* wasp (*Vespa* sp.) (see also **frelon**)

guéret *m.* 1. ploughed land not sown with any crop. 2. fallow land

guérison *f.* 1. recovery. 2. cure

guéttarde *f. Guettarda* and related spp.; **g. à fleurs rouges** *Isertia coccinea*

gueule-de-lion, gueule-de-loup *f., pl.* **gueules-de-lion (loup)** *snapdragon (Antirrhinum majus)* (= **muflier**)

gui *m.* mistletoe (*Viscum album*) and other genera, *Loranthus* etc.

guide-greffes *m. inv.* grafting tool

guigne *f.* soft juicy sweet cherry, heart cherry, gean

guignier *m.* sweet cherry tree producing **guignes**

guimauve *f.* mallow (*Althaea* sp.); **g. officinale** marsh mallow (*A. officinalis*); **g. velue** hispid mallow (*A. hirsuta*) (see also **mauve**)

gummifère *a.* gummiferous

gunnère *a. Gunnera* sp., prickly rhubarb

gur *m.* gur, brown sugar

gustatif,-ive *a.* gustative, gustatory; **préférence gustative** flavour preference

gutta-percha *f.* gutta-percha

guttation *f.* guttation

guttier *m.* gamboge (*Garcinia cambogia* and other spp.)

guttifère *a.* guttiferous

gymnoclade *m. Gymnocladus* sp. (see also **bonduc**)

gymnosperme *a.* gymnospermous; *n. m.* gymnosperm

gynandro *m. Afr.* cat's whiskers (*Gynandropsis girandra*), used as a pot-herb

gynécée *m.* gynaecium

gynérion, gynérium *m. Gynerium* (= *Cortaderia*) sp.; **g. argenté** pampas grass (*C. argentea*)

gynodioécie *f.* gynodioecy

gynœcie *f.* gynoecism

gynogénèse *f.* gynogenesis

gynogénétique *a.* gynogenetic

gynoïque *a.* gynoecious

gynophore *m.* gynophore

gynostème *m.* gynostemium (= **colonne**)

gypse *m.* gypsum

gypsophile *f. Gypsophila* sp.

gyrobroyeur *m.* rotary cultivator, rotary slasher

gyromitre *f. Fung. Gyromitra* sp.

gyroselle *f. Dodecatheon* sp.; **g. de Virginie** American cowslip, shooting star (*D. meadia*)

H

An asterisk * indicates an aspirate H

habillage *m. Hort.* pruning, trimming (before planting); **h. des racines** root pruning (before planting); **h. de la tige** heading back (before planting)
habiller *v.t.* Hort. to prune, to trim (before planting)
habitat *m.* habitat
***hache** *f.* axe; **h. d'abattage** felling axe; **h. dresse-bordure** edging tool (for lawns) (= **dresse-bordure**); **h. à main** hatchet
***hache-sarments** *m. Vit.* = **coupe-sarments** (*q.v.*)
***hachette** *f.* small axe, hatchet
***hachisch** *m.* hashish
***hachoir-broyeur** *m., pl.* **hachoirs-broyeurs** *Sug.* chopper-crusher
***haie** *f.* hedge; **h. d'abri** windbreak; **h. défensive** spiny hedge; **h. fruitière** (fruit) hedge system; **h. morte, h. sèche** dead hedge; **h. vive** live, quickset hedge
***haie-rideau** *f.* (small) windbreak
halesia *m.*, **halésie** *f. Halesia* sp., silver bell, snowdrop tree (*H. carolina*)
***hallier** *m.* thicket, copse
halophile *a.* salt-loving, halophilous
halophyte *a.* halophytic; *n. f.* halophyte
haltica *m. Ent., Vit. Haltica* sp. (= **altise**)
hamamélidacées *n. f., pl.* Hamamelidaceae
hamamélis *m. Hamamelis* sp., witch-hazel, including *H. mollis*; **h. de Virginie** *H. virginiana*
***hamelia** *m.*, ***hamélie** *f. Hamelia* sp.
hamiltonia *m. Hamiltonia* sp.
***hampe** *f. Bot.* stem, scape; **h. (florale)** spike
***hangar** *m.* open shed, lean-to
***hanneton** *m.* cockchafer (*Melolontha melolontha*); **h. bronzé** *Vit. Anomala vitis* and other spp.; **h. horticole** garden chafer, June bug (*Phyllopertha horticola*); **h. de la St. Jean** summer chafer (*Amphimallon solstitialis*); **h. vert** rose chafer (*Cetonia aurata*) (= **cétoine**)
***hannetonnage** *m.* removing cockchafers from trees and destroying them; **h. chimique** chemical control of cockchafers
***hannetonner** *v.t.* to control cockchafers
haplobiontique *a.* haplobiontic
haplodiploïdie *f.* haplodiploidy
haplodiploïdisation *f.* production of haploid plants
haploïde *a.* haploid; *n.m.* haploid
haploïdie *f.* haploidy
haplonte *m.* haplont
haplopétale *a.* haplopetalous
haplophase *f.* haplophase
***happa** *m.* (*Pineapple*) hapa, shoots produced at the base of the peduncle
***haque** *f. Vit.* planter, dibbler
***haricot** *m.* bean (*Phaseolus* sp. and other genera), usually french bean (*P. vulgaris*) when used by itself; I. **h. blanc** haricot bean (grown for ripe seed); **h. beurre** (1) wax-podded (yellow-podded) (stringless) french bean, (2) butter bean (*P. lunatus*); **h. à bouquets** = **h. à rames** (1) (*q.v.*); **h. commun** french bean, kidney bean (*P. vulgaris*); **h. à cosse jaune** wax-podded (yellow-podded) (stringless) french bean; **h. à cosse verte** green-podded (stringless) french bean; **h. à cosse violette** climbing purple-podded kidney bean; **h. à écosser** french bean type for shelling; **h. d'Espagne** = **h. à rames** (1) (*q.v.*); **h. à filets, h. "filet"** fine french bean type for eating green, needle bean; **h. mange-tout** mange-tout bean; **h. nain** dwarf french bean, bush bean; **h. à rames** (1) scarlet runner bean (*P. coccineus*), (2) climbing french bean; **h. rouge** red-seeded kidney bean; **h. sans parchemin** *P. vulgaris* group to which true **mange-tout** cvs belong; **h. vert** = **h. commun** *q.v.*; **haricots en grains** shelled beans; **haricots secs** dried beans; II. Other beans: **h. asperge** asparagus bean (*Vigna sesquipedalis*) (= **pois ficelle**); **h. doré** green gram (*P. aureus*); **h. kilomètre** = **h. asperge** *q.v.*; **h. de Lima, h. du Cap, h. du Kissi** Lima bean (*P. lunatus*) (= **pois du Cap, pois savon**); **h. (-)limaçon** snail flower (*P. caracalla*); **h. niébé, h. à œil noir** southern pea, cowpea (*Vigna unguiculata*) (= **niébé**); **h. papillon** moth bean (*Vigna aconitifolia*); **h. pistache** Bambara groundnut (*Voandzeia subterranea*) (= **voandzou**); **h. riz** rice bean (*P. calcaratus*); **h. sabre** sword bean (*Canavalia ensiformis*, *C. gladiata*) (= **pois sabre**); **h. velu** mung bean (*P. aureus*)
harmattan *m. Afr.* harmattan
***haschisch** *m.* = **hachisch** *q.v.*

***hasté,-e** *a.* hastate, spear-shaped
***hastifolié,-e** *a.* with hastate, spear-shaped leaves
***hâté,-e** *a.* forward (of season), forced (of crop etc.); **culture hâtée** (forced) early crop
***hâter** *v.t. Hort.* to force
***hâtif,-ive** *a.* early, early ripening
***hâtiveau** *m., pl.* **-eaux** early fruit or vegetable: early pear
***hâtiveté** *f.* earliness
***hauban** *m.* guy, stay
***haubanage, haubannage** *m.* guying, staying, bracing
***haubaner, haubanner** *v.t.* to guy, to stay, to brace
haustorie *f.* haustorium
haustorium *m.* (*Coconut*) haustorium, coconut apple
***haute-tige** *f., pl.* **hautes-tiges** standard, free growing tree
***hauteur** *f.* height; **h. de coupe** height of cut, cutting height
***hautin, hautain** *m. Vit.* tall growing vine: support for tall growing vine
***hautiné,-e** *a. Vit.* said of vineyard planted with (tall) espalier-trained vines
***havane** *a. inv.* light brown colour
***haworthia** *m.,* **haworthie** *f.* haworthia; **h. à bandelettes** zebra haworthia (*Haworthia fasciata*)
hechtia *m. Hechtia* sp.
hectare *m.* hectare (= 2.47 acres)
hélénie *f. Helenium* sp., helen-flower
hélianthe *m. Helianthus* sp., sunflower (*H. annuus*); **h. tubéreux** (= **topinambour** *q.v.*)
hélianthème *m. Helianthemum* sp., rock-rose
hélianthi, hélianti *m.* woodland sunflower (*Helianthus strumosus*)
hélice *f.* helix
hélichrysum *m.,* **hélichryse** *f. Helichrysum* sp., everlasting flower
hélicide *a. & n.m.* snail-killing compound
heliconia *m. Heliconia* sp.; **h. à feuilles striées de jaune** *H. aureo-striata* (= *H. bihai*); **h. illustré** *H. illustris*; **h. métallique** *H. metallica*
hélicoptère *m.* helicopter
héliographe *m.* sunshine recorder
héliophile *a.* light loving, sun loving; *n. f. Heliophila* sp., sun cress
héliotrope *m.* heliotrope (*Heliotropium* sp.); **h. d'hiver** winter heliotrope (*Petasites fragrans*); **h. du Pérou** cherry pie, Peruvian heliotrope (*H. peruvianum*) (= **fleur des dames**); *a. inv.* heliotrope-coloured
héliotropique *a.* heliotropic
héliotropisme *m.* heliotropism; **h. négatif** apheliotropism, negative heliotropism
hellébore *m.* hellebore (see **ellébore**)
helléborine *f.* winter aconite (*Eranthis hyemalis*) (see also **renoncule d'hiver**)
helminthe *m.* helminth
helvelle *f. Fung. Helvella* sp.
helxine *f.* helxine (*Helxine soleirolii*)
hémagglutinine *f.* haemagglutinin
hémanthe *m. Haemanthus* sp., blood-flower
hémérocalle *f.,* **hémérocallis** *m. Hemerocallis* sp., day-lily; **h. jaune** Chinese day-lily (*H. citrina*) (= **lis jaune**); **h. de Sibérie** *H. middendorffii*
héméropériodique *a.* long-day (of plant)
hémicellulose *f.* hemicellulose
hémicryptophyte *f.* hemicryptophyte
hémidibiose *f.* type of graft in which scion and stock develop their own foliage (in viticulture etc.)
hémiléia *m. Hemileia* sp., including *H. vastatrix*, coffee rust (see also **rouille**)
hémiparasite *a.* hemiparasitic; **plante hémiparasite** hemiparasitic plant; *n. m.* hemiparasite
hémiptère *a.* hemipterous; *n. m.* hemipterous insect, hemipteran; in *pl.* Hemiptera
hémoglobine *f.* haemoglobin
***henné** *m.* henna (*Lawsonia inermis*)
hépatique *f.* 1. liverwort (Hepaticae). 2. **h. (des jardins)** common hepatica (*Hepatica triloba* = *Anemone hepatica*)
hépiale *m. Hepialus* sp., swift moth; **h. du houblon** ghost swift moth (*H. humuli*); **h. du muguet** common swift moth, garden swift moth (*H. lupulina*)
herbacé,-e *a.* herbaceous; **bordure herbacée** herbaceous border
herbage *m. Hort.* grass, herbage
herbe *f.* herb, plant, grass; I. General: **h. potagère** green vegetable, pot herb; **fines herbes** culinary herbs; **mauvaise herbe** weed (see also under **mauvaise herbe**); **brin d'herbe** blade of grass; **h. aromatique** (aromatic) herb; **h. vivace** herbaceous plant; II. Specific: **h. amère, h. aux vers** tansy (*Chrysanthemum vulgare*); **h. d'amour** mignonette (*Reseda odorata*); **h. d'argent** silver grass (*Ischaemum aristatum*); **h. aux ânes** evening primrose (*Oenothera* sp.); **h. bassine, h. bambou** (*Mauritius*) bearded setaria (*Setaria barbata*); **h. des Bermudes** Bermuda grass (*Cynodon dactylon*); **h. bleue** (*Mauritius*) clasping heliotrope (*Heliotropium amplexicaule*); **h. bol** (*Mauritius*) = **h. tam-tam** (*q.v.*); **h. aux boucs** greater celandine (*Chelidonium majus*); **h. de Brinvilliers** West Indian spigelia, pink root (*Spigelia anthelmia*); **h. caroline** (*Mauritius*) ribwort plantain (*Plantago lanceolata*); **h. aux chantres** hedge mustard (*Sisymbrium officinale*) (= **vélar**); **h. aux charpentiers** milfoil,

yarrow (*Achillea millefolium*); **h. aux chats** catmint (*Nepeta cataria*), valerian (*Valeriana officinalis*); **h. au citron** balm (*Melissa officinalis*); **h. aux clochettes** crown imperial (*Fritillaria imperialis*); **h. à cloques** bladder-herb, winter cherry (*Physalis alkekengi*; **h. aux écus** honesty (*Lunaria annua*), creeping jenny (*Lysimachia nummularia*) (= **lysimaque nummulaire**); **h. à éléphant** elephant grass (*Pennisetum purpureum*); **h. à feu d'artifice** artillery plant (*Pilea* sp.); **h. aux goutteux** ground elder (*Aegopodium podagraria*) (= **égopode**); **h. de Guatémala** Guatemala grass (*Tripsacum laxum*); **h. aux gueux** traveller's joy (*Clematis vitalba*) (= **clématite des haies**); **h. de Guinée** Guinea grass (*Panicum maximum*); **h. à jaunir** dyer's greenweed (*Genista tinctoria*); **h. du Laos** Siam weed (*Eupatorium odoratum*); **h. à miel** molasses grass (*Melinis minutiflora*); **h. aux mites** moth mullein (*Verbascum blattaria*) (= **blattaire**); **h. à (la) ouate** silk weed (*Asclepias cornuti* = *A. syriaca*); **h. à paillottes** spear grass (*Imperata cylindrica*); **h. des Pampas** Pampas grass (*Cortaderia argentea*); **h. aux panthères** leopard's bane (*Doronicum pardalianches*); **h. de Pâques** pasque flower (*Pulsatilla vulgaris*) (= **fleur de Pâques**); **h. de Para** Para grass (*Panicum purpurascens*); **h. aux perruches** = **h. à (la) ouate** (*q.v.*); **h. à la puce** (*Canada*) poison ivy (*Rhus radicans*); **h. aux puces** psyllium (*Plantago psyllium*); **h. sacrée** vervain (*Verbena* sp.); **h. de Saint-Christophe** baneberry, herb Christopher (*Actaea spicata*) (= **actée**); **h. du Saint-Esprit** angelica (*Angelica archangelica*) (= **angélique**); **h. de Saint-Fiacre** cherry pie (*Heliotropium peruvianum*) (= **fleur des dames, héliotrope du Pérou**); **h. de Saint-Gérard** = **h. aux goutteux** (*q.v.*); **h. de (la) Saint-Jean** St. John's wort (*Hypericum* sp.), wormwood (*Artemisia* sp.) (= **armoise**), ground ivy (*Glechoma hederacea*); **h. de Saint-Martin** St. Martin's herb (*Sauvagesia erecta*); **h. solférino** (*Mauritius*) western ragweed (*Ambrosia psilostachya*); **h. aux sonnettes** crown imperial (*Fritillaria imperialis*) (= **couronne impériale**); **h. sainte** wormwood (*Artemisia absinthium*) (= **absinthe**); **h. de Sainte-Barbe** winter cress (*Barbarea vulgaris*); **h. de Sainte-Barbe à fleurs doubles** double yellow rocket (*B. vulgaris* var. *flore-pleno*); **h. des sorciers** thorn apple (*Datura stramonium*); **h. tam-tam** (*Mauritius*) *Hydrocotyle bonariensis*; **h. à taupes** caper spurge (*Euphorbia lathyrus*) (= **épurge**); **h. aux turquoises** lily turf (*Ophiopogon japonicus*); **h. villebague** (*Mauritius*) hairy beggarticks (*Bidens philosa*); **h. vivante** sensitive plant (*Mimosa pudica*), telegraph plant (*Desmodium gyrans*) (= **sainfoin oscillant**)

herbeux,-euse *a.* grassy

herbicide *a.* herbicidal; **sélection herbicide** herbicidal selection. *n.m.* herbicide, weed killer; **h. de contact** contact herbicide; **h. persistant, h. rémanent** persistent herbicide (the terms **h. résiduaire, h. résiduel** are sometimes used but are not recommended); **h. sélectif** selective herbicide; **h. systémique** systemic herbicide; **h. total** non-selective herbicide (see also **traitement**)

herbier *m.* herbal, herbarium; *Bot.* water plant community

herborisateur,-trice *n.* herborizer, botanizer

herborisation *f.* 1. herborizing, botanizing. 2. botanical excursion

herboriser *v.i.* to herborize, to botanize, to gather plants or herbs

herboriste *n.m., n.f.* herbalist

herboristerie *f.* 1. herbalist's shop. 2. herb trade

herbu,-e *a.* grassy

herbue *f. Vit.* loam

hercogamie *f.* hercogamy

héréditaire *a.* hereditary

herédité *f.* heredity; **h. cytoplasmique** cytoplasmic heredity

***hérissé,-e** *a.* bristling, bristly; *Bot.* prickly

***hérisson** *m.* 1. toothed roller. 2. chestnut husk. 3. *Onobrychis caput-galli* 4. hedgehog (*Erinaceus*); *Afr.* cane rat (see **agouti**)

héritabilité *f.* heritability

héritable *a.* heritable; **caractère héritable** heritable character

***hermannie** *f. Hermannia* sp.

hermaphrodite *a.* hermaphrodite

hermétique *a.* hermetically sealed

***herniaire** *f. Herniaria* sp., rupturewort

***hernie** (*f.*) **du chou** clubroot (*Plasmodiophora brassicae*)

***hersage** *m.* harrowing

***herse** *f.* harrow; **h. à disques** disc harrow; **h. à chaînons, h. souple** chain harrow; **h. rotative** rotary hoe

***herser** *v.t.* to harrow

***herseur** *m.* harrower

hespéridie *f.* hesperidium

hespéridine *f.* hesperidin

hétéroauxine *f.* heteroauxin

hétérocarpe *a.* heterocarpous

hétérogame *a.* heterogamous

hétérogénéité *f.* heterogeneity

hétérogreffe *f.* heterograft

hétéromorphe *a.* heteromorphous

hétéromorphie *f.* heteromorphism

hétérophylle *a.* heterophyllous

hétérophyllie *f.* heterophylly
hétéroside *m.* heteroside (= **glycoside**); **h. flavonique** flavonoid glycoside
hétérosis *m.* heterosis; **effet d'hétérosis** hybrid vigour
hétérosporiose *f.* heterosporiosis
hétérostylie *f.* heterostyly
hétérotrophe *a.* heterotrophic
hétérozygote *a.* heterozygotic; *n. m.* heterozygote
***hêtraie** *f.* beech grove, beech wood
***hêtre** *m.* beech (*Fagus* sp.); **h. commun, h. (des bois)** common beech (*F. sylvatica*); **h. pleureur** weeping beech (*F. sylvatica* var. *pendula*); **h. pourpre** purple beech (*F. sylvatica* var. *purpurea*), copper beech (var. *cuprea*); **h. blanc** hornbeam (*Carpinus betulus*) (= **charme**)
heuchère *f. Heuchera* sp., alum root
heure *f.* hour; **h. de travail** man-hour
hévéa *m.* rubber tree (*Hevea brasiliensis*)
hévéaculteur *m.* rubber grower, planter
hévéaculture *f.* rubber growing
hévéicole *a.* relating to rubber growing
hiba *m. Thujopsis*, a genus of evergreen coniferous trees
hibernacle *m.* hibernacle, hibernaculum
hibernation *f.* hibernation, dormancy (of plant) (see also **hivernement**)
hiberner *v.i.* to hibernate
hibernie *f.* mottled umber (moth) (*Erannis (Hibernia) defoliaria*) (= **phalène défeuillante**); winter moth (*Operophtera (Cheimatobia) brumata*)
hibiscus *m.* hibiscus (*Hibiscus* sp.)
hièble *f.* dwarf elder (*Sambucus ebulus*); sometimes ground elder (*Aegopodium podagraria*)
***hilaire** *m. Afr.* (long-handled) hoe; *a. Bot.* hilar
***hile** *m.* hilum
hippeastrum *m. Hippeastrum* sp. (see also **amaryllis**)
hippocratée *f. Hippocratea* sp., including beacon bush (*H. senegalensis* (= *Salacia senegalensis*))
hippocrépide *f. Hippocrepis* sp., including *H. comosa*, kidney vetch
hippomobile *a.* horse-drawn
hippophaé *m. Hippophaë* sp., sea buckthorn (*H. rhamnoides*) (= **argousier, saule épineux**)
hirsute *a.* hirsute, hairy
hispide *a.* hispid
histogenèse *f.* histogenesis
histologie *f.* histology
histologique *a.* histological
histone *f.* histone
hiver *m.* winter
hivernage *m.* 1. wintering: ploughing before winter. 2. (a) winter season, (b) tropical rainy season. 3. *Hort.* storing vegetables before forcing; **h. en tranchées** heeling in
hivernal,-e *a., m. pl.* **-aux** winter, wintry
hivernant,-e *a.* wintering
hiverné,-e *a.* overwintered
hivernement *m.* hibernation (= **hibernation**); **œufs d'hivernement** overwintering eggs (of mites etc.)
hiverner *v.i.* to winter, to overwinter; *v.t.* to plough before winter; to shelter plants for the winter
Hj see **homme-jour**
holodibiose *f.* (normal) graft in which only the scion develops leaves
holoparasite *a.* holoparasitic; *n. m.* holoparasite
holoside *m.* holoside
homme-jour, Hj *m.* man-day
homocarpe *a.* homocarpous
homogame *a.* homogamous
homogénat *m.* homogenate
homogène *a.* homogeneous
homogénéisation *f.* homogenization
homologation *f. Hort.* licensing, registration, (of new product, variety etc.)
homologué,-e *a.* approved, authorized, licenced, registered; **dose homologuée** standard (recommended) rate
homologuer *v.t.* to approve, to authorize (use of), to licence, to register (new product, variety etc.)
homophylle *a.* homophyllous
homozygote *a.* homozygous; *n. m.* homozygote
hoplie *m.* hoplia beetle (*Hoplia* sp.)
hoplocampe *m.* sawfly (*Hoplocampa* sp.); **h. commun du prunier** plum sawfly (*H. flava*); **h. noir du prunier** black plum sawfly (*H. minuta*); **h. du poirier** pear sawfly (*H. brevis*); **h. des pommes** apple sawfly (*H. testudinea*)
horizon *m.* horizon; *Ped.* (soil) horizon; **un h. B argileux** a clayey B horizon; **h. humifère** humus-bearing horizon
***hormin** *m.* dragonmouth (*Horminum pyrenaicum*)
hormonage *m.* growth regulator application
hormonal,-e *a., m. pl.* **-aux** hormonal; **équilibre hormonal** hormonal balance
hormone *f.* hormone; **h. de bouturage** rooting hormone; **h. de croissance** growth hormone, growth substance; **h. juvénile** juvenile hormone; **h. de nouaison** fruit-setting hormone; **h. d'enracinement, h. de racinage** rooting hormone; **h. de synthèse** synthetic hormone
hormoné,-e *a.* treated with, containing (plant)

hormones; **cire hormonée** grafting wax containing plant hormones

***hors saison** *a. inv.* out of season; **produits horticoles hors saison** out-of-season horticultural produce

***hors-sol** *a. inv.* soilless, outside the soil; **culture hors-sol** soilless culture; **pépinière hors-sol** container nursery, nursery of plants in pots

hortensia *m.* **(des jardins)** hydrangea (*Hydrangea macrophylla* = *H. hortensia*); **h. grimpant** climbing hydrangea (*H. petiolaris*); **h. de Virginie** *H. arborescens*

horticole *a.* horticultural

horticulteur *m.* horticulturist

horticultural,-e *a., m. pl.* **-aux** horticultural

horticulture *f.* horticulture, gardening; **exposition d'horticulture** flower show; **h. d'agrément** amenity horticulture; **h. comestible** market-gardening; **h. florale** ornamental horticulture; **h. professionnelle** commercial horticulture; **h. de serre, h. sous verre** horticulture under glass

hortillon *m.* 1. = **hortillonnage**. 2. market gardener who cultivates a **hortillonnage**

hortillonnage *m.* 1. drained marsh used for market gardening. 2. cultivation of a **hortillonnage**

hortillonneur *m.* market gardener (= **hortillon** 2)

hosta *m.* *Hosta* sp., funkia, plantain lily

hôte *m.* host (plant etc.)

***hotte** *f.* basket carried on the back; hood (in laboratory)

***hottée** *f.* contents of a **hotte**, basketful

***hotter** *v.t.* to carry in a **hotte**

H.P.D. see **hybride**

hottonie *f.* *Hottonia* sp., including water violet (*H. palustris*) (= **millefeuille aquatique**)

***houblon** *m.* hop (*Humulus lupulus*); **h. du Japon** Japanese hop (*H. japonicus*); **tige de houblon** hop bine

***houblonnier,-ière** *a.* relating to hops; **pays houblonnier** hop-growing country; *n. m.* hop-grower; *n. f.* hop garden

***houe** *f.* (i) hoe, harrow; **h. à bras** wheeled hoe; **h. à cheval** horse hoe; **h. à main** hand hoe; **h. de Paris** Paris hoe; **h. rotative** rotary hoe; **h. vigneronne** *Vit.* horse hoe; (ii) **h. à billonner** *Afr.* ridging hoe; **h. à long manche** *Afr.* long-handled hoe; **h. à manche courbe** *Afr.* curved-handled hoe; **petite houe** *Afr.* small, weeding hoe

***houement** *m.* hoeing, harrowing

***houer** *v.t.* to hoe, to harrow

***houlette** *f.* (small) garden trowel

***houlque** *f.* *Holcus* sp. grass; **h. laineuse** Yorkshire fog (*H. lanatus*)

***houppette** *f.* small tuft

***houppe** *f.* bunch, tuft

***houppier** *m.* crown (of tree)

***houssaie, houssaye** *f.* area planted with holly, holly plantation

***housse** *f.* (protecting) bag; loose cover

houstonia *f.* *Houstonia*, a genus of hardy herbaceous perennials

***houx** *m.* holly (*Ilex* sp.); **h. commun** common holly (*I. aquifolium*); **h. panaché** variegated holly; **h. du Paraguay** maté (*I. paraguensis*); **petit houx** butcher's broom (*Ruscus aculeatus*) (= **fragon**)

hoya *m.* *Hoya* sp., honey-plant, wax-flower

***hoyau** *m., pl.* **-aux** mattock, grubbing hoe

huilage *m.* *PD* topple disorder of tulips

huile *f.* oil; **h. d'abrasin** tung oil; **h. d'anthracène** anthracene oil, tar oil; **huile de ben** oil of ben (see also **ben ailé**); **h. blanche** white oil; **h. de cade** cade oil (see also **cade** (1)); **h. de coco, h. de coprah** (*Coconut*) coconut oil; **h. essentielle** essential oil; **h. de girofle** oil of cloves; **h. de lin** linseed oil; **h. de marmotte** marmotte oil (see also **abricotier**); **h. de palme** *OP* palm oil; **h. de palmiste** *OP* palm kernel oil; **h. végétale** vegetable oil; **h. vierge** virgin oil

huilerie *f.* oil-expressing factory

huileux,-euse *a.* oily

humate *m.* humate

humectation *f.* moistening, wetting

humecter *v.t.* to moisten, to wet

humide *a.* humid, moist, wet

humidificateur *m.* humidifier

humidifier *v.t.* to damp, to humidify

humidimètre *m.* moisture meter; **h. à neutrons** neutron moisture meter

humidité *f.* humidity, moisture; **"supporte l'humidité"** "tolerates waterlogging"; **h. absolue** absolute humidity; **h. atmosphérique** atmospheric humidity; **h. relative** relative humidity; **teneur en humidité** moisture content

humidomètre *m.* humidometer

humifère *a.* humus-bearing; **horizon humifère** humus-bearing layer (of soil)

humification *f.* humification

humifié,-e *a.* humified (of soil)

humique *a.* humic; **matière humique** humic matter

humus *m.* humus; **h. acide** acid humus; **h. doux** mild humus; **h. durable** stable humus

hybridation *f.* hybridization; **h. interspécifique** interspecific hybridization; **h. somatique** somatic hybridization

hybride *a.* & *n. m.* hybrid; **h. complexe** complex hybrid; **h. de greffe** graft hybrid, chimera; **h. producteur direct (H.P.D.)** *Vit.* hybrid direct producer; **h. testé** proven hybrid

hybrider *v.t.* to hybridize; **s'hybrider** *v. p.* (of plant) to be cross-fertilized
hybrideur *m.* breeder
hybridisme *m.* hybridism
hydne *m. Fung. Hydnum* sp. (see also **pied-de-mouton**)
hydrangea, hydrangelle *f. Hydrangea* sp. (see **hortensia**)
hydraste *m.* golden seal (*Hydrastis canadensis*)
hydratation *f.* hydration
hydrate (*m.*) **de carbone** carbohydrate (= **hydrocarbure**)
hydraté,-e *a.* hydrated, hydrous
hydrazide (*f.*) **maléique** maleic hydrazide, MH
hydrique *a.* hydrous; **absorption hydrique** water absorption
hydrocarboné,-e *a.* hydrocarbonic; **réserves hydrocarbonées** carbohydrate reserves
hydrocarbure *m.* hydrocarbon; **h. insaturé** unsaturated hydrocarbon
hydroculture *f.* hydroponics, water (sand) culture
hydrogène *m.* hydrogen. **h. sulfuré** hydrogen sulphide
hydrolase *f.* hydrolase
hydrologie *f.* hydrology
hydrolysable *a.* hydrolysable
hydrolysat *m.* hydrolysate
hydrolyse *f.* hydrolysis; **h. acide** acid hydrolysis
hydrolysé,-e *a.* hydrolized
hydrolyser *v.t.* to hydrolize
hydromel (vineux) *m.* mead
hydromorphe *a.* hydromorphic; **sol hydromorphe** hydromorphic soil
hydromorphie *f.* hydromorphism
hydropériodicité *f.* alternation of wet and dry periods
hydrophile *a.* hydrophilous
hydrophylle *f. Hydrophyllum* sp.
hydrophyte *f.* hydrophyte
hydroponique *a.* hydroponic; **culture hydroponique** hydroponics
hydrorétenteur *m.* water-retaining product
hydrosolubilité *f.* solubility in water
hydrosoluble *a.* soluble in water
hydrostat *m.* hydrostat
hydrotropisme *m.* hydrotropism
hydrovase *m.* container for (indoor) hydroponic culture
hydroxylation *f.* hydroxylation
hydroxyle *m.* hydroxyl
hygiène *f.* hygiene; **h. culturale** crop sanitation
hygromètre *m.* hygrometer; **h. à condensation, h. à point de rosée** dew-point hygrometer
hygrométrie *f.* hygrometry
hygrométrique *a.* hygrometric, hygrometrical; **degré hygrométrique** relative humidity
hygrophile *a.* hygrophile, hygrophilous
hyménium *m. Fung.* hymenium; *pl.* hymenia
hyménoptère *a.* hymenopterous; *n. m.* hymenopteron, hymenopterous insect; *pl.* Hymenoptera
hyperparasite *a.* hyperparasitic; *n. m.* hyperparasite
hyperparasitisme *m.* hyperparasitism
hyperplasie *f.* hyperplasia
hypersensibilité *f.* hypersensitivity
hypersensible *a.* hypersensitive
hypertonique *a.* hypertonic
hypertrophie *f.* hypertrophy
hyphe *m.* hypha; **h. d'infection** infectious hypha
hypne *f.* hypnea moss
hypoagressif,-ive *a. PD.* hypovirulent
hypobiote *m.* = **porte-greffe** (*q.v.*)
hypocotyle *m.* hypocotyl
hypoderme *m.* hypodermis
hypogé,-e *a.* hypogeal, hypogeous
hypogyne *a.* hypogynous
hyponomeute *m. Yponomeuta (Hyponomeuta)* sp., ermine moth; **h. du pommier** small ermine moth (*Y. malinella*)
hypotonique *a.* hypotonic
hypovirulent,-e *a.* hypovirulent
hysope (officinale) *f.* hyssop (*Hyssopus officinalis*)
hystéranthié,-e *a.* hysteranthous, growing leaves after flowering

I

ibéride *f.*, **ibéris** *m. Iberis* sp., candytuft
iboga *m. Iboga* sp.
icaque *f.* coco-plum
icaquier *m.* coco-plum tree (*Chrysobalanus icaco*) (= **chrysobalanier, prune coton, prune d'icaque**)
ichtyotoxique *a.* & *n.* toxic to fish, fish poison
iciquier *m. Icica viridiflora*, one of the sources of **élémi** (*q.v.*)
if *m.* yew (*Taxus baccata* and other spp.); **i. d'Irlande** Irish yew (*T. baccata stricta*); **i. puant** stinking yew (*Torreya* sp.)
igname *f.* yam (*Dioscorea* sp.); **i. ailée** white yam (*D. alata*); **i. de Chine** Chinese yam (*D. batatas*); **i. de Cayenne** Guinea yam (*D. cayenensis*)
ignatie *f. Ignatius* bean tree (*Strychnos ignatii*) (see also **fève de St. Ignace**)
ilama *m.* cherimoya of the lowlands (*Annona diversifolia*) (see also **chérimolier**)
ilang-ilang *m.* ilang-ilang (*Cananga odorata*) (= **ylang-ylang**)
illégitime *a. Hort.* rogue (of plant)
imaginal,-e *a.*, *m. pl.* **-aux** *Ent.* imaginal; **vie imaginale** imaginal life
imago *f. Ent.* imago
imantophyllum *m. Imantophyllum* (= *Clivia*) sp.
imbibition *f.* 1. soaking. 2. absorption
imbouchable *a.* non-blocking (of pipe etc.)
imbouturable *a.* unrootable (of cutting)
imbriqué,-e *a.* imbricate
imbuvable *a.* undrinkable
immangeable *a.* uneatable, inedible
immature *a.* immature; **graines immatures** immature seed
immergé,-e *a.* immersed
immortelle *f.* everlasting flower; **i. des Alpes** = **edelweiss** (*q.v.*); **i. annuelle** immortelle (*Xeranthemum annuum*); **i. de Belleville** = **i. annuelle** (*q.v.*); **i. blanche** = **i. de Virginie** (*q.v.*); **i. à bouquets** *Helichrysum orientale*; **i. (à bractées)** immortelle (*H. bracteatum*); **i. jaune** = **i. à bouquets** (*q.v.*); **i. des neiges** = **edelweiss** (*q.v.*); **i. de Virginie** pearly everlasting (*Anaphalis (Antennaria) margaritacea*)
immunité *f.* immunity
immunoélectrophorèse *f.* immunoelectrophoresis
immuno-enzymatique *a.* immunoenzymatic; **technique immuno-enzymatique** immunoenzymatic technique
imparipenné,-e *a.* imparipinnate
impatiens, impatiente *f. Impatiens* sp., balsam (= **balsamine**)
impénétrable *a.* impenetrable; **i. à l'eau** impervious to water
impérata *m.* lalang, spear grass (*Imperata cylindrica*)
impératoire *f. Astrantia* sp., masterwort
imperméabilité *f.* impermeability, imperviousness
imperméable *a.* impermeable, impervious, waterproof
impietratura *f. Cit.*, *PD* impietratura
implant *m.* inoculum, implant
implantation *f.* planting, implantation, establishment, introduction (of plant)
implanté,-e *a.* planted, implanted, introduced (of plant)
implanter *v.t.* to plant, to introduce, to establish; **s'implanter** to take root
improductif,-ive *a.* unproductive
improductivité *f.* unproductiveness
impureté *f.* impurity
imputrescible *a.* imputrescible, rot-proof
inactivé,-e *a.* inactivated
inadaptation *f.* maladjustment; **i. au climat local** lack of adaptation to the local climate
inadapté,-e *a.* maladjusted, not adapted
inaltérable *a.* undeteriorating; **i. à l'eau** unaffected by water
inbreeding *m.* inbreeding
incarvillée *f. Incarvillea* sp.
incassable *a.* unbreakable
incinération *f.* incineration
incisé,-e *a.* incised
inciser *v.t.* to incise, to cut, to score (hyacinths etc.)
inciseur *m.* tool for ringing or tapping trees, girdling tool
incision *f.* incision, incising, cut, cutting; **i. annulaire** *Hort.* the operation of (bark) ringing, girdling and the resulting cut; **i. en coin** cut for coin graft; **i. longitudinale** longitudinal incision; **i. en T** T cut; **i. transversale (au-dessus d'un œil)** notch (above a bud); **i. transversale (au-dessous d'un œil)** nick (below a bud)
incliné,-e *a.* inclined, slanted, sloping

incliner *v.t.* to incline, to slant, to slope; **s'incliner** to slant, to slope
inclus,-e *a. Bot.* included
incolore *a.* colourless
incombant,-e *a.* incumbent
incommercialisable *a.* unsaleable
incompatibilité *f. Hort.* incompatibility; **i. (faible)** uncongeniality; **i. au greffage** graft incompatibility; **i. pollinique** pollen incompatibility
incorporation *f.* incorporation
incorporé,-e *a. Hort.* incorporated
incorporer *v.t.* to incorporate
incubation *f.* incubation; *Mush.* spawn running
inculte *a.* uncultivated
inculture *f.* non-cultivation; lack of cultivation
incurvé,-e *a.* incurved
indéhiscence *f.* indehiscence
indéhiscent,-e *a.* indehiscent
indélébile *a.* indelible
indémaillable *a.* non-run, tear resistant (of fabric, netting etc.)
indemne *a.* undamaged, uninjured; *Hort.* **i. de virus (connus), i. de maladies de dégénérescence connues** virus-free
indéracinable *a.* that cannot be uprooted, ineradicable
indéterminé,-e *a.* indeterminate; **à croissance indéterminée** of indeterminate growth
index *m.* index; **i. de plugging** *Rubb.* plugging index
indexage *m.* indexing
indexation *f.* indexing; **technique d'indexation** indexing technique
indicateur,-trice *a.* indicatory; *n. m.* **i. de gelée** frost indicator; *Hort.* indicator plant (= **plante indicatrice**)
indice *m.* index (number), coefficient; **i. climatique** climatic index; **i. d'obstruction** *Rubb.* plugging index; **i. de pouvoir chlorosant (I.P.C.)** chlorosis-inducing index; **i. de salinité** salt index; **i. (de surface) foliaire** leaf area index
indifférencié,-e *a.* undifferentiated; **cellules indifférenciées** undifferentiated cells
indifférent,-e *a.* indifferent; **i. à la longueur du jour** day-neutral (of plant)
indigène *a.* indigenous, native, home-grown (of produce); **flore indigène** indigenous flora
indigo *m.* 1. indigo. 2. indigo plant (= **indigotier**)
indigofera *m. Indigofera* sp.
indigotier *m.* indigo plant (*Indigofera tinctoria*); *Fung.* blue boletus (*Boletus cyanescens*) (see also **bolet, cèpe, nonette**)
indivis,-e *a.* undivided, entire
indolique *a.* indole; **alcaloïdes indoliques** indole alkaloids
induction *f.* induction; **i. florale** floral induction
induit,-e *a.* induced
indupliqué,-e *a.* induplicate (see also **préfloraison**)
indusie *f.* indusium
induré,-e *a.* indurated, hardened
industrie *f.* industry; **i. de la conserve** preserving industry; **i. maraîchère** market-gardening (industry)
inerme *a.* inerm(ous), without spines or prickles, spineless
infécond,-e *a.* barren, sterile, unfruitful
infécondité *f.* barrenness, sterility, unfruitfulness
infecté,-e *a.* infected, contaminated
infecter *v.t.* to infect, to contaminate, to pollute
infectieux,-euse *a.* infectious
infection *f.* infection, contamination; **i. latente** latent infection
infectivité *f.* infectivity
inféodé,-e *a.* (of insect, pathogen etc.) specific to, tied to, limited to
infère *a. Bot.* inferior
infertile *a.* infertile, unfruitful
infestation *f.* infestation
infesté,-e *a.* infested
infester *v.t.* to infest
infiltration *f.* infiltration
infiltrer *v.t.* to infiltrate; **s'infiltrer** to infiltrate, to percolate
infléchir *v.t.* to bend, to curve; **s'infléchir** to bend, to become curved, to become inflexed
inflorescence *f.* inflorescence; **i. femelle** female inflorescence; **i. mâle** male inflorescence
inflorescentiel,-elle *a.* relating to the inflorescence; **bourgeons inflorescentiels** flower buds
infra(-)rouge *a.* & *n. m.* infra-red (light)
infranchissable *a.* impassable (of barrier, fence etc.)
infructueux,-euse *a.* unfruitful, unprofitable
infrutescence *f.* infructescence
infundibuliforme *a.* infundibular, funnel-shaped
infusion *f.* infusion
inhabituel,-elle *a.* unusual
inhiber *v.t.* to inhibit, to retard
inhibiteur,-trice *a.* inhibiting, inhibitory; *n. m.* inhibitor; **i. de métabolisme** metabolism inhibitor; **i. de la nitrification** nitrification inhibitor
inhibitif,-ive *a.* inhibitory
inhibition *f.* inhibition; **i. de croissance** growth inhibition
initiation *f.* initiation; **i. florale** flower initiation
injecter *v.t.* to inject
injection *f.* injection; **i. sous pression** pressure injection
innocuité *f.* innocuousness, harmlessness
inoculat *m.* inoculum

inoculation *f.* inoculation; *Mush.* spawning
inoculum *m.* inoculum; **i. primaire** primary inoculum
inocybe *m. Fung. Inocybe* sp.
inodore *a.* scentless
inondable *a.* liable to inundation, flooding
inondation *f.* inundation, flood; flood irrigation
inondé,-e *a.* inundated, flooded
inonder *v.t.* to inundate, to flood
inorganique *a.* inorganic
inquilinisme *m.* inquilinism
insaponifiable *a.* unsaponifiable
insaturation *f.* non-saturation
insaturé,-e *a.* unsaturated; **acides gras insaturés** unsaturated fatty acids
insécable *a.* indivisible
insecte *m.* insect; **i. auxiliaire** beneficial insect; **i. butineur** foraging insect; **i. nuisible** harmful insect, pest; **i. parfait** imago; **i. pollinisateur** pollinating insect; **i. rampant** crawling insect; **i. utile** useful, beneficial insect; **i. volant** flying insect
insecticide *a.* insecticidal; *n. m.* insecticide; **i. de contact** contact insecticide; **i. d'ingestion** stomach poison; **i. de translocation, i. systémique** systemic insecticide
insectifuge *a.* & *n. m.* insect repellent
insectivore *a.* insectivorous; *n. m.* insectivore
insertion *f.* insertion; **i. des feuilles** leaf insertion
insignificant,-e *a.* insignificant
insipide *a.* insipid, tasteless
insipidité *f.* insipidity, tastelessness
insolation *f.* insolation; *Hort., Arb.* sun scald; **fente d'insolation** sun crack
insoluble *a.* insoluble
instable *a.* unstable
instrument *m.* instrument, implement, tool; **instruments de jardinage** garden tools; **instruments de taille** pruning tools
insuffisance *f.* insufficiency, deficiency
intempéries *f. pl.* bad weather
intensif,-ive *a.* intensive; **culture intensive** intensive cultivation
intensification *f.* intensification
intensité *f.* intensity
interaction *f.* interaction
interbillon *m.* furrow
intercalaire *a.* intercalary; *Hort.* interplanted, grown in the interrows; (of screens etc.) placed between rows of plants
interceps *m. Vit.* row cultivator
interchangeable *a.* interchangeable
intercompatibilité *f.* intercompatibility
interfécondation *f.* inter-crossing
interfertilité *f.* interfertility
interfoliacé,-e *a.* interfoliaceous
interfoliaire *a.* interfoliar
intergénérique *a.* intergeneric; **hybride intergénérique** intergeneric hybrid
interligne *m. Hort.* interrow
intermédiaire *a.* intermediate, intervening; *n. m.* 1. *Hort.* intermediate stempiece, stembuilder, interstock. 2. *Comm.* middleman
internervaire *a.* internerval
internodal,-e *a., m.pl.* **-aux** internodal
interphase *f.* interphase
interplantation *f.* interplanting
interrang *m.* interrow
interruption *f.* interruption
interspécifique *a.* interspecific
interstérile *a.* mutually infertile; **variétés interstériles** mutually infertile varieties
interstérilité *f.* mutual infertility
intertropical,-e *a., m. pl.* **-aux** intertropical
intervalle *m.* interval, gap, space; *Hort., Plant.* space between plants in the row
intervention *f.* intervention; *Hort.* **i. printanière** spring cultivation, spring treatment
intine *m.* intine
intisy *m.* intisy (*Euphorbia intisy*)
intoxication *f.* poisoning, toxicity; **i. manganique** manganese toxicity
intracellulaire *a.* intracellular
intrant *m.* input
intratissulaire *a.* occurring in tissue
introduction *f.* introduction; **i. différée** delayed introduction
introgression *f. PB* introgression
introrse *a.* introrse
intumescence *f.* gall, wart
inuline *f.* inulin
invendable *a.* unsaleable
invendu,-e *a.* unsold
involucelle *m.* involucel
involucre *m.* involucre
involuté,-e *a.* involute
involution *f.* involution
invulnérable *a.* invulnerable; **i. au vent** windproof
iode *m.* iodine
iodure *m.* iodide
ion *m.* ion; **ions azotés** nitrogen ions; **ions minéraux** mineral ions
ionisant,-e *a.* ionizing; **rayonnements ionisants** ionizing radiation
ionisation *f.* ionization, irradiation
I.P.C. see **indice**
ipéca *m.* = **ipécacuana** (*q.v.*)
ipécacuana *m.* ipecacuanha (*Cephaelis ipecacuanha*)
ipomée *f. Ipomoea* sp., morning glory (= **volubilis**); **i. bonne-nuit** moonflower (*I. bona-nox*) (= **belle-de-nuit**)
IR = **infrarouge** (*q.v.*)
iriartée *f. Iriartea (Socratea)* sp.

iridacées *f. pl.* Iridaceae
iridoïde *m.* iridoid
iris *m.* iris; **i. barbu** bearded iris; **i. sans barbe** beardless iris; **i. (des jardins)** common iris (*Iris germanica*); **i. d'Angleterre** English iris (*I. xiphioides*); **i. deuil** mourning iris (*I. susiana*); **i. d'Espagne** Spanish iris (*I. xiphium*); **i. de Florence** *I. pallida*; **i. gigot** gladdon, stinking iris (*I. foetidissima*); **i. de Hollande** Dutch iris (*I.* × *hollandica*); **i. japonais** *I. kaempferi*; **i. (jaune) des marais, i. faux-acore** yellow flag (*I. pseudacorus*); **i. nain** *I. chamaeiris*; **i. de Provence** *I. pallida*; **i. de Sibérie** *I. sibirica*; **i. de Tanger** *I. tingitana*; **i. à tête de serpent** snake's head (*Hermodactylus tuberosus*); **i. tigré** *Belamcanda chinensis* (= **fleur de léopard)**
irradiateur (*m.*) **(de plantes)** horticultural lamp
irradiation *f.* irradiation, lighting; **i. gamma** gamma irradiation; **i. ionisante** ionizing irradiation; **i. répétée** recurrent irradiation
irradié,-e *a.* irradiated
irradier *v.i.* to radiate; *v.t.* to irradiate
irrégulier,-ère *a.* irregular
irrenversable *a.* that cannot be knocked over or upset
irrigable *a.* irrigable; **périmètre irrigable** irrigable area
irrigant,-e *a.* irrigating, *n. m.* irrigator (person)
irrigateur,-trice *a.* irrigating; *n. m.* hose, pump for watering
irrigation *f.* irrigation, flooding; **i. d'appoint** supplementary irrigation, irrigation to make up deficiency; **i. par aspersion** sprinkler irrigation; **i. de complément** supplementary irrigation; **i. au goutte à goutte** drip irrigation, trickle irrigation; **i. fertilisante** fertigation; **i. localisée** drip irrigation; **i. en planches, i. à la raie, i. par ruissellement** furrow irrigation; **i. souterraine** sub-irrigation; **i. par submersion** basin irrigation; **i. de surface** surface irrigation (see also **goutte)**
irrigatoire *a.* irrigatory
irriguer *v.t.* to irrigate, to flood
irritabilité *f.* irritability, sensitiveness
irvingie *f. Irvingia* sp. (see also **beurre de dika**)
isobare *f.* isobar
isoélectrique *a.* isoelectric
isohyète *f.* isohyet
isolat *m.* isolate
isolation *f.* insulation; **i. thermique** thermal insulation
isolé,-e *a.* isolated: insulated
isolement *m.* isolation; **distance d'isolement** *PB, PD* isolation distance
isomère *m.* isomer
isostémone *a.* isostemonous
isotherme *f.* isotherm
isotonique *a.* isotonic
isovaleur *f. Stat.* isovalue
iule *m.* 1. catkin, spike. 2. millipede
ive *f. Iva* sp.
ivoire végétal *m.* ivory nut palm (*Phytelephas macrocarpa*)
ivoirien,-enne *a.* of the Ivory Coast
ivraie (*f.*) **enivrante** darnel (*Lolium temulentum*); **i. raide** *L. rigidum*
ixia, ixie *f. Ixia* sp., African corn lily
ixora *f. Ixora* sp., West Indian jasmine
ixtle *m.* ixtle, fibre from *Agave* spp.

J

jaborandi *m. Pilocarpus jaborandi, P. pinnatifolius* and their products
jaborosa *m.* jaborosa (*Jaborosa integrifolia*)
jaboticaba *m.* jaboticaba (*Myrciaria cauliflora, M. trunciflora*); **j. de Sao Paulo** (*M. jaboticaba*)
jacapucayo *m.* monkey pod (*Lecythis ollaria*) (see also **noix**)
jacaranda *m.* jacaranda (*Jacaranda* sp.)
jacée *f.* knapweed (*Centaurea jacea* and other spp.); **j. des blés** cornflower (*C. cyanus*); **petite jacée** wild pansy (*Viola tricolor*)
jachère *f.* fallow, uncultivated land; **en jachère** fallow(ed); **j. arborée, j. arbustive** *Afr.* bush fallow
jacinthe *f.* 1. hyacinth (*Hyacinthus* sp.); **j. des bois** = **j. sauvage** (*q.v.*); **j. d'Espagne** *H. amethystinus*; **j. des fleuristes, j. d'Orient** common hyacinth (*H. orientalis*); **j. romaine** roman hyacinth (*H. orientalis* var. *albulus*); **j. parisienne** *H. orientalis* var. *provincialis*; **j. des Pyrénées** = **j. d'Espagne** (*q.v.*). 2. Other genera: **j. du Cap** summer hyacinth (*Galtonia candicans*) (= **muguet géant**); **j. d'eau** water hyacinth (*Eichhornia crassipes*); **j. musquée** musk hyacinth (*Muscari moschatum*); **j. du Pérou** Cuban lily (*Scilla peruviana*) (= **scille du Pérou**); **j. plumeuse** feather hyacinth (*Muscari comosum* var. *monstrosum*); **j. sauvage** bluebell (*Endymion non-scriptus*); **j. de Sienne** = **j. plumeuse** (*q.v.*)
jacksonie *f. Jacksonia* sp.
jacobée *f. Jacobaea* (*Senecio*) *elegans*, jacobaea
jacquier *m.* = **jaquier** *q.v.*
jacquinie *f. Jacquinia* sp., including *J. ruscifolia*, a tropical flowering shrub
jagré *m. Sug.* jaggery
jalap *m.* jalap (*Ipomoea purga* and *Ipomoea* sp.)
jalon *m.* surveyor's staff, stake, marker, sighting mark, planting peg
jalousie *f.* sweet william (*Dianthus barbatus*) (= **œillet de(s) poète(s)**)
jambe *f.* leg; **j. de force** strut; *PD* **j. noire (de la pomme de terre)** blackleg (of potato) (*Erwinia atroseptica*) (see also **pied noir**)
jambon *m.* ham; **j. des jardiniers, j. des jardins** *Coll.* evening primrose (*Oenothera biennis*) (= **onagre bisannuelle**)
jambose *f.* rose apple
jambosier *m.* rose apple tree, jambosa (*Eugenia jambos*) (= **pomme rose**); **j. de Michéli** pitanga cherry (*E. uniflora*); **j. rouge** pomerac, mountain apple (*E. malaccensis*) (= **jamelac**)
jambul *m.* jambu (*Eugenia* sp.)
jamelac, jamlac *m.* pomerac, mountain apple tree (*Eugenia malaccensis*) (= **jambosier rouge**)
jamelong *m.*, **jamelongue** *f.* java plum (*Eugenia jambolana*)
jamerose *f.*, **jamerosier** *m.* = **jambose, jambosier** (*q.v.*)
jaque *m.* jackfruit
jaquier *m.* jackfruit tree (*Artocarpus heterophyllus* = *A. integrifolia*)
jardin *m.* garden; **j. d'agrément** pleasure garden, flower garden; **j. alpin** rock garden; **j. anglais, j. à l'anglaise** landscape garden; **j. botanique** botanical garden; **j. d'eau** water garden; **j. d'essai** experimental, trial garden or plot; **j. d'exposition** show garden; **j. familial** home garden, allotment (see also **j. ouvrier**); **j. fleuriste, j. de fleurs** flower garden; **j. à la française** formal garden; **j. fruitier** (walled) fruit garden; **j. grainier** *Plant.* plot, area reserved for (hybrid) seed production; **j. d'hiver** winter garden; **petit jardin d'hiver** small indoor greenhouse; **j. japonais** (small) Japanese garden; **j. mixte** garden with vegetables and fruit trees; **j. de mousses** moss garden; **j. ouvrier** allotment (see also **j. familial**); **j. des plantes** = **j. botanique** (*q.v.*); **j. potager** vegetable garden, kitchen garden; **j. public** public garden; **j. de rocaille** = **j. alpin** (*q.v.*); **j. semencier** = **j. grainier** *q.v.*; **j. de temple** temple garden
jardinage *m.* gardening
jardiner *v.i.* to garden
jardinerie *f.* garden-centre
jardinet *m.* small garden
jardinier,-ière *a.*, **culture jardinière** garden culture; **plante jardinière** garden plant; *n.* gardener; **j.-chef** head gardener; **j. fleuriste** flower grower; **(j.) maraîcher** market gardener; **j. paysagiste** landscape gardener; **j. pépiniériste** tree and shrub nurseryman; **j. trois branches** *Coll.* flower, fruit and vegetable grower; **j. quatre branches** *Coll.*

flower, fruit, vegetable and greenhouse grower

jardinière *f.* 1. flower stand, rectangular pot (holder) (for indoor growing), window-box. 2. market gardener's cart. 3. pruning saw

jargeau *m.* (*Canada*) = **jarosse** (*q.v.*)

jarosse *f.*, **jarousse** *f.* lesser chickpea (*Lathyrus cicera*)

jarovisation *f.* vernalization

jarrah *m.* jarrah (*Eucalyptus marginata, E. rostrata*); bois de jarrah, jarrah wood

jarre *f.* (large glazed) earthenware jar

jasione *f.* *Jasione* sp., including sheep's-bit (*J. montana*)

jasmin *m.* jasmine (*Jasminum* sp. and other genera); **j. d'Arabie** Arabian jasmine (*J. sambac*); **j. en arbre** syringa (*Philadelphus coronarius*); **j. blanc** = **j. officinal** (*q.v.*); **j. du Cap** Cape jasmine (*Gardenia florida*); **j. commun** wild jasmine (*J. fruticans*); **j. d'Espagne** = **j. à grandes fleurs** (*q.v.*); **j. étoilé** Chinese jasmine (*Trachelospermum jasminoides*); **j. à fleurs nues** = **j. d'hiver** (*q.v.*); **j. à grandes fleurs** Spanish jasmine (*J. grandiflorum*); **j. d'hiver** winter jasmine (*J. nudiflorum*); **j. d'Italie** Italian jasmine (*J. humile*); **j. jaune** = **j. commun** (*q.v.*); **j. jonquille** *J. odoratissimum*; **j. officinal** common jasmine (*J. officinale*); **j. sauvage des Alpes** = **j. commun** (*q.v.*); **j. trompette** trumpet creeper (*Campsis (Tecoma) radicans*); **j. de Virginie** (= **técoma jasmin de Virginie**) = **j. trompette** (*q.v.*)

jasside *m.* *Ent.* jassid

jassidés *n. m. pl.* *Ent.* Jassidae, jassids

jatropha *f.* *Jatropha* sp., especially *J. curcas* (= **médicinier, purghère**)

jauge *f.* gauge; *Hort.* trench, including trench for heeling in; **mettre en jauge** to heel in

jaunâtre *a.* yellowish

jaune *a.* yellow; *a. inv.* **j. citron** lemon yellow; **j. paille** straw yellow; **j. serin** canary yellow

jaune d'œuf *m.* canistel, egg-fruit (*Lucuma rivicoa*)

jaunet d'eau *m.* yellow water lily (*Nuphar lutea*)

jaunir *v.t.* to colour yellow; *v.i.* to become yellow, to turn yellow

jaunisse *f.* yellowing, chlorosis (= **jaunissement**); **j. de la betterave** beet yellows; **j. des nervures** vein yellows; **j. du chou** black rot (*Xanthomonas campestris*) (= **nervation noire**); **j. des nervures** vein yellows (virus)

jaunissement *m.* yellowing, ripening, chlorosis (= **jaunisse**); **j. des asters** aster yellows; **j. du pêcher** peach yellows virus; **j. des nervures du poirier** pear vein yellows virus; **jaunissement mortel** (*Coconut*) lethal yellowing

javart *m.* chestnut canker, chestnut blight (*Endothia parasitica*)

javelage *m.* laying in heaps, laying in swaths

javeler *v.t.* to lay in heaps, to lay in swaths

javelle *f.* bundle of vine twigs, hop poles etc., swath

javellisé,-e *a.* chlorinated; **eau javellisée** chlorinated water

jeannette (blanche) *f.* pheasant's eye (*Narcissus poeticus*); **j. jaune** daffodil (*N. pseudo-narcissus*) (= **jonquille**)

jeffersonie *f.* *Jeffersonia* sp., twin-leaf

jéquirity *m.* jequirity bean (*Abrus precatorius*) (= **abrus**)

jérose *f.* rose of Jericho (*Anastatica hierochuntica*)

jersiais,-e *a.* of Jersey

jet *m.* 1. throwing; **j. de pelle** earth thrown up by spade. 2. jet, gush, stream; **j. d'eau** fountain. 3. jet, nozzle; **j. brouillard** misting nozzle (see also **buse**); **j. pinceau** pencil jet. 4. *Hort.* shoot, sprout; **j. de houblon** hop shoot; **j. (souterrain)** sucker

jeter *v.t.* to throw; *Hort.* to produce buds; **j. des racines** to grow roots, to strike

jeune clone *m.* *Cit.* young line (= **lignée nucellaire**)

jojoba *m.* jojoba (*Simmondsia californica*)

jonc *m.* rush (*Juncus* sp.) and other genera; **canne de jonc** Malacca cane; **j. à balais** reed (*Phragmites communis*); **j. des crapauds** toad rush (*J. bufonius*); **j. des chaisiers** bulrush (*Scirpus lacustris*); **j. à feuilles panachées** striped bulrush (*Scirpus tabernaemontani zebrinus*); **j. fleuri** flowering rush (*Butomus umbellatus*); **j. glauque** hard rush (*J. inflexus* = *J. glaucus*); **j. d'Inde** rattan (*Calamus* sp.); **j. japonais** glaucous bulrush (*Scirpus tabernaemontani*); **j. des jardiniers** = **j. glauque** (*q.v.*); **j. odorant** sweet flag (*Acorus calamus*)

jonchaie *f.* 1. rush bed. 2. canebrake

jonquille *a. inv.* pale yellow; *n. f.* **(petite) jonquille** jonquil (*Narcissus jonquilla*), daffodil (*N. pseudo-narcissus*) (= **faux narcisse**); **grande jonquille** campernelle (*N. odorus*) (= **campernelle**); **j. des prés** daffodil

joubarbe *f.* *Sempervivum* sp., houseleek; **j. des toits** common houseleek (*S. tectorum*); **petite joubarbe** white stonecrop (*Sedum album*)

jouet-du-vent *m.* silky apera (*Apera spica-venti*)

jour *m.* day, daylight; **de j. court** (of plant) short day; **de j. long** long day; **en jours courts** in short days; **en jours longs** in long days

journée *f.* day, daytime; **en journée(s) courte(s)** in short days; **en journée(s) longue(s)** in long days; **journée(s) d'étude** meeting of specialists or scientists, subject day(s)

juanulloa *m. Juanulloa*, a genus of tropical solanaceous plants grown as greenhouse ornamentals

jubéa *m.*, **jubée** *f.* wine palm (*Jubaea* sp.); **j. remarquable** Chile wine palm (*J. spectabilis*) (= *J. chilensis*) (= **cocotier du Chili**)

juglans *m. Juglans* sp., walnut tree

jujube *f.* jujube fruit

jujubier *m.* jujube tree (*Ziziphus* sp.); **j. de l'Afrique tropicale** Indian jujube (*Z. mauritiana*); **j. de l'Asie** = **j. commun** *q.v.*; **j. de la Berberie** lotus tree (*Z. lotus*); **j. commun** Chinese jujube (*Z. jujuba* Mill); **j. de Palestine, j. épine du Christ** (*Z. spina-Christi*)

julienne *f. Hesperis* sp., rocket; **j. des dames, j. des jardins** dame's violet, sweet rocket (*H. matronalis*); **j. jaune** double yellow rocket (*Barbarea vulgaris* var. *flore pleno*) (see also **herbe de Sainte-Barbe**); **j. de Mahon** Virginian stock (*Malcomia maritima*) (= **giroflée de Mahon, mahonille**); **j. d'Orient** *H. violacea*

jumelé,-e *a.* coupled, arranged in pairs

juniperus *m. Juniperus* sp., juniper tree (see also **genévrier**)

jus *m.* juice; **j. d'agrume** citrus juice; **j. de canne (à sucre)** (sugar) cane juice (= **vesou**); **j. de fruit** fruit juice; **j. de groseilles** currant juice; **j. de pomme** apple juice; **j. de raisin** grape juice; **j. absolu** *Sug.* absolute juice; **j. de première pression** *Sug.* first expressed juice; **j. résiduel** Sug. residual juice

jusquiame (*f.*) **(noire)** henbane (*Hyoscyamus niger*); **j. dorée** *H. aureus*

jussiée *f. Jussieua* sp., primrose willow

justicia *m.*, **justicie** *f. Justicia* sp. (see also **carmantine**)

jute *m.* jute (*Corchorus olitorius*) (= **corète potagère**); **j. bâtard** kenaf (*Hibiscus cannabinus*)

juter *v.i.* to be juicy

juteur *m. Irrig.* trickler (= **baveur**)

juteux,-euse *a.* juicy

jutosité *f.* juiciness

juvénile *a.* juvenile; *Hort.* **caractère juvénile** juvenile character; **stade juvénile** juvenile phase

juvénilité *f.* juvenility

K

kadsura *m. Kadsura* sp.
kaempferie *f. Kaempferia* sp.
kaempférol *m.* kaempferol
kafta *m.* = **khat** (*q.v.*)
kaïnite *f.* kainit
kaki *m.* persimmon (*Diospyros kaki*) (see also **plaqueminier**); **kakis âpres** varieties in which fruits require after-ripening; **kakis doux** varieties in which fruits do not require after-ripening
kaladana *m.* kaladana (*Ipomoea hederacea*)
kalanchoé *m.* kalanchoë (*Kalanchoë* sp., including *K. blossfeldiana*)
kali *m.* kali, glasswort (*Salsola kali*)
kalmia *m.*, **kalmie** *f. Kalmia* sp.; **k. à feuilles étroites** sheep laurel (*K. angustifolia*); **k. à larges feuilles** calico bush (*K. latifolia*)
kamala *m.* kamala tree (*Mallotus philippensis*)
kanya *m. Afr.* fat from butter tree (*Pentadesma butyracea*)
kaolinite *f.* kaolinite
kapok *m.* kapok
kapokier *m.* silk-cotton tree (*Ceiba pentandra*) (= **fromager**); **k. (à fleurs rouges), k. rouge** red-flowering silk-cotton tree (*Bombax costatum*)
kapulasan *m.* pulasan (*Nephelium mutabile*)
karamatsu *m.* Japanese larch (*Larix leptolepis*)
karatas, karata *m. Karatas plumieri*, an ornamental bromeliad
karé *m.* = **karité** *q.v.*
karité *m.* shea-nut tree (*Butyrospermum paradoxum*) (= *Afr.* **arbre à beurre**) (see also **beurre de karité**)
kava, kawa *f.*, **kawa-kawa** *m.* kava pepper (*Piper methysticum*) and drink prepared from it
kennédya *m.*, **kennédie** *f. Kennedya* sp., coral creeper, Australian bean flower
kentrophylle *m.* distaff thistle (*Carthamus (Kentrophyllum) lanatus*); safflower (*C. tinctorius*)
kepel *m.* **(grand) kepel** (*Tea*) fish leaf
kermès *m.* kermes, scale insect; **k. coquille** mussel scale (*Lepidosaphes ulmi*); **k. du pêcher** peach scale, brown scale (*Eulecanium (Parthenolecanium) corni*); **k. du poirier** = **cochenille rouge du poirier** (*q.v.*); **k. de la vigne** woolly (currant) scale (*Pulvinaria vitis*); **k. virgule** = **k. coquille** (*q.v.*)
kerria *m.*, **kerrie** *f. Kerria japonica* jew's mallow (= **spirée du Japon**)
ketmie *f. Hibiscus* sp.; **k. comestible** okra (*H. esculentus*) (= **gombo**); **k. des jardins** tree hollyhock (*H. syriacus*); **k. musquée** musk mallow (*H. abelmoschus*) (= **ambrette**); **k. rose de Chine** hibiscus (*H. rosa-sinensis*)
kettoul *m.* = **kitool** (*q.v.*)
khat *m.* khat (*Catha edulis*)
kiesérite *f.* kieserite
kikuyu *m.* kikuyu grass (*Pennisetum clandestinum*)
kinétine *f.* kinetin
kitaïbélie *f. Kitaibelia vitifolia*
kitool *m.* kitool (fibre from *Caryota urens*)
kiwi *m. Hort.* Chinese gooseberry, kiwifruit (*Actinidia chinensis*) (= **groseille de Chine**)
kiwiculteur *m.* Chinese gooseberry grower
klugie *f. Klugia* sp.
knautia *m. Knautia* sp., including field scabious (*K. arvensis*)
knépier *m.* honey-berry (*Melicocca bijuga*) (= **quenette**)
kochia *m. Kochia* sp.
kœlreuteria *m.* **kœlreutérie** *f. Koelreuteria* sp.; **k. paniculé** *K. paniculata*
kokam *m.* kokam (*Garcinia indica*)
koko *m. Afr. Gnetum africanum*, whose leaves are used as a pot-herb
kola *m.* kola (= **cola**); **noix de kola** kola nut
kolatier *m.* kola tree (*Cola* sp.) (= **colatier** *q.v.*)
koumquat *m.* = **kumquat** (*q.v.*)
kralen *m. pl.* cormlets, cormels
krameria *m.*, **kramérie** *f. Krameria* sp.
krasnozem *m. Ped.* krasnozem
krigeage *m. Stat.* kriging
kumquat *m.* kumquat (*Fortunella* sp.); **k. de Hong Kong** *F. hindsii*; **k. oblong, k. à fruit oblong, k. ovale, k. Nagami** oval kumquat (*F. margarita*); **k. rond, k. Maruni** round kumquat (*F. japonica*)
kyste *m.* cyst; **k. de nématode** nematode cyst
kystique *a.*, **nématode kystique** cyst nematode

L

label *m.* label, seal of approval; l. de purité purity label (of seed etc.)
labelle *m.* labellum; *pl.* labella (see also **lèvre**)
labiacées *n. f. pl.* Labiatae
labié,-e *a.* labiate, lipped; *n. f. pl.* **labiées** Labiatae
labile *a.* labile, unstable
lablab *m.* hyacinth bean (*Dolichos lablab*)
labour *m.* tilling, tillage, ploughing; **l. à la bêche** digging; **l. à plat** one-way ploughing; **l. en profondeur** deep ploughing, subsoiling
labourable *a.* arable, fit for ploughing
labourage *m.* tilling, tillage, ploughing
labourer *v.t.* to plough, to till, to cultivate; **l. en profondeur** to plough deeply, to subsoil
laboureuse *f.* 1. motor cultivator. 2. *Ent.* mole-cricket (= **courtilière**)
laburne, laburnum *m.* laburnum (*Laburnum* sp.)
laccifère *a.* lac-bearing
lâche *a.* loose, slack
lâcher 1. *v.t.* to release. 2. *n. m.* release; **l. expérimental** *PC* experimental release
lacinié,-e *a. Bot.* laciniate, jagged
lacinifolié,-e *a. Bot.* laciniate-leaved
laciniure *f. Bot.* laciniation
lacis *m.* network; **l. de racines** network of roots
lactaire *m. Lactarius* sp.; **l. délicieux** saffron milk cap (*L. deliciosus*); **l. à lait abondant** *L. volemus* (= **vachette**); **l. poivré** *L. piperatus*; **l. aux tranchées** woolly milk cap (*L. torminosus*); **l. velouté** velvet cap (*L. vellereus*)
lactifère *a.* lactiferous; **plantes lactifères** plants with milky latex
lacuneux,-euse *a.* lacunose
lagéniforme *a.* lageniform
lagerstroemia *m. Lagerstroemia* sp., including *L. indica* crape myrtle, Indian lilac, and *L. flos-reginae* queen's flower
lagerstrémie, lagerstroemie *f.* crape myrtle (*Lagerstroemia indica*)
lagunaire *a.* relating to lagoons
lagurus, lagure *m. Lagurus* sp., hare's tail grass (*L. ovatus*)
laîche *f.* sedge (Cyperaceae)
laine *f.* wool; **l. de roche** rock wool; **l. de verre** glass wool
laineux,-euse *a. Bot.* woolly, tomentose
lait *m.* milk; **l. de coco** coconut milk; **l. de chaux** whitewash
laiteron *m.* sowthistle (*Sonchus* sp.); **l. âpre, l. épineux** spiny sowthistle (*S. asper*); **l. des champs** perennial sowthistle (*S. arvensis*); **l. maraîcher, l. potager** common sowthistle (*S. oleraceus*)
laiteux,-euse *a.* milky; **suc laiteux** milky sap
laitier *m.* 1. milkwort (*Polygala* sp.) 2. slag, cinders (of iron furnace); **l. basique** basic slag
laitue *f.* 1. lettuce (*Lactuca sativa*) (cos lettuce in Switzerland); **l. beurre** standard lettuce with green leaves and firm heart; **l. à couper** cutting lettuce which does not heart, resistant to bolting; **l. d'été** summer lettuce; **l. pommée** cabbage lettuce, heading lettuce; **l. romaine** cos lettuce; **l. scarole** broad-leaved endive (*Cichorium endivia*); **l. vireuse** acrid lettuce (*L. virosa*); l. vivace perennial lettuce (*L. perennis*) (see also **batavia, romaine**). 2. **l. d'eau** water lettuce (*Pistia stratiotes*)
laize *f.* width (cloth, plastic etc.); difference between actual and nominal width
lala *m. Sug.* aerial branching
lambeau *m.* flap (e.g. of bark); **l. d'épiderme foliaire** epidermal leaf strip
lambourde *f. Arb.* fruit shoot; spur; **pommier à lambourdes** spur-type apple tree
lambrotte *f.* thin bunch of grapes
lambruche, lambrusque *f.* wild vine
lame *f.* blade; **l. de coupe** mower blade; *Bot.* lamina, blade; *Mush.* gill
lamelle *f. Mush.* gill; **lamelles décurrentes** decurrent gills
lamellé,-e *a.*, **lamelleux,-euse** *a.* lamellate
lamier *m. Lamium* sp.; **l. amplexicaule** henbit (*L. amplexicaule*); **l. blanc** white deadnettle (*L. album*); **l.galéobdolon, l. jaune** yellow archangel (*L. (Lamiastrum) galeobdolon*); **l. hybride** cut-leaved dead-nettle (*L. hybridum*); **l. panaché** spotted dead nettle (*L. maculatum*); **l. pourpre** purple dead-nettle (*L. purpureum*) (see also **ortie**)
lampe *f.* lamp; **l. infra-rouge** infra-red lamp; **l. à incandescence** incandescent lamp
lampourde *f. Xanthium* sp., cocklebur
lampsane (*f.*) **commune** nipplewort (*Lapsana communis*) (= **lapsana**)

lance *f.* lance; **l. d'arrosage, l. à eau** water-hose nozzle; **en fer de lance** *Bot.* lanceolate
lance de Christ *m.* gipsywort (*Lycopus europaeus*) (see also **lycope**)
lance-flammes *m. inv.* flame gun, flame thrower
lancéiforme = **lanciforme** (*q.v.*)
lancéolé,-e *a.* lanceolate, spear-shaped
lanciforme *a.* lanciform, lance-shaped
lande *f.* sandy moor, heath, heathland
langage *m.* language; **l. des fleurs** *m.* language of flowers
langsat *m.* langsat (*Lansium domesticum*)
langue-de-bœuf *f., pl.* **langues-de-bœuf** 1. beefsteak fungus (*Fistulina hepatica*) (= **fistuline**). 2. = **langue-de-cerf** (*q.v.*)
langue-de-cerf *f., pl.* **langues-de-cerf** hart's tongue fern (*Phyllitis scolopendrium*) (= **scolopendre**)
languette *f.* small tongue, strip; *Bot.* ligule; **fleur en languette** ligulate floret (= **fleuron ligulé**)
lanière *f.* thin strip of material, thong
lanifère *a.* laniferous
laniflore *a.* laniflorous
lanigère *a.* lanigerous; **puceron lanigère** woolly aphis (*Eriosoma lanigerum*)
lantana, lantanier *m.* *Lantana* sp., lantana; **l. épineux** lantana (*L. camara*)
lanterne *f.* lantern; **l. chinoise, l. japonaise** bladder cherry, Chinese lantern plant (*Physalis alkekengi, P. franchetii* and related spp.)
lanthane *m. Chem.* lanthanum
lapin *m.* rabbit; **l. de garenne** wild rabbit (*Oryctolagus cuniculus*)
laportea *m.* *Laportea* sp.; the leaves of some tropical species are eaten as vegetables
lapsana *m.* lapsana, nipplewort (*Lapsana communis*) (= **lampsane**)
laque *f.* lac; **gomme laque** shellac, gum lac
laqué *m. Fung.* deceiver (*Laccaria laccata*)
laquier *m.* lacquer tree (*Rhus verniciflua*); wax tree (*R. succedanea*) (= **vernis du Japon**)
lardage *m. Mush.* spawning
larder *v.t. Mush.* to spawn
lardon *m. Mush.* portion of spawn
largeur *f.* width
larme-de-Job *f., pl.* **larmes-de-Job** Job's tears (*Coix lachryma-jobi*)
larmille *f.* = **larme-de-Job** (*q.v.*)
larvaire *a.* larval; **vie larvaire** larval life
larve *f.* larva, grub; **l. cuir** leatherjacket (Tipulidae); **l. fil de fer** wireworm (Elateridae); **larves de noctuelles** cutworms, noctuid moth caterpillars (= **vers gris**)
larvicide *m.* larvicide
latania, latanier *m.* *Latania* sp.; **l. de Bourbon** Bourbon palm (*Livistona chinensis* = *Latania borbonica*)
latence *f.* latency
latent,-e *a.* latent
latéral,-e *a., pl.* **-aux** lateral; **bourgeon latéral** lateral bud
latérite *f.* laterite; *Coll.* ironstone cap
latéritisation *f.* laterization (= **ferrallitisation**)
latex *m.* latex
lathyrisme *m.* lathyrism
laticifère *a.* lacticiferous, latex-bearing; *n.m.* laticifer
latifolié,-e *a.* latifoliate, broad-leaved
latosol *m.* latosol (= **sol ferrallitique**)
latte *f.* lath, batten, slat; *Vit.* = **long bois** (*q.v.*)
lattis *m.* laths, lathing; **clôture en lattis** lath fence
lauracé,-e *a.* lauraceous; **lauracées** *n. f. pl.* Lauraceae
laurelle *f.* oleander (= **laurier-rose** *q.v.*)
lauréole *f.* *Daphne* sp., garland flower, spurge laurel
laurier *m.* laurel; **l. d'Alexandrie** Alexandrian laurel (*Danaë racemosa*); **l. (-)amandier, l. (-) amande** cherry laurel (*Prunus laurocerasus*) (= **laurier-cerise**); **l. américain** mountain laurel (*Kalmia latifolia*); **l. d'Apollon** = **l. commun** (*q.v.*); **l. benzoin** spice bush (*Lindera benzoin*); **l. des bois** spurge laurel (*Daphne laureola*); **l. de Californie** California olive (*Umbellularia californica*); **l. commun** bay laurel (*Laurus nobilis*); **l. franc** = **l. commun** (*q.v.*); **l. jaune** yellow oleander (*Thevetia peruviana* = *T. nereifolia*); **l. noble** = **l. commun** (*q.v.*); **l. des poètes** = **l. commun** (*q.v.*); **l. du Portugal** Portugal laurel (*Prunus lusitanica*); **l. de Saint-Antoine** rosebay willow herb (*Chamaenerion angustifolium*); **l. sassafras** sassafras (*Sassafras officinale* = *S. albidum*)
laurier-cerise *m., pl.* **lauriers-cerises** cherry-laurel (*Prunus laurocerasus*)
laurier-rose *m., pl.* **lauriers-rose(s)** oleander (*Nerium oleander*) (= **laurelle, laurose**)
laurier-sauce *m., pl.* **lauriers-sauce** bay laurel (*Laurus nobilis*) (used for flavouring) (= **laurier commun, l. franc**)
laurier-tin, laurier-thym *m., pl.* **lauriers-tin, lauriers-thym** laurustinus (*Viburnum tinus*) (= **viorne laurier-thym**) (see also **boule-de-neige**)
laurier-tulipier *m., pl.* **lauriers-tulipiers** laurel magnolia (*Magnolia grandiflora*)
laurifolié,-e *a.* laurel-leaved, resembling laurel
laurose *m.* oleander (*Nerium oleander*) (= **laurier-rose**)
lavage *m.* washing; scrubbing (of gas)
lavande *f.* lavender (*Lavandula* sp.); **l. aspic** spike lavender (*L. latifolia*); **l. femelle** = **l. officinale** (*q.v.*); **l. mâle** = **l. aspic** (*q.v.*); **l. de mer** sea lavender (*Limonium* sp.); **l.**

officinale common lavender (*L. vera* = *L. angustifolia*); **l. vraie** = **l. officinale** (*q.v.*); *a. inv.* **bleu lavande** lavender blue
lavanderaie, lavanderie *f.* lavender field or plantation
lavandiculteur *m.* lavender grower
lavandière *f. Hort.* lavender field or plantation
lavandin *m.* lavandin (*Lavandula* × *hybrida*)
lavatère *f. Lavatera* sp.; **l. en arbre** tree mallow (*L. arborea*) (see also **mauve**); **l. à grandes fleurs, l. rose** annual mallow (*L. trimestris*); **l. d'Hyères** tree lavatera (*L. olbia*)
lave-racines *m. inv.* = **laveur** *q.v.*
laveur *m. Hort.* machine for washing roots
lawsonia *m. Lawsonia* sp., henna
lécanium *m. Ent. Lecanium* sp.; **l. de la vigne** brown scale (*Parthenolecanium corni*)
lécithine *f.* lecithin
lécythis *m. Lecythis* spp., including sapucaia nut, paradise nut, monkey pot (*L. zabucajo*) (= **marmite de singe**)
ledum *m. Ledum* sp.
léger,-ère *a.* light (weight etc.); **sol léger** light soil
légèreté *f.* lightness
légitimité *f.* legitimacy (of seeds etc.)
légume *m.* 1. vegetable; **l. à gousse** pulse; **légumes blanchis** blanched vegetables; **légumes à bulbes, légumes bulbeux** bulbous vegetables (onion, garlic etc.); **légumes de conserve** vegetables for canning; **légumes déshydratés** dehydrated vegetables; **légumes desséchés** dried vegetables; **légumes feuilles, légumes feuillus, légumes foliacés** leafy vegetables (lettuce, spinach etc.); **légumes "fins"** unusual, exotic, gourmet, luxury-market vegetables; **légumes fleurs** inflorescent vegetables; **légumes frais** vegetables eaten soon after picking; **légumes fruits** fruit vegetables (melon, tomato etc.); **légumes "lourds"** ordinary vegetables; **légumes racines** root vegetables; **légumes secs** dried vegetables (peas, beans etc.); **légumes surgelés** frozen vegetables; **légumes tiges** stalk vegetables; **légumes tubéreux** tuberous vegetables (potato, Jerusalem artichoke etc.); **légumes verts** green vegetables, greens; **légumes vivaces** perennial vegetables (asparagus, globe artichoke etc.). 2. *Bot.* legume, pod
légumier,-ière *a.* relating to vegetables; **culture légumière** vegetable crop, vegetable growing; **jardin légumier** vegetable garden
légumineuse *f.* legume; **l. sèche** pulse; **légumineuses** *n. f. pl.* Leguminosae
légumineux,-euse *a.* leguminous
légumiste *n.* vegetable grower
leitneria *m. Leitneria floridana*, corkwood
lémonange *m.* orange × lemon hybrid
lemon-grass *m.* lemon grass (*Cymbopogon citratus*) (see also **citronnelle**)
lénacile *m.* lenacil
lenticellaire *a.* lenticular; **parasites lenticellaires** lenticel pests
lenticelle *f.* lenticel
lenticellé,-e *a.* lenticellate
lenticulaire *a.* lenticular
lentille *f.* lentil (*Lens culinaris*); **l. d'Espagne** chickling pea (*Lathyrus sativus*) (= **gesse cultivée**); **l. d'eau** duckweed (*Lemna* sp.) (= **canillée**); **l. de terre** *Afr.* Kersting's groundnut (*Kerstingiella geocarpa*)
lentisque *m.* lentisc, mastic-tree (*Pistacia lentiscus*) (= **arbre au mastic**)
léontice *f. Leontice* sp., lion's leaf
léonotis *m. Leonotis* sp., lion's ear
léontodon *m. Leontodon* sp.
léontopodium *m. Leontopodium* sp.
léonure *m.*, **léonurus** *m.* motherwort (*Leonurus cardiaca*)
léopoldinia *m. Leopoldinia* sp.
lépidoptère *a.* lepidopterous; *n. m.* lepidopter; *pl.* Lepidoptera, butterflies and moths
lépidote *a.* lepidote, scaly, scurfy
lépiote *f. Lepiota* sp.; **l. (élevée)** parasol mushroom (*L. procera*) (= **coulemelle**); **l. déguenillée** shaggy parasol (*L. rhacodes*); **l. pudique** nutshell lepiota (*L. pudica* = *L. naucina*)
lèpre *f. PD* diseases such as **blanc** (*q.v.*); algal disease (*Cephaleuros virescens*); **l. explosive** *Cit.* attack by *Brevipalpus obovatus*
lérot *m.* garden dormouse (*Eliomys quercinus*)
lésion *f.* lesion; **l. phylloxérique** *Vit.* lesion caused by phylloxera attack
lessivage *m. Ped.* leaching
lessivat *m. Ped.* leachate
lessivé,-e *a. Ped.* leached; **sol lessivé** leached soil
lessiver *v.t. Ped.* to leach; **se lessiver** to leach, to become leached
létal, léthal,-e, *a.*, *m. pl.* **-aux** lethal; **dose létale** lethal dose
létalité, léthalité *f.* lethality
letchi *m.* = **litchi** *q.v.*
léthal,-e *a.* = **létal** (*q.v.*)
leucanthème, leucanthemum *m.* **(des prés)** ox-eye daisy (*Chrysanthemum leucanthemum*) (= **grande marguerite**)
lève *f. Vit.* (*Switzerland*) = **palissage** (*q.v.*)
levée *f.* 1. raising, lifting; **l. de dormance** breaking of dormancy; **l. d'inhibition** release of inhibition. 2. *Hort.* germination, emergence, shooting, sprouting (of plant); **l. tardive de graminées** late grass emergence. 3. **l. de terre** bank-bed, earth bank
lève-gazon *m. inv.* turfing-iron

lever *v.i.* (of plants) to shoot, to germinate, to sprout
lèvogyre *a.* laevorotatory
lèvre *f. Bot.* lip, labium
lévulose *m.* (often used in *f.*) laevulose
levure *f.* yeast; **l. de bière** brewer's yeast
levuré,-e *a.* yeast-containing, with added yeast
levurer *v.t.* to add yeast to
li *m.* = **poire orientale** (*q.v.*)
liage *m.* tying, binding
liaison *f. PB* linkage
liane *f.* liana, creeper; **l. de poivrier** pepper vine (*Piper nigrum*); **l. à agoutis** sweet cup (*Passiflora maliformis*) (= **pomme calebasse**); **l. (à) réglisse** jequirity (*Abrus precatorius*) (= **abre, jéquirity**); **l. serpent** scarlet passion flower (*Passiflora coccinea*)
liant *m.* binding material, binder
liatris *m.* **liatride** *f. Liatris* sp.
liber *m.* phloem
libérien,-enne *a.* pertaining to the phloem; **cellules libériennes** phloem cells
libéro-ligneux,-euse *a.* relating to phloem and xylem; **faisceau libéro-ligneux** vascular bundle
libocèdre *m.* incense cedar (*Libocedrus decurrens*)
lichen *m.* lichen
liège *m.* cork, corky tissue; *PD* bitter pit (of apple)
liégeux,-euse *a.* corky; **racines liégeuses** *PD* corky root; **taches liégeuses** *PD* bitter pit (of apple)
lien *m.* tie; **l. pour arbres** tree tie
lierre *m.* **(commun)** ivy (*Hedera helix*) (= **l. des bois**); **couvert de lierre** covered with ivy; **l. en arbre** *H. helix arborescens*; **l. d'Irlande** *H. helix hibernica*; **l. japonais** Boston ivy (*Parthenocissus tricuspidata*) (= **vigne vierge de Veitch**); **l. terrestre** ground ivy (*Glechoma hederacea*) (= **gléchome**)
lieu(-)dit *m., pl.* **lieux(-)dits** *Vit.* etc. named place
lièvre *m.* hare (*Lepus* sp.)
liff *m.* (*Date*) fibre (= **fibrillum**)
ligature *f.* tie
lignane *m.* lignan
ligne *f.* line, row; **en ligne** in a row, in rows; **entre les lignes** between rows; **planter à la ligne** to plant with a line; **semis en ligne** sowing in line; **l. de plantation** planting line; **l. de canne** *Sug.* row of cane
ligné,-e *a. Bot.* lineate
lignée *f. PB* line, strain; **l. nucellaire** *Cit.* young line (= **jeune clone**); **vieille lignée** *Cit.* old line
ligneux,-euse *a.* ligneous, woody; **plante ligneuse** woody plant; *n. m.* woody plant; **l. d'ornement** woody ornamental; *Sug.* fibre, insoluble dry matter
lignification *f.* lignification
lignifié,-e *a.* lignified
lignifier *v.t.* to lignify; **se lignifier** to become lignified, to become woody
lignine *f.* lignin
lignoïde *m.* lignoid
lignosité *f.* woodiness
ligulaire *f.*, **ligularia** *m. Ligularia* sp.
ligule *f.* ligule
ligulé,-e *a.* ligulate, strap-shaped; **fleuron ligulé** ligulate floret, ray-floret (see also **languette**)
liguliflore *a.* liguliflorous
ligustrum *m. Ligustrum* sp., privet
lilas *a. inv.* lilac, lilac-coloured; *n. m.* 1. lilac (*Syringa* sp.); **l. de Bretschneider** = **l. velu** (*q.v.*); **l. commun** common lilac (*S. vulgaris*); **l. de fortune** *S. oblata*; **l. de Josika** Hungarian lilac (*S. josikaea*); **l. de Perse** Persian lilac (*S. persica*); **l. de Transylvanie** = **l. de Josika** (*q.v.*); **l. varin** Rouen lilac (*S.* × *chinensis*); **l. velu** *S. villosa*. 2. Other genera. **l. d'été** butterfly bush (*Buddleia* sp.) (= **buddleia**); **l. des Indes** Indian lilac (*Lagerstroemia indica*) (= **lagerstroemie**); bead tree (*Melia azedarach*)
liliacé,-e *a.* liliaceous; *n. f. pl.* Liliaceae
lilial,-e *a., m. pl.* **-aux** lily-like, lily-white
lilium *m. Lilium* sp., lily
limace *f.* slug; **l. grise** (grey) field slug (*Agriolimax (agrestis) reticulatus*); **l. horticole, l. noire** garden slug (*Arion hortensis*)
limaçon *m.* 1. snail. 2. *Hort.* snail medic (*Medicago scutellata*)
limbe *m. Bot.* limb, lamina, frond, blade
limbifère *a.* limbate
lime *f.* lime fruit; **l. (acide), l. (antillaise), l. (commune), l. (mexicaine), l. (vraie)** (common) lime fruit (= **citron vert à petits fruits**); **l. à gros fruits, l. Tahiti** Persian, Tahiti lime fruit (= **citron vert**) (see also **limettier**)
limequat *m. Cit.* limequat (lime × kumquat)
limette (*f.*) **plate** Mediterranean sweet lemon (*Citrus limetta*); **l. ronde** sweet lime of India (*C. limettioides*)
limettier *m.* lime tree; **l. (vrai), l. (mexicain)** (common) lime tree (*Citrus aurantifolia*); **l. doux** sweet lime of India (*C. limettioides*); **l. à gros fruits** Persian lime, Tahiti lime (*C. latifolia*); **l. Rangpur** Rangpur lime (*C. limonia*) (see also **lime**)
limiteur *m.* limiting device, regulator; **l. de débit d'eau** water regulator, water meter
limon *m.* 1. *Ped.* silt: loam. 2. alluvium. 3. lime fruit

limonette (*f.*) **de Marrakech** *Citrus* hybrid grown in Morocco

limoneux,-euse *a.* 1. *Ped.* silty, loamy; **limon limoneux** silty loam. 2. growing in mud (of plant). 3. alluvial

limonier, limonnier *m.* Rangpur lime tree (*Citrus limonia*) (see also **lime**)

lin *m.* (*Linum* sp.) flax; **l. de la Nouvelle Zélande** New Zealand hemp (*Phormium tenax*)

linacé,-e *a.* linaceous; *n. f. pl.* Linaceae

linaigrette *f. Eriophorum* sp., cotton grass; **l. à larges feuilles** broad-leaved cotton grass (*E. latifolium*)

linaire *f. Linaria* sp. and others, toadflax; **l. bâtarde** round-leaved fluellen (*Kickxia spuria*); **l. commune** common toadflax (*L. vulgaris*); **l. cymbalaire** ivy-leaved toadflax (*L. cymbalaria*); **l. élatine** fluellen (*Kickxia elatina*); **l. pourpre** *L. bipartita*

linéaire *a.* linear

linguiforme *a.* linguiform

lipase *f.* lipase

lipide *m.* lipid

lipidique *a.* lipid(ic); **constitution lipidique** lipidic constitution

lipochrome *m.* lipochrome

lipogenèse *f.* lipogenesis

lipolyse *f.* lipolysis

lipolytique *a.* lipolytic

lipophile *a.* lipophilic

liquidambar *m.* liquidambar (*Liquidambar* sp.)

liquide *m.* liquid; **l. de la coque (du péricarpe) de la noix de cajou** cashewnut shell liquid, CSL

liquidité *f.* liquidity

liquoreux,-euse *a.* liqueur-like, sweet and soft (of wine)

lis *m.* 1. lily (*Lilium* sp.); **l. à bandes dorées** = **l. doré du Japon** *q.v.*; **l. des Bermudes** *L. longiflorum* var. *eximium*; **l. blanc (commun)** white lily, Madonna, Bourbon lily (*Lilium candidum*); **l. brillant** *L. speciosum*; **l. de Constantinople** *L. chalcedonicum* (= **martagon écarlate**); **l. doré du Japon** golden-rayed lily of Japan (*L. auratum*); **l. isabelle** *L.* × *testaceum* (= *L. isabellinum*); **l. léopard** leopard lily (*L. pardalinum*) (= *L. roezlii*); **l. de la Madone** = **l. blanc** (*q.v.*); **l. martagon, l. des montagnes** martagon lily, Turk's cap lily (*L. martagon*) (= **martagon**); **l. orangé** orange lily (*L. (bulbiferum) croceum*); **l. de Pompone** scarlet Pompona lily (*L. pomponium*); **l. des Pyrénées** Pyrenean lily (*L. pyrenaicum*); **l. rose du Japon** Japanese lily (*L. speciosum rubrum*); **l. royal** regal lily (*L. regale*); **l. superbe** *L. superbum*; **l. soufré** *L. sulphureum*; **l. tigré** tiger lily (*L. tigrinum*); **l. tigré de Californie** = **l. léopard** (*q.v.*); **l. turban** = **l. de Pompone** *q.v.* 2. Other genera. **l. belladone** belladonna lily (*Amaryllis belladonna*) (= **amaryllis**); **l. bleu du Nil** African lily (*Agapanthus umbellatus*) (= *A. africanus*); **l. croix de Saint-Jacques** Jacobaea lily (*Sprekelia formosissima*) (= **amaryllis croix de Saint-Jacques**); **l. d'eau, l. des étangs** water lily (*Nymphaea* sp.) (= **nénufar**); **l. d'Illyrie** *Pancratium illyricum*; **l. du Japon** Guernsey lily (*Nerine sarniensis*) (= **amaryllis de Guernesey**); **l. jaune** Chinese day-lily (*Hemerocallis citrina*); **l. de mai** = **l. des vallées** *q.v.*; **l. des marais** sweet flag (*Acorus calamus*) (= **acore odorant**); **l. maritime** Mediterranean lily (*Pancratium maritimum*) (= **pancratier**); **l. matthiole** = **l. maritime** (*q.v.*); **l. du Mexique** Mexican lily (*Hippeastrum reginae*); **l. narcisse** = **l. maritime** (*q.v.*); **l. du Pérou** Peruvian lily (*Alstroemeria aurantiaca*) (= *A. aurea*); **l. des sables** = **l. maritime** *q.v.*; **l. de Saint Bruno** St. Bruno's lily (*Paradisea liliastrum*); **l. des vallées** lily-of-the-valley (*Convallaria majalis*) (= **muguet**)

liseré, liséré,-e *a.* edged, bordered; *n. m.* edge, border

liseron *m.* bindweed, convolvulus; **l. des champs** field bindweed (*Convolvulus arvensis*); **l. des haies, grand liseron** bellbine (*Calystegia sepium*); **l. scammonée** scammony (*Convolvulus scammonia*) (= **scammonée**)

lisette *f.* vine grub (*Adoxus obscurus*)

lisier *m.* liquid manure

lissage *m.* smoothing

lisse *a.* smooth

lisser *v.t.* to smooth

listère *f.* twayblade (*Listera ovata*)

lit *m.* bed; **l. de semences** seed bed

litchi, li-tchi *m.* litchi, lychee (*Litchi sinensis*); **l. chevelu** rambutan (*Nephelium lappaceum*) (= **ramboutan**); **l. ponceau** longan (*Euphoria (Nephelium) longana*) (= **longanier**)

lithocyste *m.* lithocyst

litière *f.* (stable) litter; **l. (de feuilles)** leaf litter

livèche *f.* lovage (*Levisticum officinale*) (= **ache de montagne**); **l. du Péloponèse** striped hemlock (*Molopospermum cicutarium*)

lixiviation *f.* lixiviation, leaching

lixivier *v.t.* to lixiviate, to leach

loasa *m. Loasa* sp., Chile nettle

lobe *m. Bot.* lobe

lobé,-e *a.* lobed, lobate

lobélia *m.*, **lobélie** *f.* lobelia (*Lobelia* sp.); **lobélie cardinale** cardinal flower (*L. cardinalis*); **lobélie enflée** Indian tobacco (*L. inflata*)

localisateur (*m.*) **d'engrais** attachment for fertilizer placement

localisation *f.* localization, placement (of fertilizer); **l. de la fumure** fertilizer placement
localisé,-e *a.* localized, in patches
localiser *v.t.* to localize, to locate
localité *f.* locality, place
loche *f.* slug, especially (grey) field slug (*Agriolimax reticulatus*) (see also **limace**)
locule *f.* loculus
loculicide *a.* loculicidal
locus *m., pl.* **loci** *PB* locus
locuste *f.* locust; **l. migratrice** migratory locust (see also **criquet, sauterelle**)
lodoïcée *f. Lodoicea* sp.; **l. des Seychelles** (*L. sechellarum*) (= **cocotier des Maldives**), double coconut, coco-de-mer
loess *m.* loess
loessique *a.* loessial
loganberry *m.* loganberry (*Rubus* × *loganobaccus*)
loganie *f. Logania* sp.
loge *f. Bot.* loculus, small space or cavity
logette *f. Bot.* small loculus
loir *m.* (edible) dormouse (*Glis glis*) (see also **lérot**)
loisir *m.* leasure
lombric *m.* earthworm (*Lumbricus* sp.) (= **ver de terre**)
lombricide *a.* worm-killing; *n.m.* worm-killing compound
lombricompost *m.* worm compost, vermicompost
lombricompostage *m.* worm composting, vermicomposting
lombriculture *f.* worm culture, vermiculture
longanier *m.* longan (*Euphoria (Nephelium) longana*)
long bois *m. Vit.* fruit cane (with more than two buds): arched cane, bow (= **latte**)
longévité *f.* longevity
longicorne *a.* & *n.m. Ent.* longicorn
longifolié,-e *a.* long-leaved
long-scion, lonsion *m.* elongated current year's shoot: elongated one-year-old grafted tree
longueur *f.* length; **l. du jour** length of day, day length; **l. d'onde** wavelength
lonicera *m. Lonicera* sp., honeysuckle
loofa, loofah *m.* luffa (*Luffa cylindrica*)
lophosperme *m. Lophospermum* sp. (= *Maurandya* sp.)
lopin (*m.*) **de terre** piece, plot of ground: allotment
loranthacées *n. f. pl.* Loranthaceae
loranthe *f. Loranthus* sp.
loriforme *a.* loriform, shaped like a strap
loriot *m.* oriole (*Oriolus oriolus*)
lotier *m. Lotus* sp. and related spp.; **l. corniculé** bird's foot trefoil (*L. corniculatus*); **l. cultivé** asparagus pea (*L. tetragonolobus* = *Tetragonolobus purpureus*); **l. odorant** blue fenugreek (*Trigonella caerulea*) (= **mélilot bleu**)
lotus *m.* lotus (*Nelumbium* and *Nymphaea* spp.); **l. blanc d'Egypte** Egyptian lotus (*Nymphaea lotus*); **l. bleu d'Egypte** blue lotus of the Nile (*Nymphaea caerulea*); **l. bleu des Indes** *Nymphaea stellata*; **l. des Indes, l. d'Egypte, l. égyptien, l. rose** East Indian lotus (*Nelumbium nucifera*); **l. jaune d'Amérique** American lotus (*Nelumbium lutea*) (see also **nélombo, nénufar**)
louchet *m.* draining spade
loup *m. Vit.* water shoot, sucker
loupe *f.* lens, magnifying glass; *Hort.* excrescence on tree
lourd,-e *a.* heavy; **terre lourde** heavy land
lubrifiant,-e *a.* lubricating; *n. m.* lubricant
lubrification *f.* lubrication
lubrifier *v.t.* to lubricate
lucet (*m.*) **musqué** ugni shrub, Chilean guava (*Myrtus ugni*) (= **goyavier du Chili, myrte musqué**)
lucuma, lucume *m. Lucuma* sp.; **l. à grandes feuilles** broad-leaved lucuma (*Pouteria (Lucuma) multiflora*)
luisant,-e *a.* shining, shiny, glossy
lulo (*m.*) **(de Colombie)** lulo, naranjilla (*Solanum quitoense*)
lumen *m.* lumen
lumière *f.* light; **l. artificielle** artificial light; **l. blanche** white light; **l. du jour** daylight; **l. rouge (clair)** red light; **l. rouge lointain** far red light; **l. du soleil** sunlight; **l. rouge sombre** = **l. rouge lointain** (*q.v.*); **en lumière continue** under continuous light
luminescence *f.* luminescence; **éclairage par luminescence** neon lighting
lumineux,-euse *a.* luminous
luminosité *f.* luminosity, brightness, light intensity
lunaire *f.* honesty (*Lunaria annua* = *L. biennis*) (= **monnaie du pape**)
lunettes (*f. pl.*) **de protection** protective goggles
lunetière *f. Biscutella* sp., including buckler mustard (*B. laevigata*)
lunulé,-e *a.* lunulate, lunulated
lunure *f.* crescent-shaped crack or defect in wood
lupin *m.* lupin (*Lupinus* sp.); **l. en arbre** tree lupin (*L. arboreus*); **l. blanc** *L. (nanus* var.*) albus*; **l. changeant** *L. mutabilis*; **l. grand** *L. hirsutus*; **l. d'Hartweg** *L. hartwegii*; **l. jaune** *L. luteus*; **l. polyphylle** garden lupin (*L. polyphyllus*); **l. de Russell** Russell lupin
lupulin *m.* lupulin
lupuline *f.* 1. lupulin. 2. black medick (*Medicago lupulina*)

lustrage *m.* polishing, glossing; **l. des agrumes** polishing citrus fruits
lustré,-e *a.* polished, glossy
lustrer *v.t.* to polish, to gloss
lustreuse *f.* polishing machine
lutoïde *m.* lutoid
lutte *f.* contest, struggle; *Hort.* control, protection; **l. antiacridienne** locust control; **l. antiaviaire** bird control; **l. antigel** frost protection; **l. autocide** autocidal, genetic control; **l. biologique** biological control; **l. chimique** chemical control; **l. contrôlée** regulated control; **l. culturale** cultural control; **l. dirigée** directed control; **l. génétique** genetic control; **l. intégrée** integrated control; **l. contre les mauvaises herbes** weed control; **l. mécanique** mechanical control; **l. physiologique** physiological control; **l. physique** physical control; **l. psychique** psychic control; **l. raisonnée** supervised control
lutter *v.i.* **(contre)** *Hort.* to control (pests etc.)
luxmètre *m.* lux-meter
luxuriant,-e *a.* luxuriant, abundant (of vegetation)
luzerne *f.* lucerne (*Medicago sativa*); **l. à feuilles tâchées** spotted medick (*M. arabica*); **l. minette** black medick (*M. lupulina*) (= **minette**)
luzule *f.* *Luzula* sp.
lycaste *f.* *Lycaste* sp., lycaste (orchid)
lychnide *f.*, **lychnis** *m.* *Lychnis* sp., campion; **l. diurne** red campion (*Silene dioica*) (= **compagnon rouge**); **l. des prés** ragged robin (*Lychnis flos-cuculi*) (= **fleur de coucou**)
lyciet *m.* *Lycium* sp., box thorn; **l. commun** *L. europaeum*; **l. jasminoide** *L. barbarum*
lycope *m.* *Lycopus* sp., including *L. europaeus*, gipsywort (= **lance de Christ**)
lycopène *m.* lycopene
lycopode (*m.*) **des jardiniers** *Selaginella kraussiana*
lycopside *f.*, **lycopsis** *m.* *Lycopsis* sp.; **l. des champs** lesser bugloss (*L. arvensis*)
lyda (*f.*) **du poirier** social pear sawfly (*Neurotoma saltuum* = *N. flaviventris*)
lygée *f.* albardine (*Lygeum spartum*)
lyonie *f.* *Lyonia* sp.; **l. en arbre** *L. ligustrina*
lyophilisation *f.* freeze drying, lyophilization
lyophilisé,-e *a.* freeze dried, lyophilized
lyophiliser *v.t.* to freeze dry, to lyophilize
lyre *f.* lyre; **en lyre** lyre-shaped
lyré,-e *a.* lyrate
lyse *f.* lysis
lysimachie, lysimaque *f.* *Lysimachia* sp., loosestrife; **l. nummulaire** moneywort, creeping jenny (*L. nummularia*); **l. rouge** purple loosestrife (*Lythrum salicaria*)
lysimètre *m.* lysimeter
lysimétrique *a.* relating to lysimeters; **case lysimétrique** *f.* lysimeter
lysine *f.* lysine

M

MA = **matière active** active ingredient (see also **matière**)
macabo *m.* tannia, new cocoyam (*Xanthosoma sagittifolium*)
macadamier *m.* macadamia (*Macadamia ternifolia, M. tetraphylla*)
maceron *m.* **(potager)** alexanders (*Smyrnium olusatrum*)
mâche *f.* corn salad, lamb's lettuce (*Valerianella locusta*) (= **boursette, doucette**); **m. d'Italie** Italian corn salad (*V. eriocarpa*)
mâchefer *m.* clinker, slag
machette *f.* machete, matchet
machine *f.* machine; **m. à bêcher** spading machine; **machines horticoles** horticultural machinery; **m. à effeuiller** defoliating machine; **m. à faire des mottes** soil block machine; **m. à vendanger, m. à récolter le raisin** grape harvesting machine
machinerie *f.* machinery
macis *m.* mace (= **fleur de muscade**) (see also **muscade**)
maclure *f.* *Maclura* sp.; **m. épineuse** Osage orange (*Maclura aurantiaca*)
macre *f.* **(nageante)** water chestnut (*Trapa natans*)
macrocarpe *a.* macrocarpous
macroscopique *q.* macroscopic
macule *f.* spot, stain, speak; **macules foliaires** leaf spots
maculé,-e *a.* spotted, stained, speckled
madérisation *f.* maderization, over-oxidation (in wine-making)
madi, madia *m.*, **madie** *f.* tarweed (*Madia sativa*) and *Madia* sp.
maërl *m.* (*Brittany*) calcareous algae (e.g. *Lithothamnion*) and sand for soil amelioration (= **merl**)
magasin *m.* shop; storehouse, storeroom; **m. industriel de stockage** industrial warehouse
magasinage *m.* warehousing, storing
magnésie *f.* magnesia
magnésien,-ienne *a.* magnesian; **nutrition magnésienne** magnesium nutrition
magnésium *m.* magnesium
magnolia *m.* magnolia flower, magnolia tree
magnolier *m.* magnolia tree (*Magnolia* sp.); **m. à grandes fleurs** laurel magnolia (*M. grandiflora*); **m. glauque** *M. glauca*; **m. à feuilles acuminées** cucumber tree (*M. acuminata*); **m. à fleurs pourpres** *M. liliflora*; **m. à grandes feuilles** *M. macrophylla*; **m. (en) parasol** umbrella tree (*M. tripetala*)
mahaleb *m.* mahaleb cherry (*Prunus mahaleb*) (= **cerisier (de) Sainte Lucie** or **sainte-lucie**)
mahonia *m.* *Mahonia* sp.; **m. (à feuilles de houx)** Oregon grape (*M. aquifolium*); **m. du Japon** *M. japonica*
mahonille *f.* Virginian stock (*Malcomia maritima*) (= **giroflée, julienne de Mahon**)
maïanthème *m.* May lily (*Maianthemum bifolium*) (= **fleur de mai**)
maigre *a.* thin, lean
maille *f.* mesh (of net etc.); *Hort.* female cucurbitaceous flower; *Vit.* vine bud
maillochage *m.* bruising or crushing the base of cuttings to encourage root formation
maillole *f.* 1. bud. 2. cutting
main *f.* hand; **à la main** by hand; **m. de bananes** hand of bananas; **m. de Bouddha** fingered lemon (*Citrus medica* var. *sarcodactylis*)
main-d'œuvre *f.*, *pl.* **mains-d'œuvre** labour, manpower
maïs *m.* maize (*Zea mays*); **m. doux, m. sucré** sweet corn; **m. d'eau** giant water lily (*Victoria regia*)
maîtriser *v.t.* to master, to subdue; *PC* to control
makoré *m.* *Afr.* *Tieghemella africana*
mal sec *m.* *Cit.* mal secco (*Phoma tracheiphila*)
malade *a.* diseased, ill
maladie *f.* illness, disease, disorder; **les maladies des plantes** plant diseases; I. General: **maladie bactérienne** bacterial disease (= **bactériose**); **m. de carence** deficiency disorder; **m. commune du froid** internal breakdown; **m. cryptogamique** fungus disease; **m. d'entreposage** storage disease r disorder; **m. de gazon** lawn disease; **m. inconnue** unidentified disease; **m. à mycoplasme** mycoplasma disease; **m. non-parasitaire** disorder; **m. physiologique** physiological disorder; **m. vermiculaire** disease caused by nematodes; **m. à virus** virus disease (= **virose**); II. Specific: **maladie bactérienne des gousses du haricot** bacterial blight of beans, halo blight (*Pseudomonas phaseolicola*); **m. blanche (du poirot)** white rot of leek (*Sclerotium cepivorum*); **m. du bois noir de la vigne** *Vit.* black wood disease

(a form of **flavescence dorée** *q.v.*); **m. du bout de cigare** *Ban.* cigar end (*Verticillium theobromae*); **m. bronzée** (*Coconut*) bronze leaf wilt; **m. bronzée (de la tomate)** tomato spotted wilt virus; **m. chlorotique du houblon** hop chlorotic virus; **m. du cœur noir (de la banane)** banana black heart (*Fusarium moniliforme*); **m. criblée** shot hole disease (*Stigmina carpophila* = *Clasterosporium carpophyllum*); **m. de l'écorce liégeuse** *Vit.* corky bark (virus); **m. des encoches sèches** *Rubb.* brown bast disease; **m. de l'encre (du châtaignier)** chestnut ink disease (*Phytophthora cambivora, P. cinnamomi, P. syringae*); **m. de l'encre (de l'iris)** iris ink disease (*Mystrosporium adustum*); **m. (d'entreposage) de la tache amère** (or **des taches amères**) **(de la pomme)** apple bitter pit; **m. à filament(s) blanc(s)** (*Cocoa*) etc. white thread blight (*Marasmius scandens* and related spp.); **m. (hollandaise) de l'orme** Dutch elm disease (*Ceratocystis ulmi*) (= **graphiose de l'orme**); **m. de Kaincopé** (*Coconut*) Cape St. Paul wilt; **m. de l'œil de paon de l'olivier** peacock's eye (leaf spot) disease of olive (*Cycloconium oleaginum*); **m. d'Oléron (de la vigne)** vine bacterial blight (*Xanthomonas ampelina*) (= **nécrose bactérienne**); **m. de la "petite feuille"** *Vit.* little leaf disease (due to Zn deficiency); **m. de Pierce** *Vit.* Pierce's disease (virus); **m. de la pointe de crayon** (*Coconut*) pencil point, tapering stem wilt; **m. poudreuse** mealy pod disease (of several crops) (*Trachysphaera fructigena*); **m. des proliférations du pommier** apple proliferation disease; **m. de rabougrissement (des repousses)** *Sug.* ratoon stunting disease, RSD; **m. des racines liégeuses (de la tomate)** corky root (of tomato) (*Pyrenochaeta lycopersici*); **m. des racines de Travancore** (*Coconut*) Travancore root wilt; **m. des raies noires** *Rubb.* black stripe disease (*Phytophthora palmivora*); **m. rose** pink disease (*Corticium salmonicolor*); **m. des sclérotes (de la carotte)** carrot sclerotinia rot (*Sclerotinia sclerotiorum*); **m. de Sigatoka** *Ban.* Sigatoka disease (*Cercospora musae*); **m. de la tache (de la carotte)** carrot cavity spot (*Pythium violae*); **m. des taches amères** apple bitter pit; **m. des taches annulaires du chou** cabbage black ring spot virus; **m. des taches brunes du palmier-dattier** brown-blotch of date palms (*Mycosphaerella tassiana*); **m. des taches brunes** (or **taches noires**) **du pois** pea leaf and pod spot (*Ascochyta pisi*); **m. des taches brunes (de la pomme de terre)** early blight (*Alternaria solani*); **m. des taches ligneuses** apple rough skin virus (= **rugosité de la peau**); **m. des taches noires du rosier** black spot of roses (*Diplocarpon rosae*); **m. des taches rouges du fraisier** strawberry leaf spot (*Mycosphaerella fragariae*); **m. de la toile** seedling disease caused by *Botrytis cinerea* and other fungi (see **toile**); **m. vermiculaire du muguet** caused by lily-of-the-valley nematode (*Pratylenchus convallariae*); **m. des yeux bruns** *Coff.* leaf spot (*Cercospora coffeicola*); **m. X occidentale du pêcher** peach western X virus (see also **strie**); **m. de Zelfana (du palmier-dattier)** sudden decline (of date palm) (*Fusarium oxysporum*) (see also **bayoud**)

maladif,-ive *a.* sickly

malate *m.* malate

malate déshydrogénase *f.* malate dehydrogenase

malaxage *m.* kneading, mixing

mâle-stérile *a. PB* male-sterile

malformation *f.* deformation, malformation

malherbe *f.* 1. plumbago (*Plumbago* sp.). 2. mezereon (*Daphne mezereum*)

malherbologie *f.* weed science

malherbologue *m.* weed scientist

maliforme *a.* apple-shaped

malingre *a.* sickly, puny; *Hort.* **plant malingre** leggy plant

malique *a.* malic

mal nero *m. Vit.* gummosis, bacterial blight (*Xanthomonas ampelina*) (= **nécrose bactérienne**)

malpighie *f. Malpighia* sp. (see **cerisier des Antilles**)

mal secco *m. Cit.* mal secco (*Deuterophoma tracheiphila*)

maltose *m.* maltose

malvacé,-e *a.* malvaceous; **malvacées** *n. f. pl.* Malvaceae

mamelle *f. Hort.* excrescence, protuberance; **arbre aux mamelles** (see **mammea**)

mamelon *m. Hort.* small excrescence

mamelonné,-e *a.* mamillate

mamey *m.* mammey sapote (*Calocarpum sapota*); egg fruit (*Lucuma* sp.)

mamillaire *f. Mammillaria* sp., nipple cactus

mammea *m.*, **mammée** *f. Mammea* sp.; **m. américaine, m. d'Amérique** mammey apple (*M. americana*) (= **abricot de Saint-Domingue, arbre aux mamelles**)

man *m.* cockchafer larva (= **ver blanc**)

mancenille *f.* manchineel apple

mancenillier *m.* manchineel tree (*Hippomane mancinella*)

manche *f.* sleeve; **m. à eau** (water) hose

manchon *m.* sleeve, muff; **m. de grillage** wire sleeve; **m. plastique** plastic sleeve

mancienne *f.* wayfaring tree (*Viburnum lantana*)

mancône *m.* red-water tree (*Erythrophleum guineense*)
mandarine *f.* mandarin, tangerine fruit
mandarinier *m.* mandarin, tangerine tree; **m. (commun), m. à fruits moyens** mandarin tree (*Citrus reticulata*); **m. à gros fruits** king mandarin tree (*C. nobilis*); **m. satsuma** satsuma (*C. unshiu*); **m. à très petits fruits** Cleopatra mandarin (*C. reshni*)
mandragore *f.* **(officinale)** mandrake (*Mandragora officinarum*)
mangaba *m.* mangoba (*Hancornia speciosa*), a tree with edible fruits; source of mangabeira rubber
manganèse *m.* manganese
manganeux,-euse *a.* manganous
manganique *a.* manganic; **nutrition manganique** manganese nutrition
mangeable *a.* edible, eatable
mange-maillols *m. inv. Vit.* cerambycid beetle (larva) (*Vesperus xatarti*)
mange-mil *m. inv., Afr.* quelea, red-billed weaver bird (*Quelea quelea*) (= **quéléa**)
mange-tout *n. m. inv.* type of french bean or pea with edible pods
mangle *f.* mangrove fruit; mangrove tree
manglier *m.* mangrove tree (*Rhizophora* sp.)
mango *m.* mango fruit from unimproved tree
mangottier *m. Afr.* grove of unimproved mango trees
mangoustan *m.* mangosteen (fruit)
mangoustanier *m.* mangosteen tree (*Garcinia mangostana*); **m. sauvage** santol (*Sandoricum indicum*), a tree cultivated for its edible fruits
mangue *f.* mango (fruit) from budded tree
manguier *m.* mango tree (*Mangifera indica*); **m. sauvage** *Afr.* wild mango (*Irvingia gabonensis*) (= **dika du Gabon**) (see also **pain d'Odika**)
maniabilité *f.* manageableness, handiness (of tool)
maniçoba *m.* manicoba (= **céara** *q.v.*)
maniguette *f.* malaguetta pepper, grains of paradise (*Aframomum melegueta*) (= **graines de paradis**)
manioc *m.* cassava (*Manihot utilissima*); **m. amer** bitter cassava; **m. doux** sweet cassava
maniveau *m., pl.* **-eaux** display basket, tray, punnet (see also **clayette**)
mannane *f.* mannan
manne *f.* basket, hamper, crate
mannite *f.*, **mannitol** *m.* mannitol
mannose *m.* mannose
manœuvre *m.* labourer
manomètre *m.* manometer, pressure gauge
manoque *f.* hand of tobacco leaves, bunch of esparto grass
manoquer *v.t.* to put tobacco in hands
manquant,-e *a.* missing, absent; *n. m.* gap (in nursery, plantation etc.); *Hort.* **remplacer les manquants** to fill in the gaps, to supply
manque *m.* lack, want, deficiency; **m. d'azote** nitrogen deficiency; **m. d'eau** moisture deficiency, water shortage; **m. de pluie** lack of rain
manteau *m.* cloak, mantle; *Bot.* mantle of vegetation; **m. de Notre-Dame** lady's mantle (*Alchemilla vulgaris*)
manutention *f.* 1. management. 2. handling (of materials)
maquis *m.* Mediterranean scrub, brush (type of plant community)
maraîchage *m.* market gardening, vegetable growing
maraîcher,-ère *a.* pertaining to market gardening or vegetable growing; *n.* market gardener, grower
marais *m.* 1. marshland, marsh, bog, fen; **m. salant** salt marsh; **m. tourbeux** peat bog. 2. market garden
marang *m.* marang (*Artocarpus odoratissima*)
marante *f.* *Maranta* sp., including *M. arundinacea* arrowroot
marasme *m. Fung.* *Marasmius* sp.; **m. d'oréade** fairy ring champignon (*M. oreades*) (= **faux mousseron**)
marbré,-e *a.* marbled, mottled, veined
marbrure *f. PD* mottle, mottling (virus); **m. annulaire** ring mottle; **m. chlorotique du chrysanthème** chrysanthemum chlorotic mottle virus; **m. de la fève** broad bean mottle virus; **m. foliaire** leaf mottling; **(virus de la) marbrure de l'œillet** carnation mottle virus; **m. de la vigne** fleck (of grapevine); **m. zonale** plum line pattern virus
marc *m.* **m. de raisin** marc of grapes; **m. de café** coffee grounds; **m. de pommes** pomace
marcescence *f.* marcescence, withering
marcescent,-e *a.* marcescent, withering without falling off; **feuilles marcescentes** marcescent leaves
marchand,-e *a.* commercial, saleable, marketable; **cacao marchand** marketable cocoa; **variété marchande** commercial variety
marchantie *f.* *Marchantia* sp.
marché *m.* 1. dealing, buying, contract. 2. market; **m. aux fleurs** flower market; **m. frais** fresh market; **m. local** local market; **jour de marché** market day
marcottage *m.* layering, marcotting, suckering; **m. aérien, m. en l'air** air layering, marcotting; **m. en arceau** = **m. simple** *q.v.*; **m. en arceaux** = **m. en serpenteaux** *q.v.*; **m. en archet** = **m. simple** *q.v.*; **m. avec branche ramifié** multiple layering; **m. en butte, m.**

par buttage stool layering, stooling; **m. en cépée** = **m. en butte** *q.v.*; **m. chinois** Chinese layering; **m. par couchage** simple layering, tip layering; **m. par drageons** propagation by suckers; **m. naturel** natural layering (e.g. strawberry); **m. en pannier** layering into a basket; **m. en pot** layering into a pot; **m. par provignage** *Vit.* layering (= **provignage**); **m. par racines** propagation from root suckers; **m. en serpenteaux** serpentine layering; **m. simple** simple layering

marcotte *f.* 1. layer, marcot. 2. runner, sucker

marcotter *v.t.* to layer, to marcot

marcottière *f.* layer-bed, stool-bed

mare *f.* pond

marécage *m.* swamp, bog

marécageux,-euse *a.* swampy, boggy

margarodes *m.* margarodes (*Margarodes* sp.)

marge *f.* border, edge; **m. de sécurité** safety margin

margelle *f.* curb, edge

marginal,-e *a., m. pl.* **-aux** marginal

marginé,-e *a.* marginated, marginal

margose *f.* balsam pear (*Momordica charantia*) (= **concombre amer**) (see also **momordique**)

margousier *m.* bead-tree, pride of India (*Melia azedarach*)

marguerite *f.*, **grande marguerite, m. des champs** oxeye daisy (*Chrysanthemum leucanthemum*); **(petite) marguerite** daisy (*Bellis perennis*); **m. africaine** African daisy (*Lonas inodora*); **m. d'automne** = **m. de la Saint-Michel** (*q.v.*); **m. de Chine** = **reine-marguerite** (*q.v.*); **m. dorée** corn marigold (*Chrysanthemum segetum*); **m. de la Saint-Michel** Michaelmas daisy (*Aster* sp.) (= **chrysanthème à petites fleurs**)

Marie honteuse *f.* (*West Indies*) sensitive plant (*Mimosa pudica*)

marignan *m.* eggplant (*Solanum melongena*) (= **aubergine**)

marisque *f.* 1. *Hort.* variety of large fig. 2. fen sedge (*Cladium mariscus*) (see also **cladiaie**)

marjolaine *f.* marjoram (*Origanum* sp.; **m. (à coquille)** sweet marjoram (*Origanum majorana*); **m. vivace** common, wild marjoram (*O. vulgare*) (= **marjolaine bâtarde**) (see also **origan**)

marmelade *f.* compote of fruit; **m. d'orange(s)** (orange) marmalade

marmenteau, *pl.* **-eaux** *a.* & *n.m.* **arbres marmenteaux, (bois) marmenteaux** high ornamental trees forming part of the amenities of an estate, and to be preserved

marmite (*f.*) **de singe** sapucaia nut, paradise nut, monkey pot (*Lecythis zabucajo*) and related spp.

marnage *m.* marling

marne *f.* marl; **m. calcaire** lime marl

marner *v.t.* to marl

marneux,-euse *a.* marly

marotte *f. Sug.* late tiller, sucker (= **baba**)

maroute *f.* stinking chamomile (*Anthemis cotula*) (= **camomille puante**)

marquage *m.* marking, (radioactive) labelling

marqué,-e *a.* marked, labelled; **marqué au ^{14}C** ^{14}C labelled; **élément marqué** labelled element

marqueur,-euse *n.* marker; **m. génétique** genetic marker; **pollen marqueur** marker pollen

marquise *f.* marquise tree shape

marron *m.* large sweet chestnut, from single-seeded non-septate fruit (see also **châtaigne**); **m. d'Inde** horse chestnut

marronnier *m.* (sweet) chestnut tree producing **marrons** (see also **châtaignier**); **m. (d'Inde)** horse chestnut tree (*Aesculus hippocastanum*); **m. de l'Ohio** Ohio buckeye (*A. glabra*); **m. rose** red horse chestnut (*A.* × *carnea*); **m. rouge** red buckeye (*A. pavia*) (= **pavier**)

marrube *m. Marrubium* sp., horehound; **m. aquatique** gipsywort (*Lycopus europaeus*); **m. blanc, m. commun** white horehound (*M. vulgare*); **m. noir** (black) horehound (*Ballota nigra*) (see also **ballote**)

marsault, marsaux *m.* goat willow (*Salix caprea*)

marssonina *m.* black spot of roses (*Diplocarpon rosae*) (= **maladie des taches noires**)

martagon *m.* Turk's cap lily (*Lilium martagon*) (= **lis martagon**); **m. écarlate** scarlet martagon (*L. chalcedonicum*) (= **lis de Constantinople**)

marteau-piochon *m., pl.* **marteaux-piochons** *Vit.* combined cleaver and mattock

mas *m.* farm, holding (in South of France)

masque *m.* mask (see also **anti-poussière**)

masse-cuite *f., pl.* **masses-cuites** *Sug.* massecuite

massette *f.* bulrush, reed-mace (*Typha latifolia*)

massif *m. Hort.* clump of shrubs, trees, rose bushes etc.; **m. nain** clump of small shrubs etc., clump of flowers, flower bed; **plantes à massif** bedding plants

massique *a.* pertaining to mass; **pertes massiques** losses in weight

massue *f.* club; **en forme de massue** club-shaped; **m. d'Hercule** Hercules' club (*Aralia spinosa*) (= **angélique épineuse, baton du diable**)

mastic (*m.*) **(à greffer)** grafting wax (= **cire à greffer**); **m. cicatrisant** grafting wax with wound dressing (see also **onguent**)

masticage *m. Hort.* application of grafting wax

masticatoire *a.* & *n.m.* masticatory

mastiqué,-e *a. Hort.* waxed (of graft)

mastiquer *v.t. Hort.* to apply grafting wax

maté *m.* maté (*Ilex paraguensis*) (= **thé du Paraguay**)

matelas *m.* mattress; **m. d'air, m. thermique** layer of warm air, air space (in building); **m. capillaire** capillary mat

matériau *m., pl.* **-aux** building material; **matériaux de recouvrement** covering materials; **matériaux d'origine** *Ped.* parent materials

matériel *m.* equipment, material; **m. d'arrachage** plant lifting equipment; **m. horticole** horticultural implements; **m. végétal** plant material; **m. végétal garanti sain** *Hort.* special stock

matico *m.* matico (*Piper angustifolium*)

matière *f.* matter, material; **m. active, MA** active ingredient, a.i.; **m. colorante** colouring matter; **m. élaborée** assimilate; **matières hydrocarbonées** carbohydrates; **m. organique** organic matter; **m. plastique** plastic; **m. première** raw material; **m. sèche** dry matter

matricaire (*f.*) **camomille, m. officinale** wild chamomile (*Matricaria recutita* = *M. chamomilla*) (= **petite camomille**); **m. discoïde** rayless mayweed (*M. matricarioides* = *M. discoidea*); **m. inodore** scentless chamomile (*Tripleurospermum maritimum* ssp., *inodorum*); **m. maritime** (*T. maritimum* ssp. *maritimum*) (see also **camomille**)

matthiole *f. Matthiola* sp., stock (see also **giroflée, quarantaine**)

maturation *f.* maturation, ripening; **m. complémentaire** after-ripening

maturité *f.* maturity, ripeness; **m. industrielle** industrial ripeness; **m. physiologique** physiological ripeness

maurelle *f.* dyer's croton (*Chrozophora tinctoria*) (see also **tournesol**)

mauricien,-enne *a.* of Mauritius; *n.* inhabitant or native of Mauritius

mauritie *f.*, **mauritier** *m. Mauritia* sp. (see also **sagoutier**)

mauvais pourri *m. Vit.* grey rot (*Botrytis cinerea*) (= **pourriture grise**)

mauvaise herbe *f.* weed; **m. h. annuelle, m. h. à graines** annual weed; **m. h. vivace** perennial weed (see also **adventice, plante**)

mauvaise récolte *f.* poor harvest, crop failure

mauve *f.* 1. mallow (*Malva* and other spp.) (see also **guimauve**); **m. d'Alger, m. d'Algérie** smaller tree mallow (*Lavatera cretica*); **m. en arbre** tree mallow (*Lavatera arborea*), shrubby althaea (*Hibiscus syriacus*); **m. à feuilles rondes** dwarf mallow (*M. rotundifolia* = *M. neglecta*); **m. frisée** curled mallow (*M. crispa*); **m. musquée** musk mallow (*M. moschata*); **grande mauve, m. sauvage, m. sylvestre** common mallow (*M. sylvestris*); tree mallow (= **m. en arbre**); **petite mauve** dwarf mallow (*M. neglecta*); **m. sauvage** hispid marsh mallow (*Althaea hirsuta*). 2. *a.* mauve

mayenquage *m. Vit.* first May ploughing or cultivation

mayorquin *m. Vit.* white cultivar used for raisin production

mayorquine *f. Vit.* type of veneer graft

maxillaire *f. Maxillaria* sp. (orchid)

mazout *m.* fuel oil

méat *m.* meatus; **m. intercellulaire** intercellular space

mécanisation *f.* mechanization

mécanisé,-e *a.* mechanized; **récolte mécanisée** mechanized harvesting

mécaniser *v.t.* to mechanize

médaille (*f.*) **(de Judas)** honesty (*Lunaria annua*)

médian,-e *a.* median

médicamenteux,-euse *a.* 1. medicinal. 2. medicamentous, caused by medicine(s)

médicinier *m.* physic nut (*Jatropha curcas*) (= **jatropha, purghère**)

méditerranéen,-enne *a.* Mediterranean

médocain,-e *a.* of Medoc

médullaire *a.* medullary; **rayon médullaire** medullary ray

médulleux,-euse *a.* medullary, medullated

mégachile *m.* leaf cutter bee (*Megachile* sp.) (= **abeille coupeuse de feuilles**); **m. du rosier** rose leaf cutter bee (*M. centuncularis*)

méiomère *a.* meiomerous, with a small number of parts

méiose *f.* meiosis

méiotique *a.* meiotic

mélaleuque *m. Melaleuca* sp., cajeput (= **niaouli**)

mélange *m.* 1. mixing, blending. 2. mixture; **m. huile-eau** oil-water mixture; **m. terreux** soil mixture; **"en beau mélange"** *Comm., Hort.* "a splendid mixture"

mélanger *v.t.* to mix, to blend

mélangeur *m.* mixer

mélanose *f.* melanose; **m. des agrumes** citrus melanose (*Diaporthe (Phomopsis) citri*); *Vit.* septoria leaf spot (*Septoria ampelina*)

mélasse *f.* molasses

mélastome *m. Melastoma* sp., including the ornamental *M. malabathricum*, and other plants; **m.-arbre** *Loreya arborescens*

méléagride *a.*, **fritillaire méléagride** fritillary (*Fritillaria meleagris*)

mêler *v.t.* to mix, to blend

mélèze *m.* larch (*Larix* sp.); **m. commun, m. d'Europe** European larch (*L. decidua*); **m. du Japon** Japanese larch (*L. leptolepis*)

mélia *m. Melia* sp., bead tree

mélianthe *m. Melianthus* sp., Cape honey-flower

mélicoque *m. Melicocca bijuga*, Spanish lime, honey-berry
méligèthe *m. Meligethes* sp., blossom beetle, pollen beetle
mélilot *m.* melilot (*Melilotus* sp.); **m. bleu** blue fenugreek (*Trigonella caerulea*) (= **lotier odorant**); **m. des champs** ribbed melilot (*Melilotus officinalis*)
mélinet *m.* honeywort (*Cerinthe* sp.)
mélinis *m.* molasses grass (*Melinis minutiflora*)
mélique *f.* melick (*Melica* sp.); **m. bleue** purple moor-grass (*Molinia coerulea*)
mélisse *f.* melissa, balm; **m. officinale, m. citronnelle** balm (*Melissa officinalis*); **m. de Moldavie, m. turque** Moldavian balm (*Dracocephalum moldavica*)
mélitte *f. Melittis* sp., bastard balm
mellifère *a.* melliferous, honey-bearing; *n.m. pl.* honey bees
mélocacte *m. Melocactus* sp., melon cactus
mélocoton *m.* curuba (*Sicana odorifera*)
melon *m.* melon (*Cucumis melo*); **m. brodé** netted melon, musk melon (*C. melo* var. *reticulatus*); **m. cantaloup** cantaloupe (melon); **m. d'eau** water melon (*Citrullus vulgaris*) (= **pastèque**); **m. d'hiver, m. sans odeur** winter melon (*C. melo* var. *inodorus*); **m. de Malabar** Malabar gourd (*Cucurbita ficifolia*) (see also **courge**); **m. serpent** (1) *C. melo* var. *flexuosus*, (2) snake gourd (*Tricosanthes cucumerina*); **m. sucrin** sucrin, pineapple melon (*C. melo* var. *saccharinus*)
melon-coton *m.* = **mélocoton** (*q.v.*)
mélongène *f.* aubergine, eggplant (*Solanum melongena*) (= **aubergine**)
melonné,-e *a.* resembling a melon
melonnière *f.* melon bed, melon patch
melon-poire *m.* melon pear, pepino (*Solanum muricatum*) (= **poire-melon**)
membranaire *a.* relating to membranes; **thermostabilité membranaire** membrane thermostability
membrane *f.* membrane; **m. cellulaire** cell wall
membraneux,-euse *a.* membranous
mémécyle *m. Memecylon* sp., including *M. edule*
mendélien,-enne *a.* Mendelian; **caractère mendélien** Mendelian character
mendélisme *m.* Mendelism
ménisperme *m. Menispermum* sp., moon seed; **m. du Canada** *M. canadense*
menthe *f.* mint (*Mentha* sp.) and related plants; **m. anglaise** peppermint (*M* × *piperita*); **m. aquatique** water mint (*M. aquatica*); **m. des champs** corn mint (*M. arvensis*); **m. de chat** catmint (*Nepeta cataria*); **m. crépue** round-leaved mint (*M. rotundifolia*); **m. douce** = **m. verte** (*q.v.*); **m. des jardins** garden mint (prob. *M. arvensis* × *M. viridis*); **m. poivrée** = **m. anglaise** (*q.v.*), round-leaved mint (*M. rotundifolia*); **m. poivrée blanche** white mint (*M.* × *piperita* var. *officinalis* f. *pallescens*); **m. poivrée noire** black mint (*M.* × *piperita* var. *officinalis* f. *rubescens*); **m. pouliot** pennyroyal (*M. pulegium*); **m. verte** spearmint (*M.* × *spicata* = *M. viridis*)
menthe-coq *f., pl.* **menthes-coqs** costmary, alecost (*Chrysanthemum balsamita*)
menthol *m.* menthol
mentor *m. Hort.* **rameau mentor** shoot retained as a prop for graft etc.
menuiserie *f.* carpentry
ményanthe *m. Menyanthes trifoliata*, buckbean
méplat *m.* flat part, flattening
mérangène *f.* eggplant (*Solanum melongena*)
mercure *m.* mercury
mercureux *a. m.* mercurous
mercuriale *f.* mercury; **m. annuelle** annual mercury (*Mercurialis annua*); **m. vivace** dog's mercury (*M. perennis*)
mercuriel,-ielle *a.* mercurial, relating to mercury
mercurique *a.* mercuric
mère *f.* mother; phellogen (in cork oak) (see also **pied-mère, plante mère**)
mérendère *f. Merendera* sp., Pyrenean meadow saffron
méricarp *m.* mericarp
méricIone *m.* clone obtained by meristem culture
merise *f.* wild cherry fruit; **m. de Virginie** choke cherry fruit; in Antilles, fruit of *Eugenia ligustrina*
merisier *m.* **(des bois)** wild cherry tree, gean (*Prunus avium*); **m. à grappes** bird cherry tree (*P. padus*); **m. à grappes pleureur** *P. padus pendula*; **m. doré** locust berry (*Byrsonima coriacea*) (= **bois chardon**); **m. noir** (*Antilles*) *Eugenia ligustrina*; **m. de Virginie** choke cherry (*P. virginiana*)
méristématique *a.* meristematic
méristème *m.* meristem; **m. apical** apical meristem; **m. radiculaire** root meristem
méristémoïde *m.* meristemoid
mérithalle *m.* internode (= **entre-nœud**)
merl *m.* (*Brittany*) (= **maël** *q.v.*)
merle *m.* blackbird (*Turdus merula*)
mérule *f. Fung.* dry rot (*Merulius lacrymans*)
merveille (*f.*) **du Pérou** marvel of Peru (*Mirabilis jalapa*) (= **mirabilis**)
mescal, mezcal *m.* 1. mescal drink (from *Agave* sp.). 2. mescal (*Lophophora williamsii*)
mesclun *m.* mixed salad crop (e.g. chicory, endive, lettuce, chervil, rocket)
mésembryanthème *m. Mesembryanthemum* sp. (= **ficoïde**); **m. cristallin** ice plant (*M. crystallinum*)
mésobiote *m.* interstem, interstock

mésocarpe *m.* mesocarp; **m. (fibreux)** (*Coconut*) (fibrous) husk
mésoclimat *m.* mesoclimate
mésocotyle *m.* mesocotyl
mésophile *a.* mesophyllous
mésophylle *m.* mesophyll
mésotrophe *a.* mesotrophic
mesquite *m.* mesquite (*Prosopis juliflora*)
mesure *f.* measure; **m. de lutte** control measure
métabolique *a.* metabolic
métabolisation *f.* metabolization
métabolisme *m.* metabolism; **m. acide crassulacéen** crassulacean acid metabolism; **m. azoté** nitrogen metabolism; **m. carboné** carbon metabolism; **m. phosphoré** phosphorus metabolism
métabolite *m.* metabolite; **métabolites volatils** volatile metabolites
métal *m., pl.* **métaux** metal; **m. déployé** expanded metal; **métaux lourds** heavy metals
métaldéhyde *f.* metaldehyde
métamorphose *f.* metamorphosis
métaphase *f.* metaphase
métaphloème *m.* metaphloem
métaphosphate *m.* metaphosphate
métaxénie *f.* metaxenia
méthanol *m.* methanol
méthionine *f.* methionine
méthode *f.* method, system; **m. de conduite** training system; **m. de culture** cultural system, way of growing; **m. d'établissement** method of establishment; **m. de lutte** control method; **m. de sélection** selection method, screening
méthylation *f.* methylation
méthyle *m. Chem.* methyl
métis,-isse *a.* half-bred, cross-bred; **plante métisse** hybrid plant (applied more to crosses between varieties than between species)
métissage *m.* cross breeding
métisser *v.t.* to cross, to hybridize
métobromuron *m.* metobromuron
mètre *m.* meter; **m. pliant** folding rule; **m. à ruban** tape measure; *Vit.* (large) rootstock cutting (ready for grafting) (= **bouture greffable**)
métrogreffe *m.* grafting clamp
métroxyle *m. Metroxylon* sp. (see also **sagoutier**)
mettre *v.t.* to place, to put; *Hort.* **m. en jauge** to heel in; **m. en pot** to pot, pot-on, pot up; **m. en tas** to stack, to build up a pile (e.g. manure)
meuble *a.* movable, friable; **terre meuble, sol meuble** mellow, light, friable loam, running soil
meule *f.* 1. millstone; **m. à aiguiser** grindstone. 2. *Ag.* stack, rick; **m. de foin** haystack. 3. *Hort.* hotbed, manure bed; **m. à champignons** mushroom bed
méum *m.* meum, baldmoney (*Meum athamanticum*)
meunier *m. Fung.* the miller (*Clitopilus prunulus*); *PD* applied to various white fungi attacking crops; e.g. lettuce downy mildew (*Bremia lactucae*); vine powdery mildew (*Uncinula necator) (see also* **blanc, mildiou**)
meurtri,-e *a.* bruised (e.g. fruit)
meurtrir *v.t.* to bruise
meurtrissure *f.* bruise
mézéréon *m.* mezereon (*Daphne mezereum*) (= **bois gentil**)
mi- *adv.* half, mid; **à mi-bois** to half-way mark (of cross-section of trunk of tree) (see also **plein bois**); **à mi-ombre** half shaded; **à mi-pétiole** halfway along the petiole; **radis mi-blanc, mi-rouge** half-white half-red radish
micelle *m. Bot.* micella
mi-clos,-e *a.* half-shut
micocoule *f.* fruit of **micocoulier** *q.v.*
micocoulier *m.* **(de Provence)** nettle-tree (*Celtis australis*)
miconie *f. Miconia* sp., including *M. flammea*
microanalyse *f.* microanalysis
microbe *m.* microbe
microbicide *a.* microbicidal; *n. m.* microbicide, germ-killer
microbien,-enne *a.* microbial
microbouturage *m.* micropropagation
microclimat *m.* micro-climate
microencapsulation *f.* microencapsulation
microencapsulé,-e *a.* microencapsulated
microflore *f.* microflora; **m. fongique** fungal microflora
microgreffage *m.* micrografting
microhabitat *m.* micro-habitat
microlépidoptère *m.* microlepidopter; *pl.* Microlepidoptera
micromérie *f. Micromeria* sp., including *M. douglasii*, medicinal
micromotte *f.* = **minimotte** (*q.v.*)
micromorphologie *f. Ped.* micromorphology
micro-onde *f., pl.* **micro-ondes** microwave
microorganisme *m.* microorganism; **microorganismes antagonistes** antagonistic microorganisms
microparcelle *f.* microplot
microphylle *a.* microphyllous
micropipette *f.* micropipette
microplante *f. in vitro* plant
micropropagation *f.* micropropagation
micropropagé,-e *a.* micropropagated
micropyle *m.* micropyle
microsaignée *f. Rubb.* microtapping, puncture tapping

microscope *m.* microscope; **m. électronique** electron microscope

microscopie *f.* microscopy; **m. électronique** electron microscopy; **m. électronique à balayage** scanning electron microscopy

microscopique *a.* microscopic

microsonde *f.* microprobe; **m. électronique** electron microprobe

microspore *f.* microspore

microtome *m.* microtome

microtracteur *m.* horticultural tractor

miel *m.* honey

miélat, miellat *m.* honeydew

miellée *f.* nectar flow (of plant); period of nectar flow

mignardise *f.* pink (*Dianthus plumarius*) (= **œillet mignardise**)

mignon,-onne *a.* dainty, tiny; **dahlia mignon** dwarf dahlia (= **dahlia nain**)

mignonnette *f.* mignonette (*Reseda odorata*) (= **réséda odorant**); **m. des Alpes** wood saxifrage (*Saxifraga cuneifolia*) (= **saxifrage cunéiforme**)

migrateur,-trice *a.* migratory; **criquet migrateur** migratory locust

migration *f.* 1. migration; *Bot.* translocation; **m. de l'eau** water movement; *PC* drift (of pesticide)

migratoire *a.* migratory

mi-hâtif,-ive *a.* semi-early

mil (*m.*) **chandelle, m. pénicillaire** bulrush millet (*Pennisetum typhoides*)

mildiou *m.* mildew; **m. de la betterave** beet downy mildew (*Peronospora farinosa* = *P. schachtii*); **m. de l'épinard** spinach downy mildew (*Peronospora farinosa* = *P. effusa*); **m. des giroflées** wallflower downy mildew (*Peronospora parasitica*); **m. de la laitue** lettuce downy mildew (*Bremia lactucae*) (= **meunier**); **m. en mosaïque** mosaic form of **m. de la vigne** (*q.v.*); **m. de l'oignon** onion downy mildew (*Peronospora destructor*); **m. du panais** parsnip downy mildew (*Plasmopara nivea*); **m. des pommes** phytophthora fruit rot of apples (*Phytophthora cactorum*); **m. (de la pomme de terre)** potato blight (*Phytophthora infestans*); **m. de la tomate** tomato blight (*Phytophthora infestans*); **m. de la vigne** vine downy mildew (*Plasmopara viticola*)

mildiousé,-e *a.* mildewed

milieu *m., pl.* **-ieux** 1. middle, midst. 2. medium, environment; **en m. salé** in a saline medium; **m. nutritif** nutrient medium; **m. rural** rural environment; **m. stabilisateur** stabilizing medium

millefeuille *f., pl.* **millefeuilles** milfoil, yarrow (*Achillea millefolium*); **m. aquatique** water violet (*Hottonia palustris*) (= **hottonie**)

mille-pattes *m. inv.* centipede, millipede

millepertuis *m. inv.* St. John's wort (*Hypericum* sp.); **m. fétide** stinking St. John's wort (*H. hircinum*); **m. à grandes fleurs** rose of Sharon (*H. calycinum*)

millerand *m. Vit.* small seedless berry of grape variety the berries of which normally contain seeds

millerandage *m.* 1. *Vit.* uneven development (due to poor fruit setting) of berries on a bunch; imperfect fruit set; (*USA*) shot berries. 2. partial failure of grape harvest (see also **coulure**)

millerandé,-e *a.* 1. **grain millerandé** *Vit.* small seedless berry of grape variety the berries of which normally contain seeds; (*USA*) shot berry. 2. applied to grape harvest which has partially failed

millet (*m.*) **chevelu** witchgrass (*Panicum capillare*); **m. sauvage** bristly foxtail grass (*Setaria pallide-fusca*)

mimétique *a.* & *n.m.* mimetic, mimic

mimeux,-euse *a.* sensitive (of plant); *n. f.* **m. pudique** sensitive plant (*Mimosa pudica*)

mimosa *m.* mimosa (*Mimosa* and *Acacia* spp., including *A. dealbata*); **m. des fleuristes** silver wattle (*A. dealbata*)

mimule, mimulus *m.* musk, monkey flower (*Mimulus* sp.)

mina *f. Mina lobata* (= *Ipomoea versicolor*)

minage *m. Vit.* deep ploughing, subsoiling before planting

minéral,-e *a., m. pl.* **-aux** mineral, inorganic; *n. m.* mineral; **m. de l'argile** clay mineral

minéralisation *f.* mineralization

minette *f.* black medic (*Medicago lupulina*) (= **luzerne minette**)

mineur,-euse *a. Ent.* burrowing; *n.* burrowing insect, borer, miner; **mineuse** (*f.*) **cerclée (du poirier)** pear leaf blister moth (*Leucoptera scitella*); **mineuse des feuilles de pommier** apple pygmy moth (*Stigmella (Nepticula) malella*); **mineuse des feuilles (du rosier)** rose leaf miner (*Stigmella (Nepticula) anomalella*); **mineur de racine** root miner; **mineuse sinueuse (des arbres fruitiers, du pommier)** apple leaf miner (*Lyonetia clerkella*); **mineuse** (*f.*) **de tiges** stem borer; **petite mineuse du pêcher** peach twig borer (*Anarsia lineatella*)

mini *inv. prefix* mini; **mini choux-fleurs** mini cauliflowers

miniaturisation *f.* miniaturization

mini-jardin *m.* bottle garden

minimotte *f.* miniblock, wedge (of modular-raised plant)

minirose *f.* minirose
mini-serre *f.* small indoor greenhouse for ornamentals
minium *m.* red lead
minot *m.* old measure (about 39 litres)
mi-ombre see **mi-**
mi-pétiole see **mi-**
mirabelle *f.* mirabelle plum
mirabellier *m.* mirabelle plum tree (*Prunus insititia*)
mirabilis *m.* *Mirabilis* sp., including *M. jalapa*, marvel of Peru
miride *m.* *Ent.* mirid
mirobolan see **myrobolan**
miroir *m.* mirror; **m. de Vénus** Venus' looking glass (*Specularia speculum*); *Arb.* blaze on tree for felling
mi-rustique *a.* half-hardy
mise *f.* placing, putting; **m. à fleurs** flowering; **m. à fruit** fruit-setting, fruiting; **m. en jauge** heeling in; **m. à nu du liber** baring the phloem; **m. à nu (du sol)** clean weeding, bare soil treatment; **m. en place** planting out; **m. au point** perfecting; **m. en pots** potting (up); **m. au repos** resting; **m. en route** starting up, getting under way; **m. en saignée** *Rubb.* tapping; **m. en (pleine) terre** planting (out)
misère *f.* *Hort.* *Tradescantia* sp., including spiderwort (*T. virginiana*) (= **éphémère**); silver inch plant (*Zebrina pendula*)
miste *m.* mist chamber
mite *f.* 1. mite, acarid. 2. moth (especially those damaging various products); **m. du raisin** raisin moth (*Ephestia (Cadra) figuliella*)
miticide *a.* & *n.m.* acaricide (= **acaricide**)
mitochondrial,-e *a.*, *m.pl.* **-aux** mitochondrial
mitochondrie *f.* mitochondrion; *pl.* mitochondria
mitose *f.* mitosis
mitotique *a.* mitotic; **activité mitotique** mitotic activity
mixoploïde *a.* mixoploid
moabi *m.* orere (*Baillonella toxisperma*)
mode (*m.*) **d'emploi** directions for use
modèle *m.* model; **m. mathématique aléatoire** random mathematical model
modélisation *f.* modelling, construction of a model
modification *f.* modification, alteration
module *m.* module
moelle *f.* marrow; *Bot.* pith, medulla; **m. noire (de la tomate)** *PD* black pith (of tomato)
moignon *m.* stump of amputated branch or limb
moineau *m.* sparrow (*Passer* sp.); **m. commun** house sparrow (*P. domesticus*); **m. friquet** tree sparrow (*P. montanus*)
moisi,-e *a.* mouldy, mildewed; *n. m.* mildew, mould
moisir *v.t.* to mildew, to make mouldy; *v.i.* to become mildewed, to go mouldy
moisissure *f.* mildew, mould, storage rot: mouldiness, mustiness; **moisissures bleue et verte** blue and green moulds (*Penicillium italicum, P. digitatum*); **m. grise des fruits** brown rot etc. (*Sclerotinia* spp.); **m. jaune** *Mush.* yellow mould, verdigris, mat disease (*Myceliophthora lutea*) (= **vert-de-gris**); **m. olivâtre des feuilles et des fruits** leaf mould (tomato) (*Cladosporium fulvum*); **m. rose (du pommier)** apple pink mould, pink rot, core rot (*Trichothecium roseum*); **m. rouge à lèvres** *Mush.* lipstick mould (*Sporendonema purpurascens*)
moisson *f.* harvesting, harvest (of cereals)
moissonneuse *f.* harvesting machine, harvester; **m. à pois** pea harvester
mol *a.* see **mou**
molasse *f.* molasse, sandstone
môle *f.* *Mush.* white mould, wet bubble (*Mycogone perniciosa*) (= **mycogone**); **m. sèche** dry bubble, verticillium disease (*Verticillium malthousei*)
moléculaire *a.* molecular
molécule *f.* molecule
molène *f.* mullein (*Verbascum* sp.); **m. blattaire** moth mullein (*V. blattaria*) (= **blattaire, herbe aux mites**); **m. commune** great mullein (*V. thapsus*) (= **bouillon blanc**); **m. de Phénicie** purple mullein (*V. phoeniceum*)
molle *a.* see **mou**
molluscicide *m.* molluscicide
mollusque *m.* mollusc
moly *m.* moly (*Allium moly*) (= **ail doré**)
molybdène *m.* molybdenum
mombin *m.* **(rouge)** Spanish plum (*Spondias purpurea*) (= **prunier d'Espagne**); **m. (jaune)** yellow mombin (*S. mombin*) (= **prunier mombin, p. d'or**)
momie *f.* *PD* mummy fruit
momification *f.* mummification; **m. des jeunes coings** quince fruitlet mummification (*Sclerotinia cydoniae*)
momifié,-e *a.* mummified; **cabosse momifiée** mummified pod (of cocoa etc.); **fruit momifié** mummified fruit
momifier *v.t.* to mummify; **se momifier** to become mummified
momordique (*f.*) **balsamique** balsam apple (*Momordica balsamina*) (= **pomme de merveille**); **m. à feuille de vigne** bitter gourd (*M. charantia*) (= **margose**)
monadelphe *a.* monadelphous
monarde *f.* *Monarda* sp., including sweet bergamot (*M. didyma*)
monbin *m.* see **mombin**

monder *v.t.* to clean, to hull (grain), to blanch (almonds), to stone (raisins)
moniliforme *a.* moniliform
moniliose *f. Monilia* disease, brown rot, wilt etc. (*Sclerotinia fructigena = M. fructigena, S. laxa = M. cinerea*) (see also **rot**)
monnaie (*f.*) **du pape** honesty (*Lunaria annua = L. biennis*) (= **lunaire**)
monnayère *f.* = **monnaie du pape, nummulaire** and other plants
monobase *a.* monobasic
monobasique *a.* monobasic
monocarpe *a.* monocarpous
monocarpellaire *a.* monocarpellary
monocarpien,-ienne *a.* monocarpic
monocarpique *a.* monocarpic
monocaule *a.* single-stemmed
monocéphale *a.* monocephalous
monochlamydé,-e *a.* monochlamydeous
monocotylédone *a.* monocotyledonous; *n. f.* monocotyledon
monoculture *f.* monoculture
monœcie *f.* monoecism
monœcique *a.* monoecious
monoembryonné,-e *a.* monoembryonic
monoembryonie *f.* monoembryony
monographie *f.* monograph
monohybridisme *m.* monohybridism
monoïque *a.* monoecious
monolinuron *m.* monolinuron
monopétale *a.* monopetalous, unipetalous
monophylle *a.* monophyllous
monoploïdie *f.* monoploidy
monopode *m.* monopodium
monopodique *a.* monopodial
monoroue *a.* single-wheeled; *n. m. Hort.* single-wheeled walking tractor
monosépale *a.* monosepalous
monosoc *n. m.* single plough
monosperme *a.* monospermous
montaison *f. Hort.* bolting, running to seed
montée *f. Hort.* bolting, running to seed; **bonne résistance à la montée** good resistance to bolting
monter (en graine) *v.i. Hort.* to bolt, to run to seed; **lent(e) à monter** slow to bolt
monticule *m.* hillock; *Hort.* hill (for planting etc.)
montre *f. Hort.* appearance of crop as a basis for estimating the harvest
morceler *v.t.* to cut up into small pieces
morcellement *m.* subdivision, splitting up, fragmentation; **m. des terres** fragmentation of land, holdings
mordoré,-e *a.* reddish-brown, bronze
morelle *f.* nightshade (*Solanum* sp.), Christmas cherry (*S. capsicastrum*); **m. comestible** aubergine, eggplant (*S. melongena*); **m. douce-amère** bittersweet, woody nightshade (*S. dulcamara*); **m. jaune** silverleaf nightshade (*Solanum elaeagnifolium*); **m. noire** black nightshade (*S. nigrum*); **m. de Quito** Quito orange, naranjilla (*S. quitoense*) (= **naranjille**); **m. toxique** deadly nightshade (*Atropa belladonna*) (= **belladone**)
morgeline *f.* 1. scarlet pimpernel (*Anagallis arvensis*) (= **mouron rouge**). 2. chickweed (*Stellaria* sp.) (see also **mouron**)
morille *f. Mush.* morel (*Morilla = Morchella* sp.) including common morel (*M. esculenta*)
morinde *f. Morinda* sp., Indian mulberry and related spp.
morine *f. Morina* sp., whorl-flower
morphactine *f.* morphactin
morphogène *a.* morphogenic, morphogenetic; **effet morphogène** formative effect
morphogénèse *f.* morphogenesis
morphologie *f.* morphology
morphologique *a.* morphological
morphotype *m.* morphotype
mort,-e *a.* dead; **branche morte** dead branch; **feuille morte** dead leaf, fallen leaf; *n.f.* death; **m. aux rats** rat poison
mortalité *f.* mortality; **m. larvaire** *Ent.* larval mortality
morte-saison *f., pl.* **mortes-saisons** dead season, off-season, slack season
morver *v.i. Hort.* to become affected by rot (of plant)
mosaïculture *f.* formal ornamental gardening
mosaïque *a.* mosaic; *n. f.* 1. *Hort.* formal ornamental bed. 2. *PD* mosaic (virus), mosaic disease; **m. aucuba** potato aucuba mosaic virus; **m. du chou-fleur** cauliflower mosaic virus; **m. du concombre** cucumber mosaic virus, CMV; **m. énation du pois** pea enation mosaic virus; **m. (commune) du haricot** bean (common) mosaic virus; **m. internervaire** interveinal mosaic; **m. jaune du haricot** bean yellow mosaic virus; **m. jaune du navet** turnip yellow mosaic virus; **m. des nervures** vein mosaic; **m. du pêcher** peach mosaic virus; **m. de la poire** pear ring pattern mosaic virus; **m. (commune) du pois** pea mosaic virus; **m. du pommier** apple mosaic virus; **m. rouge nécrotique du cerisier** cherry necrotic rusty mottle virus
moteur *m.* motor, engine; **m. (à) deux temps** two-stroke engine; **m. à essence** petrol engine; **m. à pétrole** oil engine; **m. (à) quatre temps** four-stroke engine
motobatteuse *f.* motor thresher
motobêche *f.* (hand) mechanical digger
motobineur *m.*, **motobineuse** *f.* rotary tiller
motoculteur *m.* power-driven cultivator, walking tractor

motoculture *f.* mechanized agriculture or horticulture; **m. de plaisance** mechanized (pleasure) gardening
motofaucheuse *f.* motor mower, motor scythe
motohoue *f.* motor hoe
motopompe *f.* movable motor pump
motorisation *f.* mechanization
motorisé,-e *a.* fitted with a motor, motorised
mototondeuse *f.* motor mower
mottage *m.* lumpiness (in plant protection mixture)
motte *f.* mound; **culture en mottes** growing in blocks, block propagation; **m. de taupe** molehill; **m. (de terre)** clod, lump of soil; *Hort.* ball of earth around roots of plants; **en motte** (of plant) wrapped in burlap with a ball of earth around the roots; **m. de gazon** sod, turf; **m. pressée** soil block; **m. de tourbe** peat block
motteuse *f.* block-making machine
mou, mol *a.m.*, **molle** *a.f.* soft; *Hort.* tender, half-hardy
mouche *f.* 1. fly (Diptera) and other insects; **m. de l'asperge** asparagus fly (*Platyparea poeciloptera*); **m. de la betterave** beet leaf miner (*Pegomya betae*); **m. blanche** whitefly (*Trialeurodes vaporariorum*) (= **aleurode**); **m. de la carotte** carrot fly (*Psila rosae*); **m. du céleri** celery fly (*Philophylla (Acidia) heraclei*); **m. des cerises** cherry fruit fly (*Rhagoletis cerasi*); **m. du chou** cabbage root fly (*Erioischia (Hylemyia) brassicae*); **m. des chrysanthèmes** chrysanthemum leaf miner (*Phytomyza atricornis*); **m. des fruits = m. méditerranéenne** (*q.v.*); **m. grise des semis** bean seed fly (*Delia platura = Hylemyia cilicrura*); **m. méditerranéenne (des fruits)** Mediterranean fruit fly, medfly (*Ceratitis capitata*) (= **cératite**); **m. mineuse serpentine américaine** American serpentine leaf-mining fly (*Liriomyza trifolii*); **m. des narcisses** large narcissus fly, bulb fly (*Merodon equestris*), small narcissus fly, lesser bulb fly (*Eumerus strigatus, E. tuberculatus*); **m. des oeillets** carnation fly (*Delia brunnescens*); **m. de l'oignon** onion fly (*Hylemyia (Delia) antiqua*); **m. de l'olive** olive fly (*Dacus oleae*); **m. de l'orange = m. méditerranéenne** (*q.v.*); **m. de la pomme** (*Canada*) apple maggot fly (*Rhagoletis pomonella*); **m. à scie** sawfly (see also **tenthrède**); **m. à scie du rosier** banded rose sawfly (*Allantus (Emphytus) cinctus*); **m. des semis** seed fly (*Phorbia (Hylemyia) platura*); **m. des vendanges, m. du vinaigre** fruit fly, vinegar fly (*Drosophila* sp.). 2. spot, speck, stain
moucher *v.t. Vit.* to prune
moucheron *m.* gnat, midge
moucheté,-e *a.* spotty, speckled
moucheture *f.* spot, speck, speckle; **m. nécrotique du haricot** *PD* bean stipple streak virus
mouillabilité *f.* wettability
mouillable *a.* wettable, absorbent; **poudre mouillable** *PC* wettable powder
mouillant *m. PC* wetter, spreader; **m. anionique** anionic wetter
mouille, mouillère *f. Hort., Vit.* wet, low-lying area (= **moyère**)
mouillé,-e *a.* moist, damp, wet
mouille-bouche *f. inv.* juicy pear
mouiller *v.t.* to wet, to moisten; *Hort.* **sans mouiller les feuilles** without wetting the leaves
moulage *m.* grinding, milling
moulin *m.* mill; **m. à vent** windmill
moulu,-e *a.* ground, powdered
mourant,-e *a.* dying
moureiller (*m.*) **des Caraïbes** golden spoon (*Byrsonima crassifolia*); **m. des jardins, m. lisse** Barbados cherry (*Malpighia glabra*) (= **acérolier, cerisier des Antilles**)
mouron (*m.*) **rouge, m. des champs** scarlet pimpernel (*Anagallis arvensis*); **m. des oiseaux, m. blanc** chickweed (*Stellaria media*) (see also **morgeline**); **m. de fontaine** blinks (*Montia verna*)
mousse *f.* 1. moss; **m. blanche** sphagnum moss; **couvert de mousse** covered with moss. 2. froth, foam; **m. plastique** plastic foam
mousseron *m.* edible mushroom (*Agaricus campestris*) and other spp., including **m. (de la Saint Georges), m. vrai** St. George's mushroom (*Tricholoma gambosum*) (= **tricholome de la Saint-Georges**); **faux mousseron** fairy ring champignon (*Marasmus oreades*) (= **marasme d'oréade**)
mousseux,-euse *a.* 1. mossy; **rose mousseuse** moss rose. 2. frothy, foaming, sparkling (of wine); **vin mousseux** sparkling wine; **vin non mousseux** still wine. 3. *n.m.* sparkling wine
mousson *f.* monsoon
moût *m.* must of grapes, wort of beer, unfermented wine
moutarde *f.* mustard; **m. blanche** white mustard (*Sinapis alba*); **m. brune** Indian mustard (*Brassica juncea*) (= **chou-moutarde**); **m. des champs** charlock, wild mustard (*S. arvensis*) (= **sanve**); **m. noire** black mustard (*Brassica nigra*) (= **chou noir**); *a. inv.* **jaune moutarde** mustard yellow (colour)
moyen,-enne *a.* middle, average, medium; *n. f.* average; *n. m.* means; **moyen de lutte** means of control
moyennement *adv.* moderately
moyère *f.* low-lying wet area (= **mouille**)
mucilage *m.* mucilage, gum

mucilagineux,-euse *a.* mucilaginous
mucron *m.* mucro
mucroné,-e *a.* mucronate
mucune *f. Mucuna* sp., mucuna
mue *f. Ent.* instar, ecdysis
muflier *m.* snapdragon (*Antirrhinum majus*) (= **gueule-de-lion**); **m. bâtard** toadflax (*Linaria* sp.); **m. rose** weasel's snout (*Misopates (Antirrhinum) orontium*)
muguet *m.* **(de mai)** lily-of-the-valley (*Convallaria majalis*); **petit muguet** sweet woodruff (*Galium odoratum*); **m. géant** summer hyacinth (*Galtonia candicans*) (= **jacinthe du Cap**); **m. des pampas** cock's eggs (*Salpichroa rhomboidea*) (= **plante aux oeufs**)
mulch *m.* mulch; **m. en plastique noir** black plastic mulch
mulching *m.* = **paillage** (*q.v.*)
mulot *m.* field mouse (*Apodemus* sp.); (*Canada*) *Microtus* sp., including *M. pennsylvanicus*) (see also **campagnol, souris**)
multicaule *a.* many-stemmed, multicauline; **taille multicaule** *Coff.* etc. multistem pruning
multicaulie *f. Coff.* etc. multistem pruning system (see also **unicaulie**)
multichapelles *a. inv.* multispan; **serre multichapelles** multispan greenhouse
multifide *a.* multifid, multifidous
multiflore *a.* multiflorous, many-flowered; multiflora, polyantha (roses)
multilobé,-e *a.* multilobate
multiloculaire *a.* multilocular
multiplicateur (de semences) *m. Hort.* propagator (of selected varieties)
multiplication *f.* multiplication; *Hort.* propagation; *Bot.* **m. asexuée, m. végétative** vegetative propagation; **m. sexuée** sexual reproduction; **m. avec brouillard artificiel** mist propagation
multiplier *v.t. Hort.* to propagate
multipot *m.* multipot
mur *m.* wall; **m. fleuri** wall decorated with flowers
mûr,-e *a.* ripe, mature; **trop mûr** over-ripe
mûraie *f.* mulberry plantation
mûre *f.* 1. mulberry. 2. **m. (sauvage), m. de ronce** blackberry; **m. géante sans épines** giant thornless blackberry; **m. bleue** dewberry; **m. des Andes** Andes berry (see also **mûrier**)
mûreraie *f.* mulberry plantation
muret *m.* low wall, dry stone wall; *Vit.* etc. small terrace: wall retaining a terrace
mûrier *m.* **(noir)** 1. (common) mulberry tree (*Morus nigra*); **m. blanc** white mulberry (*M. alba*); **m. (blanc) pleureur** *M. alba* var. *pendula*; **m. de Chine, m. d'Espagne, m. à papier** paper mulberry (*Broussonetia papyrifera*); **m. rouge** red mulberry (*M. rubra*), 2. **m. (sauvage)** bramble, blackberry bush (*Rubus* sp.) (= **ronce**); **m. géant sans épines** giant thornless blackberry bush; **m. bleu** dewberry bush (*R. caesius*); **m. des Andes** Andes berry bush (*R. glaucus*) (see also **mûre**)
muriqué,-e *a.* muricate
mûrir *v.t.* to ripen; *v.i.* to become ripe, to ripen, to mellow (of soil)
mûrissage *m.* ripening
mûrisserie *f.* ripening shed, depot
mûrisseur *m.* ripener
musa *f. Musa* sp.; **m. fétiche** *M. religiosa*
muscade *f.* nutmeg; **(noix) (de) muscade** nutmeg; **fleur de muscade** mace (= **macis**)
muscadelle *f.* musk pear
muscadier *m.* nutmeg tree (*Myristica fragrans*); **m. de calebasse** calabash nutmeg (*Monodora myristica*) (= **faux muscadier**); **m. de Californie** Californian nutmeg (*Torreya californica*); **m. de Madagascar** Madagascar nutmeg (*Ravensara aromatica*); **m. de montagne, m. de forêt** mountain nutmeg (*Myristica fatua*)
muscari *m. Muscari* sp.; **m. à grappes** *M. racemosum*; **m. musqué, m. odorant** musk hyacinth (*M. moschatum*) (= **jacinthe musquée**); **m. plumeux** feather hyacinth (*M. comosum* var. *monstrosum*); **m. raisin** grape hyacinth (*M. botryoides*); **m. à toupet** tassel hyacinth (*M. comosum*) (= **poireau roux**)
muscat *a.* & *n. m. Vit.* muscat; **raisin muscat** muscat grape, muscadine grape (*Vitis vulpina*)
musqué,-e *a.* musky, musk-scented; **poire musquée** musk pear (= **muscadelle**); **rosier musqué** musk rose bush (*Rosa moschata*)
mutagène *a.* mutagenic; **capacité mutagène** mutagenicity; **traitement mutagène** mutagenic treatment
mutagénèse *f.* mutagenesis; **m. artificielle** artificial mutagenesis
mutagénicité *f.* mutagenicity
mutant,-e *a.* & *n.* mutant; **m. induit** induced mutant
mutation *f.* mutation; **m. gemmaire** bud mutation, sport; **m. somatique** somatic mutation
muté,-e *a.* mutated; **cellule mutée** mutated cell
mutique *a.* muticate, without awns or spines
mutisie *f. Mutisia* sp.
mycélien,-enne *a.* mycelial; **croissance mycélienne** mycelial growth
mycélium *m.* mycelium; *Mush.* spawn (= **blanc**)
mycoflore *f.* mycoflora
mycogone *f. Mush.* white mould, wet bubble (*Mycogone perniciosa*) (= **môle**)
mycologie *f.* mycology

mycologue *m.* mycologist, fungologist
mycoplasme *m.* mycoplasma; **maladie à mycoplasmes** mycoplasma disease
mycoplasmose *f.* mycoplasma disease
mycorhization *f.* mycorrhization, symbiotic association of a fungus mycelium with the roots of a seed plant; **m. contrôlée** controlled mycorrhization
mycorhize *f.* mycorrhiza; **mycorhizes à vésicules et arbuscules** vesicular-arbuscular mycorrhizas
mycorhizé,-e *a.* infected with mycorrhizas
mycorhizien,-enne *a.* mycorrhizal; **association mycorhizienne** mycorrhizal association
mycorhizique *a.* mycorrhizal
mycorhizogène *a.* mycorhizogenous
mycose *f.* mycosis, fungal disease
mycothèque *f.* fungus (culture) collection
mycotoxine *f.* mycotoxin
myosotis *m.* *Myosotis* sp., forget-me-not; **m. des Alpes** alpine forget-me-not (*M. alpestris*) (see also **ne-m'oubliez-pas**)
myrica *m.* *Myrica* sp., including sweet gale (*M. gale*) (= **galé**) (see also **arbre à cire**)
myriophylle *m.* *Myriophyllum* sp., watermilfoil
myristica *m.* *Myristica* sp., nutmeg (see **muscadier**)
myrobolan *m.* 1. myrobalan (*Terminalia* sp.). 2. myrobalan plum (*Prunus cerasifera*) (= **prunier myrobolan, mirobolan**); **m. d'Amérique** cocoplum (*Chrysobalanus icaco* = *C. purpureus*) (= **icaquier**); **m. emblic** emblic myrobolan (*Phyllanthus emblica*)
myrosperme *m.* *Myrospermum frutescens*, a tree (Leguminosae) whose seeds yield a strong smelling resin
myroxyle *m.* *Myroxylon* sp. (see also **baumier**)
myrrhe *f.* myrrh, sweet cicely (*Myrrhis odorata*)
myrrhide *f.*, **myrrhis** *m.* *Myrrhis* sp.; **myrrhide odorante** myrrh, sweet cicely (*M. odorata*)
myrtaie *f.* myrtle (*Myrtus* sp.) plantation
myrte *m.* *Myrtus* and other spp.; **m. commun** common myrtle (*Myrtus communis*); **m. des marais** sweet gale (*Myrica gale*); **m. musqué** ugni shrub (*Myrtus ugni*)
myrtille *f.* bilberry, whortleberry (*Vaccinium myrtillus*); (*Canada*) sourtop blueberry (*V. myrtilloides*); **m. américaine, m. cultivée** swamp, high bush, tall blueberry (*V. corymbosum*); **m. ponctuée** cowberry (*V. vitis-idaea*); **m. des tourbières** small European cranberry (*V. oxycoccus*) (see also **airelle, bleuetier, canneberge**)
myrtillier *m.* bilberry, whortleberry bush (see **myrtille**)
myxa *f.* fruit of sebestens (*Cordia myxa*)

N

nacré,-e *a.* nacreous, pearly
nageant,-e *a. Bot.* floating (of leaf of water plant)
nain,-e *a.* dwarf; **arbre nain** dwarf tree
nanifiant *m.* dwarfing compound
nanisant,-e *a.* dwarfing
naniser *v.t.* to dwarf (plant)
nanisme *m.* dwarfing, stunting; **n. du pêcher** peach phony virus
nantais,-e *a.* of the Nantes region
napel *m.* monkshood (*Aconitum napellus*) (= **aconit napel**)
napier *m.* Napier, elephant grass (*Pennisetum purpureum*)
napiforme *a.* napiform
nappe *f.* sheet; **n. d'eau** sheet of water; n. aquifère, n. phréatique water table, ground water; **n. chauffante** heating mat; **n. d'irrigation** irrigation mat
nara *m.* narras (*Acanthosicyos horrida*)
naranjille *f.* naranjilla, Quito orange (*Solanum quitoense*) (= **morelle de Quito**)
narcisse *f. Narcissus* sp. and other plants; **faux-narcisse** = **n. des prés** (*q.v.*); **n. d'automne** winter daffodil (*Sternbergia lutea*); **n. Barrii** narcissus of Barrii group; **n. des bois** = **n. des prés** (*q.v.*); **n. à bouquets, n. de Constantinople** polyanthus narcissus (*Narcissus tazetta*); **n. double, n. à fleurs doubles** double-flowered narcissus; **n. à fleurs d'eucharis** narcissus of Leedsii group; **n. à hampes uniflores** narcissus with single-flowered stems; **n. à hampes multiflores** narcissus with multi-flowered stems; **n. incomparable, n. à couronne** chalice cup narcissus (*N. incomparabilis*); **n. à petite coupe** see **n. Barrii**; **n. des poètes** poet's narcissus, pheasant's eye (*N. poeticus*) (= **jeannette**); **n. des prés** daffodil (*N. pseudo-narcissus*) (= **jonquille**); **n. sauvage** = **n. des prés** (*q.v.*); **n. (à) trompette** daffodil, trumpet daffodil (see also **coucou, jonquille**)
nard *m.* 1. **n. raide** mat grass (*Nardus stricta*). 2. Essential oil plants: **n. agreste** *Valeriana phu*; **n. celtique** celtic spikenard (*Valeriana celtica*); **n. indien** Indian nard, ancient spikenard (*Nardostachys jatamansi*) (= **nardostachyde**); **n. de la Madeleine** *Cymbopogon nardus*; **n. de montagne** zedoary (*Curcuma zedoaria*); **n. sauvage** asarabacca (*Asarum europeum*) (= **oreillette**); **n. syriaque** = **n. de la Madeleine** (*q.v.*)
nardostachyde *f. Nardostachys* sp., including *N. jatamansi*, ancient spikenard (= **nard indien**)
nashi *m.* Oriental pear (see **poire orientale**)
nasitor, nasitort *m.* garden cress (*Lepidium sativum*) (= **cresson alénois**)
nasturce *m. Nasturtium* sp., including watercress
natte *f.* mat, matting
naturalisation *f.* naturalization
naturalisé,-e *a.* naturalized
naturaliser *v.t.* to naturalize, to acclimatize
navel *m. Cit.* navel
navet (potager) *m.* turnip (*Brassica rapa*); **n. de Suède** swede (*B. napus* var. *napobrassica*) (= **rutabaga**); **n. du diable** white bryony (*Bryonia dioica*)
naviculaire *a.* navicular, boat-shaped
ndole *m. Afr.* bitter leaf (*Vernonia amygdalina*), a leaf vegetable
nébulisateur *m.* mist blower, mister, fogger
nébulisation *f.* misting, fogging
nébulosité *f.* 1. nebulosity, haziness. 2. patch of mist
nécrophore *m.* carrion bettle (*Necrophorus* sp.)
nécrose *f.* necrosis, canker; **n. apicale** blossom-end rot (of tomato); **n. bactérienne de la vigne** *Vit.* bacterial blight (*Xanthomonas ampelina*) (= **maladie d'Oléron**); **n. du bois** coral spot (*Nectria cinnabarina*); **n. des bords des feuilles** tipburn (of lettuce); **n. du collet** *Vit.* etc. collar rot; **n. corticale** bark necrosis; **n. du liber de l'orme** elm phloem necrosis virus; **n. de la partie stylaire** *Cit.* stylar end necrosis; **n. de la tulipe** tobacco necrosis virus (on tulip); **n. de la vigne** *Vit.* dead arm disease (*Phomopsis* (*Cryptosporella*) *viticola*)
nécrosé,-e *a.* necrotic, cankered; **portion nécrosée** dead portion
nécroser *v.t.* to cause necrosis, to canker; **se nécroser** to become cankered
nécrotique *a.* necrotic
nectaire *m.* nectary, honey-cup
nectar *m.* nectar
nectarien,-ienne *a.* 1. nectarous. 2. nectariferous, nectar-bearing
nectarifère *a.* nectariferous
nectarine *f.* (freestone) nectarine (*Prunus*

persica var. *nectarina*) (see also **brugnon, pavie, pêche**)

néfaste *a.* which has a bad effect

nèfle *f.* medlar fruit; **n. d'Amérique** sapodilla fruit (= **sapotille**); **n. du Japon** loquat fruit

néflier *m.* **(commun)** medlar tree (*Mespilus germanica*); **n. africain** West African ebony (*Diospyros mespiliformis*) (= **ébénier de l'ouest africain**); **n. d'Amérique** sapodilla (*Manilkara achras* = *Achras sapota*) (= **sapotillier**); **n. de Bronvaux** *Crataegomespilus* + *dardari* (graft hybrid); **n. du Japon** loquat tree (*Eriobotrya japonica*) (= **bibacier**); **n. de Saujon** = **n. de Bronvaux** (*q.v.*)

négondo, négundo *m.*, **n. à feuilles de frêne** box elder (*Acer negundo* = *Negundo aceroides*) (= **érable negundo**)

neïroun *m.* olive bark beetle (*Phloeotribus scarabeoïdes*)

nélombo, nélumbo *m.* *Nelumbium* sp., lotus, sacred bean, Chinese water lily (see also **lotus**)

nématicide *m.* nematicide; *a.* nematicidal; **action nématicide** nematicidal action

nématode *m.* nematode (see also **anguillule**); **n. acuminé** pin nematode (*Paratylenchus* sp.); **n.-anneau** ring nematode (*Criconemoides* sp.); **n. de la carotte** carrot cyst nematode (*Heterodera carotae*); **n. cécidogène** root-knot nematode (*Meloidogyne* sp.); **n. du chrysanthème** chrysanthemum eelworm (*Aphelenchoides ritzemabosi*); **n. doré** golden nematode (*Heterodera rostochiensis*); **n. endoparasite des racines** root-knot nematode (*Meloidogyne* sp.) (= **n. cécidogène**); **n. des feuilles** leaf-blotch nematode (*Aphelenchoides fragariae*); **n. formant des kystes** = **n. kystique** (*q.v.*); **n. kystique, n. à kyste(s)** cyst nematode (*Heterodera* sp.); **n. kystique du cactus** cactus cyst nematode (*Heterodera cacti*); **n. des lésions racinaires** root-lesion nematode; **n. libre** free-living nematode; **n. de l'oignon** stem and bulb eelworm (*Ditylenchus dipsaci*) (= **n. des tiges et des bulbes**); **n. (parasite) des racines** root-lesion nematode; **n. provoquant le nanisme** stunt nematode (*Tylenchorhynchus* sp.); **n. des racines noueuses** = **n. cécidogène** (*q.v.*); **n. spiral** spiral nematode (*Helicotylenchus* sp.); **n. stylet** dagger nematode (*Xiphinema* sp.); **n. de la tige** stem eelworm; **n. des tiges et des bulbes** stem and bulb eelworm (= **n. de l'oignon** *q.v.*)

nématofaune *f.* nematode fauna

nématologie *f.* nematology

nématologique *a.* nematological

nématophage *a.* nematophagous, nematode-eating, nematode-destroying; **champignons nématophages** nematophagous fungi

nemesia *m.* *Nemesia* sp., including *N. strumosa*

némophile *f.* *Nemophila* sp., Californian bluebell

ne-m'oubliez-pas *m. inv.* forget-me-not (*Myosotis scorpioides* and other spp.) (see also **myosotis**)

nénufar, nénuphar *m.* water lily (*Nymphaea* and *Nuphar* sp.); **n. blanc** white water lily (*Nymphaea alba*); **n. commun, n. jaune, n. des étangs** yellow water lily (*Nuphar lutea*); **n. du Nil** East Indian lotus (*Nelumbium nucifera*) (= **lotus égyptien**); **n. parfumé** *Nymphaea odorata* (see also **nymphéa**)

néocalédonien,-enne *a.* from, of New Caledonia

néoformation *f.* neoformation

néoplasme *m.* neoplasm

néoténie *f.* neoteny

népenthes *m.* *Nepenthes* sp., pitcher plant

népèta *f.* *Nepeta* sp. (see also **cataire**)

néphélomètre *m.* nephelometer

néphélométrie *f.* nephelometry

néphrolepis *m.* *Nephrolepis* sp.; **n. élevé** ladder fern (*N. exaltata*)

néré see **nété**

nérine *f.* nerine (*Nerine* sp.), including Guernsey lily (*N. sarniensis*) (= **amaryllis de Guernesey**)

néroli *m.* neroli

nerprun *m.* buckthorn (*Rhamnus* sp.); **n. alaterne** = **alaterne** *q.v.*; **n. bourdaine** alder buckthorn (*Frangula alnus* = *Rhamnus frangula*) (= **aune noir, bourdaine**); **n. commun** common buckthorn (*R. cathartica*); **n. purgatif** = **n. commun** (*q.v.*); **n. des teinturiers** Avignon berry (*R. infectoria*)

nervation *f.* nervation, nervature, venation; **n. noire des crucifères** *PD* black rot of crucifers (*Xanthomonas campestris*)

nervé,-e *a.* ribbed, nervate

nervifolié,-e *a.* ribbed (of leaf)

nervure *f.* nervure, rib, vein of leaf; **n. centrale, n. médiane** midrib; **(maladie des) grosses nervures (de la laitue)** big vein disease (of lettuce); **nervures chlorotiques** *PD* vein clearing; **n. noire (du chou)** = **nervation noire des crucifères** (*q.v.*); **à nervures parallèles** parallel-veined

nervuré,-e *a.* ribbed

nété, néré *m. Afr.* West African locust bean tree (*Parkia biglobosa, P. clappertoniana*)

nettoyage *m.* cleaning; **n. des interlignes** interrow cleaning; *Vit.* preliminary pruning, (*South Africa*) clearing; *Mush.* trashing

nettoyer *v.t.* to clean

nettoyeur *m.* cleaning machine, cleaner; **n. de semences** seed cleaner

neutre *a.* 1. *Bot.* neuter, asexual. 2. (of bee) worker
néverdié *m. Afr.* horseradish tree (see **ben**)
nez coupé *m.* bladder nut (*Staphylea pinnata*) (= **faux pistachier, staphylier**)
niaouli *m. Melaleuca* sp., cajeput (= **mélaleuque**)
nicandra *m.*, **nicandre** *f. Nicandra* sp.; **n. faux-coqueret** apple of Peru (*N. physaloides*)
nichoir *m.* nesting box
nicotine *f.* nicotine
nid *m.* nest; **n. de chenilles** nest of caterpillars; **n. de fourmis** ant's nest
niébé *m.* cowpea, (*USA*) southern pea (*Vigna unguiculata*)
nigelle *f. Nigella* sp.; **n. aromatique, n. cultivée** black cumin (*N. sativa*) (= **toute-épice**); **n. des champs** *N. arvensis*; **n. de Damas** love-in-a-mist (*N. damascena*) (= **cheveux de Vénus**)
nipa *m.* nipa palm (*Nypa fruticans*)
nitralin *m.* nitralin
nitratage *m.* nitrate application
nitrate *m.* nitrate; **n. d'ammonium** ammonium nitrate; **n. d'argent** silver nitrate; **n. de chaux** calcium nitrate; **n. de potasse** potassium nitrate; **n. de soude** sodium nitrate; **n. de soude du Chili** Chilean nitrate
nitratophile *a.* nitrogen-loving (of plant)
nitre *m.* nitre, saltpetre
nitré,-e *a.* nitrated; **composé nitré** nitro-compound
nitreux,-euse *a.* nitrous
nitrifiant,-e *a.* nitrifying
nitrification *f.* nitrification
nitrifier *v.t.* to nitrify
nitrique *a.* nitric
nitrite *m.* nitrite
nitrofène *m.* nitrofen
nitrojection *f.* soil ammonia injection
nitrosation *f.* nitrosation
nivéal,-e *a., m. pl.* **-aux** *Bot.* winter-flowering; **érosion nivéale** (*Canada*) winter (snow) erosion
niveau *m.* level; **n. des basses eaux** low water mark; **n. des hautes eaux** high water mark; **n. critique** critical level; **n. de la nappe phréatique** water table level, underground water level; **n. de vie** standard of living
niveler *v.t.* to level
niveleur,-euse *a.* levelling; *n.* leveller (person); *n. m.* small harrow; *n. f.* grader (machine), implement for levelling, leveller
nivellement *m.* levelling, contouring; **repère de nivellement** bench mark
nivéole *f.* snowdrop (*Galanthus nivalis*); **n. (perce-neige)** spring snowflake (*Leucojum vernum*)
njansan *m. Afr.* wood-oil-nut tree (*Ricinodendron africanum*)
noble-épine *f.* = **aubépine** (*q.v.*)
nocivité *f.* noxiousness, harmfulness
noctiflore *a.* noctiflorous
noctuelle *f.* noctuid moth (Noctuidae); **n. de l'artichaut** globe artichoke noctuid (*Hydraecia xanthenes*); **n. des bourgeons** (*Cacao*) *Earias biplaga*; **n. du chou** cabbage moth (*Mamestra (Barathra) brassicae*); **n. gamma** silver Y moth (*Plusia gamma*); **n. potagère** tomato moth (*Lacanobia (Diataraxia) oleracea*); **n. de la tomate** tomato fruit worm (*Heliothis armigera*) (see also **ver gris**)
nodal,-e *a., m. pl.* **-aux** nodal
nodifère *a.* nodule-bearing
nodosité *f.* nodule; **n. radiculaire** root nodule; *Vit.* nodule caused by phylloxera
nodulaire *a.* nodular
nodulant,-e *a.* nodulating; **arachide nodulante** nodulating groundnut
nodulation *f.* nodulation
nodule *m.* nodule
noeud *m.* knot; knot (in wood); *Bot.* node, joint (see also **entre-noeud**)
noir,-e *a.* black, dark; *n. m. Vit.* sooty mould (*Capnodium* sp., *Cladosporium* sp., *Fumago* sp.) (= **fumagine**)
noirâtre *a.* blackish
noiseraie *f.* area planted with hazels or walnuts
noisetier *m.* 1. *Corylus* sp.; **n. (des bois), n. commun** hazel, cobnut tree (*C. avellana*) (= **avelinier, coudrier**); **n. (franc)** filbert tree (*C. maxima*); **n. de Byzance, n. de Constantinople** Turkish hazel (*C. colurna*); **n. pourpre** *C. maxima* var. *atropurpurea*; **n. tortueux** twisted hazel (*C. avellana* var. *contorta*). 2. Other genera. **n. de cajou** cashew (*Anacardium occidentale*) (= **anacardier**); **n. de Cayenne** sapotón (*Pachira macrocarpa*); **n. du Chili** Chilean hazel tree (*Gevuina avellana*); **n. du Gabon** Gaboon nut tree (*Coula edulis*); **n. de la Guyane** provision tree (*Pachira aquatica*) (= **châtaignier de la Guyane**); **n. de(s) sorcière(s)** witch-hazel (*Hamamelis* sp.)
noisette *f.* hazel nut, filbert; **n. du Chili** Chilean hazel nut; **n. de terre** chufa (*Cyperus esculentus*)
noisetterie *f.* area planted with hazel or filbert trees
noix *f.* 1. walnut; **n. écalée** shelled walnut; **n. verte** green walnut; **coquille de noix** walnut shell; **huile de noix** walnut oil, nut oil (see also **noyer**). 2. nut; **n. d'abrasin** tung fruit; **n. d'acajou, n. de cajou** cashew nut; **n. d'Afrique** African walnut, Gaboon nut; **n. d'Amérique** hickory, pecan; **n. d'arec** areca

nut; **n. de Banda** = **n. muscade** (*q.v.*); **n. des Barbades** physic nut; **n. du Brésil, n. de Para** Brazil nut; **n. de cajou** = **n. d'acajou** (*q.v.*); **n. cendrée** butternut; **n. de coco** coconut; **n. de Coyol** Paraguay palm (*Acrocomia sclerocarpa*) (= **gru-gru**); **n. du Gabon** = **n. d'Afrique** (*q.v.*); **n. de Jacapucayo** = **n. de paradis** (*q.v.*); **n. de kola** kola nut; **n. de Macassar, n. papoue** Macassar, Papua nutmeg (*Myristica argentea*); **n. muscade** nutmeg; **n. de palme** oil palm seed; **n. de Para** = **n. du Brésil** (*q.v.*); **n. de paradis** paradise nut, Sapucaia nut; **n. du Queensland** Queensland nut, macadamia; **n. de Ravensara** Madagascar nutmeg (*Ravensara aromatica*); **n. de Sapucaia** = **n. de paradis** (*q.v.*); **n. vomique** nux vomica (*Strychnos nux-vomica*). 3. **n. de galle** gall apple; **n. de terre** pignut (*Conopodium majus*)

nolane *f. Nolana* sp., Chilian bell-flower

noli-me-tangere *n. m. inv.* touch-me-not (*Impatiens noli-tangere*)

nom *m.* name; **n. commun** common name; **n. générique** generic name; **n. scientifique** scientific name; **n. spécifique** specific name

nombre *m.* number; **n. chromosomique** chromosome number

nombril *m. Bot.* hilum; eye of fruit

nomenclature *f.* nomenclature; **n. botanique** botanical nomenclature

non-autofertile *a.* cross-fertile

non-cible *a. PC, PD* non-target

non-concordance *f.* dephasing

non-culture *f.* no cultivation, no tillage (treatment in experiments)

nonette voilée *f. Fung.* slippery jack (*Boletus luteus*) (see also **cèpe, indigotier**)

non-ombragé,-e *a.* unshaded

non-taille *f.* no pruning (treatment); **n.-t. pendant une année** no pruning for a year

non-traité,-e *a.* untreated

nopal *m.* = **figuier d'Inde** (*q.v.*) and other *Opuntia* spp.

nopalerie *f.* plantation of **nopal**

noria *f.* Persian wheel, bucket chain

normalisation *f.* standardization

normaliser *v.t.* to standardize

norme *f.* standard, specification

noruron *m.* noruron

nouaison *f.* fruit setting (= **nouure**)

noué,-e *a.* set (of fruit)

nouer *v.t.* to tie; **(se) nouer** *v.i. Hort.* to set (of fruit)

noueux,-euse *a.* knotty, gnarled

nourricier,-ière *a.* nutritious, nutritive; **tissu nourricier** nutritive tissue

nourriture *f.* food

nouure *f. Hort.* fruit setting (= **nouaison**)

nouveauté *f. Hort.* mutation, sport

noyau *m.* nucleus (of cell), stone (of fruit) kernel; **n. de pêche** peach stone; **à noyau adhérent** clingstone (of peach); **à noyau libre** freestone; **à noyau semi-adhérent** semi-freestone; **n. fendu** *PD* split pit (in peach); **fruits à noyaux** stone fruits (see also **pépin**)

noyer *m.* 1. **n. (commun)** walnut tree (*Juglans regia*); **n. cendré** butternut (*J. cinerea*); **n. noir** black walnut tree (*J. nigra*) (see also **noix**). 2. nut tree; **n. d'Afrique** African walnut tree, Gaboon nut tree (*Coula edulis*) (= **noisetier du Gabon**); **n. d'Amérique** hickory, pecan tree (*Carya* spp.), black walnut (*Juglans nigra*) (sometimes applied to Brazil nut tree); **n. des Barbades** physic nut (*Jatropha curcas*); **n. du Brésil** Brazil nut tree (*Bertholettia excelsa*) (sometimes called **n. d'Amérique**); **n. à feuilles de frène** wing nut (*Pterocarya caucasica*); **n. du Gabon** = **n. d'Afrique** (*q.v.*); **n. de Jacapucayo** = **n. de paradis** (*q.v.*); **n. de Para** = **n. du Brésil** (*q.v.*); **n. de paradis** paradise nut tree, sapucaia nut tree (*Lecythis zabucajo* and other spp.); **n. du Queensland** Queensland nut tree, macadamia (*Macadamia ternifolia, M. tetraphylla*) (= **macadamier**); **n. de Sapucaia** = **n. de paradis** (*q.v.*)

noyeraie *f.* walnut plantation; plantation of other nut trees

nu,-e *a.* bare; *Bot.* naked; **sol nu** bare soil

nuancier *m.* chart of colour range

nucelle *f.* nucellus

nucellaire *a.* nucellar

nuciculteur *m.* nut grower

nuciculture *f.* nut growing

nucifère *a.* nuciferous, nut-bearing

nuciforme *a.* nuciform

nucléaire *a.* nuclear

nucléase *f.* nuclease

nucléé,-e *a.* nucleate(d)

nucléique *a.* nucleic; **acide nucléique** nucleic acid

nucléole *m.* nucleolus

nucléoprotéine *f.* nucleoprotein

nucléoside *m.* nucleoside

nucléotide *m.* nucleotide

nucléus *m.* nucleus (= **noyau**)

nuculaine *m.* nuculanium

nucule *f.* nutlet, nucule

nuculeux,-euse *a.* containing nucules

nudiflore *a.* nudiflorous

nuée *f.* large cloud, storm cloud; **n. de sauterelles** swarm of locusts

nuile *f.* **(grise), n. noire** cucumber gummosis (*Cladosporium cucumerinum*) (= **cladosporiose des cucurbitacées**); **n. rouge** cucurbit anthracnose (*Colletotrichum lagenarium*) (= **anthracnose des cucurbitacées**)

nuisibilité *f.* noxiousness
nuisible *a.* harmful, noxious; **concurrence nuisible** harmful competition; **insectes nuisibles** insect pests; **plantes nuisibles** weeds, noxious plants
nummulaire *f.* creeping jenny, moneywort (*Lysimachia nummularia*)
nutation *f.* nutation
nutritif,-ive *a.* nutritive; **valeur nutritive** food value
nutrition *f.* nutrition; **n. azotée** nitrogen nutrition; **n. équilibrée** balanced nutrition; **n. minérale** mineral nutrition
nyctanthe *m.* night-flowering jasmine (*Nyctanthes arbor-tristis*)
nycthéméral,-e *a., m. pl.* **-aux** nycthemeral
nyctipériode *f.* night period (in day/night treatments)
nyctipériodique *a.* short-day (of plant)
nymphe *f. Ent.* nymph, pupa, chrysalis
nymphéa *m.*, **nymphée** *f. Nymphaea* sp., water lily; **n. bleu du Cap** Cape blue waterlily (*N. capensis*) (see also **nénufar**)
nymphose *f.* pupation
nyssa *m. Nyssa* sp., tupelo

O

oasis *f.* oasis
oba *m. Afr.* African mango (*Irvingia gabonensis*) (= **manguier sauvage**)
obconique *a.* obconic, obconical
obier *m.* guelder rose, snowball tree (*Viburnum opulus*) (see also **boule de neige, viorne**)
oblique *a.* oblique
oblong,-ongue *a.* oblong
obovale *a.* obovate
obscurcissement *m.* darkening
obscurité *f.* darkness; **à l'obscurité** in the dark
obtenteur,-trice *n. Hort.* breeder, raiser (of a new variety)
obtention *f. Hort.* creation, production (of a new variety); **"obtention Dubois"** "bred by Dubois"
obturation *f.* obturation, closing, sealing
obtus,-e *a.* blunt, rounded
obvoluté,-e *a.* obvolute
ocelle *m. Bot.* ocellus
ocellé,-e *a.* ocellate
ocre *a. inv.* ochre
ocréa, ochréa *f.* ocrea, ochrea; *Sug.* dewlap
odeur *f.* smell, scent
odontalgique *a.* & *n.m.* odontalgic
odontoglosse *m. Odontoglossum* sp., almond-scented, violet-scented orchid
odorant,-e *a.* odorous, sweet-smelling, fragrant
odoriférant,-e *a.* odoriferous, sweet-smelling, fragrant
oedème *m.* oedema; **o. (du pélargonium)** *PD* (geranium) oedema
oeil *m., pl.* **yeux** eye; *Bot., Hort.* eye, bud (of a plant), eye (of a fruit); **cavité de l'oeil (d'une pomme)** eye cavity, stylar cavity (of apple); *Ban.* incipient sucker; **o. adventif** adventitious bud; **o. axillaire** axillary bud; **o. à bois** wood bud; **o. sur bourse** bourse bud; **o. dormant** dormant bud; bud grafted at the end of the season which develops the following spring; **o. fertile** fruitful bud; **o. à fleur, o. floral** fruit bud; **o. franc** well-developed bud; **o. fructifère** fruiting bud; **o. latent** latent bud; **o. latéral** = **o. axillaire** (*q.v.*); **o. poussant** grafted bud which develops immediately; **o. stipulaire** basal bud; **o. terminal** terminal bud
oeil-de-boeuf *m., pl.* **oeils-de-boeuf** *Hort.* oxeye daisy (*Chrysanthemum leucanthemum*), yellow chamomile (*Anthemis tinctoria*)
oeil-de-chat *m., pl.* **oeils-de-chat** *Bot.* nicker nut, bonduc seed (*Caesalpinia bonducella*)
oeil-de-Dieu *m., pl.* **oeils-de-Dieu** pasque flower (*Pulsatilla vulgaris*)
oeil-de-dragon *m., pl.* **oeils-de-dragon** longan (*Euphoria longana*)
oeil-de-faisan *m., pl.* **oeils-de-faisan** pheasant's eye narcissus (*Narcissus poeticus*)
oeil-de-paon *m., pl.* **oeils-de-paon** tiger flower (*Tigridia pavonia*) (= **belle-d'onze-heures**); **oeil-de-paon de l'olivier** *PD* olive leaf spot (*Cycloconium oleaginum*)
oeil-du-Christ *m., pl.* **oeils-du-Christ** 1. Italian starwort (*Aster amellus*) and other spp. 2. yellow hawkweed (*Tolpis barbata*)
oeillet *m.* carnation, pink (*Dianthus* sp.) and other plants; **o. pour la fleur coupée** carnation for cut flower production; **o. de bouc** see **o. sauvage; o. à bouquets** carnation (*Dianthus caryophyllus*); **o. de Chine** Chinese pink, Indian pink (*D. chinensis*); **o. des fleuristes** = **o. à bouquets** (*q.v.*); **o. giroflée** clove-pink; **o. horticole** = **o. à bouquets** (*q.v.*); **o. d'Inde** French marigold (*Tagetes patula*) (see also **rose d'Inde**); **o. jaspé** picotee; **o. maritime** sea-pink, thrift (*Armeria maritima*); **o. mignardise** pink (*D. plumarius*) (= **mignardise**); **o. des montagnes** maiden pink (*D. deltoides*); **o. de Pâques** poet's narcissus, pheasant's eye narcissus (*Narcissus poeticus*); **o. de poète, o. des poètes** sweet william (*D. barbatus*) (= **jalousie**); **o. des prés** ragged robin (*Lychnis flos-cuculi*); **o. sauvage** Deptford pink (*Dianthus armeria* and other *Dianthus* spp.); **o. tiqueté** = **o. jaspé** (*q.v.*)
oeilletiste *m.* carnation grower
oeilleton *m.* offset, offshoot; eye, bud
oeilletonnage *m.* 1. removal of surplus offsets or buds, disbudding; **o. du bananier** *Ban.* suckering, removal of unwanted offshoots from banana plant. 2. propagation by offsets etc.
oeilletonner *v.t.* 1. to remove offsets or buds. 2. to propagate by (separating) offsets etc.
oeillette *f.* oil poppy, opium poppy (*Papaver somniferum*); **huile d'oeillette** poppy seed oil
oenanthe *f. Oenanthe* sp.; **o. fistuleuse** water dropwort (*O. fistulosa*); **o. safranée** hemlock water dropwort (*O. crocata*)

oenologie *f.* oenology, science of wine
oenothère *m.* *Oenothera* sp., evening primrose (= **onagraire**); **o. bisannuelle** common evening primrose (*O. biennis*); **o. blanc** *O. tetraptera*; **o. élégant** *O. speciosa*; **o. à feuilles de pissenlit** *O. taraxifolia* (= *O. acaulis*) (= **onagraire**)
oeuf *m.* egg, ovum; **o. végétal** eggplant (*Solanum melongena*) (= **aubergine**)
officinal,-e *a., m. pl.* **-aux** officinal, medicinal
offre *f.* offer; **l'o. et la demande** supply and demand
ogival,-e *a., m. pl.* **-aux** ogival
ogive *f.*, **en ogive** often used to mean "pointed"
ognon *m.* = **oignon** (*q.v.*)
oïdié,-e *a.* infected with oidium
oïdium *m.* oidium, powdery mildew; **o. (de la vigne)** vine powdery mildew (*Uncinula necator*); **o. américain** American gooseberry mildew (*Sphaerotheca mors-uvae*); **o. des cucurbitacées** cucurbit powdery mildew (*Erysiphe cichoracearum*); **o. du framboisier** raspberry powdery mildew (*Sphaerotheca macularis*); **o. du groseillier** = **o. américain** (*q.v.*); **o. de l'hévéa** hevea powdery mildew (*Oidium heveae*); **o. du pommier** apple powdery mildew (*Podosphaera leucotricha*); **o. du rosier** rose mildew (*S. pannosa*)
oignon *m.* 1. bulb of plant (= **bulbe**); **o. à fleurs** bulb of ornamental plant. 2. onion (*Allium cepa*); **o. de Mulhouse** type of onion grown from small bulbs and transplanted; **o. patate** potato onion (*A. cepa* var. *aggregatum*); **o. à planter** onion set; **petits oignons** spring onions, pickling onions; **o. d'Egypte** tree onion (*A. cepa* var. *viviparum*); (= **rocambole**) (*q.v.*)
oignonet *m.* small onion, onion set
oignonière *f.* onion bed, land down to onions
oiseau (*m.*) **de paradis** *Hort.* bird of paradise flower (*Strelitzia reginae*); *Caesalpinia gilliesii*
oldenlandie *f.* *Oldenlandia* sp.
oléagineux,-euse *a.* oleaginous, oil-yielding; **graines oléagineuses** oil seeds; **plantes oléagineuses** oil plants; *n. m. pl.* **oléagineux** oil plants
oléandre *m.* old name for oleander (see **laurier-rose**)
oléastre *m.* wild olive (*Olea europaea* var. *oleaster*)
oléicole *a.* relating to olive growing or vegetable oils
oléiculteur *m.* olive grower
oléiculture *f.* olive growing: olive-oil industry
oléifère *a.* oil-producing, oleiferous
oléiforme *a.* oily, of oily consistency
oléocellose *f.* *Cit.* oleocellosis, rind-oil spot
oléracé,-e *a.* oleraceous
oliban *m.* Bible frankincense (*Boswellia carteri*) (see also **arbre à encens**)
oligo-élément *m., pl.* **oligo-éléments** minor element, trace element
oligotrophe *a.* oligotrophic
olivacé,-e *a.* olivaceous, olive-green
olivaie *f.* olive plantation, olive grove
olivaison *f.* olive harvest, olive crop: olive season
olive *f.* olive; **huile d'olive** olive oil; **o. de Nice** type of olive (= **cailletier**)
oliver *v.i.* & *v.t.* to pick olives
oliveraie *f.* olive plantation, olive grove
oliverie *f.* olive-oil factory
olivette *f.* 1. olive plantation, olive grove; **o. déprimée** declining olive grove; small olive. 2. name given to several grape varieties with olive-shaped fruit. 3. oval tomato
oliveur,-euse *n.* olive picker
olivier *m.* olive tree (*Olea europaea*); **o. de Bohème** oleaster (*Elaeagnus angustifolia*) (= **datte de Trébizonde, chalef à feuilles étroites**); **o. odorant** fragrant olive (*Osmanthus fragrans*)
ombelle *f.* umbel; **en ombelle** umbellate
ombellé,-e *a.* umbellate
ombellifère *a.* umbelliferous; *n. f. pl.* Umbellifers
ombelliforme *a.* umbelliform
ombellule *f.* umbellule
ombilic *m.* 1. *Bot.* hilum; *Cit.* navel (of fruit); *Vit.* mark of stigma (on grape). 2. *Umbilicus* sp.
ombiliqué,-e *a.* umbilicate
ombrage *m.* shade (of trees), shading; **o. mutuel** mutual shading; **arbre d'ombrage** shade tree; **plant d'ombrage** shade plant
ombragé,-e *a.* shaded, shady; **une cacaoyère ombragée** a shaded cacao plantation
ombrager *v.t.* to shade; **s'ombrager** to become shaded
ombre *f.* shadow, shade; **à l'ombre** in the shade (see also **mi-ombre**)
ombré,-e *a.* shaded
ombrer *v.t.* to shade; *Hort.* **paillasson à ombrer** shading mat
ombrière *f.* *Plant.* etc. shaded nursery or pre-nursery
ombrophile *a.* ombrophilous; **forêt ombrophile** tropical rain forest; *n.m.* ombrophile
omphalier (*m.*) **de la Guyane** Jamaican navel-spurge (*Omphalea diandra*)
onagraire, onagre *f.* *Oenothera* sp., evening primrose; **o. bisannuelle** common evening primrose (*O. biennis*) (= **oenothère**)
oncidie *f.*, **oncidium** *m.* *Oncidium* sp.; **o. papillon** butterfly orchid (*O. papilio*)

ondatra *m.* muskrat (= **rat musqué** *q.v.*)
ondatricide *a.* relating to muskrat control; **préparation ondatricide** preparation for muskrat control
ondulé,-e *a.* undulating, corrugated (of iron etc.) (see also **tôle**)
ongle *m.* finger nail; *Hort.* **à l'ongle** with the (thumb) nail; **ongles** (*m.pl.*) **du diable** *Martynia lutea* (= **cornaret**)
onglet *m. Bot.* claw (of petal); *Hort.* snag
onguent (*m.*) **de St. Fiacre** grafting wax, "pug" (originally two-thirds clay and one-third cow dung)
onguiculé,-e *a. Bot.* unguiculate
onopordon, onoporde *m. Onopordon* sp., cotton thistle, scotch thistle (see also **chardon**); **o. d'Arabie** *O. arabicum*; **o. d'Illyrie** *O. illyricum*
ontogenèse *f.* ontogeny, ontogenesis
ontogénie *f.* ontogeny, ontogenesis
ontogénique *a.* ontogenic
oophage *a.* egg-eating, oophagous
oosphère *f.* oosphere
opacité *f.* opacity, cloudiness (of liquid); denseness (of wood etc.)
opalin,-e *a.* opaline, opalescent
opaque *a.* opaque
ophrys *f. Ophrys* sp.; **o. abeille** bee orchid (*O. apifera*); **o. araignée** early spider orchid (*O. sphegodes*); **o. mouche** fly orchid (*O. insectifera*)
opium *m.* opium
oponce *m.* = **opunce** (*q.v.*)
opopanax *m.* opopanax (*Opopanax* sp.)
opposé,-e *a. Bot.* opposite
oppositifolié,-e *a.* opposite-leaved
opunce *m. Opuntia* sp., prickly pear; **o. robuste** *O. robusta*
opuntia *m.* = **opunce** (*q.v.*)
or *m.* gold
orange *f.* 1. orange (*Citrus*) (see also **oranger**); **o. amère** bitter, sour, Seville orange; **o. douce** sweet orange; **o. navel** Navel orange; **o. sanguine** blood orange; **écorce d'orange** orange peel. 2. Other genera: **o. du Mexique** Mexican orange flower (*Choisya ternata*); **o. de Quito** naranjilla (*Solanum quitoense*) (= **naranjille**). 3. *a.* orange-coloured
orangé,-e *a.* orange-coloured
orangeade *f.* orangeade
orangeat *m.* candied orange peel
oranger *m.* 1. **o. (doux)** (sweet) orange tree (*Citrus sinensis*) (see also **orange**); **o. amer** bitter, sour, Seville orange tree (*C. aurantium*) (= **bigaradier**); **o. trifoliolé** trifoliate orange (*Poncirus trifoliata*). 2. Other genera: **o. des Osages** Osage orange (*Maclura aurantiaca*) (= **maclure**); **o. de savetier** Jerusalem cherry (*Solanum pseudocapsicum*)
orangeraie *f.* orange grove
orangerie *f.* orangery
orangette *f.* small orange picked before maturity
orangiste *n.* orange grower
orbiculaire *a.* orbicular
orcanette *f.* alkanet (*Alkanna tinctoria*)
orchidacées *f. pl.* Orchidaceae
orchidée *f.* orchid; **o. de pleine terre** ground orchid; **o. chinoise du pauvre** *Bletia* sp.; *pl.* Orchidaceae, orchids
orchidéiste *m.* orchid grower
orchidophile *n.* orchid fancier
orchidophilie *f.* orchid fancying
orchis *m. Orchis* sp.; **o. bouffon** green-winged orchid (*O. morio*); **o. à larges feuilles** marsh orchid (*O. latifolia*); **o. militaire** soldier orchid (*O. militaris*); **o. pourpre** lady orchid (*O. purpurea*); **o. singe** monkey orchid (*O. simia*)
ordinateur *m.* computer
ordre *m.* order
ordures (*f. pl.*) **ménagères** household refuse; **ordures de ville** town refuse
oreille *f.* ear; **o. d'un butteur** wing of ridger; **o. d'une charrue** mouldboard of plough; **o. de chat** lamb's tongue (*Stachys lanata*); **o. de lièvre** (i) *Vit.* hare's ear pruning, (ii) shrubby hare's ear (*Bupleurum fruticosum*); **o. d'ours** auricula (*Primula auricula*) (= **auricule**); lamb's tongue (*Stachys lanata*) (= **o. de chat**)
oreillette *f.* asarabacca (*Asarum europeum*) (= **nard sauvage**)
oreillon *m. Comm.* peach or apricot cut in half with stone removed
orélie *f. Orelia* (*Allemanda*) sp.
organe *m.* organ; **o. excisé** excised organ; **o. de fructification** fruiting organ; **o. reproducteur** reproductive organ
organique *a.* organic
organisation *f.* organization
organisme *m.* organism
organite *m.* organelle
organochloré *m.* organochlorine compound
organogène *a.* organogenetic
organogenèse *f.* organogenesis
organographie *f.* organography
organo-halogène *a.* organo-halogen
organoleptique *a.* organoleptic
organostannique *a.* organo-tin
orge *f.* barley (*Hordeum* sp.); **o. à crinière** squirrel tail grass (*H. jubatum*)
orgyie (*f.*) **antique** vapourer moth (*Orgyia antiqua*) (= **bombyx étoilé**) (also applied to *O. gonostigma*)
orientation *f.* orientation, aspect
origan *m. Origanum sp.;* **o. vulgaire** common,

wild marjoram (*O. vulgare*) (= **marjolaine bâtarde**) (see also **dictame, marjolaine**)
origine *f.* 1. descent, origin. 2. source, derivation
orme *m.* 1. elm (*Ulmus* sp.); **o. blanc** = **o. de montagne** (*q.v.*); **o. champêtre** common elm, English elm (*U. procera*); **o. de la Chine** dwarf elm (*U. pumila*); **o. de(s) montagne(s)** wych elm (*U. glabra*); **o. lisse, o. pédonculé** European white elm (*U. laevis* = *U. pedunculata*); **o. à petites feuilles** = **o. champêtre** (*q.v.*); **o. pleureur** weeping elm (*U. glabra* var. *pendula*). 2. Other genera: **o. à trois feuilles, o. de Samarie** swamp dogwood, hoptree (*Ptelea trifoliata*); **o. de Sibérie** *Zelkova crenata* (= *Z. carpinifolia*)
ornemental,-e *a., m. pl.* **-aux** ornamental; **plante ornementale** ornamental plant
ornithogale *m.* *Ornithogalum* sp., star of Bethlehem
ornithogame *a.* pollinated by birds
ornithogamie *f.* pollination by birds
orobanche *f.* *Orobanche* sp., broomrape; **o. de Virginie** beechdrops (*Epiphegus virginiana*)
oronge *f.* *Fung.* royal agaric (*Amanita caesarea*); **o. vineuse** blusher (*A. rubescens*) (= **golmote**) (see also **amanite**)
orpin *m.* stonecrop (*Sedum* sp.); **o. âcre** yellow stonecrop (*S. acre*); **o. blanc** white stonecrop (*S. album*); **o. reprise, grand orpin** orpine (*S. telephium*) (see also **sédum**)
orthostique *m.* orthostichy, vertical rank of leaves on a stem
orthotrope *a.* orthotropous
orthotropisme *m.* orthotropism
ortie *f.* nettle; **o. brûlante** small nettle (*Urtica urens*); **o. commune, grande ortie** stinging nettle (*U. dioica*); **o. blanche** white dead nettle (*Lamium album*) (= **lamier blanc**); **o. rouge** purple dead nettle (*L. purpureum*) (= **lamier pourpre**); **o. à fleurs jaunes** yellow archangel (*Galeobdolon luteum*); **o. royale** hempnettle (*Galeopsis tetrahit*)
orvale *f.* clary (*Salvia sclarea*) (= **sclarée, toutebonne**)
oryctes *m.* *Oryctes* sp., including *O. rhinoceros* rhinoceros beetle
os *m.* bone; **o. en poudre, poudre d'os** bone meal
ose *m.* *Chem.* ose; in *pl.* monosaccharoses
oseille *f.* 1. sorrel (*Rumex* sp.); **o. commune** sorrel (*R. acetosa*); **o. crépue** curled dock (*R. crispus*) (= **parelle**); **o. épinard** herb patience (*R. patientia*) (= **patience**); **grande oseille, o. (des jardins)** = **o. commune** *q.v.*; **o. ronde** French sorrel (*R. scutatus*) (= **oseillon**); **o. sauvage** = **o. crépue** (*q.v.*); **petite oseille** sheep's sorrel (*R. acetosella*), wood sorrel (*Oxalis acetosella*). 2. **o. de Guinée** red sorrel (*Hibiscus sabdariffa*) (= **roselle**)
oseillon *m.* French sorrel (*Rumex scutatus*) (= **oseille ronde**)
osier *m.* osier; **panier d'osier** wicker basket
osmonde *f.* *Osmunda* sp.; **o. royale** royal fern (*O. regalis*) (= **fougère royale**)
osmose *f.* osmosis
osmotique *a.* osmotic
ossature *f.* frame, framework (of greenhouse etc.)
osselet *m.* *Bot.* stone of nuculanium
ostiole *m.* ostiole
osyris *m.* *Osyris* sp., including *O. alba*
otiorrhynche *m.* weevil (*Otiorrhynchus* sp.), including clay-coloured weevil (*O. singularis*); **o. des cyclamens** *O. rugostriatus*; **o. de la livèche** lovage weevil (*O. ligustici*); **o. (de la vigne)** vine weevil (*O. sulcatus*)
oudo *m.* udo (*Aralia cordata*), a herbaceous perennial with edible shoots
oued *m., pl.* **oueds, ouadi** wadi, watercourse
ouillère, ouillière *f.* *Vit.* space between vine rows used for other crops
ouratée *m.* *Ouratea* sp.
outil *m.* tool, implement; **o. à fendre** splitting tool; **outils à dents** tined implements; **outils de taille** pruning tools
outillage *m.* equipment; **o. horticole** horticultural equipment; **o. de récolte** harvesting equipment
outremer *m.* ultramarine; **bleu d'outremer** ultramarine blue
ouvert,-e *a.* open, (of plant) spreading horizontally
ouverture *f.* opening; **o. stomatique** stomatal opening
ouvrant,-e *a.* opening; *n. m.* shutter, ventilator
ouvrier,-ière *a.* & *n.* worker, workman; **o. horticole** garden worker; *a. f.* & *n. f.* *Ent.* worker bee (= **abeille ouvrière**)
ovaire *m.* ovary; **o. infère** inferior ovary; **o. supère** superior ovary
ovale *a.* oval
ovarien,-enne *a.* *Bot.* ovarian
ové,-e *a.* ovate, egg-shaped
ovicide *a.* ovicidal
oviforme *a.* ovate, oviform
ovipositeur *m.* *Ent.* ovipositor
oviscapte *m.* oviscapt
ovogenèse *f.* *Ent.* oogenesis, origin and development of ova
ovoïde *a.* ovoid
ovovivipare *a.* ovoviviparous
ovule *m.* ovule; **o. fécondé** fertilized ovule
owala *m.* *Afr.* owala oil tree (*Pentaclethra macrophylla*)
oxalate *m.* oxalate

oxalide *f.*, **oxalis** *m.* **o. blanche** wood sorrel (*Oxalis acetosella*) (= **petite oseille**); **o. crénelée** oca (*O. crenata* = *O. tuberosa*)
oxalique *a.* oxalic; **acide oxalique** oxalic acid
oxalis *m.* *Oxalis* sp.; **o. acide** wood sorrel (*O. acetosella*); **o. corniculé** yellow sorrel (*O. corniculata*)
oxycèdre *m.* prickly juniper (*Juniperus oxycedrus*)
oxychlorure (*m.*) **de cuivre** copper oxychloride
oxycoccos *f.* *Oxycoccus (Vaccinium)* sp., cranberry
oxydant,-e *a.* oxidizing; *n. m.* oxidizing agent, oxidant
oxydase *f.* oxidase
oxydation *f.* oxidation
oxydoréduction *f.* oxyreduction, redox; **potentiel d'oxydoréduction** redox potential
oxygénation, oxidation; **o. des racines** root oxygenation
oxylobe *m.* *Oxylobium* sp.

P

pacane *f.* pecan nut
pacanier *m.* pecan tree (*Carya illinoensis*)
pachire, pachirier *m.* *Pachira* sp. (see also **noisetier de Cayenne**); **p. aquatique** provision tree (*P. aquatica*) (= **châtaignier de la Guyane**)
pachypodium *m.* *Pachypodium* sp.
pachyrhizus *m.*, **pachyrhize** *f.* *Pachyrhizus* sp. (see **pois-manioc**)
pachysandre *f.* *Pachysandra* , a genus which includes ornamental shrubs
pacourier *m.* *Landolphia (Pacouria)* sp., a source of rubber
paederia *f.* *Paederia* , a genus which includes medicinal plants
pagoscope *m.* pagoscope
paillage *m.* mulch(ing); **p. plastique** plastic mulching; **p. PE** polyethylene mulching; **p. radiant** radiant mulching
paillasson *m.* *Hort.* (straw) mat used to protect fruit trees, cover greenhouses etc.; **p. en roseau** reed mat; **p. sulfaté** mat treated against rot etc. (see also **ombrer**)
paillassonnage *m.* covering, protecting with **paillassons**
paillassonner *v.t.* to cover, to protect with **paillassons**
paille *f.* straw; *Sug.* field trash; *Vit.* fruit cane
paillé,-e *a.* 1. mulched, protected with straw; *n. m.* 2. (fresh) stable litter
pailler *v.t.* to mulch, to protect with straw
paillette *f.* flake; *Bot.* palea; *PC* pellet
pailleux,-euse *a.* strawy; **fumier pailleux** strawy manure
paillis *m.* mulch (see **paillage**)
paillon *m.* wisp of straw
pain *m.* bread; **p. de laine de roche** rockwool block; **p. d'Odika** Dika bread (from *Irvingia gabonensis* = **manguier sauvage**); **p. de singe** baobab fruit
paisseau *m.* prop, stake for grapevine or other plant
paisselage *m.* propping, staking vines or other plants
paisseler *v.t.* to prop, to stake vines or other plants
pak-choï *m.* pak-choi, Chinese cabbage (*Brassica chinensis*) (= **chou de Chine**) (see also **pe-tsai**)
pal *m.* stake; **au pal** at stake; *Vit.* planting stick; **p. injecteur** lance injector, soil injector (for soil fumigants)
palais *m.* *Bot.* palate
palaque *m.* *Palaquium*, a genus which includes plants which yield gutta-percha
palava *f.* *Palava (Palaua)*, a genus which includes ornamental plants
paléacé,-e *a.* *Bot.* paleaceous
paleton *m.* *Vit.* metal protector for the end of a stake
palette *f.* pallet (of fork lift truck etc.); **en palettes** palletized
palétuvier *m.* mangrove tree; **p. blanc** white mangrove (*Avicennia officinalis*); **p. rouge** red mangrove (*Rhizophora mangle*)
palissade *f.* palisade, fence, live fence
palissadement *m.* palisading, fencing
palissader *v.t.* to palisade, to fence in, to enclose
palissadique *a.* palisade; **tissue palissadique** palisade tissue
palissage *m.* training, tying (of plant) (*Vit.* = **relevage**)
palissé,-e *a.* trained, tied (of plant)
palisser *v.t.* to train, to tie (plant)
paliure *m.* *Paliurus* sp.; **p. épineux** Christ's thorn (*P. spina-Christi*) (= **épine du Christ**)
palma-christi *m. inv.* castor (*Ricinus communis*) (= **ricin commun**)
palmatifide *a.* palmatifid
palmatilobé,-e *a.* palmatilobate
palmatinervé,-e *a.* palmately veined
palmatipartite *a.* palmatipartite
palmatiséqué,-e *a.* palmatisect
palme *f.* palm frond; **p. de cocotier** coconut leaf, frond; **palmes d'élagage** pruned fronds; **huile de palme** palm oil; **vin de palme** palm wine
palmé,-e *a.* palmate (leaf)
palmeraie *f.* palm-grove, palm plantation; **la palmeraie du Sénégal** the palm resources of Senegal
palmette *f.* 1. *Arb.* fan, palmette; **en palmette** fan-shaped; **p. à branches oblique, p. oblique** oblique palmette; **p. libre** irregular (free) palmette; **p. régulière** regular palmette. 2. palmetto (*Sabal palmetto*). 3. *Fung.* hyphae with many short ramifications (in mycorrhizal association)
palmier *m.* palm tree; **p. d'appartement** potted palm; **p. bâche** Ita palm (*Mauritia flexuosa*); **p. à cire** wax palm (*Ceroxylon andicola*); **p.**

(-)**chanvre** windmill palm (*Trachycarpus fortunei*); **p. (-)dattier** date palm (*Phoenix dactylifera*); **p. doum, p. fourchu** branching palm, dum palm (*Hyphaene thebaica*); **p. fétiche** oil palm, form *idolatrica*; **p. à huile** oil palm (*Elaeis guineensis*); **p. à huile d'Amérique** American oil palm (*Elaeis oleifera*); **p. royal de Cuba** royal palm (*Roystonea regia*); **p. royal de l'île Barbade** cabbage palm (*R. oleracea*) (= **palmiste**); **p. royal de la Jamaïque** *R. princeps*; **p. royal de Porto Rico** royal palm of Porto Rico (*R. borinquena*); **p. à sucre** sugar palm (*Arenga saccharifera* = *A. pinnata*); palmyra palm (*Borassus flabellifer*) (= **rônier**); **p. talipot** talipot palm (*Corypha umbraculifera*)
palmier-dattier *m.* see **palmier**
palmier-pêche *m.* peach palm, pejibaye (*Guilielma gasipaes*) (see also **parépou**)
palmifide *a.* palmatifid
palmiforme *a.* palmiform
palmiparti,-e, palmipartite *a.* palmatipartite
palmiste *m.* 1. cabbage palm (*Roystonea (Oreodoxa) oleracea*). 2. *OP* palm kernel; *a.* **chou** (*m.*) **palmiste** palm cabbage (growing point eaten as a vegetable)
palombe *f.* wood pigeon, ring dove (*Columba palumbus*) (= **pigeon ramier**)
palomet *m. Fung.* green russule (*Russula virescens*) (= **russule verdoyante**)
palus *m.* alluvial soil in valley (in the Bordeaux region); **vin de palus** wine from **palus** areas
pamplemousse *m.* pummelo, shaddock; often used for grapefruit (= **grapefruit**)
pamplemoussier *m.* pummelo, shaddock tree (*Citrus grandis*); often used for grapefruit tree
pampre *m.* primary vine shoot or branch with leaves
pan *m. Ped.* pan
panacée *f.* panacea
panache *m.* plume, tuft; **p. des pampas** = **gynérion argenté** (*q.v.*); *Hort.* stripe, variegation
panaché,-e *a.* parti-coloured, variegated
panacher *v.t. Hort.* to variegate; **se panacher** *v.i.* to become variegated (of flowers etc.)
panachure *f.* variegation, streak, stripe, breaking disease; **p. infectieuse** *Cit.* infectious variegation virus; **p. infectieuse de l'abutilon** abutilon mosaic virus; **p. printanière (du fraisier)** strawberry variegation, June yellows; **p. de la tulipe** tulip breaking, tulip streak virus; **p. de la vigne** vine yellow mosaic virus
panais *m.* parsnip (*Pastinaca sativa* = *Peucedanum sativum*)
pancratier, pancrais *m. Pancratium* sp.; **p. d'Illyrie** *P. illyricum*; **p. maritime** Mediterranean lily, sea daffodil (*P. maritimum*) (= **lis maritime**)
pandanus *m. Pandanus* sp., screw-pine
pangola *f.* pangola grass (*Digitaria decumbens*)
panic *m.* jungle rice (*Echinochloa colonum*); **p. pied-de-coq** barnyard grass, cockspur grass (*E. crus-galli*) (= **panisse, pied-de-coq**); **p. sanguin** crabgrass (*Digitaria sanguinalis* = *Panicum sanguinale*)
panicaut *m. Eryngium* sp.; **p. des Alpes** alpine eryngo (*E. alpinium*); **p. champêtre, p. des champs** field eryngo (*E. campestre*) (= **chardon Roland**); **p. maritime** sea holly (*E. maritimum*) (= **chardon bleu**)
panicule *f.* panicle
paniculé,-e *a.* paniculate
panier *m.* basket; **gros panier** hamper; **p. à herbes** grassbox of mower; **p. d'osier** wicker basket; **p. suspendu** hanging basket
panisse *f.* barnyard grass (*Echinochloa crus-galli*) (= **panic, pied-de-coq**)
panmictique *a. PB* panmictic
panmixie *f. PB* panmixia
panne *f.* 1. breakdown (of machine etc.). 2. peen (of hoe, hammer etc.)
panneau *m.* panel; **p. d'aération** ventilation panel; **p. (vitré)** glazed frame, light; **p. de saignée** *Rubb.* tapping panel; **p. solaire** solar panel
panouille *f.* maize ear, cob
papaïne *f.* papain
papaver *m. Papaver* sp., poppy
papaye *f.* papaw fruit
papayer *m.* pawpaw tree, papaya (*Carica papaya*); **p. de (la) montagne** mountain papaw (*C. candamarcensis*)
paperpot *m., pl.* **paperpots** paper pot (= **godet en papier**)
papier *m.* paper; **p. buvard** blotting paper; **p.-filtre** filter paper; **p. de soie** tissue paper
papilionacé,-e *a.* papilionaceous
papilionacées *n. f. pl.* Papilionaceae
papille *f.* papilla
papillé,-e *a.* papillate, papillose
papilleux,-euse *a.* papillose, papillous
papillion *m.* butterfly; **p. (de nuit)** moth; *Ent.* lepidopterous perfect insect (butterfly or moth); **p. du yucca** yucca moth (*Tegeticula* sp.)
pappe *m.* pappus
paprika *m.* paprika (*Capsicum annuum*)
papyracé,-e *a.* papery
papyrus *m.* papyrus (*Cyperus papyrus*)
pâquerette *f.* daisy (*Bellis* sp.), and other spp.; **p. bleue** blue daisy (*Felicia amelloides*); **p. d'automne** southern daisy (*B. sylvestris*); **p. vivace** common daisy (*B. perennis*)

parabiose *f.* parabiosis
paradis *m.* paradise apple (*Malus pumila* var. *paradisiaca*) (= **pommier paradis**)
paraffinage *m.* treating with paraffin or paraffin wax
paraffine *f.* paraffin, paraffin wax
paraffiné,-e *a.* treated with paraffin or paraffin wax
parage *m.* paring, trimming; *Ban.* trimming stems before removing dried floral parts (see also **épistillage**); *Vit.* cultivation before winter
paragrêle *a.* anti-hail; **filet paragrêle** net for hail protection; **fusée paragrêle** rocket to disperse hail clouds; *n.m.* **(canon) paragrêle** gun to disperse hail clouds
parallèle *a.* parallel
paramètre *m.* parameter
paraplasme *m.* paraplasm
parasitaire *a.* parasitic
parasite *m.* parasite; **p. de blessures** wound parasite; **p. obligé, p. obligatoire** obligate parasite; *pl.* pests; **parasites animaux** pests; **parasites végétaux** fungal diseases etc.; *a.* parasitic
parasité,-e *a.* parasitized, attacked, damaged by parasite
parasitisme *m.* parasitism
parasitoïde *m.* parasitoid
parasol *m.* parasol; **fleur en parasol** umbellate flower; **en forme de parasol** parasol-shaped; **magnolier en parasol** umbrella tree (*Magnolia tripetala*)
parasolier *m.* umbrella tree (*Musanga smithii*)
parastique *f.* parastichy
parc *m.* 1. park. 2. enclosure; **p. à bois** plot of trees reserved for budwood or cutting production
parcelle *f.* plot, patch of land; **p. expérimentale** experimental plot; small fragment (of plant etc.)
parche *f. Coff.* parchment
parchemin *m.* parchment
parcheminé,-e *a.* parchment-like, dried
pare-feu *m. inv.* firebreak
parelle *f.* curled dock (*Rumex crispus*) (see also **oseille, patience**)
parenchymateux,-euse *a.* parenchymatous
parenchyme *m.* parenchyma; **p. chlorophyllien** chlorophyll parenchyma; **p. foliaire** foliar parenchyma; **p. lacuneux** spongy mesophyll; **p. palissadique** palisade layer; **p. sous-épidermique** sub-epidermal parenchyma
parental,-e *a., m. pl.* **-aux** *PB* parent(al); **espèces parentales** parent species
parenté *f. PB* affinity, relationship
parépou *m.* paripi palm (*Guilielma speciosa*) (see also **palmier-pêche**)
parer *v.t. Hort.* to trim, to tidy (up)
pare-soleil *m. inv.* sun shield
parfum *m.* perfume, fragrance (of flower)
parfumé,-e *a.* scented, fragrant
parfumerie *f.* perfumery manufacture and trade
pariétaire *f.* pellitory of the wall (*Parietaria judaica*)
pariétal,-e *a., m. pl.* **-aux** parietal
parinaire *m.* = **pommier du Cayor** (*q.v.*)
paripenné,-e *a.* paripinnate
parisette (*f.*) **(à quatre feuilles)** herb paris (*Paris quadrifolia*)
parkie *f. Parkia* sp. (see also **nété**)
parkinsonie *f. Parkinsonia* sp., including Jerusalem thorn (*P. aculeata*)
parlatoria *m.*, **p. gris** *Cit.* chaff scale (*Parlatoria pergandii*)
parmentiera *f. Parmentiera* sp., including *P. cereifera* candle tree
paroi *f.* 1. (partition) wall; **p. cellulaire** cell wall. 2. lining; *Hort.* cladding
paroir *m.* scraper, broad hoe
parsemé,-e *a.* strewn (with), sprinkled (with); **parcelle parsemée de coquelicots** plot sprinkled with poppies
parsemer *v.t.* to strew, to sprinkle
partage *m.* distribution, partition
parterre *m.* flower-bed; **p. de gazon** lawn, grass plot
parthénocarpie *f.* parthenocarpy
parthénocarpique *a.* parthenocarpic
parthénogénèse *f.* parthenogenesis
particule *f.* particle; **particules du sol** soil particles; **particules virales** virus particles
partie *f.* part; **p. basale** basal part, base; **parties aériennes** aerial parts (of plant); **parties creuses** hollow parts; **parties par million, p.p.m.** parts per million, p.p.m.
partiel,-elle *a.* partial
partiteur *m. Irrig.* sluice
partition *f.* partition
parvifolié,-e *a.* parvifoliate
pasanie *f. Pasania (Lithocarpus)* sp.
pas-d'âne *m.* coltsfoot (*Tussilago farfara*) (= **tussilage**)
pas japonais *m.* stepping stones or slabs (in a garden)
passage *m.* 1. passage; **p. d'un cultivateur** cultivation. 2. way, thoroughfare
passé,-e *a.* past, gone by; **fleur passée** *Vit.* period of flower fertilization
passe-fleur *f., pl.* **passe-fleurs** pasque flower (*Pulsatilla vulgaris*); rose campion (*Lychnis coronaria*)
passer *v.i. Hort.* to cease flowering, to fade; **p.** (*v.t.*) **à la vapeur** *Mush.* to cook out
passerage *f. Lepidium* spp., including *L. draba* (= *Cardaria draba*); **p. des champs** field

pepperwort (*Lepidium campestre*); **p. cultivée** garden cress (*L. sativum*); **p. sauvage** = **cressonnette** (*q.v.*)
passerelle *f.* footbridge
passerillage *m. Vit.* raisining
passerine *f. Passerina* sp., sparrow-wort
passe-rose *f.* hollyhock (*Althaea rosea*) (= **rose trémière**)
passe-velours *m. inv.* cockscomb (*Celosia cristata*) (see also **amarante, célosie**)
passiflore *f.* passion flower (*Passiflora* sp.); **p. des berges** *Passiflora riparia*; **p. du Chili** = **p. à stipules pennées** (*q.v.*); **p. écarlate** scarlet passion flower (*P. coccinea*); **p. fétide** Tagua passion flower (*P. foetida*); **p. à feuilles brillantes** bel-apple (*P. nitida*); **p. à feuilles tripartites** *P. tripartita*; **p. à feuilles de vigne** *P. vitifolia*; **p. de Popenoe** *P. popenovii*; **p. rouge** may pops, may apple (*P. incarnata*); **p. de Seeman** *P. seemanii*; **p. à stipules pennées** *P. pinnatistipula*; **p. à tiges ailées** wingstem passion flower (*P. alata*) (see also **barbadine, grenadille, pomme-liane**)
pastel *m. Isatis* sp.; **p. des teinturiers** woad (*I. tinctoria*) (= **guède**)
pastèque *f.* watermelon (*Citrullus lanatus* = *C. vulgaris*) (= **melon d'eau**)
pasteurisation *f.* pasteurization; *Mush.* peak heat
pasteuriser *v.t.* to pasteurize
pastille *f.* pastille; **p. expansible** *Hort.* mesh support
patate *f.* **(douce)** sweet potato (*Ipomaea batatas*); *Coll.* potato; **p. aquatique** water spinach (*I. aquatica*)
patchouli *m.* patchouli (*Pogostemon cablin*)
pâte *f.* paste; **p. de dattes** date paste; **p. à praliner** root dip, slurry
paternoster *m.* crab's eye vine (*Abrus precatorius*) (= **abre, jéquirity**)
patersonie *f. Patersonia* sp.
pâteux,-euse *a.* thick, pasty; **poire pâteuse** woolly pear; **vin pâteux** wine of thick consistency
pathogène *a.* pathogenic
pathogenèse *f.* pathogenesis
pathogénésie *f.* = **pathogénie** (*q.v.*)
pathogénicité *f.* pathogenicity
pathogénie *f.* pathogenesis (= **pathogenèse**)
pathologie (*f.*) **végétale** plant pathology, phytopathology
pathotype *m.* pathotype
pathovar *m. PD* pathovar, pv
patience *f.* patience dock, herb patience (*Rumex patientia*) (= **patience officinale, grande patience**); **p. d'eau** great water dock (*R. hydrolapathum*); **p. des Alpes** monk's rhubarb (*R. alpinus*); **p. rouge** red-veined dock (*R. sanguineus*); **p. sauvage** broad-leaved dock (*R. obtusifolius*) (see also **oseille, parelle**)
pâtisson *m.* scalloped summer squash, custard marrow (*Cucurbita pepo*) (= **artichaut d'Espagne, bonnet d'électeur**) (see also **courge**)
patrimoine (*m.*) **génétique** genetical make-up
patte *f. Hort.* flattened tuber (of anemone etc.); **patte de lièvre** balsa (*Ochroma lagopus*)
pâture *f.* grazing, pasture
pâturin *m.* meadow grass; **p. annuel** annual meadow grass (*Poa annua*); **p. des bois** wood meadow grass (*P. nemoralis*); **p. commun** rough-stalked meadow grass (*P. trivialis*); **p. comprimé** flattened meadow grass (*P. compressa*); **p. des prés** smooth-stalked meadow grass (*P. pratensis*)
pauciflore *a.* pauciflorous
paullinie *f. Paullinia* sp.
paulownia *m. Paulownia* sp.; **p. impérial** *P. (tomentosa) imperialis*
pavie *f. Hort.* clingstone peach (see also **pêche**)
pavillon *m. Bot.* standard (of leguminous flower)
pavonie *f. Pavonia* sp., including *P. rosea*
pavot *m.* poppy (*Papaver* sp.); **p. argémone** long rough-headed poppy (*P. argemone*); **p. cornu** = **glaucier** (*q.v.*); **p. de Hollande** *P. somniferum* var. *nigrum*; **p. d'Islande** Iceland poppy (*P. nudicaule*); **p. à opium** = **p. somnifère** (*q.v.*); **p. d'Orient** Oriental poppy (*P. orientale*); **p. rouge** field poppy (*P. rhoeas* and other spp.) (= **coquelicot**); **p. somnifère** opium poppy (*P. somniferum*); **p. tulipe** tulip poppy (*P. glaucum*)
paxille *m. Fung. Paxillus* sp.; **p. enroulé** brown roll-rim (*P. involutus*)
payant,-e *a.* paying, profitable, (of horticultural enterprise) economic
paysage *m.* landscape, scenery
paysager,-ère *a.* arranged to give landscape effects; **jardin paysager** landscape garden
paysagiste *m. Hort.* landscape gardener; *a.* **jardinier paysagiste** landscape gardener
PE = **polyéthylène** (*q.v.*)
peau *f., pl.* **peaux** skin, (of fruit) peel; **p. brune** (*Coconut*) brown testa of parings
pêche *f.* peach, (freestone) peach (see also **pavie**); **p. à peau duveteuse** velvety-skinned peach; **p. à peau lisse** smooth-skinned peach, nectarine (= **nectarine**) (see also **brugnon**)
pêcher *m.* peach tree (*Prunus persica*); **p. d'espalier** espalier peach; **p. à fleurs** flowering peach; **p. à fleurs doubles** double-flowered peach; **p. de semis** peach tree grown from seed; **p. de vigne** subspontaneous ungrafted peach tree
pectase *f.* pectase

pectine *f.* pectin
pectiné,-e *a.* pectinate
pectinestérase *f.* pectinesterase
pédalé,-e *a.* pedate
pédicelle *m.* pedicel
pédicellé,-e *a.* pedicellate
pédiculaire *f. Pedicularis* sp., lousewort
pédicule *f.* pedicel
pédilanthe *m. Pedilanthus* sp., including *P. pavonis*
pédiluve *m.* footbath
pédobiologique *a.* pedobiological
pédoclimat *m.* soil climate; **p. thermique** soil thermal climate
pédofaune *f.* soil fauna
pédogénèse *f.* pedogenesis
pédologie *f.* pedology, soil science
pédologique *a.* pedological
pédonculaire *a.* peduncular
pédoncule *m.* peduncle, stalk, fruit stalk
pégomyie *f. PC Pegomya* sp.; **p. du framboisier** loganberry cane fly (*P. rubivora*)
peinture *f.* paint; **p. fongicide** fungicidal paint
pelage *m. Hort.* peeling of fruit
pélargonium *m. Pelargonium* sp., geranium; **p. élégant** = **p. à grandes fleurs** (*q.v.*); **p. à feuilles de lierre** ivy-leaved geranium (*P.* × *hederaefolium* = *P. peltatum*); **p. des fleuristes** regal pelargonium, show, fancy geranium (*P* × *domesticum*); **p. à grandes fleurs** *P. grandiflorum*; **p. des horticulteurs, p. à feuilles zonées** zonal geranium (*P. zonale*); **p. des jardins** garden geranium (*P.* × *hortorum*) (see also **géranium**)
peler *v.t.* to peel (off)
pelle *f.* shovel; **p. méchanique** mechanical digger; **p. plate** flat shovel
pellée *f.* shovelful
pelletage *m.* shovelling
pelletée *f.* shovelful
pelleter *v.t.* to shovel
pellicule *f.* skin, fruit skin; **p. de grain de raisin** grape skin
pelliculage *m.* film-coating of seed
pellucide *a.* pellucid
pélorie *f.* peloria, pelory, regularity in a normally irregular flower
pelouse *f.* lawn, sward; **p. de sport** playing field
pelté,-e *a.* peltate, shield-shaped with the stalk in the centre
pelu,-e *a.* hairy
pelucheux,-euse *a.* shaggy, fluffy, downy (fruit)
pelure *f.* peel, rind, skin, tunic (of bulb); **p. d'oignon** onion skin; *Vit.* a dark rosé wine
pendant,-e *a.* hanging, pendulous
pénétration *f.* penetration; **p. d'un herbicide** herbicide penetration; **p. des racines** root penetration
pénétromètre *m.* penetrometer
pennatifide *a.* pinnatifid
pennatilobé,-e *a.* pinnatilobed
pennatiséqué,-e *a.* pinnatisect
penné,-e *a.* pinnate
penniforme *a.* pinniform
pensée *f.* pansy (*Viola* × *wittrockiana* and other spp.); **p. sauvage** field pansy (*V. arvensis*); **p. tricolore** wild pansy, heartsease (*V. tricolor*)
penstémon *m. Penstemon* sp., beard tongue
pentamère *a.* pentamerous
pentanochlore *m.* pentanochlor
pentas *m. Pentas* sp.
pentatomidés *f. pl. Ent.* Pentatomidae
pente *f.* slope, incline, gradient, pitch (of roof)
pentstémon = **penstémon** (*q.v.*)
pénurie *f.* scarcity, shortage; **p. d'eau** water shortage
pépin *m.* pip, seed (of berries and other fruit); **fruit à pépins** pome fruit; **p. d'agrume** citrus seed; **p. de pomme** apple seed; **p. de raisin** grape seed; **sans pépins** seedless
pépinière *f.* nursery, seedbed for plants (including herbaceous plants) to be transplanted later; **p. d'agrumes** citrus nursery; **p. fruitière** fruit tree nursery; **p. d'ornement, p. ornementale** ornamental plant nursery; **p. de semis** nursery in which seeds are sown (as opposed to cuttings planted); **p. sylvicole** tree nursery
pépiniériste *n. m. & a.* nurseryman, nursery gardener (= **jardinier pépiniériste**)
pépon *m.* pepo
perce-neige *m.* or *f. inv.* snowdrop (*Galanthus nivalis* and others)
perce-oreille *m., pl.* **perce-oreilles** earwig (*Forficula* sp.) (= **forficule**)
perce-pierre *f., pl.* **perce-pierres** samphire (*Crithmum maritimum*)
perce-tige *f.* **(de la pomme de terre)**, *pl.* **perce-tiges** potato stem borer (*Hydraecia micacea*)
perche *f.* pole; **p. à houblon** hop pole
perchette *f.* small pole
percolation *f.* percolation
perdrigon *m.* perdrigon plum
pérennant,-e *a.* perennating
pérenne *a.* perennial
pérennité *f.* perenniality
péreskie *f. Pereskia* sp., including *P. aculeata*, Barbados gooseberry
perfolié,-e *a.* perfoliate
perforation *f.* perforation; hole, puncture
perforé,-e *a.* perforated; **carte perforée** punched card; **film perforé** perforated (plastic) film
perfusion *f.* perfusion
pergola *f.* pergola
pergulaire *f. Pergularia* sp., including *P. extense*, a medicinal plant

périanthe *m.* perianth
péricarpe *m.* pericarp
péricarpial,-e *a., m. pl.* **-aux** pericarpial
péricarpique *a.* pericarpic
péricycle *m.* pericycle
périderme *m.* periderm
périgyne *a.* perigynous
périlla, pérille *f.* perilla (*Perilla ocymoides*); **p. de Nankin** *P. frutescens* var. *nankinensis*
périmètre *m.* perimeter
période *f.* period; **p. de croissance** growth period; **p. de repos** resting, dormant period; **p. d'incubation** *PD* incubation period; **p. de grande croissance** *Sug.* boom stage
périodicité *f.* periodicity; **p. journalière** diurnal periodicity
périphérie *f.* periphery, bounding line; outskirts
périssable *a.* perishable; **produit périssable** perishable product, produce
périsperme *m.* perisperm
péritèle (*m.*) **gris** vine weevil (*Peritelus griseus* = *P. sphaeroides*) (= **grisette**)
périthèce *m.* perithecium
perlite *f.* perlite
perméabilité *f.* permeability; **p. cellulaire** cell permeability
perméable *a.* permeable, pervious, porous
permutation *f.* permutation
péronospora *m.* *Peronospora* sp., downy mildew
perovskia *m.* *Perovskia* sp.; **p. à feuilles d'arroche** Russian sage (*P. atriplicifolia*)
peroxydase *f.* peroxidase
peroxydasique *a.* relating to peroxidase; **activité peroxydasique** peroxidase activity
perroquet *m.* parrot; **systéme "perroquet"** *Coff.* "parrot perch" system (of training)
persea *m.* *Persea* sp.
persicaire *f.* common persicaria (*Polygonum persicaria*) and related spp.; **p. du Levant** kiss-me-over-the-garden-gate (*Polygonum orientale*)
persil *m.* parsley (*Petroselinum crispum*); **faux persil, p. des fous** fool's parsley (*Aethusa cynapium*); **p. à grosse racine** turnip-rooted parsley (*P. crispum* var. *tuberosum*); **p. des moissons** corn parsley (*P. segetum*); **p. sauvage** cow parsley and related sp. (*Anthriscus* sp.)
persillère *f.* clay container for indoor growing of parsley, parsley pot
persimmon *m.* persimmon tree (*Diospyros kaki, D. virginiana*) (see also **plaqueminier**)
persistant,-e *a. Bot.* persistent; **à feuilles persistantes** evergreen
personé,-e *a.* personate; **corolle personée** personate corolla
perte *f.* loss; **p. de poids, p. de masse** loss in weight
perturbation *f.* perturbation, disturbance
pervenche *f.* periwinkle (*Catharanthus (Vinca)* sp.); **grande pervenche** large periwinkle (*C. major*); **petite pervenche** lesser periwinkle (*C. minor*); **p. rose des tropiques, p. tropicale** Madagascar periwinkle, Cape periwinkle (*C. roseus*); *a. inv.* **bleu pervenche** periwinkle blue
pesée *f.* weighing
pesticide *m.* pesticide
pétale *m.* petal; **pétales tombés** fallen petals: petal-fall (stage)
pétalodé,-e *a.* petalodic
pétalodie *f.* petalody
pétaloïde *a.* petaloid
pétard antiaviaire *m.* banger for bird scaring
pétasite *m.* *Petasites* sp., winter heliotrope
pétillant,-e *a.* crackling, sparkling; semi-sparkling (of wine) (see also **mousseux**)
pétiolaire *a.* petiolar
pétiole *m.* petiole
petiolé,-e *a.* petiolate
pétiolule *m.* petiolule
petit,-e *a.* small; **(maladie de la) petite feuille, (maladie des) petites feuilles** little leaf disease; **petits fruits** small fruits, berried-fruits
petit basilic *m.* bush basil (*Ocimum minimum*)
petit-cyprès *m.* cypress spurge (*Euphorbia cyparissias*)
petit-grain *m.* small dried bitter oranges; **essence de petit-grain** petit-grain oil (made from dried citrus fruits but also leaves and twigs)
petit muguet *m.* sweet woodruff (*Galium odoratum*) (= **aspérule odorante**)
petite bourrache *f.* blue-eyed Mary (*Omphalodes verna*)
petite cerise *f. PD* cherry little cherry virus
petite mineuse du pêcher *f.* peach twig borer (*Anarsia lineatella*)
petite pêche *f. PD* peach little peach virus
petite rave *f.* = **radis** (*q.v.*)
petite tomate du Mexique *f.* tomatillo, ground-cherry (*Physalis ixocarpa*) (= **alkékenge du Mexique, coqueret du Mexique**)
petits pois *m. pl.* green peas (see also **pois**)
pétreau *m., pl.* **-eaux** sucker
pe-tsai *m.* pe-tsai (*Brassica pekinensis*) (see also **pak-choï**)
pétunia *m.* petunia (*Petunia* sp.)
peuplement *m. Arb., Hort.* 1. plantation, stand, crop; **p. serré** dense stand, plant population. 2. **p. végétal** vegetation
peuplier *m.* poplar (*Populus* sp.); **p. baumier** *P. balsamifera* (= *P. tacamahaca*) balsam poplar; **p. blanc** white poplar (*P. alba*); **p. franc** = **p. noir** (*q.v.*); **p. grisaille** grey poplar (*P. canescens*) (= **grisard**); **p. d'Italie**

Lombardy poplar (*P. nigra italica*); **p. neige** *P. alba* var. *nivea*; **p. noir** black poplar (*P. nigra*); **p. de l'Ontario** balm of Gilead (*P. candicans*); **p. pyramidal** = **p. d'Italie** (*q.v.*) (see also **tremble**)

peyote, peyotl *m.* peyote (*Lophophora williamsii*)

pézize *f. Fung. Peziza* sp.; **p. orangée** orange peel fungus (*P. (Aleuria) aurantia*)

p.f. = **produit formulé** *PC* formulated product (see also **produit**)

pH *m.* pH; **pH-mètre** pH meter

phacélia *m.*, **phacélie** *f. Phacelia* sp.

phagocytose *f.* phagocytosis

phalange *f. Bot.* phalange, bundle of stamens

phalangère *f. Phalangium* sp. (Liliaceae)

phalène *f.* geometrid moth; **p. américaine des plantes sèches** American dried-plant moth (*Idaea bonifata*); **p. anguleuse** green pug moth (*Chloroclystis rectangulata*); **p. du châtaignier** March moth (*Alsophila aescularia*); **p. défeuillante** mottled umber moth (*Erannis defoliaria*); **p. du groseiller** magpie moth (*Abraxas grossulariata*); **p. hiémale** winter moth (*Operophtera brumata*) (= **chéimatobie**)

phalénopsis *m. Phalaenopsis* sp.

phanérogame *a.* phanerogamous, phanerogamic; *n. f.* phanerogam

phanérophyte *m.* phanerophyte

pharmacognosie *f.* pharmacognosy, the study of drugs from natural sources

pharmacologie *f.* pharmacology

pharmacologique *a.* pharmacological

pharmacopée *f.* pharmacopoeia; **la p. française** the French Pharmacopoeia

phase *f.* phase, stage, period; **p. de croissance ralentie** lag phase

phasme *m.* phasmid, stick insect (Phasmidae)

phellodendron *m. Phellodendron* sp.; **p. de l'amour** *P. amurense*

phelloderme *m.* phelloderm

phellogène *a.* phellogenic; *n. m.* phellogen

phellophage *a.* phellophagic

phénicicole *a.* date-growing, relating to dates; **station phénicicole** date research station

phéniculteur *m.* date grower

phéniculture *f.* date-palm growing

phénix see **phoenix**

phénol *m.* phenol

phénolique *a.* phenolic

phénologie *f.* phenology

phénologique *a.* phenological; **stade phénologique** phenological stage

phénomène *m.* phenomenon

phénophase *f.* phenophase

phénotype *m.* phenotype

phénotypique *a.* phenotypical

phénovariant *m.* phenovariant

phéromone *f.* pheromone; **p. sexuelle** sex(ual) pheromone

phialophora *m.* phialophora wilt of carnation (*Phialophora (Verticillium) cinerescens*)

phillyrea *f. Phillyrea* sp., mock privet

philo *m. Coll.* = **philodendron** (*q.v.*)

philodendron *m. Philodendron* sp.

phloème *m.* phloem (= **liber**)

phlox *m. Phlox* sp., phlox; **p. de Drummond** *P. drummondii*; **p. d'été** *P.* × *paniculata*; **p. mousse, p. subulé** moss phlox (*P. subulata*)

phoenicicole *a.* = **phénicicole** *q.v.*

phoeniciculture *f.* = **phéniciculture** (*q.v.*)

phoenix, phénix *m. Phoenix* sp., date palm (*P. dactylifera*) and other spp.; **p. des Canaries** Canary date-palm (*P. canariensis*)

pholiote *f. Pholiota* sp.; **p. écailleuse** scaly cluster fungus (*P. squarrosa*); **p. changeante** changing pholiota (*P. mutabilis*); **p. des charbonnières** charcoal toadstool (*P. carbonaria*)

phorétique *a.* phoretic; **acarien phorétique** phoretic mite

phoride *m.* phorid fly, in *pl.* Phoridae

phormium, phormion *m. Phormium* sp., including *P. tenax* New Zealand flax

phosphatase *f.* phosphatase

phosphatasique *a.* relating to phosphatase; **activité phosphatasique acide** acid phosphatase activity

phosphate *m.* phosphate; **p. dicalcique** dicalcium phosphate; **p. naturel** rock phosphate; **p. Thomas** basic slag

phosphaté,-e *a.* phosphated; **nutrition phosphatée** phosphate nutrition

phosphore *m.* phosphorus

phosphoré,-e *a.* phosphorated, containing phosphorus; **nutrition phosphorée** phosphorus nutrition

photinia *m. Photinia* sp., Chinese hawthorn

photo-apériodique *a.* daylength-insensitive, photo-insensitive

photoassimilat *m.* photosynthate

photochimique *a.* photochemical

photodégradable *a.* photodegradable

photon *m.* photon

photopériode *f.* photoperiod

photopériodicité *f.* photoperiodicity

photopériodisme *m.* photoperiodism

photophosphorylation *f.* photophosphorylation

photorécepteur,-trice *a.* photoreceptor; *n.m.* photoreceptor

photoréception *f.* photoreception

photorespiration *f.* photorespiration

photorespiratoire *a.* photorespiratory; **métabolisme photorespiratoire** photorespiratory metabolism

photosensibilisation *f.* photosensitization

photosensibilité *f.* photosensitivity
photosensible *a.* photosensitive
photosynthèse *f.* photosynthesis; **p. apparente** net photosynthesis; **p. réelle, p. totale** gross photosynthesis
photosynthétique *a.* photosyntheric; **activité photosynthétique** photosynthetic activity
photosynthétiquement *adv.* photosynthetically
photosynthétisant,-e *a.* photosynthetizing; **bactéries photosynthétisantes** photosynthetizing bacteria
photosynthétisé,-e *a.* photosynthetized
phototropisme *m.* phototropism
phragme *m.* phragma
phragmoplaste *m.* phragmoplast
phréatique *a.* phreatic; **nappe phréatique** water table
phthiriose *f. Vit.* phthiriosis (root fungal infection caused by *Bornetina corium* owing to damage by *Planococcus citri*)
phyllanthe *m. Phyllanthus* sp. (see **groseillier** 2)
phyllobie *f. Phyllobius* sp.; **p. oblongue** brown leaf weevil (*P. oblongus*); **p. du poirier** common leaf weevil (*P. pyri*)
phyllocactus *m. Phyllocactus* sp.
phylloclade *m.* 1. celery pine (*Phyllocladus* sp.) 2. phylloclade
phyllode *f.* phyllode
phyllodie *f.* phyllody; **p. du fraisier** *PD* strawberry green petal disease
phylloïde *a.* phylloid
phyllophage *a.* phyllophagous, leaf-eating; *n.m.* leaf-eating organism
phyllophore *m.* phyllophore
phylloplan *m.* phylloplane
phyllotaxie *f.* phyllotaxis, phyllotaxy
phylloxéra, phylloxera *m.* phylloxera (*Viteus vitifoliae*)
phylloxérant,-e *a.* said of soil which favours phylloxera
phylloxéré,-e *a.* attacked or damaged by phylloxera
phylloxérien,-enne *a.* relating to phylloxera
phylloxérique *a.* relating to phylloxera; **invasion phylloxérique** phylloxera invasion; **résistance phylloxérique** resistance to phylloxera
phylloxérisation *f.* infestation with phylloxera
phylum *m.* phylum
physalis *m. Physalis* sp., including Cape gooseberry (*P. peruviana*) (see also **alkékenge, coqueret**)
physiologie *f.* physiology; **p. des semences** seed physiology
physiologique *a.* physiological
physosiphon *m. Physosiphon* sp. (orchid)
physosperme *m. Physospermum* (*Pleurospermum*) sp.
physostegia *m. Physostegia* sp., including *P. virginiana*, obedient plant
physostigma *m. Physostigma* sp. (see **fève de Calabar**)
phytéléphas *m. Phytelephas* sp., including ivory nut palm (*P. macrocarpa*) (= **corozo**)
phytiatrie *f.* plant pathology
phytine *f.* phytin
phytobiologie *f.* phytobiology, plant biology
phytobiologique *a.* phytobiological
phytocénose *f.* plant community
phytochimie *f.* phytochemistry, plant chemistry
phytochimique *a.* phytochemical
phytochrome *m.* phytochrome
phytocide *a.* phytocidal; *n.m.* phytocide, herbicide
phytoécologie *f.* plant ecology
phytoécologique *a.* phytoecological
phytogéographie *f.* phytogeography, plant geography
phytohémagglutinine *f.* phytohaemagglutinin
phytohormone (*f.*) **(de croissance)** plant growth regulator, plant growth substance, plant hormone; **p. de croissance marquée** labelled plant growth regulator; **p. de synthèse** synthetic plant hormone
phytolacca *m. Phytolacca* sp. (= **phytolaque** *q.v.*)
phytolaque *m. Phytolacca* sp., including pokeberry (*P. americana*); **p. en arbre** ombu (*P. dioica* = *P. arborea*)
phytologie *f.* phytology
phytomètre *f. Ent. Phytometra* = *Plusia* sp., plusia moth
phytoparasite *a.* plant parasitic; *n.m.* plant parasite
phytopathogène *a.* phytopathogenic; **agent phytopathogène** plant pathogen; **bactéries phytopathogènes** plant pathogenic bacteria
phytopathologie *f.* phytopathology, plant pathology
phytopathologiste *m.* & *f.* plant pathologist
phytophage *a.* phytophagous, plant-eating; *n.m.* phytophagous organism
phytopharmaceutique *a.* phytopharmacological
phytopharmacie *f.* pesticide science; **station de phytopharmacie** plant protection chemical station
phytopte *m. Phytoptus* sp. mite and other genera; **p. du cassissier** black currant gall mite, big bud mite (*Cecidophyopsis (Eriophyes) ribis*); **p. du noisetier** nut gall mite (*P. avellanae*); **p. de la tulipe** tulip mite (*Aceria tulipae*); **p. de la vigne** vine leaf blister mite (*Eriophyes (Phytoptus) vitis*) (= **érinose de la vigne**)
phytoptose *f.* attack or damage by phytoptic mites

phytorégulateur *m.* plant growth substance
phytosanitaire *a.* phytosanitary
phytosociologie *f.* plant sociology
phytotechnie *f.* phytotechny
phytothérapie *f.* phytotherapy
phytotoxicité *f.* phytotoxicity
phytotoxine *f.* phytotoxin
phytotron *m.* growth chamber, phytotron
phytotronique *f.* phytotron techniques
phytovirus *m.* plant virus
piassava *m.* piassava
pic *m.* 1. pic, pickaxe, planting tool. 2. peak; **p. de chromatogramme** chromatogram peak
picholine *f.* olive cultivar, green olive for eating raw, pickled olive
picotiane *f.* breadroot (*Psoralea esculenta*)
picride *f.* = **picris** (*q.v.*)
picris *m.* *Picris* sp.; **p. fausse épervière** hawkweed ox-tongue (*P. hieracioides*)
pièce *f.* piece; *Hort.* plot; **p. d'eau** sheet of water, ornamental lake
pied *m.* foot, (base of) trunk (of tree) (see also **plain pied**); **de p. franc** grown from seed, on its own roots; *Hort.* (stock of) plant, rootstock; **p. mâle** male plant; **p. femelle** female plant; **p. d'artichaut** artichoke plant; **p. de céleri** head of celery; **p. de laitue** head of lettuce; **p. de vigne** grapevine plant (= **cep**)
pied-bleu *m.*, *pl.* **pieds-bleus** *Fung.* blewit (*Lepista nuda*)
pied-d'alouette *m.*, *pl.* **pieds-d'alouette** larkspur (*Delphinium* sp.) (= **dauphinelle**); **p.-d'alouette des blés** forking larkspur (*D. consolida*); **p.-d'alouette des blés double impérial** *D. consolida* var. *imperiale*; **p.-d'alouette de la Chine** *D. grandiflorum chinense*; **p.-d'alouette des jardins** larkspur (*D. ajacis*); **p.-d'alouette à fleur de jacinthe** hyacinth-flowered larkspur (*D. ajacis* var. *hyacinthiflorum*); **p.-d'alouette à fleur de renoncule** rocket larkspur (*D. ajacis* var. *ranunculiflorum*)
pied-de-chat *m.*, *pl.* **pieds-de-chat** catsfoot (*Antennaria dioica*)
pied-de-coq *m.*, *pl.* **pieds-de-coq** barnyard grass (*Echinochloa crus-galli*) (= **panic, panisse**)
pied d'éléphant *m.*, *pl.* **pieds d'éléphant** elephant's foot (*Testudinaria* (*Dioscorea*) *elephantipes*)
pied-de-mouton *m.*, *pl.* **pieds-de-mouton** *Fung.* wood hedgehog (*Hydnum repandum*) (see also **hydne**)
pied-droit *m.*, *pl.* **pieds-droits** 1. pier (of arch). 2. jamb, pier (of window)
pied(-)mère *m.*, *pl.* **pieds(-)mères** mother-tree, mother-plant, stock, parent stock, stool
pied-noir *m.* *PD* black leg; **p.-n. de la betterave** beet black leg (*Pleospora betae* = *Phoma betae*); **p.-n. du châtaignier** = **maladie de l'encre** (*q.v.*) (see also **jambe noire**)
piège *m.* trap; **p. lumineux** light trap; **p. à mulots** trap for field mice; **p. à succion** suction trap (for arthropods etc.)
piège-abri *m.*, *pl.* **pièges-abris** shelter trap
piégeage *m.* trapping (of insects etc.); **p. alimentaire** trapping with food; **p. sexuel** sex trapping
piège-appât *m.*, *pl.* **pièges-appâts** bait trap
piéride *f.* *Pieris* sp. butterfly; **grande piéride, p. du chou** cabbage white, large white (*P. brassicae*); **p. du navet** green-veined white (*P. napi*); **p. de la rave** small white (*P. rapae*)
pierraille *f.* broken stones, rubble, gravel; ballast
pierre *f.* stone; *Hort.* grit (in pear)
pierreux,-euse *a.* stony, gravelly, gritty (of pear)
piétinement *m.* trampling, treading; **p. artificial** artificial trampling (see also **foulage**)
piétiner *v.t.* to trample, to tread on; to tread (light soil)
pieu *m.* stake, post; **p. de défibrage** (*Coconut*) husking blade
pigamon *m.* meadow-rue (*Thalictrum* sp.)
pigeon *m.* pigeon; **p. ramier** wood pigeon (*Columba palumbus*) (= **palombe**)
pigment *m.* pigment
pigmentation *f.* pigmentation
pigne *f.* pine cone, (edible) pine seed (especially of stone pine)
pignon *m.* 1. gable, gable end. 2. (a) (edible) pine seed, piñon, (b) **(pin) pignon** stone pine (*Pinus pinea*) (= **pin parasol**) (see also **pin**); **p. d'Inde** physic nut (*Jatropha curcas*) (= **purghère**)
pilée *f.*, **pilea** *m.* *Pilea* sp., artillery plant
pileux,-euse *a.* hairy, pilose
pilifère *a.* piliferous
pilocarpe, pilocarpus *m.* *Pilocarpus* sp.; **p. à grandes fleurs** *P. pennatifolius*
pilocère, pilocereus *m.* *Pilocereus* (*Cephalocereus*) sp.
piloselle *f.* mouse-ear hawkweed (*Hieracium pilosella*)
pilosité *f.* hairiness
pimélée *f.* *Pimelea* sp.
piment *m.* 1. capsicum, pepper (*Capsicum annuum*); **p. à bouquets** cluster peppers (*C. annuum* var. *fasciculatum*); **p. caraïbe** = **p. enragé**; **p. cerise** cherry pepper (*C. annuum* var. *cerasiforme*); **p. du Chili** = **p. enragé** (*q.v.*); **p. doux** sweet pepper; **p. enragé** chilli pepper (*C. annuum* var. *acuminata*); **p. à gros fruits, gros piment** green pepper (*C. annuum* var. *grossum*) (= **poivron**). 2. allspice (= **piment-poivre (de la Jamaïque), pimenta** *q.v.*); **p. âcre, p. couronné** bayberry

(*Pimenta acris*). 3. **p. royal** bog myrtle (*Myrica gale*); **faux piment** winter cherry (*Physalis alkekengi, P. franchetii*) (see also **coqueret**)

pimenta *m.* allspice (*Pimenta dioica* = *P. officinalis*)

piment-poivre (*m.*) **(de la Jamaïque)** = **pimenta** *q.v.*

pimprenelle *f.* burnet (*Sanguisorba* sp.); **grande pimprenelle, p. des prés** great burnet (*S. officinalis*); **petite pimprenelle, p. commune** salad burnet (*S. minor*)

pin *m.* pine (*Pinus* sp.); **pomme de pin** pine cone; **pin d'Alep** Aleppo pine (*P. halepensis*); **p. de Banks** jack pine (*P. banksiana*); **p. à bois lourd** ponderosa pine (*P. ponderosa*); **p. des Canaries** Canary pine (*P. canariensis*); **p. cembro** arolla pine (*P. cembra* = **arole, tinier**); **p. de Coulter** big-cone pine (*P. coulteri*); **p. à crochets** *P. mugorostrata*; **p. à l'encens** frankincense pine (*P. taeda*); **p. à feuilles tombantes** spreading-leaved pine (*P. patula*); **p. à graine comestible** pinyon pine (*P. edulis*); **p. de l'Himalaya** = **p. pleureur** (*q.v.*); **p. jaune** = **p. à bois lourd** (*q.v.*); **p. de Jeffrey** *P. jeffreyi*; **p. de Jérusalem** = **p. d'Alep** (*q.v.*); **p. du Labrador** = **p. de Banks** (*q.v.*); **p. de Lambert** sugar pine (*P. lambertiana*); **p. laricio de Calabre** *P. nigra calabarica*; **p. laricio de Corse** Corsican pine (*P. nigra corsicana*) (see also **p. noir**); **p. du Lord, p. de Lord Weymouth** = **p. Weymouth** (*q.v.*); **p. des marais** long-leaf pine (*P. palustris*); **p. Napoléon** lacebark pine (*P. bungeana*); **p. maritime** maritime pine (*P. pinaster*); **p. du Mexique** Mexican pinyon pine (*P. cembroides*); **p. de(s) montagne(s)** mountain pine (*P. mugo*); **p. de Monterey** Monterey pine (*P. radiata*); **p. noir (d'Autriche)** Austrian pine (*P. (laricio) nigra austriaca*); **p. parasol** = **p. pignon** (*q.v.*); **p. pignon** stone pine (*P. pinea*) (= **pignon**); **p. pinier** = **p. pignon** (*q.v.*); **p. pleureur (de l'Himalaya)** Bhutan pine (*P. wallichiana* = *P. excelsa* = *P. griffithii*); **p. résineux** red pine (*P. resinosa*); **p. rigide** Northern pitch pine (*P. rigida*); **p. rouge du Japon** Japanese red pine (*P. densiflora*); **p. de Sabine** digger pine (*P. sabiniana*); **p. silvestre, p. sylvestre** Scots pine (*P. sylvestris*); **p. de Virginie** scrub pine (*P. virginiana*); **p. Weymouth** Weymouth pine (*P. strobus*)

pinçage *m.* pinching, pinching-off, nipping off (buds), stopping, topping

pince *f.* 1. pincers, pliers; **p. à inciser** *Hort.* girdling, (bark-) ringing tool. 2. **p. (à soulever)** crowbar; **p. à rejet** (*Date*) offshoot cutter (= **ciseau à rejet**)

pincé,-e *a.* pinched off, nipped off, stopped, topped; **p. court** pinched off short

pinceau *m., pl.* **-eaux** paint brush; *Vit.* brush, bundle of vascular strands remaining on pedicel when berry is pulled off

pincement *m.* = **pinçage** (*q.v.*); **p. d'Argenteuil (du figuier)** Argenteuil method of fig pruning; **p. chimique** chemical pinching

pincer *v.t. Hort.* to pinch off, to nip, to stop, to top

pince-sève *m. inv.* girdling tool, (bark-) ringing tool

pinet *m. Coll.* = **lactaire délicieux** (*q.v.*)

pinnatifide *a.* pinnatifid

pinnatiséqué,-e *a.* pinnatisect

pinné,-e *a.* pinnate, pennate

pinnule *f.* pinnule

pinot *m.* cabbage palm (*Euterpe oleracea*)

pinotière *f.* grove, stand of **pinots**

pioche *f.* pickaxe, pick, mattock, hoe; **p. à arracher, p. coupante** mattock

pioché,-e *a.* hoed, dug

piocher *v.t.* to dig with a mattock or pick

piochetout *m. Vit.* row cultivator

piocheur (*m.*) **multiple** multi-tine hand hoe

piocheur-vibrateur *m.* spring-tine cultivator (= **cultivateur canadien, canadienne**)

pipéricole *a.* relating to pepper (*Piper*)

pipériculteur *m.* pepper (*Piper*) grower

pipériculture *f.* pepper (*Piper*) growing, culture

pipérine *f.* piperine

pipette *f.* pipette

piquant,-e *a.* 1. prickling, prickly, stinging, pungent, sharp. 2. *n. m.* prickle, sting (of plant)

pique-bouton (*m.*) **du pommier** (*Canada*) eye-spotted bud moth (*Spilonota ocellana*)

pique-fleurs *m. inv.* flower holder

piqué,-e *a.* mould-spotted, attacked by mould (see also **tulipe**); attacked by insect (of seed etc.)

piquet *m.* peg, stake, post; *Vit.* trellis post; **piquets intermédiaires** posts between the end posts; **p. de protection** protective stake; **p. de tête** post at the end of a row

piquetage *m.* marking out, staking, pegging out

piqueté,-e *a.* 1. marked out, staked out, pegged out. 2. spotted, dotted

piqueter *v.t.* 1. to mark out, to stake out, to peg out. 2. to spot, to dot

piqueur *a.* & *n. m. Ent.* sucking, piercing insect; **papillons piqueurs de fruits** fruit-piercing moths

piqûre *f.* 1. prick, sting, bite. 2. puncture, small hole, bird peck in fruit

pirées *f. pl.* pome fruits (= **pomées, pomacées**)

piriforme *a.* pyriform, pear-shaped
pirola *m.*, **pirole** *f.* *Pyrola* sp., wintergreen (= **pyrole**)
pisciculture *f.* fish farming
piscidie *f.* *Piscidia* sp.
pisiforme *a.* pea-shaped
pissenlit *m.* dandelion (*Taraxacum officinale*) (= **dent de lion**)
pissette *f.* *Chem.* wash bottle
pissevin *m.* *Vit.* supplementary fruiting cane left at pruning
pistache *f.* pistachio nut
pistachier *m.* pistachio tree (*Pistacia vera*); **p. de l'Atlas** Mount Atlas mastic (*P. atlantica*)
pistil *m.* pistil
pistillaire *a.* relating to the pistil
pistillé,-e *a.* pistillate
pistolet *m.* pistol; **p. d'arrosage** pistol nozzle; *Ban.* sword sucker; *Vit.* fruit cane on spur-pruned vine
pistou *m.* basil (see **basilic**)
pitaya *m.* mandacaru cereus (*Cereus jamacaru*)
pite *f.* American aloe (*Agave americana*)
pittone (hirsute) *f.* Trinidad tournefortia (*Tournefortia hirsutissima*), a liana with edible fruits
pittospore *m.* *Pittosporum* sp.
pivoine *f.* 1. peony (*Paeonia* sp.); **p. arbustive, p. en arbre** Moutan peony, tree peony (*P. suffruticosa*); **p. de Chine** *P. lactiflora* (= *P. albiflora*); **p. coralline** *P. corallina* (= *P. mascula*); **p. à feuilles menues** *P. tenuifolia*; **p. officinale** *P. officinalis*. 2. *Coll.* bullfinch (see **bouvreuil**) 3. *a. inv.* peony-red
pivot *m.* *Bot.* taproot
pivotant,-e *a.* *Bot.* taprooted; **racine pivotante** taproot
pivoter *v.i.* to form a tap root (of plant)
placage *m.* 1. (veneer) grafting. 2. laying turf. 3. glazing, capping (of soil)
place (*f.*) **gélive** frost pocket
placement *m.* placement
placenta *m.* *Bot.* placenta
placentaire *a.* placental
placentation *f.* placentation
placette *f.* small experimental plot
plagiotrope *a.* plagiotropic
plagiotropie *f.* plagiotropy
plagiotropisme *m.* plagiotropism
plaie *f.* wound; **plaie de taille** pruning wound; *Rubb.* tapping cut
plain pied, in **de plain pied** *adv. phr.* on a level (with), on one floor
plan,-e *a.* even, flat, level, plane; *n. m.* plan
plancha *f.*, **greffe à la plancha** (*Morocco*) = **greffe en placage** (*q.v.*)
planche *f.* *Hort.* rectangular bed. **p. d'arrosage** *Irrig.* border strip; **p. de légumes** vegetable bed; **p. de semis** seed bed; **p. de stolons** runner bed. 2. board, plank; **p. de recouvrement** plank for shading etc.
plancher *m.* floor; **p. de séchage** drying floor
plançon *m.* 1. sapling. 2. set, slip, cutting, truncheon (in olive cultivation)
plane *m.* plane tree (*Platanus* sp.); Norway maple (*Acer platanoides*) (= **érable plane**)
plant *m.* 1. nursery plant (tree or vegetable), young plant, sapling, set, slip; **p. d'un an** one-year-old plant; **p. en arrachis** bare-root-lifted plant; **p. borgne** blind plant; **p. de bouture** plant derived from cutting; **p. certifié** certified stock; **p. filé** etiolated plant; **p. franc** seedling, plant on its own roots; **p. (de) frigo** cold-stored runner (strawberry); **p. greffé** (successfully) grafted plant; **p. issu de graine** seedling; **p. malingre** drawn plant, etiolated plant; **p. en motte** plant with ball of earth around the roots; **p. raciné** rooted cutting, rooted plant; **p. ramassé** compact plant; **p. repiqué** bedded plant, transplanted plant; **p. à repiquer** plant for bedding-out; **p. sain** healthy plant; **p. de semis** seedling; **p. de vigne** young grapevine. 2. bed, field of plants; **p. d'asperges** asparagus bed; **p. de choux** cabbage patch
plantable *a.* capable of being planted
plantage *m.* 1. planting. 2. cultivated patch of ground: plantation
plantain *m.* 1. plantain (*Plantago* sp.); **grand plantain** great plantain (*P. major*); **p. lancéolé** ribwort plantain (*P. lanceolata*), 2. plantain banana (*Musa* sp.) (= **banane plantain**); **p. corne** Horn plantain type; **p. créole** French plantain type, 3. **p. d'eau** water plantain (*Alisma plantago-aquatica*) (= **flûteau**)
plantanier *m.* (banana) plantain (*Musa* sp.)
plantard *m.* 1. sapling. 2. set, slip, cutting
plantation *f.* 1. planting, method of planting (= **système de plantation**); **p. d'automne** autumn planting; **p. en blocs** block planting; **p. en carré** square planting; **p. en chaintre** *Vit.* vine growing at margins of fields, marginal vine culture; **p. à demeure** planting direct, planting at stake; **p. à la fiche** *Vit.* etc. planting in hole made by a metal bar; **p. en mottes** planting with a root ball (of earth); **p. progressive** *Vit.* etc. planting as soil preparation continues; **p. en quinconce** staggered planting, triangular planting; **p. en rangées** row planting; **p. dans le(s) sable(s)** *Vit.* planting in sand (to avoid phylloxera damage); **p. en tranchées** *Vit.* planting in open trench; **p. en triangles** = **p. en quinconce** (*q.v.*); **p. par trous** planting in holes. 2. plantation (of cocoa, rubber etc.); **p. de citrons** citrus grove; **p. familiale** family

holding; **p. de fraisiers** strawberry field; **p. de rapport** commercial plantation

plante *f.* plant; **jeunes plantes** young plants; **p. aberrante (dans un lot)** rogue; **p. accumulatrice** accumulator plant; **p. adventice** adventitious plant, weed; **p. alimentaire** food plant; **p. alpine** alpine, rock plant; **p. améliorante** soil-improving plant; **p. d'appartement** house plant; **p. aquatique** aquatic plant, water plant; **p. aquatique nuisible** water weed; **p. aromatique** aromatic plant; **p. en bac** tub plant; **p. à balcon** balcony plant; **p. de bordure** border plant; **p. boussole** compass plant (*Silphium laciniatum*) (*Lactuca scariola*); **p. bulbeuse, p. à bulbe** bulbous plant; **"p. caoutchouc"** india rubber plant (*Ficus elastica*); **p. carnivore** carnivorous plant; **p. à cire** wax plant; **p. condimentaire** culinary (herb) plant; **p. de couverture** cover crop; **p. à épice** spice plant; **p. féculente** starchy plant; **p. au feu d'artifice** artillery plant (*Pilea muscosa*); **p. à feuillage (ornemental)** (ornamental) foliage plant; **p. à feuilles persistantes** evergreen; **p. fleurie** flower-covered plant, flowering plant; **p. à fleurs, p. florale** flowering plant, plant grown for flowers; **p. frileuse** tender plant, non-hardy plant; **p. fruitière** fruit plant; **p. de grande culture** field crop, crop grown on a field scale; **p. grasse** succulent plant; **p. grimpante** climbing plant; **p. héméropériodique** long-day plant; **p. herbacée** herbaceous plant; **p. à huile essentielle** essential oil plant; **p. indésirable** unwanted plant, weed; **p. indicatrice** indicator plant; **p. industrielle** industrial plant, industrial crop; **p. insectivore** insectivorous plant; **p. d'intérieur** houseplant; **p. jardinière** garden plant; **p. de jour(s) court(s)** short-day plant; **p. de jour(s) long(s)** long-day plant; **p. lampion** = **lanterne japonaise** (*q.v.*); **p. à latex** rubber plant; **p. légumière** vegetable; **p. ligneuse** woody plant; **p. de lumière** light-loving plant; **p. de marécage** marsh plant; **p. à massif** plant for planting in clumps; **p. masticatoire** masticatory plant; **p. médicinale** medicinal plant; **p. mère** mother plant, stock plant, stool; **p. molle** non-hardy, tender plant; non-woody plant; **p. qui monte en graine, p. montant en graine** bolting plant, bolter; **p. en motte** plant with a ball of earth around roots; **p. nyctipériodique** short-day plant; **p. aux oeufs** eggplant (= **aubergine** *q.v.*); **p. aux oeufs (de coq)** cock's eggs (*Salpichroa rhomboidea*) (= **muguet des pampas**); **p. oléagineuse** oil plant; **p. d'ombrage** shade plant; **p. d'ombre** shade(-loving) plant; **p. d'ornement, p. ornementale** ornamental plant, decorative plant; **p. parasite** parasitic plant; **p. à parfum** essential oil plant; **p. punk** club rush (*Scirpus cernuus*) grown on a tree-fern stump; **p. qui porte des chatons** catkin-bearing plant; **p. en pot** potted plant, pot plant; **p. potagère** (1) herb, (2) vegetable; **p. psychédélique** psychedelic plant; **p. de ramassage** *Afr.* etc. gathered plant; **p. rampante** creeping plant; **p. retombante** weeping plant; **p. rince-bouteilles** bottle-brush (*Callistemon* sp.); **p. de rocaille** rock plant; **p. saine** healthy plant; **p. de semis** seedling; **p. de serre** greenhouse plant; **p. de serre chaude** hothouse plant, stove plant; **p. spontanée** native plant, self-sown plant; **p. stimulante** stimulant beverage or masticatory plant (coffee, cocoa etc.); **p. sucrière** sugar plant; **p. supérieure** higher plant; **plantes en surplus (à l'éclaircissage)** thinnings; **p. tapissante** ground-cover plant, plant forming a cover; **p. témoin** test plant; **p. test** test plant, plant for testing or indexing; **p. textile** fibre plant; **p. tinctoriale** dye plant; **p. toxique** poisonous plant; **p. tubéreuse, p. tuberculifère** tuberous plant; **p. verte** ornamental foliage plant; **p. vivace** (herbaceous) perennial; *pl.* perennials (see also **mauvaise herbe**); **p. vulnéraire** healing plant; **"p. en zinc"** *Billbergia vittata*

plante-cobaye *f.*, *pl.* **plantes-cobayes** test plant

plante-hôte *f.*, *pl.* **plantes-hôtes** host plant, food plant (of insect etc.)

plante-piège *f.*, *pl.* **plantes-pièges** trap crop, trap plant

planter *v.t.* to plant, to set (seeds etc.); **p. à demeure, p. en place** to plant direct, to plant at stake; **p. en ligne** to line-out; **p. les manquants** to gap-up, to fill the gaps

plante(-)test *f.*, *pl.* **plantes(-)tests** see **plante**

planteur *m.* planter, grower; **p. de café** coffee planter; **p. d'hévéa** rubber planter; **p. de pommes de terre** potato grower

planteuse *f.* planting machine, planter; **p. automatique** mechanical planter; **p. (de pommes de terre)** potato planter; **p. tous-légumes** multi-purpose planter

plantier *m.* *Vit.* young shoot

plant-thermo *m.*, *pl.* **plants-thermos** virus-free plant (e.g. strawberry) obtained by heat-treatment, heat-treated plant

plantoir *m.* dibble, dibber, planting stick; **p. à bulbes** bulb planter

plantule *f.* small plant, small seedling, plantlet; **p. d'adventice** weed seedling

plaque *f.* plate, sheet, slab, patch; **p. de gazon** patch of turf; **p. de verre** plate of glass; module (for plant raising)

plaquemine *f.* persimmon fruit

plaqueminier *m.* *Diospyros* sp. (see also **kaki**);

p. du Caucase, p. d'Italie date plum (*D. lotus*); **p. du Japon** Japanese persimmon (*D. kaki*) (= **kaki**); **p. (de Virginie)** American persimmon (*D. virginiana*)
plasmalemme *m.* plasmalemma
plasmode *f.* plasmodium
plasmodesme *m.* plasmodesm
plasmolyse *f.* plasmolysis
plasmolysé,-e *a.* plasmolysed
plaste *m.* plastid
plasticité *f.* plasticity
plasticulture *f.* use of plastic in horticulture, growing under plastic, plastic mulching etc.
plastide *m.* plastid
plastidial,-e *a., m. pl.* **-aux** related to plastids
plastidome *m.* plastidome, the total plastids in a cell
plastifié,-e *a.* plasticized
plastique *n. m.* & *a.* plastic
plastisemeuse *f.* plastic-laying and sowing machine
plastisson *m.* plastic mat (to protect fruit trees, cover greenhouses etc.)
plastochrone *m.* plastochrone
plastomanie *f. PD* flat limb virus
plat,-e *a.* flat
platane *m.* plane tree (*Platanus* sp.); **p. à feuilles d'érable, p. hybride, p. du Midi** London plane (*P.* × *acerifolia*); **p. d'occident** buttonwood (*P. occidentalis*); **p. oriental, p. d'orient** oriental plane (*P. orientalis*)
plateau *m.* 1. tray; **p. à pots** pot tray; *Bot.* short stem of bulb; **p. criblé** *Bot.* sieve plate; **p. radiculaire** *OP* bulbous base (= **bulbe**). 2. plateau, tableland
plate-bande *f., pl.* **plates-bandes** grass border; bed, flower bed
plate-forme *f., pl.* **plates-formes** platform; **p.-f. automotrice** self-propelled platform
plâtrage *m.* plastering
plâtre *m.* plaster; **(pierre à) plâtre** gypsum; **p. blanc** *PD, Mush.* white plaster mould (*Scopulariopsis fimicola*); **p. brun, p. rouge** *PD, Mush.* brown plaster mould (*Papulospora byssina*)
platycérium (à corne d'élan) *m.* staghorn fern (*Platycerium alcicorne*); **p. de deux formes** *P. biforme*; **grand platycérium** *P. grande*
platystémon *m. Platystemon* sp., including cream cups (*P. californicus*)
plectenchyme *m.* plectenchyma
plein,-e *a.* full, entire, whole, full (of flower); **en plein air** in the open, outside; **à plein bois** right across the rootstock (of grafting split) (see also **à mi-bois**); **en plein champ** in the open; **en pleine lumière** in full light; **en plein soleil** in full sunlight; **de pleine terre** hardy (of plant); **en pleine saison** at the height of the season; **en pleine terre** in the open; **au stade pleine fleur** at full flowering stage; **plein** *Ban.* full (maturation grade); **¾ plein** three-quarters full
plein-vent *a. inv.* & *n. m., pl.* **pleins-vents** standard fruit tree, hardy fruit tree, isolated tree
pléiomère *a.* pleiomerous, with many parts or organs
pléione *f. Pleione* sp.
pleïotrope *a. PB* pleiotropous
pléomorphe *a.* pleomorphic
pléophage *a.* pleophagous, capable of attacking several species of host-plant
plérome *m.* plerome
pleurard,-e *a. Hort.* weeping, drooping; **rameau pleurard** drooping branch
pleureur,-euse *a.* Hort. weeping; **saule pleureur** weeping willow
pleurote *m. Fung. Pleurotus* sp.; **p. en coquille, p. huitré** oyster mushroom (*Pleurotus ostreatus*); **p. en corne d'abondance** *P. (ostreatus* var.*) cornucopiae*; **p. du panicaut** *P. eryngii*
pleuroticulteur *m.* oyster mushroom grower
pleurs *m. pl. Vit.* bleeding
pleyon *m.* 1. tied back fruit branch, fruit cane. 2. osier tie
pliant,-e *a.* folding; **scie pliante** folding saw
plissé,-e *a.* pleated; **plissée** *Vit.* applied to leaf folded at the point of attachment to the petiole
ploïdie *f.* ploidy
plomb *m.* lead; *PD* **p. (des arbres fruitiers)** silver leaf disease (*Stereum purpureum*)
plombage *m. Hort.* rolling, ramming, tamping (soil)
plombago *m.* (see **plumbago**)
plombé,-e *a. Hort.* rolled, rammed, tamped (of soil)
plomber *v.t. Hort.* to roll, to ram, to tamp (soil)
plombique *a.* relating to lead; **contamination plombique** lead contamination
plongé,-e *a.* submerged (of plant)
plugging *m. Rubb.* plugging (see also **index**)
pluie *f.* rain; **p. battante** beating rain; **p. fine** drizzle; **les pluies** *Afr.* etc. the rainy season, the rains (= **hivernage**)
pluie d'or *f.* laburnum, golden chain (*Laburnum anagyroides*)
plumbago *m.* plumbago, leadwort (*Plumbago* sp.)
plume(s) du Kansas *f.* gay feather (*Liatris spicata*), Kansas feather (*L. pycnostachya*)
plumet *m.* plume; **p. des pampas** = **gynérion argenté** (*q.v.*)
plumeux,-euse *a.* feathery, plumose
plumule *f.* plumule
pluriannuel *a.* pluriannual
pluricellulaire *a.* pluricellular, multicellular

pluriflore *a.* multiflorous
plurifolié,-e *a.* multifoliate
pluriloculaire *a.* multilocular, plurilocular
pluvial,-e *a., m.pl.* **-aux** pluvial, rainy; **culture pluviale** rain-fed crop; **eau pluviale** rainwater
pluvieux,-euse *a.* rainy, wet (weather)
pluviographe *m.* self-recording rain gauge
pluviolessivage *m.* rain washing
pluviomètre *m.* rain gauge
pluviométrie *f.* (meteorological) rainfall; **p. mal répartie** badly distributed rainfall
pluviosité *f.* rainfall, precipitation
PM = **poudre mouillable** *PC* w.p. wettable powder
pneu *m.* tyre; **p. gazon** mower tyre
pneumatophore *a.* pneumatophorous; *n. m.* pneumatophore
PO = **pression osmotique** OP osmotic pressure
poche *f.* pocket; **p. d'air** air pocket; **p. à manche** sleeved pocket (for date harvesting)
pochette *f.* small pocket; **pochettes du prunier** *PD* pocket plums (*Taphrina pruni*)
podagraire *f.* ground elder (*Aegopodium podagraria*)
podophylle *m.* *Podophyllum* sp., May apple
podsol, podzol *m.* podsol, podzol
podsolisation *f.* podsolization
podsolique *a.* podsolic
podsolisé,-e *a.* podsolized
pogostémon *m.* *Pogostemon* sp.; **p. suave** patchouli (*P. cablin*) (= **patchouli**)
poids *m.* weight; **p. atomique** atomic weight; **p. brut** gross weight; **p. frais** fresh weight; **p. de 1000 grains** 1000-grain weight; **p. moléculaire** molecular weight; **p. net** net weight; **p. normal** *Sug.* etc. normal weight; **p. sec** dry weight
poil *m.* *Bot.* hair, in *pl.* down, pubescence; **p. absorbant** = **p. radical** (*q.v.*); **p. foliaire** leaf hair; **p. glanduleux, p. sécréteur** glandular hair; **p. radical, p. radiculaire** root hair; **p. urticant, p. vésicant** stinging hair
poilu,-e *a.* hairy, pilose; **non poilu** smooth, glabrous (see also **glabre**)
poinciane *f.* *Poinciana* sp., including flamboyant (*P. regia*)
poinsettia *m.* poinsettia (*Euphorbia pulcherrima*) (= **étoile de Noël**)
point *m.* 1. stitch; **points de tapisserie** *Vit.* "tapestry stitches", applied to mosaic form of downy mildew (*Plasmopara (Peronospora) viticola*). 2. point; **points bruns (de la chair de la pomme)** *PD* (apple) bitter pit; **p. de compensation du CO_2** CO_2 compensation point; **p. de croissance** growing point; **p. d'ébullition** boiling point; **p. de flétrissement** wilting point; **p. de fusion** melting point; **p. de greffe** union (of graft); **points liégeux** bitter pit (of apple); **p. pétiolaire** petiolar junction; **p. de plantation** (planting) hill; **p. de saturation** saturation point; **p. de soudure** = **p. de greffe** (*q.v.*); **p. végétatif** shoot apex, meristem; **p. de végétation** = **p. de croissance** (*q.v.*)
pointe *f.* point, sharp end, peak; **pointes d'asperges** asparagus tips; **p. d'un pépin** beak of seed; **p. de production** production peak; **p. de racine** root tip
pointillé,-e *a.* dotted, stippled, spotted
pointu,-e *a.* sharp-pointed, pointed
poire *f.* 1. pear (see also **poirier**); **p. asiatique** see **p. orientale**; **p. d'avocat** avocado pear (= **avocat**); **p. blanche de Chine** Chinese pear (*Pyrus ussuriensis*); **p. channe** (*Switzerland*) type of pear which is dried before eating; **p. japonaise** sand pear; **p. à poiré** perry pear; **p. d'Anchois** anchovy pear (*Grias cauliflora*); **p. de chardon** (*Antilles*) night-blooming cereus (*Hylocereus undulatus*); **p. orientale** oriental pear, sand pear (= **nashi**); **p. pierreuse** *f.* *PD* (pear affected by) stony pit (virus) (see also **gravelle**); **p. de terre** Yacon strawberry (*Polymnia edulis*) (see also **polymnia**)
poiré *m.* perry
poireau *m.* leek (*Allium porrum*); **p. roux** tassel hyacinth (*Muscari comosum*) (= **muscari à toupet**)
poirée *f.* spinach beet; **p. à cardes** Swiss chard (*Beta vulgaris* var. *cicla*) (see also **bette**)
poire-melon *f.* melon pear, pepino (*Solanum muricatum*) (= **melon-poire**)
poire-pomme *f.* round oriental pear, sand pear (see **poire**)
poirette *f.* small pear, immature pear
poirier *m.* pear tree (*Pyrus* sp.) (see **poire**); **p. commun** wild pear (*P. communis*); **p. japonais** sand pear tree (*P. serotina* = *P. pyrifolia*)
pois *m.* 1. pea (*Pisum sativum*); **petits pois** green peas; **pois cassés** split peas; **pois congelés** frozen peas; **pois de casserie** peas for splitting; **p. de conserve** peas for canning; **p. à écosser** garden pea, green pea (with non-edible pod); **p. frais** green pea; **p. en gousses** unshelled peas; **p. à grain ridé** wrinkle-seeded pea; **p. à grain rond** smooth-seeded pea; **p. mange-tout** edible-podded pea; **p. nain** dwarf pea; **p. à parchemin** = **p. à écosser** (*q.v.*); **p. potager** garden pea; **p. à rames** climbing pea; **p. sans parchemin** = **p. mange-tout** (*q.v.*); **p. sec** dried pea. 2. Other genera: **p. ailé** = **p. carré** (*q.v.*); **p. d'Ambrevade, p. d'Angole** pigeon pea (*Cajanus cajan*) (= **ambrevade**); **p. Azuki** Adzuki bean (*Phaseolus angularis*); **p. bâtard** centro (*Centrosema pubescens*); **p. du Brésil** cowpea

(*Vigna unguiculata*) (= **niébé**); **p. cajan** = **p. d'Ambrevade**; **p. du Cap** lima bean (*Phaseolus lunatus*) (= **haricot de Lima**); **p. carré** asparagus pea, Goa bean (*Psophocarpus tetragonolobus*); **p. chiche** chick pea (*Cicer arietinum*); **p. de Chine** = **p. vivace** (*q.v.*); **p. de coeur** heart pea (*Cardiospermum halicacabum*); **p. cornu** = **p. chiche**; **p. doux** sackysac (*Inga laurina*); **p. ficelle** asparagus pea, yardlong bean (*Vigna sesquipedalis*) (= **haricot asperge**); **p. indien** hyacinth bean (*Dolichos lablab*); **p. mascate** Florida velvet bean (*Stizolobium deeringianum*); **p. musqué** = **p. de senteur** (*q.v.*); **p. pigeon** = **p. d'Angole** (*q.v.*); **p. sabre** sword bean (*Canavalia ensiformis, C. gladiata*) (= **haricot sabre**); **p. savon** = **p. du Cap** (*q.v.*); **p. de senteur** sweet pea (*Lathyrus odoratus*); **p. sucré** food inga (*Inga edulis*), Madras thorn (*Pithecellobium dulce*); **p. de terre** Bambara groundnut (*Voandzeia subterranea*); **p. violon** = **p. bâtard** (*q.v.*); **p. vivace** everlasting pea (*Lathyrus latifolius*) (= **gesse à large feuilles**)

pois-manioc *m.* yam bean (*Pachyrhizus erosus*) (= **pomme de terre du Mossi**)

poison *m.* poison; **p. par ingestion** stomach poison; **p. de piste** contact rodenticide

poivre *m.* pepper (*Piper* and other genera) (see also **poivrier**); **p. (noir)** (black) pepper; **p. des Achantis** Ashanti pepper (*P. guineense*); **p. de Cayenne** Cayenne pepper; **p. d'eau** water pepper (*Polygonum hydropiper*); **p. d'Espagne** = **piment** (*q.v.*); **p. de Guinée** (1) = **piment** (*q.v.*), (2) grains of Paradise (= **graines de paradis, p. maniguette** (*q.v.*), (3) Guinea pepper; **p. de la Jamaïque** allspice, pimento (= **pimenta, toute-épice**); **p. du Kissi** = **p. des Achantis** (*q.v.*); **p. long (vrai)** long pepper (*Piper chaba* = *P. officinarum*); **p. long du Bengale** Jaborandi pepper (*Piper longum*); **p. maniguette** grains of Paradise (*Aframomum melegueta*); **p. des murailles** yellow stonecrop (*Sedum acre*); **p. à queue** cubeb (*Piper cubeba*) (= **cubèbe**) (see also **bétel, kava, matico**)

poivré,-e *a.* peppery, pungent

poivrier *m.* pepper vine, pepper plant (*Piper nigrum*) (see also **poivre**); **p. d'Amérique** pepper tree (*Schinus molle*) (= **faux-poivrier**); **p. de Guinée** Guinea pepper tree (*Xylopia aethiopica*); **p. de la Jamaïque** allspice, pimento tree (*Pimenta dioica*); **p. du Japon** Japan pepper (*Zanthoxylum piperitum*) (= **clavalier poivrier**); **p. des religieux** chaste tree (*Vitex agnus-castus*) (= **arbre au poivre**)

poivrière *f.* 1. pepper (*Piper*) garden. 2. pepper pot

poivron *m.* large sweet *Capsicum* pepper, bell pepper (*Capsicum annuum* var. *grossum*) (= **gros piment**)

poix *f.* pitch

pol *m. Sug.* pol

polarité *f.* polarity

polémonie *f.* *Polemonium* sp.; **p. bleue, p. grecque** Jacob's ladder, Greek valerian (*P. caeruleum*)

polianthe *m.* *Polianthes* sp., tuberose

pollen *m.* pollen; **p. marqueur** marker pollen

pollination *f.* pollination (= **pollinisation** *q.v.*)

pollinide *m.*, **pollinie** *f.* pollinium

pollinifère *a.* polliniferous

pollinique *a.* relating to pollen; **boyau pollinique, tube pollinique** pollen tube; **viabilité pollinique** pollen viability

pollinisateur,-trice *a.* pollinating; **insectes pollinisateurs** pollinating insects; **variétés pollinisatrices** pollinating varieties; *n. m.* pollinator

pollinisation *f.* pollination; **p. assistée** assisted pollination; **p. croisée, p. indirecte** cross-pollination; **p. directe** self-pollination; **p. libre** open pollination

polliniser *v.t.* to pollinate

polluant *m.* agent of pollution, pollutant

pollué,-e *a.* polluted; **eaux polluées** polluted waters

pollution *f.* pollution; **p. atmosphérique** air pollution; **p. de l'eau** water pollution; **p. urbaine** urban pollution

polyadelphe *a.* polyadelphous

polyamide *m.* polyamide

polyandre *a. Bot.* polyandrous

polyantha *m.* polyantha rose

polyanthe *a.* polyanthous, many-flowered

polycarpique *a.* polycarpic

polycarpon (*m.*) **à quatre feuilles** four-leaved allseed (*Polycarpon tetraphyllum*)

polychrome *a.* polychrome

polyculteur *m.* grower who grows several different crops

polyculture *f.* polyculture, mixed cropping

polyembryoné,-e *a.* polyembryonic

polyembryonie *f.* polyembryony

polyester *m.* polyester

polyéthylène *m.* polyethylene (= **PE**); **p. noir** black polyethylene

polygala, polygale *m.* *Polygala* sp., milkwort; **p. de Virginie** Seneca snake-root (*P. senega*)

polygame *a. Bot.* polygamous

polygènes *m. pl. PB* polygenes

polygénique *a. PB* polygenic

polymère *m.* polymer

polyméthacrylate (*m.*) **de méthyle** polymethyl methacrylate (plastic)

polymnia *m.*, **polymnie** *f.* *Polymnia* sp., including Yacon strawberry (*P. edulis*) (= **poire de terre**)
polymorphe *a.* polymorphic
polymorphie *f.*, **polymorphisme** *m.* polymorphism
polyoside *m.* polysaccharide
polypétale *a.* polypetalous
polyphage *a.* polyphagous
polyphénol *m.* polyphenol
polyphénolique *a.* polyphenolic
polyphosphate *m.* polyphosphate
polyphylétique *a.* polyphyletic
polyphylétisme *m.* polyphyletism
polyphylle *a.* polyphyllous
polyploïde *a.* & *n. m.* polyploid
polyploïdie *f.* polyploidy
polyploïdisation *f.* polyploidization
polypode *m.* *Polypodium* sp., polypody fern; **p. commun** wall fern (*P. vulgare*); **p. à feuilles de bananier** *P. musifolium*; **p. à feuilles étroites** *P. angustifolium*; **p. à feuilles de frêne** *P. fraxinifolium*; **p. à feuilles glauques** *P. glaucophyllum*; **p. à feuilles de lycopode** *P. lycopodioides*
polypore *m.* polypore (*Polyporus* sp.); **p. en ombelle** *P. umbellatus* (see also **poule des bois**)
polysaccharide *m.* polysaccharide
polysépale *a.* polysepalous
polysoc *m.* multiple plough
polysome *m.* polysome
polysomie *f.* *PB* polysomy
polysperme *a.* polyspermous, containing many seeds
polyspermie *f.* polyspermy
polystachié,-e *a.* polystachyous
polystémone *a.* polystemonous
polystyrène *m.* polystyrene; **p. expansé** expanded polystyrene
polysulfure *m.* polysulphide
polytric *m.* polytrichum moss
polyvalence *f.* versatility (of machine, implement, chemical etc.)
polyvalent,-e *a.* polyvalent, versatile, multipurpose
polyvinyle *m.* polyvinyl; **chlorure de polyvinyle** polyvinyl chloride
pomacé,-e *a.* pomaceous; *n. f. pl.* pome fruits (= **pirées, pomées**)
pomaison *f.* period when cabbages, lettuces etc. form hearts
pomelo *m.* pomelo, pummelo (*Citrus grandis*), often used for grapefruit (*C. paradisi*) (see also **grapefruit, pamplemousse**)
pomées *f. pl.* pome fruits (= **pirées, pomacées**)
pomiculteur *m.* orchardist, pome fruit grower
pomifère *a.* pomiferous
pomiforme *a.* pomiform, apple-shaped
pommade *f.* pomade, ointment
pomme *f.* 1. (a) apple (see also **pommier**); **p. à cidre** cider apple; **p. à couteau** eating apple, dessert apple; **p. sauvage** crab apple; **pommes vitreuses** *PD* (apples affected by) water core, glassiness; (b) **p. de terre** potato (*Solanum tuberosum*); **p. de terre de consommation** ware potato; **p. de terre demi-hâtive** second early potato; **p. de terre hâtive** (first) early potato; **p. de terre potagère** garden potato; **p. de terre de semence** seed potato; **p. de terre de l'air** air potato (*Dioscorea bulbifera*); **p. de terre céleri** Peruvian parsnip (*Arracacia xanthorhiza*); **p. de terre de Madagascar** edible *Coleus* spp. including Hausa potato (*C. dazo*); **p. de terre du Mossi** yam bean (*Sphenostylis stenocarpa*); (c) Other genera: **p. d'acajou** = **p. cajou**; **p. d'amour** love apple; **p. cajou** cashew apple; **p. calebasse** sweet calabash, sweet cup (*Passiflora maliformis*); **p.-cannelle** sugar apple; **p. cannelle du Sénégal** *Afr.* wild custard apple; **p. du Cayor** *Afr.* gingerbread plum fruit; **p.-cythère** golden apple; **p. épineuse** thorn apple (*Datura stramonium*); **p. étoilée** star apple (*Chrysophyllum cainito*) (= **caïmitier**); **p. de mai** May apple (*Podophyllum* sp.); **p. de merveille** balsam apple (*Momordica balsamina*); **p. mexicaine** white sapote fruit; **p.(-)rose** rose apple; **p. sauvage** *Afr.* wild mango fruit; **p. de savon** soapberry (*Sapindus* sp.). 2. pome. 3. (a) heart, head (of plant); **p. de chou** head of cabbage; **p. de chou de Bruxelles** sprout of brussels sprouts; **p. de chou-fleur** curd of cauliflower; **p. de chou-rave** swollen stem of kohl-rabi; **p. d'endive** blanched head of witloof chicory; **p. de laitue** heart of lettuce; **p. de pin** pine cone (see also **cône**); (b) knob; **p. d'arrosoir** rose of watering can
pommé,-e *a.* rounded, hearted (of cabbage, lettuce etc.); **chou bien pommé** cabbage with a fine head; **laitue pommée** cabbage lettuce, heading lettuce
pomme de terre *f.* see **pomme** 1. (b)
pomme-liane *f.*, *pl.* **pommes-lianes** water lemon (*Passiflora laurifolia*); **p.-l. collante** wild passion flower (*P. foetida*); **p.-l. violette** purple granadilla (*P. edulis*) (= **grenadille**)
pommer *v.i.* & *p.* to form a round head, to heart (of cabbage, lettuce etc.)
pommier *m.* 1. apple tree (*Malus* sp.) (see also **pomme**); **p. baccifère** Siberian crab apple (*M. baccata*); **p. de la Chine** *M. spectabilis*; **p. à cidre** cider apple tree; **p. doucin** type of wild apple used as rootstock (*M. pumila* var. *(acerba =) sylvestris*) (= **doucin**); **p. à**

fleurs = **p. d'ornement** (*q.v.*); **p. franc** seedling apple tree, apple tree on its own roots; **p. du Japon** = **cognassier du Japon** (*q.v.*); **p. à fleurs, p. d'ornement** flowering crab-apple tree (*M. floribunda*) and others; **p. paradis** paradise apple (*M. pumila* var. *paradisiaca*) (= **paradis**); **p. pourpre** purple-leaved *Malus*; **p. sauvage** crab-apple tree (*M. pumila*); **p. de Sibérie** *M. baccata*. 2. Other genera: **p. d'amour** Jerusalem cherry (*Solanum pseudo-capsicum*); **p. cajou** cashew tree (*Anacardium occidentale*) (= **anacardier**); **p. (-)cannelle** sugar apple tree (*Annona squamosa*); **p. (-)cannelle du Sénégal** *Afr.* wild custard apple (*Annona senegalensis*); **p. du Cayor** *Afr.* gingerbread plum tree (*Parinari macrophylla*); **p. (-) cythère** golden apple tree (*Spondias cytherea*); **p. de Goa** carambola tree (*Averrhoa carambola*) (= **carambolier**); **p. mexicain** white sapote (*Casimiroa edulis*) (= **sapote blanche**); **p. rose** rose apple tree (*Eugenia jambos*) (= **jambose**); **p. sauvage** wild mango (*Irvingia gabonensis*)

pomoculture *f.* pome fruit culture

pomoïdées *f. pl.* Pomoideae (section of the Rosaceae); **pomoïdées ornementales** ornamental Pomoideae

pomologie *f.* pomology

pomologique *a.* pomological

pomologiste, pomologue *m.* pomologist

pompage *m.* pumping

pompe *f.* pump; **p. aspirante** suction pump; **p. à bras** hand pump; **p. centrifuge** centrifugal pump; **p. pulvérisatrice** spraying pump; **p. rotative** rotary pump; **p. rotative à engrenages** rotary gear pump

pomper *v.t.* to pump (up)

pompon *m.* pompon flower; **dahlia pompon** pompon dahlia; **rose pompon** pompon rose (= **pompon**); **p. de Saint-François** *Rosa gallica parvifolia* (see **rosier**); *Ban., Coll.* bracts and male flowers at end of inflorescence

ponceau *m., pl.* **-eaux** 1. culvert. 2. corn poppy (*Papaver rhoeas, P. dubium*); *a. inv.* poppy red

ponctuation *f.* spotting, speckling; **p. nécrotique** necrotic spots

ponctué,-e *a.* dotted, spotted, speckled

ponctuel,-elle *a.* pinpoint (source of light etc.); **forage ponctuel** well-point

pondeuse *f. Coll.* = **aubergine** (*q.v.*)

ponère *f. Ponera* sp. (orchid)

pontédérie *f. Pontederia* sp., pickerel weed

populage *m.* marsh marigold (*Caltha palustris*) (= **caltha des marais, souci d'eau**)

population *f.* population

populiculture *f.* poplar growing

poquet *m.* planting hole, seed hole; **semis en poquets** planting in holes

poral,-e *a., m. pl.* **-aux** relating to pores; **espace poral** pore space

porc-épic *m.* porcupine (*Hystrix* and other genera); *Afr.* brush-tailed porcupine (*Atherurus africanus*) (= **athérure**)

pore *m.* pore; **p. germinatif (de la noix de coco)** eye (of coconut)

poreux,-euse *a.* porous

poromètre *m.* porometer; **p. à diffusion** diffusion porometer

porosimétrie *f.* porosimetry

porosité *f.* porosity; *Ped.* pore space

porquet *m. Ent.* noctuid larva, including *Agrotis* sp.

port *m. Arb., Hort.* habit; **p. de la plante** habit of the plant; **p. en boule** with rounded crown (of tree); **p. buissonnant** bushy habit; **p. compact** compact habit; **p. diffus** open habit; **p. dressé** upright habit; **p. élevé** upright habit; **p. érigé** upright habit; **p. fastigié** with narrow conical crown, fastigiate habit; **p. pleureur** weeping habit; **p. rampant** creeping habit; **p. volubile** climbing habit

porte-bonheur *m. inv.* (lucky) charm, mascot; **fleur porte-bonheur** lucky flower

porte-bouture(s) *a.* reserved for cuttings (of plant)

porte-chapeau *m. inv. Bot.* Christ's thorn (*Paliurus spina-Christi*) (= **épine du Christ**)

portée pratique *f. Stat.* practical range

porte-graine(s) *n. m. inv.* & *a. inv.* reserved for seed, seed plant; **carottes porte-graines** carrots grown for seed

porte-greffe, porte-greffes *m. inv.* rootstock, stock (= **sujet**); **p.-g. affaiblissant** dwarfing rootstock; **p.-g. fort** vigorous rootstock; **p.-g. nanisant** dwarfing rootstock; **p.-g. semi-fort** semi-vigorous rootstock

porte-noix *m. inv.* butter nut (*Caryocar nuciferum*)

porte-outil(s) *m. inv.* toolbar

porte-pots *m.* flower pot stand

porte-rampe *m. inv.* boom carrier

porteur,-euse *a. Hort.* bearing; **branche porteuse, rameau porteur** bearing branch

portlandie *f. Portlandia*, a neotropical genus which includes flowering shrubs such as *P. grandiflora*

portuaire *a.* relating to a port, harbour

portulaca *m. Portulaca* sp., purslane (see also **pourpier**)

post-buttage *m.* post-earthing up, after earthing-up

postdormance *f.* postdormancy

poste *m.* post, station; **p. météorologique** weather station
post-émergence *f.* post-emergence (= **post-levée**)
post-floral,-e *a., pl.* **-aux** post-floral
post-levée *f.* post-emergence; **herbicide de post-levée** post-emergence herbicide
post-maturation *f.* after-ripening
post-plantation *f.* post-planting
post-reprise *f.* post-regrowth
post-semis *m.* post-sowing; **application en post-semis** post-sowing application
pot *m.* pot; **p. à fleurs** flower pot; **p. de fleurs** flowering plant in a pot, pot of flowers; **pots serrés côte à côte** pot-thick arrangement; **p. de tourbe** peat pot
potager,-ère *a.* edible, culinary, for the pot; **herbe potagère** pot herb; **jardin potager** kitchen garden; **plante potagère** herb, vegetable; *n. m.* kitchen garden; **potager-fruitier** fruit and vegetable garden
potalia *m.*, **potalie** *f. Potalia* sp.
potamot, potamogéton *m. Potamogeton* sp., pondweed
potasse *f.* potash
potassé,-e *a.* containing potash, combined with potassium
potassique *a.* potassic, containing potassium; **nutrition potassique** potash nutrition
potassium *m.* potassium; **p. assimilable** available potassium
poteau *m., pl.* **-aux** post, pole, stake
potée *f.* potful; **p. fleurie** potted (flowering) plant, plant in pot
potelet *m.* (small) post, strut, prop
potentialité *f.* potentiality
potentiel,-ielle *a.* potential; *n. m.* potential; **p. capillaire** capillary potential; **p. hydrique foliaire** leaf water potential; **p. infectieux** infectious potential; **p. d'inoculum** inoculum potential
potentille *f. Potentilla* sp., cinquefoil; **p. ansérine, p. argentée** silverweed (*P. anserina*)
pothos *m.* golden pothos (*Scindapsus (Pothos) aureus*)
potiron *m.* winter squash (*Cucurbita maxima*)
potomètre *m.* potometer
poto-poto *m. inv.* black organic soil (in tropics) associated with tidal rivers (e.g. mangrove soil)
potyvirus *m. PD* potyvirus
pou *m., pl.* **poux** louse; *Hort.* scale, aphid; **poux (collants)** scales; **p. de Floride** *Cit.* Florida red scale (*Chrysomphalus aonidum*); **p. des Hespérides** *Cit.* soft brown scale (*Coccus hesperidum*) (= **cochenille plate**); **p. du laiteron** currant-sowthistle aphid (*Hyperomyzus lactucae*); **p. rouge (des agrumes)** *Cit.* dictyospermum scale (*Chrysomphalus dictyospermi*); **p. (rouge) de Californie** *Cit.* citrus red scale (*Aonidiella aurantia*); **p. de San José** San José scale (*Quadraspidiotus perniciosus*); **p. des serres** long-tailed mealybug (*Pseudococcus adonidum*)
pouce *m.* 1. thumb. 2. inch
poudrage *m.* dusting, application of powder
poudre *f.* powder; **p. antiparasitaire** pesticide powder; **p. insecticide** insecticide powder; **p. mouillable** *PC* wettable powder; **p. d'os** bone meal; **p. d'os dégélatinisée** steamed bone flour; **p. soluble dans l'eau, PS** *PC* water-soluble powder
poudrer *v.t.* to dust
poudrette *f.* 1. fine powder; dried powdered night soil or household waste used as fertilizer. 2. *Vit.* young vine plant
poudreuse *f.* duster (machine), powder blower; **p. à dos** knapsack duster; **p. à grand rendement** field duster; **p. à moteur** power-driven duster
poudreux,-euse *a.* powdery
poudroyer *v.t.* to cover with dust; *v.i.* to form clouds of dust
poule des bois *Fung. Polyporus frondosus* (see also **polypore**)
poule pondeuse *f.* = **aubergine** (*q.v.*)
poupartia *m. Afr.* cat thorn (*Sclerocarya birrea*)
poupartier (*m.*) **d'Amazonie** *Poupartia amazonica*, a tree with edible fruits
pourcentage *m.* percentage
pourette *f.* (*Algeria*) olive nursery plant
pourghère *m.* = **purghère** (*q.v.*)
pourpier *m.* purslane; **p. en arbre** *Portulacaria afra*; **p. commun** common purslane (*Portulaca oleracea*); **p. à grandes fleurs** sun plant (*Portulaca grandiflora*); **p. d'hiver** winter purslane (*Claytonia perfoliata*) (= **claytone de Cuba**); **p. marin d'océan** sea purslane (*Halimione (Obione) portulacoides*); **p. maritime** sea purslane (*Sesuvium portulacastrum*); **p. de mer** Mediterranean saltbush (*Atriplex halimus*); **p. potager** cultivated, salad purslane (= **p. commun** *q.v.*) (see also **grand**)
pourpre *a.* purple, crimson; **p.-noir** *a. inv.* dark purple
pourri,-e *a.* rotten, decayed; *n. m.* rotten, decayed part; **p. noble** *Vit.* noble rot (*Botrytis cinerea*) (= **pourriture noble**)
pourridié *m.* (root) rot; **p.-agaric, p. des racines (des arbres fruitiers)** armillaria root rot (*Armillaria mellea*); **p. du noyer** walnut root rot (*Armillaria mellea* and others); **p. des racines (des végétaux)** fungal root rot (*Dematophora necatrix*) (= **blanc des racines**) (see also **p.-agaric**); **p. des vergers** fomes heart

rot (*Fomes pomaceus*); **p. de la vigne** grapevine root rot (*A. mellea, D. necatrix*) (see also **pourriture**)

pourrir *v.i.* to rot, to decay

pourriture *f.* rot, decay; **p. bactérienne (du chou)** cabbage soft rot (*Erwinia carotovora*); **p. blanche (des alliacées) (des liliacées)** (Alliaceae) white rot (*Sclerotium cepivorum*) (= **maladie blanche**); **p. blanche (des vergers)** (= **p. des vergers**) fomes heart rot (*Fomes pomaceus*); **p. bleue** blue mould (*Penicillium italicum*); **p. brune (des cabosses du cacaoyer)** (*Cacao*) cacao black pod rot (*Phytophthora palmivora*); **p. du coeur (de la betterave)** beet heart rot and dry rot (B deficiency and/or *Pleospora betae*); **p. du collet** collar rot (*Phytophthora* spp.); *Vit.* collar rot (caused by several fungi); **p. du collet (des alliacées)** (Alliaceae) neck rot (*Botrytis allii, B. byssoidea*); **p. grise** (1) grey mould (*Botrytis cinerea*), (2) **p. grise de l'oignon** onion neck rot (*B. allii, B. byssoidea*), (3) **p. grise de la tulipe** tulip grey bulb rot (*Sclerotium tuliparum*); **p. molle (du glaïeul** etc.) dry rot (of gladiolus etc.) (*Sclerotinia gladioli*); **p. maculée** *f. Vit.* black rot (*Guignardia bidwelli*); **p. molle (de la pomme de terre)** potato soft rot (*Erwinia carotovora* etc.); **p. noble (sur raisins)** *Vit.* noble rot (*B. cinerea*) (= **pourri noble**); **p. noire (de la carotte)** carrot black rot (*Stemphylium radicinum*); **p. noire des racines** black root rot (of peas etc.) (*Thielaviopsis basicola*); **p. pédonculaire (des agrumes)** *Cit.* stem-end rot (*Phomopsis citri*); **p. pédonculaire de l'avocat** avocado stem-end rot (*Diplodia natalensis*); **p. du point stylaire** *Cit.* stylar-end breakdown; **p. des racines** root rot (caused by various fungi on various crops) (= **pourridié**); **p. rouge de la racine du fraisier** strawberry red core disease (*Phytophthora fragariae*); **p. sèche (de la pomme de terre)** potato dry rot (*Fusarium caeruleum*); **p. sèche du coeur** (*Coconut*) dry bud rot, stem necrosis; **p. de la tomate** tomato fruit rot (*Didymella lycopersici* and *Phoma destructiva*); **p. des vergers** (= **p. blanche**) fomes heart rot (*Fomes pomaceus*); **p. verte** green mould (*Penicillium digitatum*)

pousse *f.* 1. growth; **p. en ortie** *Vit.* "nettle-leaf growth"—applied to deformation caused by infectious degeneration virus; **seconde pousse, p. d'août** second flush, growth. 2. shoot, sprout, bud; **p. de l'année** current year's shoot; **p. apicale** apical shoot; **p. d'asperge** asparagus spear; **p. feuillée** leafy shoot; **p. latérale** *Hort.* feather, lateral shoot; **p. principale** leader, leading shoot; **p. souterraine** underground shoot; **p. terminale** terminal shoot

poussée *f.* sprouting, growth; **p. d'août** second growth in mid or late August; **p. de croissance** flush; **p. radiculaire** root pressure; **p. de (la) sève** rising of sap; **p. végétative** flush

pousser *v.t.* to push; *v.i.* to grow, to shoot; **p. des rejetons** to sucker, to stool

poussière *f.* dust; **p. de tourbe** granulated peat, peat dust

poussiéreux,-euse *a.* dusty

pouteria *m. Pouteria*, a genus which includes several neotropical trees with edible fruit (see also **lucuma**)

pouvoir *m.* power, force; **p. émissif** emissivity; **p. germinatif (des graines)** (seed) viability; **p. infectieux** infectivity; **p. inhibiteur** inhibitory capacity; **p. mouillant** *PC* spreadability; **p. pathogène** pathogenicity, virulence; **p. réducteur** reducing ability; **p. tampon** buffering capacity

pouzzolane *f.* pozzolana

p.p.m. = **parties** (*f. pl.*) **par million** p.p.m., parts per million

pralin *m.* slurry

pralinage *m. Hort.* dipping in slurry

praliner *v.t.* to puddle, to dip roots in slurry prior to planting

prays *f. Ent. Prays* sp.

précédent,-e *a.* preceding; **culture précédente** preceding crop; *n. m.* precedent; **p. cultural** (previous) crop sequence

préchauffé,-e *a.* preheated

préchauffer *v.t.* to preheat

préchauffeur *m.* preheater

précipitation *f. Chem.* precipitation; **p. atmosphérique** rainfall

précipité *m. Chem.* precipitate

précoce *a.* early, early maturing; **floraison précoce** early flowering

précocité *f.* earliness

préconisé,-e *a.* recommended

précultural,-e *a., pl.* **-aux; travail précultural** pre-cultivation

préculture *f.* pre-cultivation; *PD* sprouting potatoes to determine their virus status before planting

précurseur *m.* precursor

prédateur,-trice *a.* predatory; *n. m.* predator

prédatisme *m.* predatism

prédébourrement *m.* pre bud-burst

prédisposé,-e *a.* susceptible

prédose *f.* preliminary dose

pré-emballage *m.* (pre)packing

pré-emballé,-e *a.* (pre)packaged

pré-emballer *v.t.* to (pre)pack

pré-émergence *f.* pre-emergence, pre-em (= **pré-levée**)

préfeuille *f.* prophyll, bracteole; pale
préfleuraison, préfloraison *f. Bot.* aestivation, prefloration (= **estivation, vernation**); **p. induplicative** induplicate aestivation (see also **valvaire**)
préfloral,-e *a., m. pl.* **-aux** prefloral
préfoliation, préfoliaison *f. Bot.* aestivation
prégermination *f.* pre-germination
prêle, prèle *f. Equisetum* sp., horsetail; **p. des champs** common horsetail (*E. arvense*); **p. d'hiver, p. des ébénistes, p. des tourneurs** Dutch rush (*E. hyemale*); **p. des marais** marsh horsetail (*E. palustre*)
prélevée *f.* pre-emergence (= **pré-émergence**); **traitement de prélevée** pre-emergence treatment
prélèvement *m.* sample, sampling
prélever (*v.t.*) **(un échantillon)** to sample
prématuration *f.* pre-ripening
prématuré,-e *a.* premature, early
premier-bourgeon *m., pl.* **premiers-bourgeons** first bud; **p.-b. (à la levée)** chit
prémunition *f. PD* protective inoculation; **p. (croisée)** cross protection
prendre *v.i. Hort.* to root; **p. racine** to take root; **plante prenant difficilement racine** shy-rooting plant
préparasitaire *a.* preparasitic; **stade préparasitaire** preparasitic stage
préparation *f.* preparation, processing; **p. du café** coffee processing
pré(-)pépinière *f.* pre-nursery
pré-plantation *f.* pre-planting
prépotent,-e *a.* prepotent
prépuse *f. Prepusa* spp., ornamental Gentianaceae
préréfrigération *f.* pre-cooling; **p. par le vide** vacuum pre-cooling
préséchage *m.* pre-drying
présélection *f.* preselection
présemis *m.* pre-sowing; **application en présemis** pre-sowing application; **en présemis incorporé** by pre-sowing incorporation
présence *f.* presence; **p. naturelle** natural occurrence
présentation *f. Comm.* (method of) displaying object for sale
présentoir *m.* 1. stand for vases etc. 2. *Comm.* display unit
préservatif,-ive *a. & n.m.* preservative
presse-mottes *m. Hort.* block-making machine
pression *f.* pressure; **p. atmosphérique** atmospheric pressure; **à p. constante** at constant pressure; **à p. entretenue** with manual pressure (of sprayer); **p. osmotique** osmotic pressure; **à p. préalable** with automatic pressure (of sprayer)
pressoir *m.* press for grapes, apples, oilseeds etc.
prêt,-e *a.* ready, prepared; **p. à consommer** ready to eat
prétaille *f.* preliminary pruning
prétailleuse *f. Vit.* machine for preliminary pruning
prétraitement *m.* pre-treatment
préventif,-ive *a.* preventive
pré-verger *m., pl.* **prés-vergers** pasture land planted with fruit trees, meadow-orchard
prévision *f.* forecasting, forecast (weather)
primefleur *f.* first bloom
primerose *f.* hollyhock (*Althaea rosea*) (= **rose trémière**)
primeur *f. Hort.* early horticultural produce; **pommes de terre primeurs** early potatoes
primeuriste *m., f.* early fruit and vegetable grower
primevère *f.* primula, primrose, cowslip (*Primula* sp.); **p. acaule** (common) primrose (*P. vulgaris* = *P. acaulis*); **p. d'Arabie** Arabian primrose (*Arnebia cornuta*); **p. auricule** auricula (*P. auricula*) (= **oreille d'ours**); **p. de Chine** Chinese primula (*P. sinensis*); **p. commune** cowslip (*P. veris* = *P. officinalis*); **p. élevée** oxlip (*P. veris* × *vulgaris*); **p. farineuse** bird's eye primrose (*P. farinosa*); **p. à grandes fleurs** primrose (*P. vulgaris*) or oxlip (*P. veris* × *vulgaris*); **p. des jardins** polyanthus primrose (*P. variabilis*); **p. obconique** *P. obconica* (see also **p. de serre**); **p. officinale** = **p. commune** (*q.v.*); **p. de serre** *P. obconica, P. sinensis* (= **p. de Chine**)
primine *f.* primine, outer integument of ovule
primordial,-e *a., m. pl.* **-aux** primordial
primordium *m.* primordium; *pl.* primordia; **p. floral** floral primordium; *Mush.* pinheads (= **grain**)
primula *f. Primula* sp.
primulacé,-e *a.* primulaceous
primulacées *n. f. pl.* Primulaceae
printanier,-ière *a.* vernal, early (of vegetable), relating to spring
printanisation *f.* vernalization
prise *f.* something taken; **p. du blanc** *Mush.* spawn run; **prise** (*f.*) **d'essai** sample; **prise** (*f.*) **de force** power-take-off, PTO
pritchardie *f. Pritchardia* sp. (= *Eupritchardia*) fan palm
prix *m.* value, cost, price; **p. courant** current price, market price; **p. garanti** guaranteed price; **p. moyen** average price; **p. de revient** cost price
procambium *m.* procambium
procédé *m.* process, method of working; **p. d'essai** experimental design

processionnaire *a.* & *n. f.* processionary caterpillar
processus *m.* method, process; **p. d'absorption** absorption process; **p. de fermentation** fermentation process; **p. d'union** union (graft) process
producteur,-trice *a.* productive; **arbre producteur** bearing tree; *n.* producer, bearer; **p. irrégulier** irregular bearer (fruit tree); **p. de légumes** vegetable grower
producteur-direct *m., pl.* **producteurs-directs** *Vit.* direct producer
productif,-ive *a.* productive, fertile; **vigne productive** bearing vine
production *f. Ag., Hort.* production, yield; produce, crop; **p. de matière sèche** dry matter production; **productions de plein champ** field crops; **productions maraîchères** market-garden produce
productivité *f.* productivity, yield capacity
produit *m.* product, produce; **p. antiparasitaire** plant protection product; **p. cuprique** copper product; **p. formulé** formulated product (= **p.f.**); **produits de pépinière** nursery stock; **p. périssable** perishable produce; **p. de photosynthèse** photosynthate; **p. répulsif** repellent; **p. toxique** toxicant; **p. transformé** processed product
profichi *m. inv.* profichi, caprifig June crop
profil *m.* profile; **p. hydrique** moisture profile; **p. de sol, p. pédologique** soil profile
profond,-e *a.* deep; **peu profond** shallow
profondeur *f.* depth; **p. de drainage** depth of drainage
programmateur *m.* automatic control; **p. d'irrigation** automatic irrigation control device
programmation *f.* programming
proie *f.* prey
prolifération *f.* proliferation; **p. du pommier** *PD* apple proliferation virus
prolifère *a.* proliferous
proliférer *v.t., v.i.* to proliferate
prolificité *f.* prolificacy
proline *f.* proline
prolongement *m.* prolongation, extension; *Bot.* **p. médullaire** medullary ray; *Hort.* leader
proméristème *m.* promeristem
prometteur,-euse *a.* promising; **hybride prometteur** promising hybrid; **résultat prometteur** promising result
prompt-bourgeon *m., pl.* **prompts-bourgeons** accessory, secondary shoot or bud: bud giving rise to feather (see also **bourgeon anticipé**); *Vit.* (small) lateral (shoot) arising from bud close to dormant bud (= **entre-coeur**)
propachlore *m.* propachlor
propagateur,-trice *a.* propagating; *n.* propagator; *n.m. Hort.* propagator
propagation *f.* propagation
propager *v.t.* to propagate
propane *m.* propane
prophase *f.* prophase
prophylaxie *f.* prophylaxis, preventive treatment of disease; **p. biologique** biological control
propre *a.* clean, weed-free (of crop)
propriété *f.* 1. property, characteristic. 2. property, estate, holding
prosenchyme *m.* prosenchyma
protandre *a.* protandrous
protandrie *f.* protandry
protéagineux,-euse *a.* rich in protein; *n. m. pl.* protein plants
protection *f.* protection; **p. croisée** *PD* cross protection; **p. raisonnée** *PC* managed control, rational control
protée *m. Protea* sp.
protéine *f.* protein; **p. végétale** plant protein
protéique *a.* protein, proteinic; **fraction protéique** protein fraction
protéogenèse *f.* protein synthesis
protéolyse *f.* proteolysis
protéosynthèse *f.* protein synthesis
prothalle *m.* prothallus
protide *m.* protid
protocole *m.* protocol; **p. d'essai** experimental design
protocorme *m.* protocorm
protogyne *a.* protogynous
protogynie *f.* protogyny
protophylle *f.* protophyll
protoplasma, protoplasme *m.* protoplasm
protoplaste *m.* protoplast; **fusion de protoplastes** protoplast fusion
protoplastique *a.* protoplastic
protozoaire *m.* protozoan
provenance *f.* source, origin
provignage, provignement *m. Vit.* layering
provigner *v.t. Vit.* to layer (= **marcotter**)
provin *m.* layer
pruche *f.* (*Canada*) eastern hemlock (*Tsuga canadensis*)
pruine *f.* waxy bloom of plums, grapes, mushrooms, cabbage leaves etc.
pruiné,-e *a.*, **pruineux,-euse** *a.* pruinose, covered with bloom
pruinosité *f.* pruinescence
prune *f.* 1. plum (see also **prunier**); **p. domestique** cultivated, European, garden plum; **p. d'Agen** d'Agen plum, a prune-producing variety; **p. de Damas** damson; **p. d'Ente** = **p. d'Agen; p. européenne** = **p. domestique**. 2. Other plants with plum-like fruit; **p. d'Amérique** (= **p. mombin**); **p. coton, p. d'icaque** coco-plum (*Chrysobalanus*

icaco) (see **chrysobalanier, icaquier**); **p. de Madagascar** botoko plum, governor's plum; **p. de Malabar** rose apple; **p. de mer** seaside plum; **p. mombin, p. d'or** yellow mombin; **p. de pierre** Mexican hawthorn, stone plum (*Crataegus mexicana*) (= **aubépine américaine**); **p. rouge** Spanish plum

prune-abricot *f., pl.* **prunes-abricots** plumcot

pruneau *m., pl.* **-eaux** prune

pruneautier *m.* plum tree of a variety used for prune production

prunées *f. pl.* stone fruits

prunelle *f.* 1. sloe. 2. *Prunella* sp., including *P. vulgaris*, self-heal

prunellier *m.* blackthorn, sloe (*Prunus spinosa*) (= **prunier sauvage**); **p. à fleurs doubles** *P. spinosa* var. *plena*

prunicole *a.* relating to plums; **région prunicole** plum-growing region

pruniculteur *m.* plum grower

prunier *m.* 1. plum tree (*Prunus* sp.) (see also **prune**); **p. commun, p. domestique** cultivated, European, garden plum tree (*P. domestica*); **p. de Damas, p. damas** damson tree (*P. damascena*); **p. épineux** = **p. sauvage** (*q.v.*); **p. japonais** Japanese plum tree (*P. salicina* = *P. triflora*); **p. myrobolan** myrobalan plum tree (*Prunus cerasifera*) (= **myrobolan**); **p. d'ornement** ornamental plum tree; **p. pourpre** *P. cerasifera* var. *atropurpurea* (= *P. pissardii*); **p. Saint-Julien** bullace (*P. insititia*); **p. sauvage** blackthorn, sloe (*P. spinosa*) (= **prunellier**); **p. à trois lobes** *P. triloba*. 2. Other plants with plum-like fruit; **p. d'Amérique** = **p. mombin** (*q.v.*); **p. de Chine** *Flacourtia rukam*; **p. d'Espagne** = **p. mombin** (*q.v.*); **p. de Madagascar** botoko plum tree, governor's plum tree (*Flacourtia ramontchi*); **p. de Malabar** rose apple tree (*Eugenia jambos*) (= **jambosier**); **p. de la Martinique** Martinique plum tree (*Flacourtia inermis*); **p. de mer** seaside plum tree (*Ximenia americana*) (= **citron de mer, ximénie**); **p. mombin, p. d'or** yellow mombin tree (*Spondias mombin* = *S. lutea*); **p. du Pérou** tree tomato (= **tomate en arbre** *q.v.*); **p. rouge** Spanish plum tree (*S. purpurea*); **p. de savane** West African plum (*Vitex doniana*); **p. sébeste** scarlet cordia (*Cordia sebestena*) (= **sébestier**)

pruniforme *a.* shaped like a plum

P.S. = **poudre soluble dans l'eau** *PC* water soluble powder

P.S.C. = **pourriture sèche du coeur** (*Coconut*) (*q.v.*)

psalliote *m.* or *f.* *Psalliota* (*Agaricus*) sp.; **p. champêtre** field mushroom (*A. (Psalliota) campestris*) (see also **champignon**); **p. jaunissante** yellow-staining mushroom (*A. (Psalliota) xanthoderma*)

psammophyte *a.* psammophytic

pseudo-bulbe *m., pl.* **pseudo-bulbes** pseudo-bulb

pseudocarpe *m.* pseudocarp

pseudocarpien,-ienne *a.* pseudocarpous

pseudogamie *f.* pseudogamy

pseudo-tronc *m. Ban.* pseudostem

psoralée (*f.*) **bitumineuse** pitch trefoil (*Psoralea bituminosa*)

psorose *f. Cit., PD* psorosis virus; **p. alvéolaire** concave gum virus; **p. écailleuse** scaly bark virus; **p. en poches** blind pocket virus

psyché *f. Ent.* psychid moth (Psychidae)

psychromètre *m.* psychrometer, wet and dry bulb hygrometer

psychrophile *a.* psychrophilous, growing best at a low temperature

psylle *m. Psylla* sp.; **p. asiatique des agrumes** oriental citrus psylla (*Diaphorina citri*); **p. des agrumes** citrus psylla (*Trioza erytreae*); **p. de la carotte** (*T. apicalis* = *T. viridula*); **p. de l'olivier** olive psyllid (*Euphyllura olivina*); **p. du poirier** pear psylla, pear sucker (*Psylla pyri*); **p. du pommier** apple sucker (*P. mali*)

psyllium *m.* **(noir)** psyllium (*Plantago indica* = *P. psyllium*) and its seed (= **herbe aux puces**)

ptélée *m. Ptelea* sp., including *P. trifoliata* hop tree (see **orme**)

ptéride *f. Pteridium* sp., including *P. aquilinum* bracken (see also **fougère**)

pteris *m. Pteris* sp.; **p. de Crète** *P. cretica*; **p. dentelé** spider fern (*P. multifida* = *P. serrulata*); **p. ombreux** *P. umbrosa*

ptérocarpe *a.* pterocarpous; *n. m. Pterocarpus* sp.; **p. officinal** dragonblood (*P. draco*)

pterocaryer *m. Pterocarya* sp., including wing nut (*P. caucasica*)

ptérostyle *m. Pterostylis* sp. orchid

pubérulent,-e *a.* puberulent

pubescence *f.* pubescence

pubescent,-e *a.* pubescent, downy

puccinia *m. Puccinia* sp., rust

puce *f.* flea; **p. de terre** flea bettle (see **altise**); **p. de la vigne** grapevine flea beetle (= **altise de la vigne, pucerotte** *q.v.*)

puceron *m.* aphid; **p. cendré du chou** cabbage aphid (*Brevicoryne brassicae*); **p. cendré (du pommier)** rosy apple aphid, bluebug (*Dysaphis plantaginea*); **p. de l'échalote** shallot aphid (*Myzus ascalonicus*); **p. du fraisier** strawberry aphid (*Chaetosiphon* (= *Pentatrichopus*) *fragaefolii*); **p. du houblon** hop aphid (*Phorodon humuli*); **p. jaune du groseillier** red-currant blister aphid (*Cryptomyzus ribis*); **p. de la laitue** lettuce root aphid (*Pemphigus bursarius*); **p. lanigère (du**

pommier) woolly aphid (*Eriosoma lanigerum*); **p. noir des agrumes** citrus aphid (*Toxoptera citricidus*); **p. noir de la fève** black bean aphid, blackfly (*Aphis fabae*); **p. noir du pêcher** (*Brachycaudus persicaecola* = *Anuraphis persicae-niger*); **p. noir du poirier** pear-grass aphid (*Longiunguis pyrarius*); **petit p. du noyer** European walnut aphid (*Chromaphis juglandicola*); **p. de l'oranger** tea aphid (*Toxoptera aurantii*); **p. de l'orme et du groseillier** currant root aphid (*Eriosoma (Schizoneura) ulmi*); **p. de l'oseille** permanent dock aphid (*Aphis rumicis*); **p. du pois** pea aphid (*Acyrthosiphon pisum*); **p. de la pomme de terre** glasshouse potato aphid (*Aulacorthum solani*), potato aphid (*Macrosiphum euphorbiae*); **p. rose du pommier** = **p. cendré (du pommier)** (*q.v.*); **p. vert farineux du prunier** mealy plum aphid (*Hyalopterus pruni* = *H. arundinis*); **p. vert du framboisier** raspberry aphid (*Aphis idaei*); **p. vert du pêcher** peach-potato aphid (*Myzus persicae*); **p. vert du pommier** green apple aphid (*Aphis pomi*); **p. vert du rosier** rose aphid (*Microsiphum rosae*)

puceronnière *f.* powder bellows

pucerotte *f.* 1. aphid. 2. grapevine flea beetle (*Haltica ampelophaga, H. lythri, H. chalybea*) (= **altise de la vigne**)

puisard *m.* drain tank, catch pit, sump

puisoir *m.* ladle, scoop

puits *m.* well; **p. artésien** artesian well; *Physiol.* sink; **relations source-puits** source-sink relationships

pullulation *f.*, **pullulement** *m.* pullulation, rapid multiplication

pulmonaire *f.* *Pulmonaria* sp., lungwort

pulpe *f.* pulp; **p. de fruit** fruit pulp; **p. jaune** *Ban.* yellow pulp disorder

pulpeux,-euse *a.* pulpy, pulpous

pulque *m.* pulque

pulsatille *f.* pasque flower (*Pulsatilla vulgaris*)

pulvérisable *a.* 1. pulverizable. 2. (of liquid) capable of being sprayed

pulvérisage *m.* pulverization (= **pulvérisation**); **p. à disques** discing

pulvérisateur *m.* 1. pulverizer, disc harrow; **p. vigneron** *Vit.* disc harrow. 2. sprayer; **p. pour cultures** crop sprayer; **p. à deux roues** pedestrian-operated sprayer; **p. à dos** knapsack sprayer; **p. logarithmique** logarithmic sprayer; **p. à main** hand sprayer; **p. à moteur** motor sprayer; **p. portatif** portable sprayer; **p. porté** tractor-mounted sprayer; **p. sur prise de force** PTO-driven sprayer

pulvérisation *f.* 1. pulverization (= **pulvérisage**). 2. spraying; **p. curative** remedial spray; **p. hydraulique avec réduction de débit d'eau** low volume spraying; **p. pneumatique** mist blowing

pulvériser *v.t.* 1. to pulverize. 2. to spray

pulvériseur *m.* disc harrow

pulvérulent,-e *a.* powdery

punaise *f.* 1. (hemipterous) bug; **p. du chou, p. potagère** *Eurydema* spp.; **p. du cocotier** coconut bug (*Pseudotheraptus* sp.); **p. terne** (*Canada*) tarnished plant bug (*Lygus lineolaris*); **p. verte** *Vit.* green bug (*Lygus spinolai*). 2. drawing pin

pupe *f.* pupa case, pupa, chrysalis; **pupes parasitées** parasitized pupae

purée *f.* puree, mash, paste; **p. de pois** pease pudding; **p. de pommes** apple sauce; **p. de tomates** tomato paste

pureté *f.* purity; **p. variétale** varietal purity; **p. apparente** *Sug.* purity; **p. clerget** *Sug.* gravity purity; **p. réelle** *Sug.* true purity

purghère, purgheire, pourghère *m.* physic nut (*Jatropha curcas*) (= **jatropha** or **médicinier)**

purin *m.* liquid manure; **fosse à purin** liquid manure pit

purpurin,-e *a.* (nearly) crimson, purplish; *n.f.* madder purple (dye)

pustule *f.* pustule, blister; **p. chancreuse** cankerous pustule

puya *m.* *Puya* sp.

PVC *m.* = **chlorure de polyvinyle** polyvinyl chloride (PVC)

pyrale *f.* pyralid moth; **p. du caféier** coffee pyralid (*Dichocrocis crocodora*); **p. des caroubes, p. du caroubier** carob moth (*Myelois ceratoniae*) (see also **ver**); **p. des dattes** date pyralid (*M. ceratoniae*); **p. des greffons** *Cit.* graft pyralid (*Ephestia rapidella*); **p. du palmier à huile** oil palm pyralid (*Pimelephila ghesquierei*); **p. de la pomme, p. des pommes** codling moth (*Cydia (Carpocapsa) pomonella*) (= **carpocapse**); **p. des prunes** cherry bark tortrix moth (*Enarmonia formosana*); **p. de la vigne** vine pyralid (*Sparganothis pilleriana*)

pyramidal,-e *a.*, *pl.* **-aux** pyramidal

pyramide *f.* pyramid; *Hort.* pyramid, central-leader tree; **p. modifiée** modified leader tree; **p. naine** dwarf pyramid

pyranomètre *m.* pyranometer; **p. linéaire** linear pyranometer

pyrénoïde *m.* pyrenoid

pyrèthre *m.* pyrethrum (*Pyrethrum* = *Chrysanthemum*) sp., including garden pyrethrum (*C. coccineum*); **p. d'Afrique** Mount Atlas daisy (*Anacyclus* sp.); **p. de Dalmatie** *C. cinerariaefolium*; **p. matricaire** feverfew (*C. parthenium*); **poudre de pyrèthre** pyrethrum powder

pyriforme *a.* pear-shaped (= **piriforme**)

pyrole *f.* *Pyrola* sp., wintergreen (= **pirola, pirole**)

pyxide *f.* pyxidium

Q

quadrangulaire *a.* quadrangular
quadrifide *a.* quadrifid
quai (*m.*) **de chargement** loading platform; **quai de déchargement** unloading platform
qualité *f.* quality; **q. gustative** organoleptic quality
quarantaine *f.* 1. quarantine. 2. annual stock (*Matthiola incana* var. *annua*) (= **giroflée quarantaine**)
quassia *m.* *Quassia* sp., especially bitter wood (*Q. amara*)
quassier *m.* = **quassia** (*q.v.*)
quaterné,-e *a.* quaternate
quatre *num. a.* four; **marchand(e) des quatre saisons** street fruit and vegetable trader, "barrow boy"; *Hort.* **des quatre saisons** everbearing
quatre-épices *f.* black cinnamon (*Pimenta acris*) (= **piment âcre, cannelier sauvage**); allspice (*Pimenta officinalis*) (= **toute-épice**); black cumin (*Nigella sativa*) (= **nigelle aromatique, toute-épice**)
quéléa *m. Afr.* quelea, red-billed finch (*Quelea quelea*)
quenette *f.* honey-berry (*Melicocca bijuga*) (= **knépier**)
quenouille *f. Arb., Hort.* pyramid
quercitron *m.* black oak (*Quercus velutina* = *Q. tinctoria*)
quetsche *f.* Quetsche plum
quetschier *m.* Quetsche plum tree
queue *f. Hort.* stalk (of flower, fruit)
queue-de-lézard *f., pl.* **queues-de-lézard** American swamp lily (*Saururus cernuus*)
queue-de-lièvre *f., pl.* **queues-de-lièvre** hare's tail grass (*Lagurus ovatus*)
queue-de-lion *f., pl.* **queues-de-lion** lion's ear (*Leonotis leonurus*)
queue-de-paon *f., pl.* **queues-de-paon** *Hort.* fan (= **éventail**)
queue-de-renard *f., pl.* **queues-de-renard** 1. love-lies-bleeding (*Amaranthus caudatus*) (= **amarante queue-de-renard**). 2. foxtail grass (*Alopecurus* sp.)
queue-de-souris *f., pl.* **queues-de-souris** mouse-tail (*Myosurus minimus*)
queue-fourchue *f., pl.* **queues-fourchues** puss moth caterpillar (*Dicranura vinula*)
quillaja *m.* *Quillaja* sp.; **q. savonneux** soap bark tree (*Q. saponaria*)
quinconce *m.* quincunx; **plantation en quinconce** staggered planting
quiné,-e *a. Bot.* quinate
quinine *f.* quinine
quinoa *m.* **(blanc)** quinoa (*Chenopodium quinoa*) (= **ansérine quinoa**)
quinone *f.* quinone
quino-quino *m.* balsam of Peru (*Myroxylon pereira*)
quinquéfolié,-e *a.* quinquefoliolate
quinquina *m.* quinine (*Cinchona* sp.)
quintefeuille *f.* cinquefoil (*Potentilla reptans* and other spp.)
quisqualis *m.* Rangoon creeper (*Quisqualis indica*)
quotient *m.* quotient; **q. respiratoire** respiratory quotient

R

rabane *f.* **(de raphia)** raffia matting, grass mat
rabattage *m. Arb., Hort.* cutting back, heading back, tipping; **r. de charpente** cutting back the framework (of tree); **r. de la végétation** cutting back the vegetation
rabattement *m.* = **rabattage** (*q.v.*)
rabattre *v.t. Arb., Hort.* to cut back, to head back, to tip
raboteux,-euse *a.* rough, uneven
rabougri,-e *a.* stunted
rabougrir *v.t.* to stunt; *v.i.* to become stunted
rabougrissement *m.* stunting, dwarfing; **r. buissonneux de la tomate** *PD* tomato bushy stunt virus; **r. du chrysanthème** chrysanthemum stunt virus; **(virus du) r. du pêcher** *PD* peach stunt virus; **r. du prunier** *PD* plum dwarf virus; **r. des repousses** *Sug.* ratoon-stunting disease (virus), RSD
raccourcir *v.t.* to shorten; *Hort.* to tip; *v.i.* to become shorter
raccourcissement *m.* shortening, tipping; **r. des coursonnes** spur shortening
race *f. Bot.* strain; **de race pure** true to type
racème *m.* raceme
racémeux,-euse *a.* racemose
rachilla *m.* rachilla
rachis *m. Bot.* rachis
racinage *m.* (*Collective*) edible (vegetable) roots; rooting
racinaire *a.* relating to roots; **lésion racinaire** root lesion; **système racinaire** root system; **traitement racinaire** root treatment
racine *f.* 1. root; **à racines superficielles** shallow-rooted; **sur ses propres racines** (of fruit tree etc.) on its own roots, own-rooted; **prendre racine** (of plant) to root; **r. absorbante** absorbing root; **r. adventive** adventitious root; **r. aérienne** aerial root; **r. d'ancrage** anchor root; **r. comestible** edible root; **r. coronaire** crown root; **r. fasciculée** fibrous root; **r. fourchue** forked root, fanging root; **racines liégeuses** *PD* corky root; **r. pivotante** taproot; **r. primaire** primary root; **r. principale** main root; **r. secondaire** secondary root; **r. traçante** creeping root, lateral root; **r. tuberculeuse** tuberous root. 2. **r. du Saint-Esprit** = **angélique** (*q.v.*)
raciné,-e *a.* rooted; *n.m.* rooted cutting, rooted (nursery) plant
racine-asperge *f., pl.* **racines-asperges** air root, pneumatophore
racineau *m., pl.* **-eaux** plant stake, support
racine-échasse *f., pl.* **racines-échasses** stilt root
racinement *m.* rooting
raciner *v.i.* to root
raclage *m.* raking, scraping (of soil); *Vit.* shallow hoeing
racle *m.* hand hoe
raclée *f. Hort.* superficial hoeing
racler *v.t.* to rake, to scrape (soil)
raclette *f.* 1. scraper. 2. hoe, tine; **r. en patte d'oie** duck-foot tine
racleur,-euse *n.* raker
racloir *m. Hort.* rake
radiaire *f.* great masterwort (*Astrantia major*)
radiateur *m.* radiator
radiation *f.* radiation
radical,-e *a., m. pl.* **-aux** *Hort.* radical
radicant,-e *a.* radicant, radicating
radication *f.* radication
radicé,-e *a.* radicate
radicellaire *a.* radicellose
radicelle *f.* rootlet
radicicole *a.* radicicolous
radiciflore *a.* radiciflorous
radiciforme *a.* radiciform
radicigène *a.* root-forming; **hormone radicigène** rooting hormone
radicivore *a.* radicivorous, rhizophagous
radiculaire *a.* radicular
radicule *f.* radicle
radiculeux,-euse *a.* radiculose
radié,-e *a.* radiate
radioactif,-ive *a.* radioactive
radioactivité *f.* radioactivity
radio(-)induit,-e *a.* radiation-induced
radiométrie *f.* radiometry; **r. infrarouge** infrared radiometry
radiotraceur *m.* radioactive tracer
radis *m.* radish (*Raphanus sativus*); **r. noir** black radish; **r. sauvage** wild radish (*R. raphanistrum*) (= **ravenelle**)
rafale *f.* squall
raffermir *v.t.* to harden (once more), to make firmer
raffinage *m.* refining
raffiné,-e *a.* refined
raffinement *m.* refining
raffinerie *f.* refinery

raffinose *m.* raffinose
rafflesia *m.*, **rafflésie** *f. Rafflesia* sp.
rafle *f.* stalk of currants, grapes, oil palm bunch etc., cob of maize
rafraîchi,-e *a.* 1. cooled, refreshed. 2. *Hort.* pruned (again), re-pruned, pruned (of roots before transplanting), recultivated (of soil)
rafraîchir *v.t.* 1. to cool, to refresh. 2. *Hort.* to prune again, to re-prune, to prune roots (before transplanting), to recultivate (soil)
rafraîchissement *m.* 1. cooling, refreshing; **r. par aspersion** sprinkler cooling. 2. *Hort.* pruning again, re-pruning; **r. (des racines)** pruning roots (before transplanting); recultivation (of soil)
raidisseur *m.* (wire) strainer
raie *f.* drill, row, interrow, furrow, line, stroke; **r. de canne** *Sug.* row of canes; **irrigation à la raie** row irrigation, furrow irrigation
raifort *m.* horseradish (*Armoracia rusticana* = *Cochlearia armoracia*) (= **cran, cranson**); **r. sauvage** = **cochléaria** (*q.v.*)
rainure *f.* groove, sunken channel; *Hort.* drill (row), furrow
raiponce *f.* rampion (*Campanula rapunculus*), sometimes applied to **mâche** (*q.v.*)
raisin *m.* 1. grape (see also **vigne**); **grain de raisin** grape; **grappe de raisin** bunch of grapes; **r. de cuve** wine grape; **r. fertile** large-seeded berry on normally seedless variety, buck currant; **r. sans pépins** seedless grape; **r. sec** raisin; **r. (sec) de Corinthe** currant; **r. de serre** hothouse grape; **r. de Smyrne** sultana; **r. de table** table grape, dessert grape; **r. teinturier** teinturier grape (with red pulp and juice); **r. de treille** = **r. de table** (*q.v.*); **r. de vigne** = **r. de cuve** (*q.v.*). 2. Other genera: **r. d'Amazonie** *Pourouma cecropiaefolia*, a tree with edible fruits; **r. d'Amérique** Virginian poke weed (*Phytolacca decandra*) (= **teinturier**); **r. d'ours** bearberry (*Arctostaphylos uva-ursi*) (= **busserole**); **r. pahouin** *Afr. Trichoscypha ferruginea*
raisinier *m.* sea-grape (*Coccolobis uvifera*); **r. acide** *Afr. Lannea acida*; **r. à petits fruits** *Afr. L. microcarpa*
rajanie *f.* bihi (*Rajania cordata*)
rajeunir *v.t.* to rejuvenate; *Hort.* to cut back, to prune
rajeunissement *m.* rejuvenation; *Hort.* cutting back, pruning
ralentisseur (*m.*) **de croissance** growth retardant
ramassage *m.* collecting, gathering, picking up; **r. d'herbe** (cut) grass collecting (by mower)
ramassé,-e *a.* 1. stocky, compact. 2. collected, gathered
ramasser *v.t.* to collect, to gather, to pick up
ramasseur,-euse *n.* collector, gatherer; **équipe de ramasseurs** gang of harvesters (for fallen fruit etc.); pick-up attachment (of implement), loading machine (often *f.*) (see also **arracheuse**), (pick-up) harvester, lifter
ramboutan *m.* rambutan (*Nephelium lappaceum*) (= **litchi chevelu**)
rame *f. Hort.* stake, stick, pole, prop; **r. à pois** pea stick; **variétés à rames (de haricot)** climbing varieties (of bean)
ramé,-e *a.* supported with sticks (e.g. peas), staked, propped (of plant)
raméaire *a. Bot.* ramal, rameous, growing on a branch
raméal,-e *a.*, *pl.* **-aux** = **raméaire** (*q.v.*)
rameau *m.*, *pl.* **-eaux** shoot, bough, branch, lateral, offshoot; **r. adventif** adventitious shoot; **r. aérien** sucker; **r. d'un an** maiden lateral (shoot); **r. anticipé** feather, lateral shoot on the current year's extension growth; **r. axillaire** axillary shoot; **r. à bois** maiden lateral, long shoot; **r. (-)bouture**, *pl.* **rameaux(-)boutures** shoots for cuttings; **r. courson** = **courson** (*q.v.*); **r. (-)greffon**, *pl.* **rameaux(-)greffons, r. de greffe** bud-stick; **r. herbacé** shoot, green shoot; **r. latéral** lateral (branch) (= **branche latérale**); **r. ligneux** woody shoot; **r. mixte** shoot bearing both fruit and wood buds; **r. d'olivier** olive branch; **r. porteur** bearing shoot; *Vit.* fruiting cane; **r. primaire** *Vit.* primary shoot (= **pampre**); **r. principal** *Vit.* main shoot (= **sarment-maître**); **r. de rajeunissement** replacement shoot; **r. de remplacement, r. de renouvellement** replacement shoot; **r. secondaire** *Vit.* secondary shoot (= **entre-cœur**); **r. terminal** terminal shoot
rameau d'or *m.* yellow wallflower (*Cheiranthus cheiri*) (= **giroflée jaune, ravenelle**)
ramée *f.* 1. cut branch with green leaves. 2. arbour
raméen,-enne *a.* rameous
ramentacé,-e *a.* ramentaceous
ramer *v.t. Hort.* to support climbing plants with sticks, props etc., to stake plants
rameux,-euse *a.* ramose, much branched
ramie *f.* ramie (*Boehmeria nivea*)
ramière *f.* border of shrubs
ramifère *a.* ramiferous
ramification *f.* ramification, branching; *Sug.* aerial branching
ramifié,-e *a.* ramified, branched
ramifier *v.t.*, *v. p.* to ramify; **se ramifier** to ramify
ramiflore *a.* ramiflorous
ramille *f.*, **ramillon** *m.* twig, shoot; *pl.* small branches, small wood

ramollir *v.t.* to soften; **se ramollir** to soften, to become soft
ramondie *f.* *Ramondia* sp.; **r. des Pyrénées** rosette mullein (*R. pyrenaica*)
rampant,-e *a.* (of plant) creeping, procumbent; **légumineuse rampante** creeping legume
rampe *f.* **(d'arrosage)** *Irrig., PC* boom, spray-line
ramule *m.* branchlet
ramuleux,-euse *a.* ramulous, ramulose, having many small branches
ramure *f.* branches, branch system, boughs; **(à) r. souple** "easily trained" (of shrub etc.), with pliable branches
ramuscule *m.* ramulus, small branch
rance *a.* rancid
rancir *v.i.* to become rancid, to grow rancid
rand *m.* *PD* tipburn (of lettuce) (= **brunissement marginal des feuilles**)
randie *f.* *Randia* sp.; some spp. have edible fruits
rang *m.* row, line; **sur le rang** within the row; **r. d'oignons** row of onions
rangée *f.* row, line
râpe *f.* rasp, grater
raphé *m.* raphe
raphia *m.* 1. raphia palm (*Raphia* sp.). 2. raffia
raphide *f.* raphide
rapistre rugueux *m.* *Rapistrum rugosum*
rapport *m.* 1. return, yield, profit. 2. report. 3. ratio, relation(ship); **r. C/N** C/N ratio; **r. longueur/largeur** length/width ratio
rapprochement *m.* *Hort.* = **rabattage** (*q.v.*)
rapprocher *v.t.* *Hort.* = **rabattre** (*q.v.*)
rappuyé,-e *a.* settled, stale (of soil)
raquette *f.* *Bot.* 1. flat cactus cladode. 2. prickly pear (*Opuntia* sp.); **(grosse) raquette, r. commune** common prickly pear (*O. vulgaris*) (see also **figuier de Barbarie, nopal**). 3. *Vit.* long shoot
rare *a.* rare; **r. en culture** rarely grown
ras,-e *a.* close-cropped; *prep. phr.* **au ras du sol** at ground level
rasette *f.* *Hort.* skim coulter; **r. latérale** angle blade
rassembleur *m.* "lumper" (in systematics) (see **désintégreur**)
rassir *v.i.* to become stale
rassis,-e *a.* stale
rat *m.* rat; **r. d'égout** brown rat (*Rattus norvegicus*); **r. fruitier** = **lérot** (*q.v.*); **r. musqué** muskrat (*Ondatra zibethicus*) (= **ondatra**); **r. noir** black rat (*R. rattus*); **r. surmulot** brown rat (= **surmulot**); **r. taupier** ground vole (*Arvicola terrestris*) (= **campagnol terrestre**)
ratanhia *m.* rhatany (*Krameria* sp.)
ratatiné,-e *a.* shrivelled, shrunken
ratatiner *v.t.* to shrivel, to shrink; **se ratatiner** to become shrivelled, shrunken
râteau *m., pl.* **-eaux** rake; **r. (genre) américain** garden rake; **r. à feuilles** leaf rake; **r. à gazon** lawn rake
râtelage *m.* raking
râteler *v.t.* to rake
râtelures *n. f. pl.* rakings
raticide *m.* rodenticide
ratissage *m.* raking
ratisser *v.t.* to rake, to hoe
ratissoir *m.*, **ratissoire** *f.* hoe, light rake, scraper; **r. à pousser** Dutch hoe; **r. à tirer** draw hoe
ratissure *f.* rakings
rattrapage *m.* compensation (for), catching up; **traitement de rattrapage** late treatment (e.g. herbicide) to make up for earlier failure etc.
ravageur,-euse *a.* ravaging; *n. m.* pest
ravalement *m.* *Hort.* cutting (hard) back, trimming
ravaler *v.t.* *Hort.* to cut (hard) back, to trim
rave *f.* *Hort.* (short-rooted) turnip, (long-rooted) radish (= **petite rave**); **céleri-rave** celeriac (*Apium graveolens* var. *rapaceum*); **chou-rave** kohl-rabi (*Brassica oleracea* var. *caulorapa*)
ravenala *m.* ravenala, traveller's tree (*Ravenala madagascariensis*)
ravenelle *f.* 1. wallflower (*Cheiranthus cheiri*) (= **giroflée, rameau d'or**). 2. wild radish (*Raphanus raphanistrum*)
ravière *f.* turnip field; radish bed
ravinement *m.* gullying, gully erosion (= **érosion en ravins**)
raviner *v.t.* (of rainwater) to gully, to channel
ravineux,-euse *a.* gullied
ray-grass *m. inv.* rye-grass (*Lolium* sp.); **r.-g. anglais** perennial rye-grass (*L. perenne*); **r.-g. d'Italie** Italian rye-grass (*L. multiflorum*)
rayon *m.* 1. ray; **rayons gamma** gamma rays; *Bot.* pedicel forming part of umbel; ray floret; **r. médullaire** medullary ray; *Hort.* drill, furrow (for sowing), row. 2. = **demi-fleuron** *q.v.*
rayonnant,-e *a.* radiant, radiating; **énergie rayonnante** radiant energy
rayonnement *m.* radiation; **rayonnements gamma** gamma rays; **r. solaire** solar radiation; **r. ultra-violet** ultra-violet radiation
rayonner *v.i.* to radiate; to beam, to shine; *v.t.* *Hort.* to drill, to furrow
rayonneur *m.* marker (for planting rows etc.); **r. multiple** multi-row marker; **r. simple** single-row marker
rayure *f.* *Hort.* stripe, streak
réactif,-ive *a.* *Chem.* etc. reactive; *n.m.* reagent
réaction *f.* reaction; **r. de défense** defence reaction; **r. de Hill** Hill reaction

rebiochage *m. Vit.* lateral shoot removal
rebiot *m. Vit.* lateral shoot
reboisement *m.* reforestation
reboiser *v.t.* to afforest
rebouchage *m.* filling in
reboucher *v.t.* to fill up (again), to fill in
rebourgeonner *v.i.* to grow new buds
rebut *m.*, **(article de) rebut** reject; **dattes de rebut** rejected dates
rebutter *v.t.* to ridge up again
recépage, recèpement *m.* close pruning, cutting back; *Coff.* stumping; *Vit.* (i) cutting back the top of a vine in readiness for grafting, (ii) cutting back the stock after successful side grafting
recéper *v.t.* to prune closely, to cut back; *Coff.* to stump; *Vit.* (i) to cut back the top of a vine in readiness for grafting, (ii) to cut back the stock after successful side grafting
réceptacle *m. Bot.* receptacle; **r. charnu** fleshy receptacle
réceptif,-ive *a.* receptive
réceptivité *f.* receptivity
récessif,-ive *a.* recessive
receveur *m.* receiver (see also **donneur**)
réchaud (*m.*) **(antigelées)** small portable stove; **réchaud** *Hort.* bank of manure or earth placed around a bed or frame (= **accot**); new layer of manure on hotbed
réchauffage *m.* reheating, warming up; **r. d'un verger** heating an orchard
réchauffer *v.t.* to reheat, to warm up; *Hort.* to add fresh manure to hotbed
rechaussage, rechaussement *m. Hort.* banking up or replacing soil at foot of tree, plant etc.
rechausser *v.t. Hort.* to bank up or replace soil at foot of tree, plant etc.
recherche *f.* search, research; **r. caféière** coffee research; **r. fruitière** fruit research; **r. maraîchère** vegetable crop research (see also **chercheur**)
récipient *m.* container, receptacle, vessel
recolonisation *f.* recolonization
récolte *f.* 1. harvesting (of crops), vintaging (of grapes); **à la récolte** at harvest; **après-récolte** post-harvest; **faire la récolte** to harvest; **r. mécanique** mechanical harvesting. 2. harvest, crop, vintage; **r. améliorante** soil-improving crop; **r. dérobée** catch crop; **mauvaise récolte** crop failure; **r. principale** main crop
récolter *v.t.* to harvest, to reap
récolteuse *f.* harvester, lifter; **r. de bulbes** bulb lifter; **r. de choux de Bruxelles** brussels sprouts harvesting machine; **r. de choux pommés** cabbage harvester; **r. d'épinards en branches** spinach harvester; **r. de haricots verts** bean harvester
recombinaison *f. PB* recombination, recombining
reconstitution *f. Hort.* replanting
recontamination *f.* recontamination
recontaminer *v.t.* to recontaminate
recourage *m. Sug.* supplying
recourbé,-e *a.* bent again, bent back, bent round
recouvrement *m.* re-covering, covering again; soil cover given by plant(s)
recouvrir *v.t.* to cover again, to cover, to overlap; **se recouvrant** (of petals etc.) overlapping
recroquevillé,-e *a.* shrivelled up, cockled
recrû, recru *m.* regrowth; **r. naturel d'herbacées** natural grass regrowth
rectinervé,-e *a.* parallel-veined
recuire *v.t. Mush.* to cook out
récupérable *a.* recoverable, salvageable
récurvé,-e *a.* recurved
recyclage *m.* recycling
recyclé,-e *a.* recycled; **solution recyclée** recycled solution
redistribution *f.* redistribution
redoul *m. Coriaria* sp.; **r. (à feuilles de myrtes)** Mediterranean coriaria (*C. myrtifolia*)
redouter *v.t.* to fear, to dread; **plante qui redoute l'humidité stagnante** plant that cannot tolerate waterlogging
redoux *m.* rise in temperature (in meteorology)
redressé,-e *a.* upright, erect, tilted up
redressement *m.* re-erecting, rectification; **fumure de redressement** corrective fertilization
réductase *f.* reductase
réducteur (*m.*) **de croissance** growth inhibitor
réduction *f.* reduction; **r. de croissance** growth reduction; **r. du jour** daylight reduction; **r. des jeunes pousses (d'un arbre)** shoot-pruning
réduire *v.t.* to reduce
réensemencement *m.* resowing
réensemencer *v.t.* to resow; **se réensemençant facilement** easily self-sown, self-perpetuating
refend *m. Hort.* **mur de refend** espalier wall
réfléchi,-e *a.* reflected; **lumière réfléchie** reflected light; *Bot.* reflexed
réflecteur *m.* reflector
refleurir *v.i.* to flower again, to reflower
refleurissement *m.* second flowering
réflexion *f.* reflection
refloraison *f.* = **refleurissement** (*q.v.*)
reflorescent,-e *a.* remontant, perpetual, repeat-flowering (= **remontant)**
refluer *v.i.* to flow back, to sweep back
réfractaire *a.* refractory; **r. au bouturage** difficult to root from cuttings
réfracté,-e *a. Bot.* refracted

réfractométrique *a.* refractometric; **index réfractométrique** refractometric index
réfrigération *f.* refrigeration, cooling, chilling; **r. par eau glacée** hydrocooling; **r. sous vide** vacuum cooling
réfrigérer *v.t.* to refrigerate
refroidir *v.t.* to cool
refroidissement *m.* cooling, refrigeration; **r. par eau glacée** hydrocooling; **r. par le vide** vacuum cooling
refroidisseur *m.* cooler; **r. à air, r. d'air** air cooler
regain *m.* second growth, regrowth, aftermath
regazonnement *m.* re-turfing
regazonner *v.t.* to re-turf, to sow again with grass seed
régence *f.* Italian corn salad (*Valerianella eriocapa*) (= **mâche d'Italie**)
régénération *f.* regeneration, rejuvenation
régénérer *v.t.* to regenerate, to rejuvenate; **r. une vieille plantation** to rejuvenate an old plantation
régie *f.* management; **r. des cultures** crop management
régime *m.* 1. regimen; **r. hydrique du sol** soil water economy. 2. bunch, cluster; **r. de bananes** bunch, stem of bananas; **r. de dattes** bunch of dates
région *f.* region, area; **r. bulbicole** bulb-growing region, area; **r. horticole** horticultural region, area
réglable *a.* adjustable
réglage *m.* regulating, adjusting, adjustment
règle (*f.*) **à encoches** planting rod (with planting distances indicated on it); **r. à planter** planting rule
réglisse *f.* liquorice (*Glycyrrhiza glabra*)
réglisserie *f.* liquorice factory
règne (*m.*) **végétal** plant, vegetable kingdom
regreffage *m.* regrafting, reworking
regreffer *v.t.* to regraft, to rework
régression *f.* regression; **coefficient de régression** regression coefficient
régularité *f.* regularity, consistency
régulateur (*m.*) **de croissance** growth regulant, growth regulator
régulation *f.* control, regulation; **r. des populations d'insectes** insect population control
régulier,-ière *a. Bot.* regular
réhumectation *f.* rewetting
réhumidification *f.* rewetting
réhydratation *f.* rehydration
réimplantation *f.* planting again, replanting, resowing
réimplanter *v.t.* to plant again, to replant, to resow
reine *f.* queen; *Ap.* queen bee; **r. secondaire** *Ent.* secondary queen (termite); **r. des Alpes** Alpine eryngo (*Eryngium alpinum*); **r. des bois** sweet woodruff (*Galium odoratum* = *Asperula odorata*) (= **aspérula, petit muguet**); **r. de la nuit** queen of the night, night-blooming cereus (*Selenicereus grandiflorus*); **r. des prés** queen of the meadows, meadow sweet (*Filipendula ulmaria*); **r. des Pyrénées** Pyrenean saxifrage (*Saxifraga longifolia*)
reine-claude *f., pl.* **reines-claudes** greengage
reine-claudier *m.* greengage tree (*Prunus insititia* var. *italica*)
reine-marguerite *f., pl.* **reines-marguerites** China aster (*Callistephus chinensis*) (= **aster de Chine, callistèphe, marguerite de Chine**); **r.-m. à fleurs de chrysanthème** chrysanthemum-flowered China aster; **r.-m. à fleurs de pivoine** peony-flowered China aster
reinette *f. Hort.* reinette apple
rejet *m. Hort.* sucker, offshoot, shoot, sprout, ratoon; **r. aérien** *Sug.* aerial branching; **r. de bananier** banana sucker; **rejets de dattiers** date palm offshoots; **r. sauvageon** sucker on rootstock; **rejets de souche** new shoots from a stump, stool shoot; **r. tardif** *Sug.* late tiller or sucker
rejeter *v.t. Hort.* to grow (new) shoots, to produce offshoots, to sucker
rejeton *m. Hort.* sucker, offshoot, shoot, sprout, ratoon; **émettre des rejetons** to sucker, to ratoon
rejetonnage *m.* suckering, shooting, sprouting, ratooning, production of offshoots
rejetonner *v.i.* to sucker (of plant)
réjuvénilisation *f.* rejuvenation
relargage *m. Chem.* salting out
relarguer *v.t. Chem.* to salt out
relation *f.* relation, connection; **r. linéaire inverse** inverse linear relation(ship)
relevage *m. Hort.* 1. tying up (vines etc.). 2. potting for the winter. 3. *Sug.* = **dessouchage** (*q.v.*)
relevé *m.* 1. abstract, summary, account, statement. 2. *a. Hort.* potted for the winter
relever *v.t. Hort.* 1. to tie up vines etc. 2. to put in pots for the winter. 3. *Sug.* see **dessoucher**
reliquat *m.* remainder, residue; **r. azoté** nitrogen residue; **reliquats d'herbicide** herbicide residues
rémanence *f. PC* persistence; **r. du DDT** persistence of DDT; **r. insuffisante** insufficient persistence
rémanent,-e *a.* residual
remblai *m.* 1. earth, material (for filling in). 2. embankment, bank
remblaiement *m.* earth, manure etc. for filling in; filling in
remblayé,-e *a.* filled up, banked up

remblayer *v.t.* to fill up, to bank (up)
rembourrage *m.* lining, padding (of box, crate); lining or padding materials
remède *m.* remedy
remembrement (*m.*) **(des terres)** consolidation, re-allocation (of land, holdings)
remijia *m. Remijia*, neotropical genus, a source of quinine
remise *f.* shed; **r. à outils** tool shed
remontance *f.* perpetual flowering ability
remontant,-e *a. Hort.* everbearing (strawberry), perpetual, remontant; **framboisier remontant** autumn-fruiting raspberry (which also fruits in spring); **rosier remontant** repeat-flowering rose (see also **reflorescent**)
remontée *f.* rise (after downward movement), tendency to be everbearing (of strawberry etc.)
remonter *v.i. Hort.* to flower again
remorque *f.* trailer; **r. autochargeuse** self-loading trailer
remplacement *m.* replacing, replacement; **rameau de remplacement** replacement shoot; *Hort.* gapping up, filling gaps, supplying; *Vit.* supplying, uprooting dead vines and replacing them with young vines
remplacer (*v.t.*) **les manquants** to gap-up, to fill the gaps
remplir *v.t.* to fill up, to refill
remplissage *m.* filling, filling up; **terre de remplissage** soil for filling (pots etc.)
rempotage *m.* potting on, repotting (see also **empotage**)
rempoter *v.t.* to pot-on, to repot
rempoteuse *f.* potting machine, repotting machine
rempotoir *m.* (enclosed) area for potting-on, repotting
rémunérateur,-trice *a.* remunerative, paying, profitable, economic
rémusatie *f. Remusatia* sp., including *R. vivipara*
renanthère *f. Renanthera* sp. (orchid)
rendement *m. Hort.* yield, output, produce; **r. au cassage** shelling (out) percentage (of nuts etc.); **r. moyen** average yield
rendre *v.t. Ag., Hort.* to yield, to produce
rendzine *f.* rendzina
renflé,-e *a.* swollen, enlarged; **tronc renflé à la base** trunk swollen at the base
renflement *m.* swelling, bulge, enlargement
renfler *v.t., v.i.* to swell, to enlarge
réniforme *a.* reniform, kidney-shaped
renonculacées *f. pl.* Ranunculaceae
renoncule *f. Ranunculus* sp., buttercup, and related genera: **r. des champs** corn buttercup (*R. arvensis*); **r. cultivée** = **r. des jardins** (*q.v.*); **r. double** yellow bachelor's buttons (*R. acris flore-pleno*); **r. à feuilles d'aconit** white bachelor's buttons (*R. aconitifolius*) (= **bouton d'argent**); **r. à fleurs de pivoine** *R. africanus*; **r. des fleuristes** = **r. des jardins** (*q.v.*); **r. flottante** water crowfoot (*R. aquatilis* and related spp.); **r. d'hiver** winter aconite (*Eranthis hyemalis*) (= **ellébore d'hiver**); **r. des jardins** garden ranunculus (*R. asiaticus*); **r. des marais** (1) hairy buttercup (*R. sardous*), (2) marsh marigold (*Caltha palustris*); **r. rampante** creeping buttercup (*R. repens*) (see also **bouton d'or**)
renonculier *m.* double-flowered cherry (*Prunus avium* var. *flore-pleno*)
renouée *f.* knotgrass (*Polygonum* sp.); **r. amphibie** amphibious bistort (*P. amphibium*); **r. liseron** black bindweed (*P. convolvulus*); **r. des oiseaux, r. traînasse** common knotgrass (*P. aviculare*) (= **traînasse**); **r. persicaire** redshank, persicaria (*P. persicaria*); **r. du Turkestan** Russian vine (*P. baldschuanicum*)
renouvellement *m.* renovation, renewal
rénovation *f.* renovation, renewal; **r. variétale** variety renewal
rénover *v.t.* to renovate, to restore
renseignement *m.* (piece of) information; **renseignements (techniques)** (technical) data; **renseignements chiffrés** numerical data
rentabilité *f.* profit-earning capacity, profitability, economics (of enterprise)
rentable *a.* profitable; **(économiquement) rentable** economic
répartition *f.* distribution; **régularité de répartition** *PC* evenness of spraying
repassage *m.* repetition (of operation); *Hort.* recultivation
repercement *m. Hort.* = **reprise** (*q.v.*); replacement shoot
repercer *Hort.* = **reprendre** (*q.v.*)
repère *m.* reference (to mark etc.); **r. topographique** landmark
répétition *f. Stat.* replication
repiquage, repiquement *m.* transplanting (of herbaceous plants), planting out, pricking-out (see also **transplantation**); subculturing (in *in vitro* culture)
repiqué,-e *a.* transplanted, planted out, pricked-out
repiquer *v.t.* to transplant, to plant out, to prick-out; **plant à repiquer** bedding plant
repiqueur,-euse *n.* planter out (person); *n.f.* transplanting machine
replantage *m.*, **replantation** *f.* replanting, re-establishing, supplying
replanter *v.t.* to replant, to re-establish, to supply
replat *m.* shoulder, shelf (of hill, mountain, terrace etc.)

replié,-e *a.* folded again, doubled up
réponse *f. Hort.* response (to fertilizers etc.)
repos *m.* rest, dormancy; **r. hivernal** winter dormancy, winter rest; **r. végétatif** dormant or resting stage, dormant period; **au repos** in the dormant stage, in the resting stage; **terre au repos** fallow land
repousse *f.* regrowth, self-sown plant (of annuals etc.); *Sug.* etc. ratoon
repousser *v.i.* to grow again; *Sug.* etc. to ratoon
reprendre *v.i. Hort.* to strike, to take, to take root again, to regrow
reprise *f.* 1. *Hort.* take (of graft), strike, rooting (of cutting etc.), growth (of offshoot, ratoon etc.); regrowth, establishment (after transplanting); **r. des pluies** return of the rains; **pourcentage de reprise, taux de reprise** percentage take. 2. orpine, live-for-ever (*Sedum telephium*) (= **sédum reprise**)
reproducteur,-trice *a.* reproductive
reproductif,-ive *a.* reproductive
reproduction *f.* reproduction; **r. asexuée** asexual reproduction; **r. sexuée** sexual reproduction; *Hort.* propagation
répulsif,-ive *a. PC* repellent; **produit répulsif** repellent (product); *n.m.* repellent (product)
répulsion *f.* repulsion
réseau *m., pl.* **-eaux** network, system; **r. de drainage** drainage system; **r. expérimental** research network; **r. d'irrigation** irrigation system; **r. racinaire** root system; **r. doré du pêcher** *PD* peach golden net virus
réséda *m. Reseda* sp.; **r. (fausse) raiponce** rampion mignonette (*R. phyteuma*); **r. jaunâtre** dyer's rocket (*R. luteola*); **r. jaune** wild mignonette (*R. lutea*); **r. odorant** mignonette (*R. odorata*); **r. des teinturiers** dyer's rocket (= **gaude**)
resemer *v.t.* to sow again, to resow
resemis *m.* resowing
réservoir *m.* reservoir, tank, cistern; **r. d'eau** water tank
résidu *m.* residue; **r. de pesticide** pesticide residue
résiduaire *a.* residual, waste; **diquat résiduaire** residual diquat
résiduel,-elle *a.* residual; **herbicide résiduel** residual herbicide
résille *f.* lattice, network
résine *f.* resin
résineux,-euse *a.* resinous; *n. m. Arb.* conifer
résinifère *a.* resiniferous
résistance *f.* resistance; **r. au détachement** resistance to detachment (of fruit); **r. au gel** frost resistance
résistant,-e *a.* resistant; **r. au froid** hardy; **extrêmement résistant(e) au manque de soins** (of plant) "thrives on neglect"
résorber *v.t.* to reabsorb
respiration *f.* respiration; **r. à l'obscurité** dark respiration
ressemer *v.t.* to sow again, to resow; **se ressème seul(e)** self-seeding (of plant); **se ressemer** to seed itself (of plant)
ressemis *m.* resowing
resserre *f.* storing, holding (of fruits etc.); store; **r. à légumes** vegetable rack; **r. de transit** cold storage room on quayside etc.
ressuyage *m.* 1. drying, drying out. 2. cleaning vegetables for market
ressuyer *v.t.* to dry; **se ressuyer** *v.i.* to dry out (of soil etc.)
restauration *f.* restoration; **r. des sols** soil conservation
restaurer *v.t.* to restore
résultat *m.* result; **r. expérimental** trial result
résupiné,-e *a.* resupinate
rétablissement *m.* recovery
retailler *v.t. Hort.* to prune again
retard *m.* delay; **r. de croissance** delayed growth
retardant (*m.*) **de croissance** growth inhibitor, growth retardant
retardateur (*m.*) **de croissance** growth inhibitor, growth retardant
retardation *f.* retardation, delay
retarder *v.t.* to retard, to delay; *v.i.* to be late, to be slow
rétenteur,-trice *a.* retaining; *n.m.* **r. d'eau** water-retaining product (= **hydrorétenteur**)
rétention *f.* retention (see also **capacité**)
reterçage, retersage *m.* fourth cultivation, recultivation (of vineyard)
retercer, reterser *v.t.* to give a fourth cultivation, to re-cultivate (vineyard)
réticulaire *a.* reticular
réticulation *f.* reticulation
réticule *m. Bot.* reticulum
réticulé,-e *a.* reticulated, reticulate
réticulum *m.* reticulum; **r. endoplasmique** endoplasmic reticulum
rétinacle *m. Bot.* retinaculum
retombant,-e *a.* dropping, pendent, hanging, pendulous
retombée *f.* 1. overhanging growth (ivy etc.). 2. fall; **retombées annuelles de feuilles** annual falls of leaves; fall-out; **retombées acides** acid rain; **r. radioactive** radioactive fall-out
retournement *m.* turning, turning over; **r. de sol** turning over soil, ploughing (under)
retourner (*v.t.*) **le sol** to turn over the soil
retrait *m.* shrinkage, contraction
rétrécir *v.t.* to contract, to shrink; *v.i.* to become contracted, to shrink
rétrécissement *m.* narrowing, contracting, shrinking, shrinkage

rétrocroisement *m.* *PB* back-crossing; back-cross
rétrodiffusion *f.* backscatter; **r. gamma** gamma backscatter
rétroinoculation *f.* back inoculation
rétus,-e *a.* retuse
réussite *f.* *Hort.* take (of graft)
revalorisation *f.* revalorization, revaluation
réveil *m.* waking, awakening; **r. printanier** spring growth, budburst; *Hort.* growth after aestivation or dormancy
réveil-matin *m.* sun spurge (*Euphorbia helioscopa*)
revente *f.* resale
reverdir *v.i.* to become green again (of plants, citrus fruits etc.)
reverdissement *m.* regreening
réversibilité *f.* reversibility
réversible *a.* reversible
réversion *f.* reversion; **r. du cassis** *PD* black currant reversion virus
revêtement *m.* facing, coating, cladding, sheathing
reviviscence *f.* reviviscence
reviviscent,-e *a.* reviviscent
révoluté,-e *a.* revolute
révolution *f.* revolution; **r. verte** green revolution
révulsif,-ive *a.* & *n.m.* revulsive, counter-irritant
rhabdovirus *m.* rhabdovirus
rhamnus *m.* *Rhamnus* sp., buckthorn
rhapis *m.* *Rhapis* sp., including *R. excelsa*, ground rattan cane
rhapontic, rhapontique *m.* *Rhaponticum* sp. (= *Centaurea* sp.)
rhexia *m.*, **rhexie** *f.* *Rhexia* sp., meadow beauty, including *R. virginica*
rhinocéros *m.* *Ent.* rhinoceros beetle (*Oryctes rhinoceros*) (= **oryctes**)
rhipsalis *m.*, **rhipsalide** *f.* *Rhipsalis* sp., mistletoe cactus
rhizobium *m.* *Rhizobium* sp., nodule bacterium
rhizoctone *m.*, **rhizoctonie** *f.* *Rhizoctonia* sp.; **r. de la pomme de terre** potato black scurf and stem canker (*Rhizoctonia solani* = *Corticium solani*); **r. violet de l'asperge** asparagus violet root rot (*R. crocorum*); **r. violet de la pomme de terre** potato violet root rot (*R. crocorum*)
rhizoderme *m.* rhizodermis, piliferous layer
rhizogène *a.* rhizogenic, rhizogenous
rhizogenèse *f.* rhizogenesis, root formation
rhizoglyphe commun *m.* bulb mite (*Rhizoglyphus echinopus*)
rhizome *m.* rhizome
rhizomorphe *m.* *Fung.* rhizomorph
rhizophage *a.* rhizophagous
rhizophora, rhizophore *m.* *Rhizophora* sp., mangrove
rhizosphère *f.* rhizosphere
rhizosphérique *a.* relating to the rhizosphere; **mycoflore rhizosphérique** rhizosphere mycoflora
rhodanien,-ienne *a.* of the Rhône
rhodante *m.* *Rhodanthe* sp. (= *Helipterum* sp.), Australian everlasting
rhododendron *m.* rhododendron (*Rhododendron* sp.); **r. du Caucase, r. pontique** *R. ponticum*; **r. de Virginie** *R. catawbiense* (see also **rose des Alpes**)
rhodora *m.* *Rhodora* (= *Rhododendron*) sp.
rhomboïdal,-e *a.*, *m. pl.* **-aux** rhomboidal
rhubarbe *f.* rhubarb (*Rheum rhaponticum* and other spp.); **r. de Chine** = **r. palmée** (*q.v.*); **r. des Indes** = **r. palmée** (*q.v.*); **r. du Népaul** *R. emodi*; **r. officinale** medicinal rhubarb (*R. officinale*); **r. ondulée** *R. undulatum*; **r. palmée** Chinese rhubarb, East Indian rhubarb (*R. palmatum*); **r. des pauvres, r. des paysans** meadow rue (*Thalictrum* sp.); **r. de Tartarie** = **r. palmée** (*q.v.*)
rhume *m.* cold (medical); **r. des foins** hay fever
rhus *m.* *Rhus* sp.
rhynchite *m.* *Rhynchites (Coenorrhinus)* sp.; **r. coupe-bourgeon, r. conique** grapevine weevil (*R. coeruleus* = *R. conicus*)
rhynchophore *m.* *Rhynchophorus* sp., palm weevil
rhytidome *m.* rhytidome
ribes *m.* *Ribes* sp. (see **groseillier**)
ribosomal,-e *a.*, *m. pl.* **-aux** ribosomal; **RNA ribosomal** ribosomal RNA
ribosome *m.* ribosome
richardia *m.*, **richardie** *f.* *Richardia* sp. (= *Zantedeschia*); **r. d'Afrique** arum lily (*R. africana* = *Z. aethiopica*) (= **arum, calla d'Éthiopie**)
richardsonia *m.* *Richardsonia*, a neotropical genus with medicinal species
riche *a.* rich; **r. en sève** sappy
richesse *f.* *Hort.* richness, fertility; *Sug.* percentage sugar in the cane (often percentage pol)
ricin *m.* **(commun)** castor plant (*Ricinus communis*) (= **palma-Christi**); **r. sanguin** *R. communis* var. *sanguineus*; **r. d'Amérique** physic nut (*Jatropha curcas*) (= **médicinier**)
rickettsie *f.* rickettsia
rickettsoïde *m.* rickettsia-like organism
ride *f.* *Bot.* girdle, scar, scale leaf scar, longitudinal fold in bark
ridé,-e *a.* 1. wrinkled, rugose. 2. ribbed, fluted
rideau *m.*, *pl.* **-eaux** screen, curtain; **r. d'arbres** curtain of trees
rider *v.t.* to wrinkle, to shrivel; **se rider** to become wrinkled, to shrivel up
rièble *m.* cleavers, goosegrass (*Galium aparine*)

rigolage *m. Hort.* channelling, trenching, furrowing
rigole *f.* channel, trough, (small) trench, (small) furrow drain
rigoler *v.t. Hort.* to make channels, to trench, to furrow
rigoleur,-euse *a.* & *n. Hort.* trenching, channelling; **équipement rigoleur** channelling equipment
rimier *m.* (seeded form of) breadfruit (*Artocarpus altilis*) (= **châtaignier** in West Indies)
rinçage *m.* rinsing (out)
rincer *v.t.* to rinse (out)
riparia *m. Vit. Vitis riparia*
ripper *m.* ripper
ris de veau (végétal) *m.* (*West Indies*) akee (*Blighia sapida*) (= **fisanier**)
rita *m.* soapnut tree (*Sapindus mukorossi*)
ritte *f.* ridge-type plough
rivina *m. Rivina* sp., including *R. humili* bloodberry
R.N.A. =**acide ribonucléique** *m.* RNA, ribonucleic acid; **R.N.A. messager** messenger RNA
rob *m.* pharmaceutical syrup
robinet *m.* tap, cock; **r. de vidange** drain cock
robinier *m. Robinia* sp.; **r. commun, r. faux-acacia, r. de Robin** locust tree, false acacia (*R. pseudo-acacia*); **r. visqueux** clammy locust (*R. viscosa*)
robustaculture *f.* robusta coffee growing
robusticité *f.* robustness
rocaille *f. Hort.* **(jardin de) rocaille** rock garden, rockery
rocambole *f.* rocambole, sand-leek (*Allium scorodoprasum*) (= **ail d'Espagne, échalote d'Espagne, oignon d'Egypte**), but also sometimes applied to *A. sativum* cultivars with coiled stems
roche *f.* rock, boulder
rochea *m.*, **rochée** *f. Rochea* sp.
roche-mère *f.* parent rock
rocher *m.* rock, boulder; **'r. monstrueux'** rock cactus (*Cereus peruvianus* var. *monstruosus*) (see also **cierge**)
rocou, roucou *m.* annatto, rocou
rocouyer *m.* annatto tree (*Bixa orellana*) (= **bixa, roucouyer**)
roëlle *f. Roella* sp., South African harebell
rognage, rognement *m.* pruning, trimming, tipping, topping
rogner *v.t.* to prune, to trim, to tip, to top
rogneuse *f.* machine for pruning, trimming, tipping, topping
rollinia *m. Rollinia* sp. (see **cachiman**)
romaine *f. Hort.* cos lettuce
romarin *m.* rosemary (*Rosmarinus officinalis*); **r. sauvage** wild rosemary (*Ledum palustre*)
ronce *f.* bramble, blackberry (*Rubus* sp.) and other plants; **ronces américaines** American blackberries; **r. d'Amérique** West Indian gooseberry, Barbados gooseberry (*Pereskia aculeata*); **r. bleue** = **r. à fruits bleus** (*q.v.*); **r. commune** common blackberry (*Rubus fruticosus*); **r. sans épines** thornless blackberry; **r. à fruits bleus** dewberry (*R. caesius*); **r. du Japon** Japanese wineberry (*R. phoenicolasius*); **r. odorante** flowering raspberry (*R. odoratus*)
ronce-framboise *f., pl.* **ronces-framboises** loganberry (*Rubus* × *loganobaccus*)
ronceraie *f.* area overgrown with brambles, bramble patch
ronceux,-euse *a.* full of brambles, brambly
roncier *m.*, **roncière** *f.* bramble bush
ronciné,-e *a.* runcinate
rond,-e *a.* round, *n. m.* ring, circle; *Plant.* circle around individual tree; **r. désherbé** weeded circle; **r. noir des feuilles (de rose)** *PD* black spot of roses (*Actinonema rosae* = *Diplocarpon rosae*); **r. de sorcière** *Fung.* fairy ring
rondelle *f.* small round disc, section; **r. de concombre** slice of cucumber
ronger *v.t.* to gnaw, to nibble
rongeur *m.* rodent; **rongeurs nuisibles** rodent pests
rônier *m.* palmyra palm (*Borassus flabellifer*) (see also **palmier**)
roquette *f.* rocket (*Eruca sativa*); **r. bâtarde** hoary mustard (*Hirschfeldia incana*); **r. blanche** white wall rocket (*Diplotaxis erucoides*); **r. des jardins** garden cress (*Barbarea vulgaris*) (= **barbarée vulgaire**); **r. maritime, r. de mer** sea rocket (*Cakile maritima*)
rosacé,-e *a.* rosaceous
rosacées *n. f. pl.* Rosaceae; **rosacées fruitières** rosaceous fruit trees
rosage *m.* rhododendron (*Rhododendron* sp.)
rosarium *m.* rose garden, rosarium
rose *f.* 1. rose (see also **rosier, églantier**); **r. capucine** *Rosa foetida bicolor* (= *R. punicea*); **r. gallique double** *R. gallica duplex*; **r. gallique pleine** *R. gallica flore-pleno* (see also **rosier de France**); **r. à grande tige** long-stemmed rose; **r. hybride de thé** hybrid tea rose; **r. du Kamtchatka** Ramanas rose (= **rosier rugueux**); **r. mousseuse, r. moussue** moss rose; **r. musquée** musk rose; **r. parfumée** fragrant rose; **r. pompon** pompon rose (= **pompon**); **r. thé** tea rose; **r. verte** *R. chinensis* var. *semperflorens viridiflora*. 2. Other genera: **r. des Alpes** alpine rose (*Rhododendron ferrugineum*); **r. de Chine** hibiscus, shoeflower (*Hibiscus rosa-sinensis*); **r. d'hiver** = **r. de Noël** (*q.v.*); **r. d'Inde** African marigold (*Tagetes erecta*) (see also

tagetes, oeillet d'Inde); **r. de Jéricho** rose of Jericho (*Anastatica hierochuntica*); **r. de Noël** Christmas rose (*Helleborus niger*) (= **ellébore noir**); **r. de Notre-Dame** peony (*Paeonia* sp.); **r. trémière** hollyhock (*Althaea rosea*); **r. trémière annuelle, r. trémière de Chine** annual form of hollyhock; *a.* pink; **une fleur rose** a pink flower

rosé,-e *a.* rose-coloured, rosy; **vin rosé** rosé wine

roseau *m., pl.* **-eaux** reed (*Phragmites communis*); **r. aromatique** sweet flag (*Acorus calamus*); **r. de la Passion** reedmace (*Typha* sp.)

rosé-des-prés *m., pl.* **rosés-des-prés** field mushroom (*Agaricus campestris*) (see also **champignon**)

rosée *f.* dew; **r. du soleil** sundew (*Drosera* sp.) (= **rossolis**); **point de rosée** dew-point

roselier,-ière *a.* reed-producing; *n. f.* reed bed

roselle *f.* red sorrel, roselle (*Hibiscus sabdariffa*) (= **oseille de Guinée**)

roseoeillet *m.* hybrid of French marigold (*Tagetes patula*) and African marigold (*T. erecta*)

roseraie *f.* rose garden

rosette *f.* small rose; rosette (of leaves); *PD* rosette; **r. du pommier** apple rosette virus

rosier *m.* rose tree, rose bush (*Rosa* sp.) (see also **rose, églantier**); **r. arbuste** tree rose plant; **r. de Banks** Banksian rose bush (*R. banksiae*); **r. du Bengale** crimson China rose bush (*R. chinensis* var. *semperflorens*) (see also **r. de Chine**); **r. blanc** *R.* × *alba*; **r. bourbon** Bourbon rose bush (*R.* × *bourboniana*); **r. buisson, r. buissonnant** bush rose plant; **r. camellia** cherokee rose bush (*R. laevigata*); **r. canin** = **r. des chiens** (*q.v.*); **r. capucine** Austrian briar bush (*R. foetida*); **r. capucine bicolore** Austrian copper bush (*R. foetida* var. *bicolor*); **r. cent(-)feuilles** cabbage, Provence rose bush (*R. centifolia*); **r. des champs** trailing rose bush (*R. arvensis*); **r. châtaigne** *R. roxburghii* = *R. microphylla*; **r. des chiens** dog rose (*R. canina*); **r. de (la) Chine** China rose, monthly rose bush (*R. chinensis* = *R. indica*); **r. de Damas** Damask rose (*R. damascena*); **r. demi-tige** half standard rose bush (see **r. tige**); **r. églantier** = **églantier** (*q.v.*); **r. à feuilles simples** *R. persica* (= *R. berberifolia*); **r. à fleurs jaunes, r. jaune** Austrian briar (*R. foetida* = *R. lutea*); **r. de France** French rose bush (*R. gallica*); **r. géant** *R. gigantea*; **r. grimpant** rambler rose bush; **r. groseillier (à maquereau)** *R. mirifica*; **r. hybride** hybrid rose bush; **r. hybride de thé** hybrid tea rose bush; **r. de l'île Bourbon** *R. gallica* × *R. chinensis*; **r. incarnat** = **r. de Damas** (*q.v.*); **r. du Japon** = **r. rugueux** (*q.v.*); **r. de Lady Banks** Banksian rose bush (*R. banksiae*); **r. Macartney** Macartney rose bush (*R. bracteata*); **r. miniature** miniature rose bush; **r. de Miss Lawrance** fairy rose bush (*R. chinensis minima* = *R. lawranceana*); **r. mousseux** moss rose bush (*R. centifolia* var. *muscosa*); **r. multiflore** *R. multiflora* = *R. polyantha*; **r. musqué** musk rose bush (*R. moschata*); **r. nain** bush rose plant; **r. noisette** noisette rose bush (*R.* × *noisettiana*); **r. à nombreuses épines** burnet rose, Scots rose (*R. spinosissima*); **r. à odeur de thé** tea rose bush (= **r. thé**); **r. (en) parasol** umbrella-trained rose bush; **r. perpétuel** repeat-flowering, perpetual rose bush (*R. gallica* × *R. odorata*) (= **r. remontant**); **r. pleureur** weeping rose bush; **r. polyantha** polyantha rose bush (see also **r. multiflore**); **r. pompon** rose de Meause bush (*R. centifolia* var. *pomponia*); **r. de Provins** = **r. de France** (*q.v.*); **r. des quatre saisons** = **r. de Chine** (*q.v.*); **r. remontant** repeat-flowering rose bush (= **r. perpétuel**); **r. rouletti** *R. rouletti* (a form of **r. de Miss Lawrance**); **r. rugueux** Ramanas rose bush (*R. rugosa*) (= **rose du Kamtchatka**); **r. sarmenteux** climbing, rambler rose bush; **r. sauvage** (i) wild rose bush (in general), (ii) **r. des chiens** (*q.v.*); **r. soyeux** *R. serica*; **r. thé** tea rose bush (*R. odorata*); **r. tige** standard rose bush (see also **r. demi-tige**); **r. toujours vert** *R. sempervirens*; **r. des Turcs** *R. hemisphaerica* (= *R. sulphurea*); **r. de Wichura** *R. wichuraiana*

rosiériste *n.* rose-grower

rosir *v.t.* to turn rosy; *v.i.* to become rosy

rossolis *m.* sundew (*Drosera* sp.) (= **droséra, rosée du soleil**)

rostellum *m.* rostellum

rostre *m. Bot.* rostrum; **en forme de rostre** beak-shaped

rot *m. PD* rot; **rot amer** *Vit.* bitter rot (*Melanconium fuligineum*); **r. blanc** *Vit.* white rot, ripe rot (*Coniothyrium (Coniella) diplodiella*) (= **coître**); **r. brenner** *Vit.* brenner, red fire disease (*Pseudopeziza tracheiphila*) (= **brenner, rougeot parasitaire**); **r. brun (de la pomme)** apple brown rot and spur canker (*Sclerotinia (Monilia) fructigena*); **r. brun du cognassier** quince brown rot (*Sclerotinia (Monilia) fructigena*) (= **moniliose du cognassier**); **r. brun** *Vit.* brown rot form of mildew (*Plasmopara viticola*); **r. noir** *Vit.* black rot of grapes (*Guignardia bidwellii*) (= **black-rot**)

rotacé,-e *a.* rotate

rotang *m.* = **rotin** (*q.v.*)

rotation *f. Hort.* rotation (see also **assolement**)
roténone *f.* notenone
rotin, rottain *m.* rattan (*Calamus rotang*)
rotobêche *f.* rotary spade
rotobineuse *f.* rotary hoe
rotundifolié,-e *a.* rotundifoliate
roucou = **rocou** (*q.v.*)
roucouyer *m.* annatto tree (*Bixa orellana*) (= **bixa, rocouyer**)
roue *f.* wheel; **r. squelette** cage wheel
rouge *a.* & *n. m.* red; **r. brique** brick red; **r. brun** brownish red; **r. carmin** carmine red; **r. clair** light red, *Physics*: red light; **r. lointain** *Physics* far red light; **r. sang** blood red; **r. sombre** dark red, *Physics*: far red light; **r. vif** bright red; **r. violacé** violet red
rougeâtre *a.* reddish
rougeau *m. Vit., PD* = **rougeot** (*q.v.*)
rougeot *m. Vit.* reddening disorder (= **flavescence**); **r. parasitaire** brenner, red fire disease (*Pseudopeziza tracheiphila*) (= **brenner, rot brenner**)
rougeur *f.* redness, blush
rough lemon *m.* rough lemon (*Citrus limon*)
rougir *v.i.* to redden, to turn red; **feuillage rougissant à l'automne** foliage turning red in autumn
rougissement *m.* reddening, turning red
roui *m.* retting
rouille *f. PD* rust; **r. blanche (des crucifères)** white (blister) rust of crucifers (*Albugo candida*) (= *Cystopus candidus*); **r. blanche du salsifis et de la scorsonère** white salsify rot (*Cystopus tragopogi*); **r. du caféier** (common) coffee rust (*Hemileia vastatrix*); **r. du chrysanthème** chrysanthemum rust (*Puccinia chrysanthemi*); **r. couronnée** crown rust (*Puccinia coronata*); **r. écidienne (du groseillier)** gooseberry cluster cup rust (*Puccinia pringsheimiana*, a form of *P. caricina*); **r. farineuse du caféier** grey coffee rust (*Hemileia coffeicola*); **r. du framboisier** raspberry rust (*Phragmidium rubi-idaei*); **r. du géranium** geranium rust (*Puccinia pelargonii-zonalis*); **r. du haricot** bean rust (*Uromyces appendiculatus*); **r. noire (de la vigne)** grapevine anthracnose (*Elsinoë ampelina* = *Gloeosporium ampelophagum*); **r. (orangée) du caféier** = **r. du caféier** (*q.v.*); **r. du poirier** pear rust (*Gymnosporangium fuscum*); **r. du rosier** rose rust (*Phragmidium mucronatum*); *a. inv.* rust-coloured, reddish-brown
rouillé,-e *a.* rusty, rust-coloured; *PD* attacked by rust
rouir *v.t.* to ret
rouissage *m.* retting
roulage *m.* rolling
roulaison *f. Sug.* harvest time
rouleau *m., pl.* **-eaux** 1. roller; **r. brise-mottes** clod crusher; **r. à dents** toothed roller; **r. à gazon, r. pour gazon** garden roller; **r. "hérisson"** spiked roller; **r. marqueur** marking roller; **r. plombeur** flat roller; **r. squelette** Cambridge roller, ring roller. 2. roll; **r. de gazon** roll of turf
rouler *v.t.* to roll
roulette *f.* small wheel; **r. à découper le gazon** edging wheel (for lawns) (see **dresse-bordure**)
roussâtre *a.* reddish
roussette *f.* fruit bat (*Rousettus* sp. & other spp.)
roussir *v.t.* to make brown, russet, reddish; *v.i.* to become brown, russet, reddish
roussissement *m.*, **roussissure** *f.* russeting, turning brown, russet, reddish
rouvraie *f.* Durmast oak plantation or grove
rouvre *m.* Durmast oak (*Quercus petraea*) (= **chêne rouvre**)
roux, rousse *a.* russet, reddish, reddish-brown
ruban *m.* ribbon, band; **r. de bergère** ribbon grass, gardener's garters (*Phalaris arundinacea variegata*); **r. plastique adhésif** adhesive plastic band
rubané,-e *a.* ribboned, striped
rubanier *m.* bur-reed (*Sparganium* sp.)
rubigineux,-euse *a.* rubiginous, rust-coloured
rubus *m. Rubus* sp. (see **ronce**)
ruche *f.* beehive
rucher *m.* apiary
rudbeckia, rudbeckie *f.* cone flower (*Rudbeckia* sp.)
rude *a.* rough
rudéral,-e *a., m. pl.* **-aux** ruderal, growing on rubbish or in waste places
rudgéa *m. Rudgea* sp., including the tropical flowering shrub *R. macrophylla*
rue *f.* rue, herb of grace (*Ruta graveolens*) and other *Ruta* spp.; **r. de(s) chèvre(s)** goat's rue (*Galega officinalis*)
rue-des-murailles *f., pl.* **rues-des-murailles** wall rue (*Asplenium ruta-muralis*) (see also **asplenium**)
ruée *f.* straw added to dunghill
ruellage *m. Vit.* trenching
rueller *v.t. Vit.* to trench (vineyards)
ruellia *m. Ruellia* sp., ornamental Acanthaceae
rugosité *f.* russeting; **r. de la peau** (apple) rough skin virus (= **maladie des taches ligneuses**)
rugueux,-euse *a.* rugose, rough, wrinkled
ruine de Rome *f.* ivy-leaved toadflax (*Cymbalaria muralis*) (= **cymbalaire**)
ruissellement *m. Ped.* run-off; **terrain de ruissellement** alluvium
ruizia *m. Ruizia* sp.
rumex *m. inv. Rumex* sp.; **r. crépu** curled dock

(*R. crispus*) (= **parelle**); **r. oseille** sorrel (*R. acetosa*) (see also **oseille, patience**)
ruminé,-e *a. Bot.* ruminate, mottled
rupestre *a.* rupestral, rupicolous
rupture *f.* breaking, rupture; **r. des tiges** stem breakage
russelie *f. Russelia* sp., including *R. juncea* and *R. sarmentosa*
russeting *m.* russeting (= **rugosité**)
russule *f. Fung. Russula* sp.; **r. cyanoxante** *R. cyanoxantha* (= **charbonnier**); **r. verdoyante** green russule (*R. virescens*) (= **palomet**)
rusticité *f. Hort.* hardiness
rustique *a. Hort.* hardy
rutabaga *m.* swede (*Brassica napus* var. *napobrassica*) (= **chou-navet**)
rutine *f.* rutin
rythme *m.* rhythm; **r. de croissance** growth rhythm
rythmique *a.* rhythmic, rhythmical; *n. f.* rhythm; **r. de croissance** growth rhythm

S

sabal *m. Sabal* sp., including fan palm (*S. blackburniana*) and cabbage palm (*S. palmetto*) (= **palmette**)
sabia *m. Sabia* sp.
sabicée *f. Sabicea* sp.
sabine *f.* savin (*Juniperus sabina*)
sablage *m.* sand application
sable *m.* sand; **s. côtier** coastal sand; **s. fin** fine sand; **s. grossier** coarse sand; **s. de rivière** river sand
sablé,-e *a.* sanded, gravelled
sabler *v.t.* to cover with sand, to spread sand
sableux,-euse *a.* sandy
sablier *m.* sand-box tree (*Hura crepitans*)
sabline *f.* sandwort; **s. à feuilles étroites** *Minuartia tenuifolia*; **s. à feuilles de serpolet** thyme-leaved sandwort (*Arenaria serpyllifolia*)
sablonneux,-euse *a.* sandy, gritty (of fruit); **lande sablonneuse** sandy heath
sabot (*m.*) **de Vénus, s. de Marie** lady's slipper (*Cypripedium* sp.) (= **cypripède**)
sac *m.* bag; **s. d'autofécondation** selfing-bag; **s. boudin** growing bag; **s. de cueillette** picking bag; **s. embryonnaire** embryo sac; **s. à fruits** fruit bag; **s. à herbe** grass bag (of mower); **s. d'isolement** isolation bag; **s. de plastique** plastic bag, polybag
saccadé,-e *a.* jerky, abrupt; **croissance saccadée** growth by flushes
saccharifère *a.* sacchariferous, containing sugar
saccharimètre *m.* saccharimeter
saccharose *m.* sucrose, cane sugar
saccharum *m. Saccharum* sp.
saccolabion, saccolabium *m. Saccolabium* sp. (orchid)
sachée *f.* sackful, bagful
sachet *m.* small bag
safoutier *m.* bush butter tree (*Dacryodes (Pachylobus) edulis*), cultivated for its edible fruits and seeds
safran *m.* 1. saffron and related spp. (see also **crocus**);**s. d'automne** autumn crocus (*Crocus nudiflorus* and other spp.); **s. cultivé, s. du Gâtinais** saffron crocus (*Crocus sativus*); **s. officinal** = **s. cultivé** (*q.v.*); **s. des prés** meadow saffron (*Colchicum autumnale*) (= **veilleuse**); **s. printanier** spring crocus (*Crocus vernus* and other spp.) 2. Other genera: **s. bâtard** safflower (*Carthamus tinctorius*) (= **carthame**); **s. des Indes** turmeric (*Curcuma longa*); **s. marron** Indian shot (*Canna indica*)
safrané,-e *a.* saffron, saffron-coloured
safraneraie *f.* saffron plantation
safranier *m.* saffron grower
safranière *f.* saffron plantation
sagesse (*f.*) **des chirurgiens** flixweed (*Descurainia sophia*) (= **sisymbre sagesse**)
sagine *f. Sagina* sp., pearlwort; **s. courbée** procumbent pearlwort (*S. procumbens*)
sagittaire *f. Saggittaria* sp., arrowhead, including *S. sagittifolia*, common arrowhead; **s. subulée** *S. subulata*
sagitté,-e *a.* sagittate
sagou *m.* sago
sagoutier *m.* sago palm (*Metroxylon sagu* and other spp.); **s. de l'Amérique** alta palm (*Mauritia flexuosa*)
sahélien,-enne *a.* relating to the Sahel, Sahelian
saignable *a. Rubb.* tappable, capable of being tapped
saignée *f.* shallow water furrow; *Rubb.* tapping; **s. par piqûre** *Rubb.* puncture tapping; *Vit.* type of green pruning
saigner *v.t. Rubb.* to tap
saigneur *m. Rubb.* tapper
saillant,-e *a.* projecting, jutting out, prominent
saillie *f.* 1. spurt, spring. 2. protrusion; **en saillie** jutting out
sain,-e *a.* healthy, sound, wholesome; **plante saine** healthy plant
sainbois *m.* garou bush (*Daphne gnidium*) (= **garou**)
sainegrain *m.* fenugreek (*Trigonella foenum-graecum*) (= **fenugrec, sénegré**)
sainfoin *m.* sainfoin (*Onobrychis sativa*); **s. à bouquets, s. d'Espagne** French honeysuckle (*Hedysarum coronarium*); **s. oscillant** telegraph plant (*Desmodium gyrans*) (= **herbe vivante**)
sainte-lucie *m.* mahaleb cherry (*Prunus mahaleb*) (= **mahaleb**)
saintpaulia *m.* African violet (*Saintpaulia ionantha*) (= **violette du Cap**)
saison *f.* season; **de première saison** early (= **hâtif**); **de (moyenne) saison** mid-season; **des quatre saisons** everbearing, perpetual; **la belle saison** summer months, dry season (in tropics); **la mauvaise saison, la saison morte** winter months; **la saison des pluies** wet

season, rains (in tropics); **la saison sèche** dry season (in tropics)

saisonnement *m.* the bearing of abundant fruit (by tree)

saisonner *v.i.* to bear abundant fruit, to fruit well (of tree)

saisonnier,-ière *a.* seasonal

salade *f.* 1. salad; **s. de fruits** fruit salad. 2. salad plant (lettuce, chicory etc.); **s. de blé, s. de chanoine** corn salad, lamb's lettuce (*Valerianella locusta*) (= **doucette, mâche**)

salage *m.* salting (of roads etc.)

salaire *m.* wages

salant *a. m.* salt, saline; *n.m.* salt pan; **marais salant** salt marsh

salarié,-e *a.* salaried; *n.* employee

sale *a.* dirty, weed-infested (of crop)

salé,-e *a.* salted, saline

salep *m.* salep, food prepared from dried orchid tubers

salicaire *f.*, **s. commune** common loosetrife (*Lythrum salicaria*)

salicole *a.* halophytic; **plante salicole** halophytic plant

salicorne *f.* glasswort (*Salicornia* sp.)

salifère *a.* saliferous

salinité *f.* salinity

salissement *m. Hort.* weed invasion (of crop); **s. rapide** rapid weed invasion, rapid weed growth

salissure *f.* stain, dirty mark

salpêtre *m.* saltpetre; **s. du Chili** Chile saltpetre

salpiglossis *m. Salpiglossis* sp.; **s. à fleurs changeantes** scalloped tube tongue (*S. sinuata*)

salsepareille *f.* sarsaparilla, prickly ivy (*Smilax glauca* and other spp.); **s. d'Europe** *S. aspera*

salsifis *m.* salsify (*Tragopogon porrifolius*); **s. d'Espagne, s. noir** = **scorsonère** (*q.v.*); **s. des prés, s. sauvage** goat's beard (*T. pratensis*)

salsola *f. Salsola* sp., saltwort

salure *f.* saltness, salinity

salvia *f. Salvia* sp. (see also **sauge**)

samare *f.* samara

samole *m. Samolus* sp., including Tasmanian water pimpernel (*S. repens*)

sang (*m.*) **desséché, s. soluble** dried blood, blood meal

sang-dragon, sang-de-dragon *m.* 1. dragon's blood (from *Dracaena draco, Pterocarpus draco* and others). 2. red-veined dock (*Rumex sanguineus*) (= **patience rouge**)

sanguin,-ine *a.* relating to blood; **orange sanguine** blood orange; *n. f.* blood orange; *n. m.* dogwood (*Cornus (Swida) sanguinea*) (= **sanguinelle**)

sanguinaire *f. Sanguinaria* sp.; **s. du Canada** blood-root (*S. canadensis*)

sanguinelle *f.* dogwood (*Cornus (Swida) sanguinea*) (= **sanguin**)

sanguisorbe *f. Sanguisorba* sp., burnet

sanicle, sanicule *f.* sanicle (*Sanicula europaea*)

sanitaire *a. Hort.* (phyto) sanitary

sans *prep.* without; **sans bourgeons** budless; **sans épines** thornless; **sans feuilles** leafless; **sans virus** virus-free

sansevière *f. Sansevieria* sp., mother-in-law's tongue

sansonnet *m.* starling (*Sturnus vulgaris*)

santal *m. Santalum* sp.; **s. blanc** sandal-wood (*S. album*)

santoline *f. Santolina* sp.; **s. petit cyprès** lavender cotton (*Santolina chamaecyparissus*) (= **aurone femelle**)

sanve *f.* charlock (*Sinapis arvensis*)

saperde *f. Ent. Saperda* sp.; **grande saperde du peuplier** large poplar longhorn (*S. carcharias*); **petite saperde** poplar borer (*S. populnea*)

sapidité *f.* sapidity

sapin *m.* fir (*Abies* sp.) and other genera; **s. d'Algérie** = **s. de Numidie** (*q.v.*); **s. d'Asie Mineure** = **s. de Cilicie** (*q.v.*); **s. baumier** balsam fir, balm of Gilead (*A. balsamea*) (= **baumier**); **s. blanc** = **s. commun** (*q.v.*); **s. blanc du Colorado** Colorado fir, white fir (*A. concolor*); **s. bleu** Colorado spruce (*Picea pungens* var. *glauca*); **s. du Caucase** Caucasian fir (*A. nordmanniana*); **s. de Céphalonie** *A. cephalonica*; **s. de (la) Cilicie** *A. cilicica*; **s. commun** European silver fir (*A. alba* = *A. pectinata*); **s. de la Corée** Korean fir (*A. koreana*); **s. de Crimée** = **s. du Caucase** (*q.v.*); **s. de Douglas** Douglas fir (*Pseudotsuga taxifolia*); **s. élevé** noble fir (*A. procera* = *A. nobilis*); **s. d'Espagne** Spanish fir (*A. pinsapo*); **s. géant** giant fir (*A. grandis*); **s. gracieux** red silver fir (*A. amabilis*); **s. de Grèce** Grecian fir (*A. cephalonica*); **s. de Lobb** Rocky Mountain fir (*A. lasiocarpa*); **s. de Nikko** Nikko fir (*A. homolepis*); **s. noble** = **s. élevé** (*q.v.*); **s. de Noël** Christmas tree; **s. de Nordmann** = **s. du Caucase** (*q.v.*); **s. de Numidie** Algerian fir (*A. numidica*); **s. d'Orégon** = **s. géant** (*q.v.*); **s. pectiné** = **s. commun** (*q.v.*); **s. rouge de Californie** Californian red fir (*A. magnifica*); **s. de Vancouver** = **s. géant** (*q.v.*); **s. de Veitch** Veitch's silver fir (*A. veitchii*); **s. des Vosges** = **s. commun** (*q.v.*)

sapinette *f.* spruce (*Picea* sp.); **s. blanche** white spruce (*P. glauca*) (= **épinette blanche**); **s. de l'Himalaya** Himalayan spruce (*P. morinda* = *P. smithiana*); **s. noire** black spruce (*P. mariana*) (= **épinette noire**); **s. d'Orient** Oriental spruce (*P. orientalis*); **s.**

rouge red spruce (*P. rubens*) (= **épinette rouge**); **s. de Sakhaline, s. de Glehn** Saghalien fir (*P. glehnii*) (see also **épicéa**)
sapinière *f.* fir plantation or grove
saponaire *f.* *Saponaria* sp.; **s. à bouquets** *S. vaccaria* (= *Vaccaria pyramidata*); **s. officinale** soapwort (*S. officinalis*); **s. des vaches** = **s. à bouquets** (*q.v.*)
saponifiable *a.* saponifiable; **fraction non saponifiable** unsaponifiable fraction
saponification *f.* saponification
saponine *f.* saponin
sapote *f.* sapodilla plum; **s. blanche** white sapote (*Casimiroa edulis*) (= **pomme mexicaine**); **s. mamey** mammee sapote (*Calocarpum sapota*) (= **sapotille mamey**); **s. nègre, s. noire** *Afr.* black sapote
sapotier *m.* sapodilla tree (*Achras sapota* = *Manilkara achras*) (= **sapotillier**); **s. blanc** see **sapote blanche**; **s. brésilien** *Quararibea cordata*, a tree with edible fruits; **s. nègre, s. noir** *Afr.* black sapote tree (*Diospyros ebenaster*)
sapotille *f.* sapodilla fruit (= **nèfle d'Amérique**); **s. mamey** mammee sapote (*Calocarpum sapota*) (= **sapote mamey**)
sapotillier *m.* sapodilla tree (*Achras sapota* = *Manikara achras*) (= **néflier d' Amérique, sapotier**)
saprophyte *m.* saprophyte
saprophytique *a.* saprophytic
saprophytisme *m.* saprophytism
sarclage *m.* weeding, weeding with a hoe; **s. à blanc** clean weeding
sarclé,-e *a.* weeded, weeded with a hoe
sarcler *v.t.* to weed, to hoe; **plantes sarclées** weeded, hoed plants
sarclet *m.*, **sarclette** *f.* weeding hoe
sarcleur,-euse *n.* 1. weeder. 2. *n. f.* weeding implement, hoe; **s. à patte de canard** A-blade tine
sarcloir *m.* weeding hoe, spud
sarclure *f.* weeds removed by hoeing, weedings
sarcocarpe *m.* sarcocarp
sarcocolle *f.* sarcocolla
sarcocollier *m.* sarcocolla tree (*Penaea sarcocolla*)
sarcotest *m.* sarcotesta
sariette *f.* see **sarriette**
sarment *m.* 1. grapevine shoot or branch. 2. woody climbing stem, bine
sarmentacé,-e *a.* sarmentaceous
sarmenter *v.i.* *Vit.* collecting the shoots after pruning
sarmenteux,-euse *a.* sarmentous, climbing; **rosier sarmenteux** rambler rose; **vigne sarmenteuse** climbing vine
sarment-greffon *m.* *Vit.* budwood
sarment-maître *m.* *Vit.* main shoot (= **rameau principal**)
sarracénie *f.* *Sarracenia* sp., North American pitcher plant; **s. pourpre** *S. purpurea*
sarrète, sarrette *f.* = **serratule** (*q.v.*)
sarriette *f.* savory (*Satureja* sp.); **s. commune, s. des jardins** summer savory (*S. hortensis*); **s. de montagne, s. vivace** winter savory (*S. montana*)
sas *m.* sieve, screen, riddle
sassafras *m.* sassafras (*Sassafras albidum*)
satiné,-e *a.* satiny, satin-like
saturation *f.* saturation
saturé,-e *a.* saturated
saturer *v.t.* to saturate
satyrion *m.* *Satyrium* sp. (orchid)
saucissonnier *m.* *Afr.* sausage tree (*Kigelia africana*)
sauge *f.* sage, salvia (*Salvia* sp.); **s. argentée** *S. argentea*; **s. écarlate, s. éclatante** red salvia (*S. splendens*); **s. officinale, s. des jardins** common sage (*S. officinalis*); **s. des prés** meadow clary (*S. pratensis*); **s. sclarée** clary (*S. sclarea*) (= **toute bonne**)
saule *m.* willow (*Salix* sp. and other spp.); **s. de Babylone** = **s. pleureur** (*q.v.*); **s. blanc** white willow (*S. alba*); **s. cendré** grey willow (*S. cinerea*); **s. épineux** sea buckthorn (*Hippophaë rhamnoides*) (= **argousier**); **s. fragile** crack willow (*S. fragilis*); **s. laurier** bay-leaved willow (*S. pentandra*); **s. marsault** sallow (*S. caprea*); **s. pleureur** weeping willow (*S. babylonica*); **s. rampant** creeping willow (*S. repens*)
saulée *f.* row of willows
saumâtre *a.* brackish
saumon *a. inv.* salmon-coloured, salmon-pink
saumure *f.* brine, pickle
saupoudrage *m.* dusting, powdering
saupoudrer *v.t.* to dust, to powder
saut-de-loup *m.*, *pl.* **sauts-de-loup**; *Hort.* ha-ha, sunk fence
saute *f.* sudden change (of temperature etc.); **s. de vent** change of wind
sautelle *f.* layered vine shoot
sauterelle *f.* grasshopper, locust, bush cricket; **s. des serres** greenhouse grasshopper (*Tachycines asynamorus*) (see also **criquet, locuste**); *Hort.* aerial part of layer
sauvageon *n. m.*; **sauvageonne** *n. f.* wild stock, wilding, seedling
sauve-vie *f. inv.* wall rue (*Asplenium ruta-muraria*) (see also **doradille**)
savane *f.* savanna; (*Canada*) swamp; **s. arborée** *Afr.* wooded savanna; **s. verger** *Afr.* orchard bush
saveur *f.* savour, taste, flavour
savinier *m.* = **sabine** (*q.v.*)

savonnier *m.* 1. soapberry (*Sapindus* sp.); **s. saponaire** southern soapberry (*S. saponaria*). 2. pride of India (*Koelreuteria paniculata*)
savorée *f.* savory (*Satureja* sp.)
saxatile *a.* saxatile
saxicole *a.* saxicolous, growing among rocks
saxifrage *f.* saxifrage (*Saxifraga* sp.); **s. d'automne** yellow saxifrage (*S. aizoides* = *S. autumnalis*); **s. cunéiforme** wood saxifrage (*S. cuneifolia*) (= **mignonnette des Alpes**); **s. éclatante de fleurs** *S. florulenta*; **s. à feuilles de benoîte** *S.* × *geum*; **s. granulée** meadow saxifrage, fair maids of France (*S. granulata*); **s. hypnoïde, s. mousse, s. mousseuse** Dovedale moss, Eve's cushion, mossy saxifrage (*S. hypnoides*) (= **gazon turc**); **s. ombreuse** London pride (*S. umbrosa*) (= **désespoir du peintre**); **s. sarmenteuse** mother-of-thousands (*S. stolonifera* = *S. sarmentosa*); **s. tridactyle, s. à trois doigts** rue-leaved saxifrage (*S. tridactylites*)
scab *m. inv. PD* scab (see also **tavelure**); **s. du bigaradier** sour orange scab, citrus scab (*Elsinoë fawcettii*)
scabieuse *f.* scabious (*Scabiosa* sp. and other genera); **s. des Alpes** *Cephalaria alpina*; **s. du Caucase** *S. caucasica*; **s. des jardins, s. à fleurs pourpre-noir, s. fleur de veuve, s. maritime** sweet scabious (*S. atropurpurea*); **s. succise, s. tronqué** *S. succisa*
scabre *a.* scabrous
scabreux,-euse *a.* scabrous, rough to the touch
scalariforme *a.* scalariform
scammonée *f.* scammony (*Convolvulus scammonia*) (= **liseron scammonée**)
scandix *m.* *Scandix* sp.; **s. peigne de Vénus** shepherd's needle (*S. pecten-veneris*)
scape *m.* scape
scarieux,-euse *a.* scarious, scariose
scarifiage *m.* scarification
scarificateur *m.* scarifier, harrow
scarification *f.* 1. scarification. 2. incising
scarifier *v.t.* to scarify; *Hort.* to incise
scarole *f.* (broad-leaved) endive (*Cichorium endivia*) (= **chicorée scarole**)
sceau *m.* seal; **s. de Notre-Dame, s. de la Vierge** black bryony (*Tamus communis*); **s. de Salomon** Solomon's seal (*Polygonatum* sp.)
schinus *m.* *Schinus* sp. (see also **poivrier**)
schisteux,-euse *a.* schistous; **sol schisteux** schistous soil, schist soil
schistocarpe *a.* schizocarpous
schizandre *m.* *Schizandra* , a genus containing hardy deciduous aromatic climbing shrubs
schizanthe *m.* *Schizanthus* sp., butterfly, fringe flower
schizée *f.*, **schizea** *m.* *Schizaea* sp., comb or rush fern
schotie *f.* *Schotia* sp., including *S. africana*
schuélage, schuellage *m.* tamping watercress plants into the substrate
schuèle *f.* watercress (wooden) tamper
schuéler, schueller *v.t.* to tamp watercress
sciaphile *a.* shade-loving, sciophyllous; **plante sciaphile** shade-loving plant, sciophyte
scie *f.* saw; **s. à chaine** chain saw; **s. égoïne, s. d'élagage** pruning saw, small hand saw (= **égoïne**); **s. de jardinier** pruning saw; **s. à main** hand saw; **s. mécanique** mechanical saw (see also **tronçonneuse**)
science (*f.*) **du sol** soil science (= **pédologie**)
scientifique *a.* scientific; *n.* scientist
scille *f.* *Scilla* sp., squill and other genera; **s. d'automne** autumn squill (*S. autumnalis*); **s. azurée** = **s. de Sibérie** (*q.v.*); **s. blanche** sea daffodil (*Pancratium maritimum*); **s. à deux feuilles** *S. bifolia*; **s. maritime** sea onion (*Urginea maritima*); **s. du Pérou** Cuban lily (*S. peruviana*) (= **jacinthe du Pérou**); **s. de Sibérie** Siberian squill (*S. sibirica*)
scintillateur *m.* *Physics* scintillation counter
scion *m.* young shoot; young grafted plant (see also **greffon**); **s. d'un an** maiden tree, one-year-whip (= **greffe d'un an**); **s de deux ans** two-year-old grafted tree; **s. de deux ans avec anticipés** two-year feathered tree; **s. ramifié** feathered maiden
scirpe *m.* *Scirpus* sp.; **s. des lacs** bulrush (*Scirpus (Schoenoplectus) lacustris*)
scissiparité *f.* scissiparity
sciure *f.* **(de bois)** sawdust
sclarée *f.* clary (*Salvia sclarea*) (= **orvale**)
scléranthe, scleranthus *m.* *Scleranthus* sp.; **s. annuel** annual knawel (*S. annuus*); **s. vivace** perennial knawel (*S. perennis*)
sclérenchyme *m.* sclerenchyma
scléreux,-euse *a.* sclerous
sclérifié,-e *a.* sclerosed, sclerified
sclérite *f.* sclerite
scléroderme *m.* *Fung.* *Scleroderma* sp.
sclérophylle *a.* sclerophyllous; *n. m.* sclerophyll
sclérote *m.* sclerotium; *pl.* sclerotia
sclérotinia *m.* *PD* *Sclerotinia* sp.
sclérotiniose *f.* sclerotinia disease; **s. du chou** cabbage sclerotinia rot (*Sclerotinia sclerotiorum*); **s. du cognassier** quince leaf blotch (*S. cydoniae*); **s. du glaïeul** gladiolus dry rot (*S. gladioli*); **s. de la salade et de l'endive** lettuce and chicory sclerotinia disease (*S. sclerotiorum* and other spp.); **s. de la tomate** tomato sclerotinia disease (*S. sclerotiorum*)
scolopendre *f.* 1. centipede. 2. hart's tongue fern (*Phyllitis scolopendrium*) (= **langue-de-cerf**)
scolyme *m.* *Scolymus* sp., including **s. d'Espagne** Spanish oyster plant (*S. hispanicus*) (=

cardouille); **s. taché** spotted Spanish oyster plant (*S. maculatus*)

scolyte *m.* bark beetle (*Scolytus* sp.); **s. de l'orme** large elm bark beetle (*S. scolytus*); **s. du pommier** fruit bark beetle (*S. rugulosus*); **s. des cerises, s. des drupes** *Coff.* berry borer (*Stephanoderes (Hypothenemus) hampei*); **s. du grain** *Coff.* berry borer (*Stephanoderes coffeae*); **s. des rameaux** *Coff.* shot hole borer (*Xyleborus compactus = X. morstatti*)

scopolia *m. Scopolia* sp., including *S. carniolica*, with medicinal properties

scorie *f.* usually *pl.*, slag; **scories de déphosphoration, scories Thomas** basic slag; **scories vitreuses** clinker (see also **laitier** 2)

scorpioïde *a.* scorpioid

scorsonère, scorzonère *f.* scorzonera (*Scorzonera hispanica*) (= **salsifis d'Espagne, s. noir**)

scotch *m.* self-adhesive tape, scotch tape

scrobiculé,-e *a.* scrobiculate, pitted

scrofulaire, scrophulaire *f. Scrophularia* sp.

scrofulariacées *n. f. pl.* Scrophulariaceae

scrubber *m.* scrubber

scutellaire *f. Scutellaria* sp., helmet flower, skull cap

scutelle *f.*, **scutellum** m. Bot. scutellum

scutigérelle *f.* greenhouse centipede (*Scutigerella immaculata*)

seaforthia *m. Seaforthia (Ptychosperma)* sp., including Australian feather palm (*P. elegans*)

seau *m.*, *pl.* **seaux** bucket; **s. de plantation** *Vit.* container for young vines at planting; **s. de vendange** *Vit.* picking bucket

sébeste *m.* scarlet cordia fruit

sébestier *m.* scarlet cordia tree (*Cordia sebestena*) (= **prunier sébeste**); **s. blanc** white cordia (*C. alba*); **s. à odeur d'oignon** onion cordia (*C. alliodora*)

sébifère *a.* sebiferous

sec, sèche *a.* dry; **la saison sèche** *Afr.* etc. the dry season

sécateur *m.* secateurs, pruning shears; **s. à ficelle = échenilloir** (*q.v.*)

séchage *m.* drying

séché,-e *a.* dried (up), cured (of tobacco); **s. à l'air** air-dried

sécher *v.t.* to dry, to dehydrate; **se sécher** *v.i.* to dry, to become dry, to wilt (of plant)

sécheresse *f.* dryness, drought

séchium *m. Sechium* sp., chayote (*S. edule*)

séchoir *m.* 1. drying place; **s. à houblon** oast-house. 2. drier, desiccator; (*Coconut*) copra kiln; **s. solaire** (*Cacao*) sun drier

secondaire *a.* secondary; **forêt secondaire** secondary forest

seconde pousse *f. Hort.* second flush, growth (= **pousse d'août**)

secondine *f.* secundine

secouage, secouement *m.* shaking; **vendange mécanique à secouage** grape mechanical harvesting by shaking

secouer *v.t.* to shake, to vibrate

secoueur *m.*, **secoueuse** *f.* shaker, vibrator; **s. de fruits** fruit (tree) shaker

sécréteur,-trice *a.* secretory; **tissu sécréteur** secretory tissue

sécrétion *f.* secretion

sécrétoire *a.* secretory; **activité sécrétoire** secretory activity

sectoriel,-elle *a.* sectorial

sédiment *m.* sediment, deposit

sédum, sedum *m. Sedum* sp., stonecrop (= **orpin**); **s. reprise** orpine, live-for-ever (*S. telephium*) (= **reprise**); **s. rougeâtre** red sedum (*S. rubens*)

seedling *m.* seedling

segment *m.* segment

ségrégation *f. PB, Ped.* segregation

séguia *f.* irrigation trench or ditch (in North Africa)

seigle *m.* rye (*Secale cereale*); **paille de seigle** rye straw

sel *m.* salt; **s. d'amine du 2,4-D** 2,4-D amine (salt); **s. gemme** rock salt; **s. marin** sea salt

sélaginelle *f. Selaginella* sp.; **s. argentée** *S. argentea*; **s. changeante** *S. serpens*; **s. à feuilles écailleuses** resurrection plant (*S. lepidophylla*) (= **fougère de la résurrection**); s. **à pied rouge** *S. erythropus*; **s. en vrille** *S. viticulosa*

sélectif,-ive *a.* selective

sélection *f.* selection; *PB* breeding, selection; s. **généalogique** single plant selection; s. **massale** mass selection; **s. par mutation** mutation breeding; **s. naturelle** natural selection; **s. nucellaire** *Cit.* nucellar selection

sélectionner *v.t.* to select; *PB* to breed, to improve

sélectionneur *m. PB* plant breeder

sélectivité *f.* selectivity

seller, se seller *v. p. Ped.* to cap, to seal (of soil)

semailles *f. pl.* 1. sowing; **l'époque des semailles** sowing time. 2. seeds

semé,-e *a.* sown, strewn, sprinkled

semelle (*f.*) **(de labour)** plough sole, plough pan

semence *f.* seed; *Sug.* etc. seed-piece; **semences d'adventices** weed seeds; **s. calibrée** graded seed; **s. enrobée** pelleted seed; **s. féconde** fertile seed; **s. potagère** vegetable seed; **semences horticoles** horticultural seeds; **semences souterraines viables** viable underground seeds

semenceau *m. Sug.* etc. seed-piece; *Plant.* nursery plant, seedling
semencier,-ière *a.* seed-bearing (of tree); *n. m.* tree reserved for seed (see also **champ**)
semen-contra *m.* santonica (*Artemisia maritima*) (see also **armoise**)
semer *v.t.* to sow; **s. clair** to sow thinly; **s. dru** to sow thickly (see also **semis**)
semeur,-euse *n.* sower
semi-automatique *a.* semi-automatic
semi-déterminé,-e *a.* semi-determinate (of plant)
semi-flosculeux,-euse *a.* semiflosculous, semi-floscular
semi-forçage *m.* semi-forcing
séminal,-e *a., m. pl.* **-aux** seminal; **dormance séminale** seed dormancy
sémination *f.* semination
séminifère *a.* seminiferous, seed-bearing
séminule *f.* spore
semi-persistant,-e *a.* semi-persistent
semis *m.* 1. sowing (see also **semer**); **multiplication par semis** propagation by seed; **pommier de semis** seedling apple; **s. en caissettes** sowing in seed boxes; **s. sur couche** sowing on a hotbed; **s. direct** direct drilling; **s. à distance constante** space sowing; **s. en godets** sowing in small pots, sowing in containers; **s. en lignes** sowing in drills, sowing in rows; **s. en lignes interrompues** space sowing, station sowing; **s. naturel** (1) natural seeding, (2) (natural) seedling, wilding (see 3 below); **s. en place** sowing in situ, sowing at stake; **s. en poquets** sowing in pockets, sowing (in groups) in holes; **s. en pots** sowing in pots; **s. en terrines** sowing in seed pans; **s. à la volée** broadcasting. 2. seedbed. 3. seedlings; **fonte des semis** *PD* damping off disease
semi-tardif,-ive *a.* semi-late
semoir *m.* sowing drill, seed drill, seeder; **s. à bras** hand seeder; **s. de précision** seed spacing drill, precision seeder
semper virens *a. inv.* evergreen (of plant)
sempervirent,-e *a.* evergreen (of plant)
sempervivum *m. inv.* sempervivum, house-leek (*Sempervivum* sp.)
séné *m.* senna (*Cassia* sp.)
senebière, senebiérie *f. Senebiera* sp.; **s. corne de cerf** swine cress (*S. coronopus* = *Coronopus squamatus*); **s. double** lesser swine-cress (*S. didyma* = *Coronopus didymus*)
seneçon, séneçon *m. Senecio* sp., groundsel; **s. jacobée** ragwort (*S. jacobaea*); **s. des oiseaux** (common) groundsel (*S. vulgaris*); **s. en arbre** groundsel tree (*Baccharis halimifolia*); **s. géant** ligularia (*Ligularia clivorum*)
sénegrain *m.* = **sénegré** (*q.v.*)
sénegré *m.* fenugreek (*Trigonella foenum-graecum*) (= **fenugrec, sainegrain**)
senelle = **cenelle** (*q.v.*)
sénescence *f.* senescence
sénescent,-e *a.* senescent
sénevé *m.* black mustard (*Brassica nigra*) (= **moutarde noire**)
sénile *a.* senile
sénilité *f.* senility
sens (*m.*) **de la spirale** direction of the spiral (in phyllotaxy)
sensibilisation *f.* sensitizing, sensitization
sensibilité *f.* sensitivity, susceptibility; **s. aux meurtrissures** susceptibility to bruising; **s. variétale** varietal sensitivity; **s. au vent** wind susceptibility
sensible *a.* susceptible; **s. à la gelée** susceptible to frost
sensitive *f.* sensitive (plant) (*Mimosa pudica*)
senteur *f.* aroma, perfume; **pois de senteur** sweet pea (*Lathyrus odoratus*)
sentier *m.* path
sep *m.* sole (of plough)
sépalaire *a.* sepaline
sépale *m.* sepal
sépaloïde *a.* sepaloid
séparable *a.* separable
séparation *f.* separation; **couche de séparation** abscission layer
séparer *v.t.* to separate
septal,-e *a., m. pl.* **-aux** septal
septicide *a.* septicidal
septoriose *f.* disease caused by *Septoria* sp.; **s. des agrumes** citrus septoria spot (*S. citri*); **s. du céleri** celery leaf spot (*S. apii* and *S. apii-graveolentis*); **s. du chrysanthème** chrysanthemum blotch (*S. chrysanthemella*); **s. du framboisier et de la ronce** blackberry septoria spot (*S. rubi*); **s. du poirier** pear leaf fleck (*Mycosphaerella sentina*); **s. de la tomate** tomato leaf spot (*S. lycopersici*)
septum *m.* septum
séquence *f.* sequence; **s. de sols** *Ped.* soil series
sequoia *m. Sequoia* sp. and related spp.; **s. géant** giant sequoia, wellingtonia (*Sequoiadendron giganteum*); **s. toujours vert** redwood (*S. sempervirens*)
sérapias *m. Serapias* sp., tongue-flowered orchid
serfouage *m.* = **serfouissage** (*q.v.*)
serfouette *f.* combined hoe and fork
serfouir *v.t.* to hoe
serfouissage *m.* hoeing
sériculture, séricilculture *f.* sericulture, silkworm breeding
sérié,-e *a.* seriate
seringa, seringat *m.* syringa, mock orange (*Philadelphus coronarius*); **s. doré** *P. coron-*

arius aureus; **s. à grandes fleurs** *P. grandiflorus*
seringage *m.* syringing, spraying
seringue *f.* syringe; **s. d'arrosage** syringe for watering; **s. bruineuse** syringe with a fine spray; **s. de jardin** garden syringe; **s. pulvérisatrice** (small) sprayer
seringuer *v.t. Hort.* to spray, to water leaves by spraying, to syringe
serjanie *f. Serjania* sp., including *S. reticulata*
sernamby *m. Rubb.* coagulated rubber on the tapping cut
serpe *f.* billhook
serpent *m.* snake; **s. végétal** snake gourd (*Tricosanthes anguina*)
serpentaire *f.* dragon arum (*Arum dracunculus* (= *Dracunculus vulgaris*))
serpentine *f. Bot.* snake wood (*Strychnes colubrina*)
serpette *f.* small billhook; pruning knife
serpolet *m.* wild thyme (*Thymus serpyllum*) (see also **thym**)
serradelle *f.* serradella (*Ornithopus sativus*)
serratule *f. Serratula* sp.; **s. des teinturiers** sawwort (*S. tinctoria*)
serre *f.* greenhouse, glasshouse; **grande serre** winter garden; **s. adossée** lean-to greenhouse; **s. "baby"** plastic cloche; **s. à trois chapelles** three-span greenhouse; **s. à châssis démontables, s. à châssis mobiles** Dutch light structure, mobile greenhouse; **s. chaude, s. chauffée** heated house, hothouse; **s. démontable** collapsible greenhouse; **s. à deux versants, s. à double pente** span-roofed greenhouse; **s. de forçage, s. à forcer** forcing house; **s. froide** cool greenhouse, cool house; **s. gonflable** inflatable greenhouse; **s. gonflée** bubble house; **s. mobile** = **s. roulante** (*q.v.*); **s. multichapelles** multispan greenhouse; **s. à multiplication** propagating house; **s. à palmiers** palm house; **s. à une pente** = **s. adossée** (*q.v.*); **s. plastique** plastic greenhouse; **s. roulante** mobile greenhouse; **s. tempérée** temperate greenhouse; **s.-tour** tower greenhouse; **s. tropicale** tropical house; **s.-tunnel** (plastic) tunnel greenhouse; **s.-verre** glasshouse; **s. à vigne** grape house, vinery
serré,-e *a.* tight, close, dense; **fleurs serrées** dense flowers; **peuplement serré** dense stand; **en rangs serrés** in close rows
serricole *a.* relating to greenhouse culture; **culture serricole** greenhouse culture
serriculture *f.* glasshouse culture, greenhouse culture
serriste *m.* glasshouse grower, greenhouse grower
serrulé,-e *a.* serrulate
service *m.* service; **s. de contrôle (des produits)** (produce) inspection service; **s. de vulgarisation** advisory, extension service
sesbania, sesbanie *f. Sesbania* sp.
séséli *m. Seseli* sp., including moon carrot (*S. libanotis*)
sésie *f.* clearwing moth; **s. du groseillier** currant clearwing moth (*Synanthedon tipuliformis* = *Aegeria tipuliformis*); **s. du pommier** pear clearwing moth (*S. myopaeformis*)
sesquioxyde *m.* sesquioxide
sessile *m.* sessile
sétacé,-e *a.* setaceous, bristly
sétaire, sétaria *f. Setaria* sp.; **s. d'Italie** Italian millet (*S. italica*); **s. verticillée** rough bristlegrass (*S. verticillata*)
séteux,-euse *a.* setose, pubescent, bristly
sétifère *a.* setiferous, setigerous
sétiforme *a.* setiform, bristle-shaped
sétigère *a.* = **sétifère** (*q.v.*)
sétule *f.* seta, bristle
seuil *m.* threshold; **s. de croissance** growth threshold; **s. de flétrissement** wilting point; **s. d'intervention** treatment threshold; **s. de nuisibilité** damage threshold; **s. de sensibilité** sensitivity threshold; **s. de tolérance** tolerance threshold; **s. de toxicité** toxicity threshold
sève *f.* sap; **s. ascendante, s. brute** crude sap; **s. descendante, s. élaborée** elaborated sap; **plein de sève** full of sap; **en sève** non-dormant; **sans sève** sapless
séveux,-euse *a.* sappy
sevrage *m.* weaning; *Hort.* 1. separation of a marcot or layer from the mother plant, separation of the scion when approach grafting; transferring an *in vitro* plant to soil. 2. suppression of scion roots. 3. hardening off (misted cuttings etc.)
sevrer *v.t.* to wean; *Hort.* 1. to separate a marcot or layer from the mother plant, to separate a scion in approach grafting; to transfer an *in vitro* plant to soil. 2. to suppress scion roots. 3. to harden off (misted cuttings etc.)
sexe *m.* sex
sex-ratio *m. Ent.* sex-ratio
sexualisation *f.* sexual differentiation; reproductive stage
sexualité *f.* sexuality
sexué,-e *a.* sexed; **à reproduction sexuée** sexually propagated
sexuel,-elle *a.* sexual
sharka *f.* (**du prunier**) plum pox virus
shérardie *f. Sherardia* sp.; **s. des champs** field madder (*S. arvensis*)
sialagogue *a.* & *n.m.* sialagogic, stimulating flow of saliva, sialagogue
siccatif,-ive *a.* siccative

siccité *f.* dryness
sicyos *m. Sicyos*, a cucurbit genus
sida *m. Sida* sp., including Queensland hemp (*S. rhombifolia*)
sidalcea *m.*, **sidalcée** *f. Sidalcea*, a genus containing ornamental hardy perennial herbs
sidération *f. Hort.* green manuring
sidéritis *m. Sideritis*, a genus containing perennial dwarf sub-shrubby plants
sidéroxyle, sidéroxylon *m. Sideroxylon* sp. (see also **fruit**)
significatif,-ive *a.* significant
silène *m. Silene* sp., catchfly; **s. à bouquets** sweet william catchfly (*S. armeria*); **s. conoïde** *S. conoidea*; **s. de Crète** *S. pendula*; **s. de France** small-flowered catchfly (*S. gallica*); **s. nocturne** night-flowering campion (*S. noctiflora* = *Melandrium noctiflorum*); **s. d'Orient** *S. compacta*
silex *m. inv.* flint
silice *f.* silica
siliceux,-euse *a.* siliceous; **terrain siliceux** siliceous soil
silicicole = **silicole** (*q.v.*)
silicifié,-e *a.* silicified
silicium *m.* silicon
silicole *a.* silicicolous
silicone *f.* silicone
silicule *f.* silicle, silicula
silique *f.* siliqua
sillon *m.* furrow; *Sug.* bud furrow
sillonnage *m.* furrowing
sillonné,-e *a.* furrowed
sillonneur *m.* drill plough; type of hoe
silo *m.* silo, clamp
silybe, silybum *m. Silybum* sp.; **s. de Marie** blessed thistle (*S. marianum*) (= **chardon-Marie**)
simaba *m. Simaba* sp.; **s. ferrugineux** calunga bark (*S. ferruginea* and others)
simaruba, simarouba *m. Simaruba* sp.; **s. amer, s. officinal** bitter damson, mountain damson (*S. amara*)
simazine *f.* simazine
simple *a.* simple; *Hort.* single (of flower, as opposed to double); *n. m. pl.* simples, medicinal herbs
sinué,-e *a.* sinuate, wavy
sinus *m. Bot.* sinus
sirop *m.* syrup; (*Coconut*) toddy syrup, jaggery
sirupeux,-euse *a.* syrupy
sisal *m.* sisal (*Agave sisalana*); **fibre de sisal** sisal fibre
sisaleraie *f.* sisal plantation
sisalier,-ière *a.* relating to sisal; **expérimentation sisalière** sisal trials
sissongo *m.* elephant grass (*Pennisetum purpureum*) (= **herbe à éléphant**)
sisymbre, sisymbrium . *Sisymbrium* sp., rocket; **s. iris** London rocket (*S. irio*); **s. officinal** hedge mustard (*S. officinale*) (= **vélar**); **s. d'Orient** *S. orientale*; **s. sagesse** flixweed (*Descurainia sophia*) (= **sagesse des chirurgiens**)
sitone (*m.*) **des pois** pea and bean weevil (*Sitona* spp. including *S. lineatus*)
sium *m. Sium* sp. (see **berle, chervis**)
skimmie *f. Skimmia* sp.
smilax *m. Smilax* sp.; **s. de Chine** China root (*S. china*) (see also **salsepareille**)
sobole *f.* sobole, creeping underground stem which produces leaf-buds and roots
soc *m.* ploughshare
sodium *m.* sodium
soie *f.* bristle
soin *m.* care; **soins culturaux** cultivations, herbicide applications, pest control etc.
soja *m.* soybean (*Glycine max*)
sol *m.* soil, earth; **s. d'alluvions** alluvial soil; **s. argileux** clay soil; **s. asphyxiant** soil liable to waterlogging; **s. battant** 'capping', sealing soil; **s. creux** puffy soil; **s. fatigué** soil affected by soil sickness; **s. fertile** fertile soil; **s. ferrallitique** ferrallitic soil; **s. gonflant** swelling soil; **s. humide** wet soil; **s. marécageux** marshy soil; **s. marneux** marly soil; **s. nu** bare soil; **s. nu non travaillé** undisturbed bare soil; **s. nu travaillé** cultivated bare soil; **s. propre** clean soil; **s. soufflé** = **s. creux** (*q.v.*); **s. tourbeux** peat soil
solanacé, solané,-e *a.* solanaceous
solanacées, solanées *n. f. pl.* Solanaceae; **solanées maraîchères** market-garden solanaceous crops
solanum *m. Solanum* sp.
solarisation *f.* solarization
soldanelle *f. Soldanella* sp., blue moonwort
sole *f. Hort.* course (of a rotation); **s. en jachère** fallow break
soleil *m.* 1. sun; **au soleil** in the sun, in sunlight. 2. *Hort.* **s. (des jardins)** sunflower (*Helianthus annuus*) (= **tournesol**); **s. du Mexique** Mexican sunflower (*Tithonia tagetiflora*); **s. miniature** *H. cucumerifolius* (= *H. debilis*)
solénostemme *m. Solenostemon* sp., including the West African *S. ocymoides*, with medicinal uses
solitaire *a.* solitary; **fleur solitaire** solitary flower
solubilité *f.* solubility; **s. dans l'eau** water solubility
soluble *a.* 1. soluble, dissolvable. 2. solvable
soluté *m.* solute; aqueous solution
solution *f.* solution; **s. d'épanouissement** (flower) opening solution; **s. nutritive** nutrient solution; **s. à pulvériser** spray liquid
solvant *m.* solvent; **s. organique** organic solvent

somatique *a.* somatic; **mutation somatique** somatic mutation
sombre *a.* dark, sombre; **vert sombre** dark green
somme *f.* sum; **s. des températures** temperature sum
sommet *m.* top, summit, apex; **au sommet** at the top; **s. touffu du bananier** *PD* banana bunchy top virus
sommité *f.* summit, top; *Hort.* apex, tip (of branch, plant etc.); **s. fleurie** flowering top; **s. fructifère** fruiting top
son *m.* bran
sondage *m.* 1. sampling, survey; **s. (d'un moût)** determining must density with a saccharimeter (in winemaking) 2. borehole, drill hole
sonde *f.* 1. probe, sampler; **s. agrologique** soil probe, auger; **s. gammamétrique de profondeur** gamma-ray depth probe; **s. à neutrons** neutron probe. 2. boring machine, drill; **trou de sonde** borehole. 3. saccharimeter (in winemaking)
sonder *v.t.* 1. to sample, to investigate, to test, to scan. 2. to sound, to make borings
sophora *m.* *Sophora* sp.; **s. du Japon** Japanese pagoda tree (*S. japonica*)
sorbe *f.* fruit of service tree
sorbier *m.* *Sorbus* sp. (see also **alisier**); **s. domestique** service tree (*S. domestica*) (= **cormier**); **s. des oiseaux, s. des oiseleurs, s. sauvage** rowan, mountain ash (*S. aucuparia*)
sore *m.* sorus
sorgo, sorgho *m.* sorghum (*Sorghum* sp.); **s. d'Alep** Johnson grass (*S. halepense*)
sorose *f.* sorosis, fleshy fruit formed from many crowded flowers
sorption *f.* sorption
sortie *f.* going out, coming out; *Ag., Hort.* output; *Ent.* emergence; **s. des germes** *Hort.* chitting, germination; *Vit.* inflorescence emergence; early berry development
souche *f.* underground part of stem of perennial plant, (rootstock of) plant, stool, stump; base of tree; *Sug.* stubble; *PB, PD* strain; **s. faible** (de virus) mild strain (of virus); **s. sévère (de virus)** severe strain (of virus)
souche-mère *f., pl.* **souches-mères** mother plant (for layer, marcot etc.)
souchet *m.* 1. sedge (*Cyperus* sp.); nutgrass (*C. rotundus*); **s. domestique** chufa (*C. esculentus*) (= **amande de terre, choufa**); **s. long, s. odorant** galingale (*C. longus*). 2. large woody olive cutting (= **souquet**)
souci *m.* *Calendula* sp.; **s. des champs** wild marigold (*C. arvensis*); **s. (des jardins)** marigold (*C. officinalis*); **s. d'eau, s. des marais** marsh marigold (*Caltha palustris*) (= **caltha des marais, populage**); **s. pluvial** rain daisy (*Dimorphotheca pluvialis*)
soude *f.* 1. saltwort (*Salsola* sp.), including prickly saltwort (*S. kali*); **s. ligneuse** shrubby seablite (*Suaeda fruticosa*); 2. soda; **s. caustique** caustic soda
soudé,-e *a.* joined, united; *Hort.* (successfully) grafted, united (of graft); *n. m.* (successfully) grafted plant (= **greffé-soudé**)
souder *v.t.* to join, to unite; **se souder** (of graft) to unite, to grow together
soudure *f.* soldering; *Hort.* graft union; *Afr.* **faire la soudure** to bridge the 'hungry gap' before the new harvest
soufflerie *f.* blower
soufflé,-e *a.* *Ped.* puffy (see **sol**)
soufflet *m.* bellows, blower
souffleur *m.*, **souffleuse** *f.* blower
souffreteux,-euse *a.* in poor health, sickly
soufrage *m.* dusting with sulphur, sulphuring
soufre *m.* sulphur; **fleur(s) de soufre** flowers of sulphur; **s. mouillable** wettable sulphur
soufré,-e *a.* treated with sulphur; **bouillie soufrée** lime sulphur
soufrer *v.t.* to dust or treat with sulphur
soufreur,-euse *n.* person who treats vines etc. with sulphur; *n. f.* bellows, puffer (for sulphur dusting)
souki *m.* vegetable marrow (*Cucurbita pepo*)
soulamé *m.*, **soulamée** *f.* *Soulamea* sp.
soulever *v.t.* *Hort.* to lift (root crop)
souleveuse *f.* lifter; **s. de pommes de terre** potato lifter
soumbala *m.* paste made from **nété** (*q.v.*)
soump *m.* desert date (*Balanites aegyptiaca*) (= **balanites, dattier du désert**)
soupape *f.* valve; **s. à bille** ball valve; **s. de sécurité** safety valve
souple *a.* supple, flexible
souplesse *f.* flexibility, pliability
souquet *m.* olive cutting (= **souchet** 2)
source *f.* spring (water), source; **s. alimentaire** food source; **s. d'infection** source of infection; **s. lumineuse** light source
souris *f.* mouse (*Mus* sp.) (see also **campagnol, mulot**); **s. végétale** Chinese gooseberry (*Actinidia chinensis*) (= **yang-tao**)
sous-arbrisseau *m., pl.* **sous-arbrisseaux** subshrub
sous-bois *m. inv.* undergrowth, underwood
sous-bourgeon *m.* bud developed on another bud
sous-charpentière *f.* *Hort.* secondary branch (see also **branche**)
sous-culture *f.* underplanted crop
sous-épidermique *a.* sub-epidermal
sous-espèce *f.* sub-species
sous-étage *m.* understorey (of trees)

sous-frutescent,-e *a.* suffrutescent; **plante sous-frutescente** sub-shrub (= **sous-arbrisseau**) (= **suffrutescent**)
sous-jacent,-e *a.* subjacent, underlying
sous-ligneux,-euse *a.* semi-woody
sous-mère *f. Hort.* secondary branch (see also **branche**)
sous-oeil *m., pl.* **sous-yeux** basal bud (= **oeil stipulaire**)
sous-produit *m.* by-product
sous-sol *m. Hort.* subsoil
sous-solage *m.* subsoiling
sous-solé,-e *a.* subsoiled
sous-soler *v.t.* to subsoil
sous-soleuse *f.* subsoiler
sous-tribu *f.* subtribe
soutènement *m.* supporting, propping up; **mur de soutènement** supporting wall
souterrain,-e *a.* underground, subterranean; **tige souterraine** underground stem
soya *m.* soyabean, soybean (*Glycine max*)
soyeux,-euse *a.* silky; **n. f.** milkweed (*Asclepias* sp.)
spadice *m.* spadix
spadicé,-e *a.* spadiceous
spadiciflore *a.* spadicifloral
sparganier *m. Sparganium* sp., including *S. racemosum* bur-reed
spargoule, spargoute *f.* spurrey (*Spergula* sp.) (= **spergule**)
sparmannia *m.* African hemp (*Sparmannia africana*) (= **tilleul d'appartement**)
sparte *m.* albardine (*Lygeum spartum*) (see also **alfa**)
spartine *f. Spartina* sp.; **s. maritime** cordgrass (*S. maritima*)
spathe *f.* spathe
spatule *f.* spatula
spatulé,-e *a.* spatulate
spécificité *f.* specificity
spécifique *a.* specific
spectre *m.* spectrum; **s. d'activité** spectrum of activity
spectrographie *f.* spectrography; **s. de masse** mass spectrography
spectrométrie *f.* spectrometry; **s. de masse à étincelles** spark source mass spectrometry
spectrophotométrie *f.* spectrophotometry; **s. infra-rouge** infra-red spectrophotometry
spectroscopie *f.* spectroscopy; **s. d'absorption** absorption spectroscopy; **s. infrarouge** infrared spectroscopy
spéculaire *f. Specularia* sp.; **s. miroir** Venus's looking glass (*S. speculum*) (= **miroir de Vénus**)
spergulaire *f.*, **spergularia** *m. Spergularia* sp., spurrey; **s. rouge** red spurrey (*S. rubra*)
spergule *f.* spurrey (*Spergula* sp.) (= **spargoule**); **s. des champs** corn spurrey (*S. arvensis*)
spermoderme *m.* spermoderm, seed coat, testa
spermophile *m.* suslik, ground squirrel (*Citellus* sp.)
SPG *m.* = **sujet porte-greffe** rootstock
sphagnum *m.*, **sphaigne** *f.* sphagnum, peat moss
sphinx *m. Ent.* sphinx moth, sphingid; **s. du laurier-rose** oleander sphinx (*Daphnis nerii*); **s. tête de mort** death's head hawk moth (*Acherontia atropos*); **s. du troène** privet hawk moth (*Sphinx ligustri*); **s. de la vigne** grape sphinx moth (*Deilephila elpenor*)
spiciflore *a.* spiciflorous
spiciforme *a.* spiciform, shaped like a spike
spigélie *f. Spigelia* sp.; **s. anthelmia** Demerara pink root (*S. anthelmia*); **s. de Maryland** Carolina pink, Maryland pink root (*S. marilandica*)
spilanthe *m. Spilanthes* sp., including *S. oleracea* (see also **cresson**)
spinescence *f.* spinescence
spinescent,-e *a.* spinescent
spinule *f.* small spine, spinule
spiquenard *m.* spikenard (*Nardostachys jatamansi*)
spiral,-e *a.* spiral; *n. f.* spiral; **s. entière** *Rubb.* full spiral
spiralé,-e *a.* spiral
spiranthe *m. Spiranthes* sp., lady's tresses
spirée *f. Spiraea* sp.; **s. arbustive** tree spiraea; **s. du Japon** (1) = **kerria** (*q.v.*), (2) *S. japonica*; **s. printanière** *S.* × *arguta, S.* × *vanhouttei* (see also **filipendula, reine des prés**)
spiroplasme *m.* spiroplasma
spiruline *f. Spirulina* sp., blue-green alga, including *S. platensis* and *S. geitleri*
spondias *m. Spondias* sp. (see also **pommier**)
spongieuse *f.* gypsy moth (*Lymantria dispar*)
spongieux,-euse *a.* spongy (see also **tache**)
spongiole *f.* spongiole
spontané,-e *a.* (of plants) self-sown, volunteer, (by extension) wild, indigenous; **arbre spontané** seedling tree
spontanéité *f.* spontaneity, indigenousness (of plant)
sporange *m.* sporangium
spore *f.* spore
sporocarpe *m.* sporocarp
sporogenèse *f.* sporogenesis
sporophyte *m.* sporophyte
sport *m. PB* sport, mutation
sporulation *f.* sporulation
spray *m.* spray (sometimes sprayer)
sprékélie *f. Sprekelia* sp., Jacobean lily
squameux,-euse *a.* squamous
squamiforme *a.* squamiform, scale-like
squamule *f.* squamule

squelettique *a.* skeletal
squelettisation *f.* skeletonizing
squelettiser *v.t.* to skeletonize (leaves etc.)
stabilisant *m.* stabilizer
stabilisateur,-trice *a.* stabilizing; **milieu stabilisateur** stabilizing medium; *n. m.* stabilizer
stabilisation *f.* stabilization
stabilité *f.* stability; **s. structurale** *Ped.* structural stability
stachys *m.*, **stachyde** *f.* *Stachys* sp.; **s. tubéreux** = **crosne** (*q.v.*) (see also **épiaire**)
stade *m.* stage; **s. de croissance, s. de développement** stage of growth; **s. 2 feuilles** two-leaf stage; **stades de la germination** stages of germination; **s. juvénil** juvenile phase; **s. de maturité** stage of maturity; **s. repère** stage for reference
stagnation *f.* stagnancy (of water)
stagner *v.i.* to stagnate (of water)
staminé,-e *a.* staminate
stamineux,-euse *a.* stamineous, stamineal
staminifère *a.* staminiferous
staminode *m.* staminode
standardisation *f.* standardization
stangerie *f.* *Stangeria* sp., including *S. paradoxa*
stanhopée *f.* *Stanhopea* sp. (orchid)
stapélia *m.*, **stapélie** *f.* *Stapelia* sp., carrion flower; **s. géante** *S. gigantea*
staphisaigre *f.* stavisacre (*Delphinium staphisagria*)
staphylier *m.* *Staphylea* sp., bladder nut; **s. penné, s. pinné** St. Anthony's nut, bladder nut (*S. pinnata*) (= **faux pistachier**)
statenchyme *m.* statenchyma
statice *m.* *Statice* sp., sea lavender (*Limonium* sp.), thrift (*Armeria* sp.)
station *f.* station; **s. horticole** horticultural research station; **s. de pompage** pumping station
statistique *a.* statistical; *n. f.* statistics
statocyte *m.* statocyst
stauntonie *f.* *Stauntonia* sp., ornamental evergreen climbing shrubs
stèle *f.* stele
stellaire *f.* *Stellaria* sp., chickweed
sténie *f.* *Stenia* sp. (orchid)
sténocarpe *m.* *Stenocarpus* sp., including the fire tree (*S. sinuatus*)
sténohalin,-e *a.* stenohaline
stéphanotis *m.* *Stephanotis* sp., Madagascar jasmine
sterculier *m.*, **sterculie** *f.* *Sterculia* sp., including *S. urens* Kataya gum
stérigmate *m.* *Fung.* sterigma, *pl.* sterigmata
stérile *a.* sterile; *PB* **s. femelle** female sterile; **s. mâle** male sterile
stérilisation *f.* sterilization; **s. à la vapeur** steam sterilization
stérilisé,-e *a.* sterilized
stériliser *v.t.* to sterilize, to disinfect
stérilité *f.* sterility; **s. mâle** male sterility
sternbergie *f.* *Sternbergia* sp., including *S. lutea* winter daffodil (= **narcisse d'automne**)
sternutatoire *a.* & *n.m.* sternutatory, snernutative
stéroïde *m.* steroid
stéroïdique *a.* relating to steroids
stérol *m.* sterol
stérolique *a.* sterol; **fraction stérolique** sterol fraction
stigmate *m.* *Bot.* stigma
stigmatique *a.* stigmatic
stigmule *m.* stigmula
stigomose *f.* *PC* foliage mottling caused by insect punctures
stillingie *f.* *Stillingia* sp. (see also **suif végétal**)
stimulant-e *a.* stimulating; *n. m.* stimulant
stimule *m.* stinging hair
stipa *m.*, **stipe** *f.* *Stipa* sp., feather grass
stipe *m.* stipe, culm, trunk (of palm)
stipelle *f.* stipel
stipellé,-e *a.* stipellate
stipité,-e *a.* stipitate, stalked
stipulaire *a.* stipular
stipule *f.* stipule
stipulé,-e *a.* stipulate
stock *m.* stock; **s. de graines** seed bank
stockage *m.* storage
stocker *v.t.* to stock, to store
stolbur *m.* *PD* stolbur (virus)
stolon *m.*, **stolone** *f.* stolon, runner (strawberry)
stolonifère *a.* stoloniferous
stolonnage *m.* stolon, runner formation (in strawberry etc.)
stomachique *a.* & *n.m.* stomachic
stomate *m.* stoma; *pl.* stomata
stomatique *a.* stomatal; **fréquence stomatique** stomatal frequency
store *m.* (window) blind; **s. (de serre)** greenhouse shade
stramoine *f.*, **stramonium** *m.* thorn apple (*Datura stramonium*) (= **pomme épineuse**); **s. en arbre** angel's trumpet (*D. arborea*); **s. odorante** angel's trumpet (*D. suaveolens*); **s. sanguine** *D. sanguinea*
strangulation *f.* strangulation
stranvésia *m.* *Stranvaesia* sp., including *S. davidiana*
strate *f.* stratum, layer; **s. végétale** plant layer
stratification *f.* *Hort.* stratification
stratifié,-e *a.* *Hort.* stratified; *n. m.* laminated plastic etc.
stratifier *v.t.* *Hort.* to stratify
stratiote *f.* *Stratiotes* sp.; **s. faux-aloès** water soldier (*S. aloides*)
strélitzie *f.*, **strélitzia** *m.* *Strelitzia* sp., bird of

Paradise flower (= **oiseau de paradis**); **s. à feuilles de jonc** *S. parvifolia* var. *juncea*
streptocarpe *m.* *Streptocarpus* sp., cape primrose
stress *m.* stress; **s. salin** saline stress
strie *f.* 1. score, scratch, streak (of colour); **stries sur (le) bois** stem pitting; **strie(s) chlorotique(s)** *Sug.* chlorotic streak (virus); **maladie des stries rouges** *Sug.* red stripe (*Xanthomonas rubrilineans*). 2. rib, ridge
strié,-e *a.* 1. striated, striped; **bois strié** *PD* stem pitting. 2. grooved
strobile *m.* strobile
strobiliforme *a.* strobiliform, cone-shaped
stroma *m.* stroma
strophante, strophanthus *m.* *Strophanthus* sp., including *S. hispidus*, source of strophanthin
strophiole *m.* strophiole
structural,-e *a., m.pl.* **-aux** structural; **porosité structurale** *Ped.* structural porosity
structure *f.* structure; **s. dégradée (du sol)** *Ped.* soil structure breakdown, deterioration; **s. granulaire** *Ped.* granular structure; **s. grumeleuse** *Ped.* crumb structure
structuré,-e *a.* structured
strychnine *f.* strychnine
strychnos *m.* *Strychnos* sp.; **s. vomiquier** strychnine tree (*S. nux-vomica*)
stubborn *m.* *PD, Cit.* stubborn (*Spiroplasma citri*)
style *m.* *Bot.* style
stylet *m.* stylet
styrax *m.* *Styrax* sp.; **s. benzoin** benzoin laurel (*S. benzoin*); **s. officinal** styrax (*S. officinalis*) (= **aliboufier**)
subalpin,-e *a.* subalpine
sub-carence *f.* sub-deficiency
subcaulescent,-e *a.* subcaulescent
suber *m.* suber, cork
subéreux,-euse *a.* suberose, suberous
subériculture *f.* cork-oak growing
subérification, subérisation *f.* suberisation
subérine *f.* suberin
subérisation *f.* suberisation
submergé,-e *a.* submerged
submersion *f.* submersion; *Hort.* 1. irrigation. 2. inundation to destroy pests (e.g. phylloxera in vines)
subraclette *f.* (see **sarcleur** 2)
subspontané,-e *a.* subspontaneous
substance *f.* substance; **s. de croissance** growth substance
substrat *m.* substrate
subterminal,-e *a., m. pl.* **-aux** subterminal
subtropical,-e *a., m. pl.* **-aux** subtropical
subulé,-e *a.* subulate, awl-shaped
suc *m.* juice; *Bot.* sap; **s. cellulaire** cell sap
succédané,-e *a.* & *n. m.* substitute
succession *f.* succession; **s. végétale** plant succession
succion *f.* suction
succulence *f.* succulence
succulent,-e *a.* succulent; **plante succulente** succulent
suceur,-euse *a.* sucking, suctorial; *n. m.* sucking insect; **insectes suceurs** sucking insects
suçoir *m.* sucking organ; *Ent.* sucker; (*Coconut*) *OP* etc. haustorium (= **haustorium**)
sucre *m.* sugar; **s. artisanal** sugar produced by local industry; **s. brut** jaggery, gur; **s. de canne** cane sugar; **s. raffiné** refined sugar; **s. réducteur** reducing sugar
sucré,-e *a.* sugared, sweet
sucrerie *f.* *Sug.* sugar factory
sucrier,-ière *a.* relating to sugar; **l'industrie sucrière** the sugar industry
sudation *f.* sudation
sudorifique *a.* sudorific
suffrutescent,-e *a.* suffrutescent (= **sous-frutescent**)
sugi *m.* Japanese cedar (*Cryptomeria japonica*)
suie *f.* soot; *PD* sooty mould (= **fumagine**)
suif (*m.*) **végétal** vegetable tallow; **arbre à suif** Chinese tallow tree (*Sapium (Stillingia) sebiferum*)
suintant,-e *a.* oozing, dripping (of rocks, walls etc.)
sujet *m.* *Hort.* stock, rootstock (= **porte-greffe**); **s. intermédiaire** intermediate stock, stem-builder (see also **SPG**); **s. de semis** seedling rootstock
sulfatage *m.* *Vit.* treating (vines) with Cu sulphate or Fe sulphate
sulfate *m.* sulphate; **s. d'ammoniaque, s. d'ammonium** ammonium sulphate, sulphate of ammonia; **s. de cuivre** copper sulphate; **s. ferreux** ferrous sulphate; **s. de magnésie** magnesium sulphate; **s. de potasse** sulphate of potash
sulfater *v.t.* *Vit.* to treat (vines) with Cu or Fe sulphate
sulfhydrique *a.*, **acide sulfhydrique** hydrogen sulphide, H_2S
sulfocalcique *a.*, **bouillie sulfocalcique** lime sulphur spray
sulfurage *m.* *Vit.* soil treatment with carbon disulphide
sulfure *m.* sulphide; **s. de carbone** carbon disulphide
sulfurer *v.t.* *Vit.* to treat soil with carbon disulphide
sulfurique *a.* sulphuric
sulla *m.* French honeysuckle (*Hedysarum coronarium*)
sumac *m.* sumac (*Rhus* sp.); **s. des corroyeurs** Sicilian sumac (*R. coriaria*); **s. fustet** smoke

tree (*R. cotinus*); **s. odorant** *R. aromatica*; **s. vénéneux** poison oak (*R. toxicodendron*), poison ivy (*R. radicans*); **s. de Virginie** stag's horn sumac (*R. typhina*) (see also **vinaigrier**)
super *m.* = **superphosphate** (*q.v.*)
supère *a. Bot.* superior; **ovaire supère** superior ovary
superficie *f.* surface; area
superficiel,-elle *a.* superficial
supermarché *m.* supermarket
superphosphate *m.* superphosphate; **s. concentré** concentrated superphosphate; **s. simple** single superphosphate
supplément *m.* supplement, addition; **s. de rendement** increase in yield
supplémentaire *a.* supplementary, additional; **frais supplémentaires** extra charges
support *m.* support, prop; *PC* carrier
suppression *f.* suppression, removal; **s. des drageons** desuckering; **s. (partielle) des fleurs** deblossoming; **s. (partielle) des fruits** defruiting
supprimer *v.t.* to suppress, to remove
sur *prep.* on; **sur pied** *Hort.* etc. on the plant and, by extension, in the field
sur,-e *a.* acid, sour, tart; **pomme sure** sour apple
surabondance *f.* superabundance, glut
surabondant,-e *a.* superabundant, superfluous
surchaulage *m.* overliming
surcroissance *f.* excessive growth, overgrowth
surdosage *m.* overdose
sureau *m.*, **s. (noir)** elder (*Sambucus nigra*); **baie de sureau** elderberry; **s. doré** golden elder(-berry) (*S. (nigra)aurea*); **s. à grappes** scarlet-berried elder (*S. racemosa*); **s. hièble, petit sureau** danewort, dwarf elder (*S. ebulus*) (see also **yèble**); **s. rameux** = **s. à grappes** (*q.v.*)
surélever *v.t.* to heighten, to raise
surelle, surette *f.* sheep's sorrel (*Rumex acetosella*); wood sorrel (*Oxalis acetosella*); upright yellow sorrel (*O. stricta*)
suret,-ette *a.* slightly sour
suretière *f.* apple tree nursery
surette *f.* 1. see **suret**. 2. = **surelle** (*q.v.*). 3. Otaheite gooseberry (*Phyllanthus acidus*) (= **cerisier de Tahiti**)
surfaçage *m. Hort.* replacing the surface soil (of a potted plant), adding fresh soil (to a potted plant)
surface *f.* surface; **s. foliaire** leaf surface; **s. vitrée** glazed surface
surfactant *m.* surfactant
surfermenté,-e *a.* overfermented
surfondre *v.t.*, *v.i.* to supercool
surfusion *f.* supercooling
surgélation *f.* **(rapide)** deep freezing, quick freezing
surgeon *m.* sucker, offshoot; **pousser des surgeons** (of plant) = **surgeonner** (*q.v.*)
surgeonner *v.i.* to sucker, to grow offshoots
surgreffage *m.* double-grafting, top grafting, topworking
surgreffe *f.* double-graft, top graft
surgreffer *v.t.* to double-graft, to top-graft, to topwork, to regraft
surin *m.* young ungrafted apple tree
surmaturation *f. Vit.* overripening
surmulot *m.* brown rat (*Rattus norvegicus*)
surnageant,-e *a.* supernatant; *n. m.* supernatant (liquid)
surnombre *m.* number above that required, excess; **en surnombre** in excess
surpeau *f. Bot.* epidermis
sursaturation *f.* supersaturation; **s. en eau** waterlogging
sursemer *v.t.* to sow again (land already sown), to re-sow, to oversow
survie *f.* survival; **s. dans le sol** survival in the soil
susceptibilité *f.* susceptibility
suspenseur *m.* suspensor
suspension *f.* suspension; **s. de cellules, s. cellulaire** cell suspension
suture *f. Bot.* suture; **s. ventrale** ventral suture
swertie *f. Swertia* sp., including marsh felwort (*S. perennis*)
swiétėnie *f. Swietenia* sp., including mahogany (*S. mahagoni*)
swollen shoot *m.* (*Cacao*), *PD* swollen shoot (virus)
sycomore *m.* **(érable) sycomore** sycamore (*Acer pseudoplatanus*)
sycone *m.* syconium, syconus
sylvestre *a.* sylvan, growing in woods
sylvicole *a.* 1. silvicolous. 2. relating to arboriculture; **pépinière sylvicole** tree nursery
sylviculteur *m.* sylviculturist
sylviculture *f.* sylviculture
sylvie *f. Bot.* wood anemone (*Anemone nemorosa*) (= **anémone des bois**)
symbiose *f.* symbiosis
symbiote *m.* symbiont; **s. mycorhizien** mycorrhizal symbiont
symbiotique *a.* symbiotic
symétrique *a.* symmetrical; **fleur symétrique** symmetrical flower
sympatrie *f.* sympatry
sympatrique *a.* sympatric
symphorine *f. Symphoricarpos* sp.; **s. boule-de-cire, s. boule-de-neige, s. à fruits blancs, s. à grappes** snowberry (*S. racemosus*)
symplaste *m.* symplast
symploque *m. Symplocos* sp., including sweet-leaf. (*S. tinctoria*)
sympode *m.* sympodium

sympodique *a.* sympodial
symptomatologie *f.* symptomatology
symptôme *m.* symptom; **s. de carence** deficiency symptom
synanthé,-e *a.* synanthous
synanthéré,-e *a.* synantherous
syncarpe *m.* syncarp
syncarpé,-e *a.* syncarpous; **fruit syncarpé** = **syncarpe** (*q.v.*)
synchrone *a.* synchronous
syndrome *m.* syndrome
synécologie *f.* synecology
synergides *f. pl. Bot.* synergidae
synergie *f.* synergy
synergique *a.* synergic, synergistic
synergisme *m.* synergism
syngonium *m. Syngonium* sp., including *S. auritum*, five fingers
synonymie *f.* synonymy
synthèse *f.* synthesis
syrphe *m.* syrphid, hover-fly (Syrphidae)
systématique *f.* systematics
système *m.* system; **s. aérien** aerial parts (of plant); **s. d'avertissements horticoles** horticultural warning system; **s. de culture** land use system; **s. d'irrigation** irrigation system; **s. de plantation** planting system; **s. racinaire, s. radiculaire** root system; **s. de taille** pruning system
systémique *a.* systemic; *n. m. pl.* systemic pesticides

T

tabac *m.* tobacco plant (*Nicotiana* sp.); **t. blanc odorant** *N. alata* var. *grandiflora*; **t. des Vosges** mountain tobacco (*Arnica montana*); **t. de cape** wrapper tobacco; **t. de cigare** cigar tobacco; **t. flue-cured** flue-cured tobacco; **t. industriel** ommercial tobacco; **t. d'Orient** Turkish tobacco

tabacole *a.* relating to tobacco

tabaculture *f.* tobacco growing

tabernamontane *m.* *Tabernaemontana* sp., including forbidden fruit (*T. dichotoma*)

table *f.* table; **t. de données** data table; **t. à empoter** potting table; **t. de cueillette** (*Tea*) plucking table

tableau *m.* list, table; **t. de mélange** table of compatibilities (of fertilizers, pesticides etc.)

tablette *f.* greenhouse bench, staging, shelf; *Mush.* shelf for growing

tablier *m.* apron; **t. de la feuille** *Vit.* auricle

tabouret (*m.*) **des champs** field pennycress (*Thlaspi arvense*); **t. perfolié** perfoliate pennycress (*T. perfoliatum*)

tacaco *m.* tacaco (*Polakowskia tacaco*), a neotropical creeping plant whose fruits are eaten cooked

tacca *m.* *Tacca* sp.

tache *f.* stain, spot; *PD* (often in *pl.*) blotch, spot; **en tache d'huile** in concentric circles; **t. amère** bitter pit; **taches angulaires du concombre** cucumber angular leaf spot (*Pseudomonas lachrymans*); **tache(s) annulaire(s)** ring spot; **taches annulaires cendrées du coing** sooty ring spot virus (of quince); **taches annulaires du cerisier** cherry ring mottle virus; **taches annulaires du chou** see **taches noires du chou**; **taches annulaires (de Henderson)** (apple) ring spot virus; **taches annulaires nécrotiques du cerisier, des prunus** prunus necrotic ring spot virus; **taches bactériennes** *Mush.* brown blotch (*Pseudomonas tolaasi*); **taches blanches du fraisier** strawberry leaf spot (*Mycosphaerella fragariae*); **taches bronzées de la tomate** tomato spotted wilt virus; **taches brunes** *Mush.* brown blotch (*Verticillium malthousei*); **taches brunes de la laitue** lettuce ring spot (*Marssonina panattoniana*) (= **anthracnose**); **taches chlorotiques** apple chlorotic leaf spot virus; **t. du collet** collar rot; **taches en couronne** spraing (of potato); **taches étoilées (asteroïdes) du pêcher** peach asteroid spot virus; **taches foliaires** leaf spot (caused by *Alternaria, Septoria, Phyllosticta* spp. etc.); **taches foliaires du chou** cabbage ring spot (*Mycosphaerella brassicicola*); **taches foliaires du glaïeul** botrytis rot and core rot (*Botrytis gladiolorum*); **taches foliaires de la laitue** lettuce leaf spot (*Pleospora herbarum*); **taches foliaires du marronnier d'Inde** horse-chestnut leaf blotch (*Guignardia aesculi*); **taches foliaires (taches des feuilles) du poirier et du cognassier** *Stigmatea mespili* (= *Entomosporium maculatum*); **taches foliaires du rosier** rose leaf scorch (*Sphaerulina rehmiana*) (*Phyllosticta rosarum* = *Sphaceloma rosarum*); **taches des fruits et des feuilles (du cerisier)** (cherry) shot-hole disease (*Stigmina carpophila* = *Coryneum beijerinckii*) (= **criblure**); **tache(s) graisseuse(s)** *Cit.* greasy spot, black melanose (*Mycosphaerella horii*); **tache d'huile** *Cit.* oil spot; *Vit.* oil spot (early symptom of downy mildew); **taches liégeuses** bitter pit (of apple) (= **tache(s) spongieuse(s)**; **taches noires de l'ananas** pineapple fruitlet core rot (*Penicillium funiculosum*); **taches noires du chou** cabbage black ring spot virus; **taches noires (de l'ellébore)** hellebore leaf spot (*Coniothyrium hellebori*); **taches noires du framboisier** raspberry spur blight (*Didymella applanata*); **taches noires du lilas** bacterial blight of lilac (*Pseudomonas syringae*); **taches noires du pois** pea leaf and pod spot (*Ascochyta pisi*); **taches noires du rosier** rose black spot (*Diplocarpon rosae*) (= **rond noir des feuilles**); **tache phylloxérique** *Vit.* patch attacked by phylloxera; **taches pourpres (du fraisier)** strawberry leaf spot (*Mycosphaerella fragariae*); **taches rouges du concombre** cucumber blotch (*Corynespora cassiicola* = *C. melonis*); **tache(s) spongieuse(s) de la pomme** apple bitter pit (= **taches liégeuses**); **taches sur tige (du rosier)** rose canker and dieback (*Griphosphaeria corticola*)

taché,-e *a.* spotted; **fruit taché** bruised fruit

tacheture *f.* spots, speckle, mottle; *PD* **t. chlorotique** chlorotic flecks; **t. de la peau** Jonathan spot

tachinaire *m.* sometimes *f.* tachinid fly (Tachinidae)
tactisme *m.* tactism
tagetes, tagète, tagette *m.* *Tagetes* sp., including striped Mexican marigold (*T. tenuifolia*), sweet-scented Mexican marigold (*T. lucida*) (see also **oeillet d'Inde, rose d'Inde**)
taille *f.* 1. *Hort.* pruning, trimming, clipping, training; **t. en chaintre** *Vit.* training to a creeping habit; **t. en cordon** cordon pruning; **t. à coursons, ta. à coursonnes** spur pruning; **t. courte** *Vit.* short pruning; **t. en créneaux alternés** alternate crenellated pruning; **t. en crochet** fan training (e.g. for peach); **t. de dégagement** *Coff.* centering; **t. d'entretien** maintenance pruning; **t. en espalier** espalier pruning; **t. d'été** summer pruning (= **t. en vert**); **t. en éventail** palmette pruning, fan pruning (= **t. en palmette**); **t. de formation** training, preliminary pruning; **t. en formation à plusieurs étages** *Coff.* capping; **t. de fructification** regular pruning, routine pruning (= **t. de production**); **t. en gobelet** bush pruning, goblet pruning; **t. en gobelet élevé** *Vit.* head pruning; **t. en haie** hedge pruning; **t. d'hiver** winter pruning (= **t. en sec, t. ligneuse**); **t. ligneuse** = **t. d'hiver** (*q.v.*); **t. longue** *Vit.* long pruning; **t. mixte** *Vit.* cane and spur pruning; **t. d'Oeschberg** Oeschberg system; **t. ornementale des arbres** topiary work; **t. en palmette** palmette, fan pruning (= **t. en éventail**); **t. préparatoire** preliminary pruning; **t. de production** = **t. de fructification** (*q.v.*); **t. des prolongements** leader pruning; **t. en quenouille** *Vit.* cone pruning; **t. des racines** root pruning; **t. de rajeunissement** rejuvenation pruning; **t. par rapprochement** pruning by shortening the branches; **t. de reconstitution** remedial pruning after accidental damage (see also **t. de renforcement, t. de restauration**); **t. de régénération** regeneration pruning; (*Tea*) medium/hard pruning; **t. de renforcement** remedial pruning (see also **t. de reconstitution, t. de restauration**); **t. de renouvellement** renewal (pruning) system; **t. de restauration** remedial pruning (see also **t. de reconstitution, t. de renforcement**); **t. riche** *Vit.* light pruning; **t. sèche, t. en sec** = **t. d'hiver** (*q.v.*); **t. sévère** heavy pruning; **t. en tête de saule** *Vit.* short head pruning; **t. à une tige** *Coff.* single-stem pruning; **t. en vert** summer pruning, green pruning (= **t. d'été**) (*Tea*) skiffing. 2. size; **t. des chromosomes** chromosome size
taille-bordures *m. inv.* (lawn) edger
taille-haies *m. inv.* hedge cutter, trimmer
taille-herbes *m. inv.* grass cutter
taille-legumes *m. inv.* vegetable cutter, slicer
tailler *v.t.* to prune, to trim, to clip, to train; **"taillez tôt ou taillez tard, rien ne vaut la taille de mars"** "prune early or prune late, pruning in March is best"
tailleur *m. Hort.* pruner, trimmer, clipper, trainer
tailleuse de haie *f.* hedge cutter, hedge trimmer
taillis *m.* copse, coppice; **bois taillis** underwood, brushwood
taka *m.* (*Date*) date mite (*Oligonychus afrasiaticus*) and the damage it causes
talc *m.* talc
talé,-e *a.* bruised (of fruit etc.); **zone talée** bruised zone (of fruit)
talin *m.*, **taline** *f.* *Talinum* sp., including waterleaf (*T. triangulare*) (see **grassé**)
tallage *m.* 1. tillering, suckering; *Sug.* etc. stooling. 2. tillers, suckers
talle *f.* tiller, sucker
taller *v.i.* to tiller, to sucker, to stool
tallipot, talipot *m.* talipot palm (*Corypha umbraculifera*)
talon *m. Hort.* heel (of cutting); **bouture à talon** heel cutting
talonnage *m. Hort.* heeling-in (of plant)
talure *f.* bruise, graze on fruit or branch; **t. de l'écorce** graze on the bark
talus *m.* 1. slope. 2. bank, embankment; **t. gazonné** turfed bank
talweg *m.* thalweg
tamano *m.* Alexandrian laurel (*Calophyllum inophyllum*)
tamarin *m.* tamarind fruit and pulp; tamarind tree
tamarinier *m.* tamarind tree (*Tamarindus indica*)
tamaris, tamarisc, tamarix *m.* *Tamarix* sp.; **t. de France** French tamarisk (*T. gallica*); **t. de printemps** spring-flowering tamarisk (*T. tetrandra*)
tambour *m.* drum (for hose etc.)
tamier, taminier *m.* black bryony (*Tamus communis*)
tamis *m.* sieve, riddle
tamisage *m.* sieving, riddling
tamisé,-e *a.* sieved, riddled, subdued (of light); **ensoleillement tamisé par la végétation** sunlight filtered by vegetation; **terre tamisée** sifted soil
tamiser *v.t.* to sieve, to screen, to strain, to filter
tamiseur,-euse *n.* sifter, screener, strainer, filter
tampico *m.* tampico fibre, ixtle fibre (from *Agave falcata*)
tampon *m.* buffer
tan *a. inv.* tan, tan-coloured; *n.m.* tan, (tanner's) bark; second seed-coat of chestnut fruit

tanacetum *m. Tanacetum* sp.
tanaisie *f.* tansy (*Tanacetum vulgare*); **t. balsamite** costmary (*Chysanthemum balsamita*) (see also **baume**)
tangelo *m. Cit.* tangelo (*Citrus reticulata* × *C. paradisi*)
tangent,-e *a.* tangential, tangent
tanghin, tanghen *m.* tanghin poison or tree (*Tanghinia venenifera*)
tanghinia *m. Tanghinia* sp.
tangue *f.* calcareous sea-mud from Channel coast used as fertilizer
tanin, tannin *m.* tannin
tannée *f.* tan, tanner's waste; **couche de tannée** bark bed
tapioca *m.* tapioca
tapis *m.* carpet; **t. de fleurs** flower carpet; **t. de gazon** sward; **t. à godets** bucket chain; **t. magique** fig marigold (*Mesembryanthemum pyropaeum*); **t. roulant** endless belt, conveyor belt
tapissant,-e *a.* (of plant) forming a carpet, sward
tapissé,-e *a.* carpeted by (of bank, soil etc.), covered by (of wall etc.); **t. de fleurs** carpeted with flowers
tarare *m.* winnowing machine
taravelle *f. Vit.* foot planter
taraxacum *m. Taraxacum* sp., dandelion
tardif,-ive *a.* late; **floraison tardive** late flowering
tardiflore *a.* late-flowering
tardiveté *f.* lateness
tare *f.* depreciation, loss in value (due to damage or waste); defect, taint
taré,-e *a.* spoilt, tainted, damaged
tarière *f.* auger, borer; *Ent.* ovipositor
taro *m.* taro, dasheen, 'old' cocoyam (*Colocasia antiquorum*)
tarsonème *m.* mite (*Steneotarsonemus* sp., including *S. laticeps*, bulb scale mite), **t. du fraisier** strawberry mite (*S. pallidus*)
tas *m.* heap, pile; **t. de fumier** manure heap
tas-piège *m., pl.* **tas-pièges** heap of organic matter as a trap for *Oryctes rhinoceros* or other pest
tasse *f.* cup; **qualité à la tasse** *Coff.* etc. cup quality
tassement *m.* compressing, ramming, settling, consolidating; **t. de terre battante** compaction of capping soil
tasser *v.t.* to compress, to ram, to firm; *v.i.* to grow thickly (of plants); **se tasser** to settle, to subside, to become consolidated
taupe *f.* mole (*Talpa* sp.); **stade "taupe" (du haricot** etc.) pre-emergence stage (of bean etc.) when cotyledons are just appearing above ground
taupe-grillon *m., pl.* **taupes-grillons** mole cricket (*Gryllotalpa gryllotalpa*) (= **courtilière**)
taupicide *m.* mole poison
taupière *f.* mole trap
taupin *m. Ent.* click-beetle, elaterid (Elateridae); **larve de taupin** wireworm
taupinière, taupinée *f.* molehill
taux *m.* price, rate, proportion, content; **t. de fémininité** *Ent.* sex ratio; **t. d'humidité** moisture content; **t. humique** humus content; **t. net d'assimilation** net assimilation rate
tavelage *m.* spotting, speckling of fruit; *PD* scab (see **tavelure**)
tavelé,-e *a.* spotted, speckled; **fruit tavelé** speckled fruit
taveler *v.t.* to spot, to speckle; **se taveler** to become spotted, speckled (of fruit)
tavelure *f. PD* scab; **t. du poirier** pear scab (*Venturia pirina*); **t. du pommier** apple scab (*V. inaequalis*)
taxon *m.* taxon
taxonomie, taxinomie *f.* taxonomy
taxus *m. Taxus* sp. (see also **if**)
tchernozem *m. Ped.* chernozem
technicité *f.* technicalness, technicality
technique *f.* technique; **t. culturale** growing technique
technologie *f.* technology; **t. végétale** plant technology
technologique *a.* technological
técoma, técome *m. Tecoma (Campsis)* sp.; **t. à grandes fleurs** *C. grandiflora*; **t. jasmin de Virginie** trumpet creeper (*Campsis radicans*) (= **jasmin de Virginie**)
tegmen *m. Bot.* tegmen
tegminé,-e *a. Bot.* provided with a tegmen
tégument *m.* tegument, integument; **t. séminal** seed-coat
tégumentaire *a.* tegumental, tegumentary
teigne *f.* 1. tineid moth (Tineidae); **t. de azalées** azalea leaf miner (*Caloptilia (Gracillaria) azaleella*); **t. du cassis** currant shoot borer (*Lampronia capittella*); **t. du cerisier** cherry fruit moth (*Argyresthia curvella* = *A. nitidella*), *A. ephippella*; **t. du chou** = **t. des crucifères** (*q.v.*); **t. des citronniers** = **t. des fleurs** (*q.v.*); **t. des crucifères** diamond-back moth (*Plutella maculipennis* = *P. xylostella*); **t. des feuilles de pommier** apple leaf skeletonizer (*Eutromula (Simaethis) pariana*); **t. des fleurs** *Cit.* citrus flower moth (*Prays citri*); **t. des fleurs du cerisier** cherry flower moth (*Argyresthia ephippella*); **t. du groseilier** = **t. du cassis** (*q.v.*); **t. des lilas** lilac leaf miner (*Caloptilia (Gracillaria) syringella*); **t. du noyer** walnut moth (*Gracillaria juglandella*); **t. de l'olivier** olive moth (*Prays oleellus*); **t. du panais** parsnip moth (*Depressaria heracliana* = *D. pastinacella*); **t. du**

pêcher peach twig borer (*Anarsia lineatella*); **t. du poireau** leek moth (*Acrolepiopsis assectella*); **t. du pois** pea moth (*Laspeyresia nigricana*) (= **tordeuse du pois**); **t. des pommes, t. du pommier** apple fruit moth, apple fruit miner (larva) (*Argyresthia conjugella*), lesser bud moth (*Recurvaria nanella* and *R. leucatella*); **t. du prunier** small ermine moth (*Yponomeuta padella*); **t. de la vigne** grape moth (*Clysiana (Clysia) ambiguella*). 2. *Coll.* dodder (*Cuscuta* sp.); bur of burdock (*Arctium* sp.)

teinturier *m.* Virginian poke week (*Phytolacca decandra*) (= **raisin d'Amérique**); *Vit.* teinturier grape cultivar

télédétection *f.* remote sensing

télékie *f.*, **telekia** *m.* *Telekia (Buphthalmum)* sp., yellow oxeye

telfairia *m.* *Telfairia* sp., including *T. occidentalis*, oyster nut vine

tellurique *a.* telluric, or from the soil; **champignon tellurique** soil fungus; **faune tellurique** soil fauna; **maladie d'origine tellurique** *PD* soil-borne disease

télophase *f.* telophase

témoin *m.* *Stat.* control, (*USA*) check; **t. biné** hoed control (plot); **t. non traité** untreated control (plot); **plantes témoins** control plants

température *f.* temperature; **températures cumulées** accumulated temperatures; **t. ambiante** ambient temperature; **t. de conservation** storage temperature; **t. maximum** maximum temperature; **t. minimum** minimum temperature; **t. seuil** threshold temperature

temporaire *m.* *Hort.* filler (in orchard) (= **arbre temporaire**)

temporisation *f.* *AM* timing, fitting with a time-delay system

ténacité *f.* *PC* adhesiveness, stickiness (of pesticide)

tendance *f.* tendency, trend

tendéromètre *m.* tenderometer

tendérométrie *f.*, **indice de tendérométrie** tenderometer index

tendérométrique *a.*, **indice tendérométrique** tenderometer index

tendeur *m.* wire strainer

tendre *a.* 1. tender, soft, not hardy, not hardened (of plant). 2. *v.t.* to stretch, to tighten

tendreté *f.* tenderness (of vegetables etc.)

tendrille *f.* tender shoot (see also **vrille**)

tendron *m.* *Bot.* tender shoot

tendu,-e *a.* stretched, taut, tight

ténébrion *m.* *Ent.* tenebrionid

teneur *f.* amount, content; **t. en chaux** lime content; **t. en eau** water, moisture content; **teneurs en éléments nutritifs des feuilles** leaf nutrient contents; **t. en humus** humus content; **t. en sel** salt content; **t. en sucre** sugar content

tensimètre, tensiomètre *m.* tensiometer

tensio-actif,-ive *a.* surface active; *n.m.* surfactant

tension *f.* tension; **t. d'eau, t. hydrique** water pressure

tenthrède *f.* sawfly; **t. des feuilles du rosier** leaf-rolling rose sawfly (*Blennocampa pusilla*); **t. (jaune) du groseillier** common gooseberry sawfly (*Nematus (Pterodinea) ribesii*); **t. limace** pear slug sawfly, pear and cherry slug-worm (larva) (*Caliroa cerasi* = *C. limacina*); **t. de l'oseille** dock sawfly (*Ametastegia glabrata*); **t. des poires** pear sawfly (*Hoplocampa brevis*); **t. des pommes** apple sawfly (*H. testudinea*); **t. des prunes** plum sawfly (*H. flava*); **t. de la rave** turnip sawfly (*Athalia rosae*); **t. du rosier** large rose sawfly (*Arge ochropus*), banded rose sawfly (*Emphytus cinctus*); **t. de la tige du rosier** rose stem sawfly (*Ardis brunniventris*)

ténu,-e *a.* tenuous, thin, slender

tenue *f.* habit (of plant, fruit etc.), comportment; **t. à l'étalage** shelf life; **t. en vase** vase life

tenure *f.* holding, estate

tépale *m.* tepal

tepary *m.* tepary bean (*Phaseolus acutifolius*)

téphrosia *m.*, **téphrosie** *f.* *Tephrosia* sp., including *T. candida*

téraspic *m.* candytuft (*Iberis* sp.)

tératologie *f.* teratology

tératologique *a.* teratological

tératome *m.* teratoma

térébinthe *m.* terebinth (*Pistacia terebinthus*)

terminal,-e *a.*, *m. pl.* **-aux** terminal

termite *m.* termite, white ant

termitière *f.* termite nest, mound

ternaire *a.* *Chem.* ternary

terne *a.* dim, dull, tarnished; **de couleur terne** dull-coloured

terné,-e, ternifolié,-e *a.* ternate, trifoliate

terpène *m.* terpene

terpénique *a.* terpenic

terrade *f.* urban waste fertilizer, town refuse compost

terrage *m.* *AM* depth of ploughing

terrain *m.* ground, soil, earth; **t. de golf** golf course; **t. infesté** infested soil; **t. de sport** playing field, sports field; **sur le terrain** in the field

terrarium *m.* terrarium; *Hort.* pitcher (of bromeliad)

terrasse *f.* terrace, bank; patio; **culture en terrasses** terrace cultivation

terrassement *m.* embankment, banking

terre *f.* earth, soil, land; **en pleine terre** in soil;

t. arable arable land; **t. de barre** *Afr.* clayey soil (near the coast); **t. battante** capping or glazing soil; **t. de bruyère** heath soil; **t. calcaire** calcareous soil; **t. fibreuse** orchid peat; **t. fine** fine soil; **t. de forêt** forest soil; **t. forte** heavy soil; **t. franche** loam; **t. de gazon** loam (decayed turves) (see also **gazon**); **t. glaise** clay; **t. de gobetage** *Mush.* casing soil; **t. grasse** rich soil; **t. graveleuse** gravelly soil; **t. inculte** waste land; **t. de jardin** garden soil; **t. légère** light soil; **t. meuble** loose, mellow, easily-worked soil; **t. naturelle** loam; **t. propre** clean land, weed-free land; **t. de remplissage** soil for filling; **t. de rempotage** potting soil; **t. sablonneuse** sandy soil; **t. sale** dirty land, weed-infested land; **t. soufflée** puffy soil; **t. superficielle** topsoil; **t. de surface** topsoil; **t. susceptible à l'érosion** erosive soil; **t. de taupinière** mole-hill soil; **t. tourbeuse** peat soil; **t. végétale** mould

terreau *m., pl.* **-eaux** vegetable mould, garden mould, compost; **t. de couche, t. de fumier** well-rotted manure; **t. de feuilles** leaf mould

terreautage *m.* composting, top-dressing with compost or mould

terreauter *v.t.* to treat with mould, to apply compost

terre-noix *f. inv.* earthnut, pignut (*Conopodium majus*) (= **châtaigne de terre**)

terre-plein *m.* earth platform; terrace; raised strip of ground (with trees etc.)

terrer *v.t.* to earth up (plant), to spread mould over bed etc.

terrestre *a.* terrestrial

terrette *f.* ground ivy (*Glechoma hederacea*)

terreux,-euse *a.* earthy

terrière *f. Hort.* area from which (top-)soil is obtained

terril *m.* spoil heap, tip, dump

terrine *f. Hort.* (earthenware) pot; **t. à semis** seed pan; **semis en terrines** sowing in pans

terroir *m. Vit.* etc. soil; **goût de terroir** tang of the soil

tertre *m.* hillock, mound, knoll; **t. de fleurs** bank of flowers

tessellé,-e *a.* tessellated

tesson *m.* crock, shard; **tessons de pots** pot fragments

test *m.* 1. *Bot.* testa. 2. test, trial; **t. préliminaire** preliminary test

testa *m.* testa

testacé,-e *a.* testaceous

têtard *m. Hort.* pollarded tree; **en têtard** pollarded (of tree)

tête *f.* head; **t. d'un arbre** summit, top of tree; **t. de ligne** end of a row (of vines etc.); **têtes plates** *PD* bullheads (of roses); **t. de rotation** first crop in a rotation; **t. de saule** *Hort.* bunch of vertical shoots or suckers; **t. de vieillard** *f.* old man cactus (*Cephalocereus senilis*) and similar cacti

têteau *m., pl.* **-eaux** (cut) extremity of main branch

tétrachlorure *m.* tetrachloride; **t. de carbone** carbon tetrachloride

tétrade *f.* tetrad

tétragone, tétragonie *f. Tetragonia* sp.; **t. (cornue), t. (étalée)** New Zealand spinach (*T. expansa*) (= **épinard d'été, é. de la Nouvelle Zélande**)

tétramère *a.* tetramerous

tétranème *m. Tetranema* sp., including *T. mexicanum*, Mexican foxglove

tétranyque *m. Tetranychus* sp.; **t. à deux points** = **t. tisserand** (*q.v.*); **t. rouge du pommier** fruit tree red spider mite, European mite (*Panonychus ulmi*); **t. tisserand** (greenhouse) red spider mite (*T. urticae*) (see also **grise**)

tetrapleura *m. Afr. Tetrapleura tetraptera*, a tree whose fruits are used as condiments

tétraploïde *a.* & *n. m.* tetraploid

tétraploïdie *f.* tetraploidy

tétraspore *m.* tetraspore

tétravalent,-e *a.* tetravalent, quadrivalent

teucrium *m. Teucrium* sp. (see **germandrée**)

texture *f.* texture; **t. du sol** soil texture

thalamiflore *a.* thalamifloral

thalle *m.* thallus

thallophytes *m. pl.* Thallophytes

thapsie *f.*, **thapsia** *m. Thapsia* , a genus of Mediterranean umbellifers, some of which have medicinal properties

thé *m.* tea; **arbre à thé** tea bush, tea tree (*Camellia sinensis*) (= **théier**); **t. noir** black tea; **t. vert** green tea; **t. des bois** (*Canada*) = **t. du Canada** (*q.v.*); **t. de Bourbon** Bourbon tea (*Angraecum fragrans*); **t. du Canada** (*Canada*) spicy wintergreen checkerberry (*Gaultheria procumbens*); **t. d'Europe** (1) common speedwell (*Veronica officinalis*) (= **véronique officinale**), (2) blue gromwell (*Lithospermum purpuro-coeruleum*); **t. de France** balm (*Melissa officinalis*); **t. des Jésuites** = **t. du Paraguay** (*q.v.*); **t. du Labrador** Labrador tea (*Ledum groenlandicum*); **t. du Mexique** Mexican tea (*Chenopodium ambrosioides*) (= **ambroisie**); **t. du Paraguay** maté (*Ilex paraguensis*)

théicole *a.* relating to tea; **région théicole** tea-growing region

théiculture *f.* tea growing

théier *m.* tea bush (*Camellia sinensis*) (see also **thé**)

théligone, thélygone *m. Theligonum (Thely-*

gonum) sp., including *T. cynocrambe*, dog's cabbage

thélymitre *f. Thelymitra* sp. (orchid)

théobroma, théobrome *m. Theobroma* sp. (see **cacao**)

théobromine *f.* theobromine

thérapeutique *a.* therapeutic; *n.f.* therapy

thermocouple *m.* thermocouple

thermographe *m.* thermograph

thermographie *f.* thermography

thermolabile *a.* thermolabile

thermomètre *m.* thermometer; **t. de couche** soil thermometer

thermométrographe *m.* self-recording thermometer

thermonébulisation *f.* (thermo)fogging

thermopériode *f.* thermoperiod

thermopériodisme *m.* thermoperiodism

thermophile *a.* thermophile, heat-loving; **plante thermophile** thermophile, heat-loving plant

thermorésistant,-e *a.* heat-resistant

thermosiphon *m.* thermosiphon

thermostable *a.* thermostable

thermostat *m.* thermostat

thermothérapie *f.* heat treatment, heat therapy

thermotraité,-e *a.* heat-treated; **bourgeon thermotraité** heat-treated bud

thermotropisme *m.* thermotropism

thérophyte *f.* therophyte

thigmotropisme *m.* thigmotropism

thlaspi *m.* 1. *Thlaspi* sp., pennycress. 2. *Hort. Iberis* sp., candytuft (see also **corbeille d'argent**); **t. blanc julienne** *I. amara* var. *hesperidifolia*; **t. odorant** *I. pinnata*; **t. de Perse** *I. semperflorens*; **t. très nain blanc** *I. pectinata* (= *I. affinis*); **t. violet** *I. umbellata*

thrips *m.* thrips (*Thrips* and other spp.); **t. de l'olivier** olive thrips (*Liothrips oleae*); **t. des pois** pea thrips (*Kakothrips robustus* = *K. pisivorus*); **t. du rosier** rose thrips (*Thrips fuscipennis*); **t. des serres** greenhouse thrips (*Heliothrips haemorrhoidalis*)

thuia, thuja *m.* = **thuya** *q.v.*

thunbergia *m.*, **thunbergie** *f. Thunbergia* sp.; **t. ailée** black-eyed Susan (*T. alata*)

thuya *m. Thuja (Thuya)* sp., arbor-vitae; **t. du Canada, t. d'occident** American arbor-vitae (*T. occidentalis*); **t. du Canada pyramidal** *T. occidentalis* var. *pyramidalis*; **t. de (la) Chine** Chinese arbor-vitae (*T. orientalis*); **t. de (la) Chine doré** *T. orientalis* var. *aurea*; **t. de Corée** Korean arbor-vitae (*T. koraiensis*); **t. du Japon** Japanese arbor-vitae (*T. standishii* = *T. japonica*); **t. géant (de Californie), t. de Lobb** Western arbor-vitae (*T. plicata*); **t. d'orient** = **t. de (la) Chine** (*q.v.*)

thylakoïde *m.* thylakoid

thylle *f.* tylose, tylosis

thym *m.* thyme (*Thymus* sp.); **t. pileux** *T. lanuginosus*; **t. à thymol** *T. zygis*; **t. vulgaire** common thyme, garden thyme (*T. vulgaris*) (see also **serpolet**)

thymélée (*f.*) **des Alpes** garland flower (*Daphne (Thymelaea) cneorum*)

thyrse *m. Bot.* thyrsus

thysanoptères *m. pl.* Thysanoptera, thrips

tiarella *m.*, **tiarelle** *f. Tiarella* sp., foam flower

tiède *a.* lukewarm, tepid

tierçage *m. Vit.* third cultivation

tige *f.* stalk, stem; **(haute) tige, arbre (à haute) tige** (tall) standard (tree) (about 2 m high) (see also **basse-tige, demi-tige**); **t. dressée** upright stem; **t. feuillée** leafy stem; **t. florale** peduncle; **t. herbacée** shoot; **t. latérale** side shoot; **t. primaire** *Sug.* etc. primary shoot; **t. principale** main stem; **t. sarmenteuse** climbing stem; **t. secondaire** *Sug.* etc. secondary shoot; **t. volubile** climbing stem; **taille à une tige** *Coff.* single-stem pruning

tigé,-e *a.* stemmed

tigelle *f.* little stalk: tigellum

tigre (*m.*) **du poirier** pear lace bug (*Stephanitis pyri*); **t. du rhododendron** rhododendron bug (*S. rhododendri*)

tigré,-e *a.* striped, speckled; **lis tigré** tiger lily (*Lilium tigrinum*)

tigridie *f. Tigridia* sp., tiger flower, tiger iris; **t. oeil-de-paon** common tiger flower (*T. pavonia*)

tilleul *m.* 1. lime, linden (*Tilia* sp.); **t. d'Amérique** basswood (*T. americana*); **t. argenté** silver lime (*T. tomentosa*); **t. argenté pleureur** pendent silver lime (*T. petiolaris*); **t. des bois** = **t. à petites feuilles** (*q.v.*); **t. de Carpentras** variety of small-leaved lime (*T. cordata*); **t. commun** common lime (*T.* × *europaea*); **t. de Hollande, t. à grandes feuilles, t. à larges feuilles** large-leaved lime (*T. platyphyllos*); **t. de Hollande** = **t. à grandes feuilles** (*q.v.*); **t. de Hongrie** = **t. argenté** (*q.v.*); **t. de Mongolie** Mongolian lime (*T. mongolica*); **t. à petites feuilles** small-leaved lime (*T. cordata*). 2. Other genera: **t. d'Afrique** African linden (*Mitragyna ciliata*); **t. d'appartement, t. de chambre, t. nain** African hemp (*Sparmannia africana*) (= **sparmannia**)

tinctorial,-e *a.*, *m. pl.* **-aux** tinctorial

tinier *m.* arolla pine (*Pinus cembra*) (= **pin cembro**)

tipule *f.* crane-fly (Tipulidae), crane-fly larva, leather-jacket

tiquet *m. Ent.* flea-beetle (see **altise**)

tirant *m.* tie

tire-bouchon *m.*, *pl.* **tire-bouchons** corkscrew; **en tire-bouchon** in a corkscrew

tire-bouchonné,-e *a.* in corkscrew curls
tire-racines *m. inv.* weeder
tire-sève *n. m. inv.* sap-drawer (= **appel-sève**); (*Tea*) lung branch
tisserin *m.* weaver (bird); **t. gendarme** = **gendarme** (*q.v.*)
tisserand *m. PC* spider mite (*Tetranychus* sp.); **t. commun** red spider mite (*T. urticae*); **t. du Midi** *T. turkestani* (see also **acarien, tétranyque**)
tissu *m.* tissue; **t. cambial** cambium tissue; **t. cellulaire** cellular tissue; **t. cellulosique** cellulose tissue; **t. conducteur** conducting tissue; **t. libérien** phloem tissue; **t. palissadique** palisade tissue; **tissus tumoraux** tumorous tissues; **t. végétal** plant tissue
tissulaire *a.* relating to tissue; **culture tissulaire** tissue culture; **différenciation tissulaire** tissue differentiation
tocoférol, tocophérol *m.* tocopherol
toddy *m.* toddy
tofieldie *f. Tofieldia* sp., including Scottish asphodel (*T. pusilla*)
toile *f.* 1. linen cloth, canvas; **t. cirée** oilcloth; **t. d'emballage** packing cloth; **t. métallique** wire gauze, wire netting; **t. à ombrer** shading cloth. 2. *PD* damping off caused by *Phythium, Phytophthora, Botrytis, Rhizoctonia* etc. (see also **fonte**); **t. (du champignon de couche)** mushroom cobweb disease (*Dactylium dendroides*). 3. webs made by certain insect larvae etc. e.g. *Neurotoma saltum;* **t. d'araignée** cobweb
toilettage *m. Hort.* (remedial) pruning (see also **taille**)
toit *m.* roof; **t. vitré** glazed roof
toiture *f.* roofing, roof
tôle *f.* sheet metal; **t. ondulée** corrugated iron
tolérance *f.* tolerance
toluifera *m. Toluifera* sp. (= *Myroxylon*) (see also **baumier**)
tomate *f.* tomato (*Lycopersicon esculentum*); **tomates d'abri** tomatoes grown under cover; **t. cerise** cherry tomato (var. *cerasiforme*); **t. de conserve** canning tomato; **t. de consommation** salad tomato; **t. d'arbre, t. en arbre, t. de la Paz** tree tomato (*Cyphomandra betacea*); **t. poire** pear tomato (var. *pyriforme*); **t. de serre** greenhouse tomato
tomatillo *m.* tomatillo (*Physalis ixocarpa*)
tombereau *m., pl.* **-eaux** tip-cart
tomenteux,-euse *a.* tomentose
tomentum *m.* tomentum
tondeuse *f.* **(à gazon)** (lawn) mower; **t. autoportée** mechanical mower; **t. autotractée** self-propelled mower; **t. électrique** electric mower; **t. à main** hand mower; **t. à moteur** motor mower; **t. à moteur à essence** petrol-driven mower; **t. à siège** mower with a seat for the driver
tondobroyeuse *f.* mower-crusher
tondre *v.t.* to trim, to clip, to mow (lawn), to cut
tonique *a.* & *n.m.* tonic
tonka *m.* tonka bean plant (*Dipteryx odorata*); **fève tonka** tonka bean (= **coumarou**)
tonne *f.* large cask, barrel; **t. à eau de pluie** rainwater butt
tonneau *m., pl.* **-eaux** barrel, cask; **t. d'arrosage** water cart; **t. pour l'eau de pluie** rainwater butt
tonnelle *f.* arbour, bower
tonte *f. Hort.* clipping, mowing; **t. du gazon** lawn mowing
tontinage *m.* balling, packing straw and sacking around the roots of a plant prior to transport
tontine *f. Hort.* straw and sacking packed around the roots of a plant prior to transport; a plant so treated
tontiner *v.t.* to ball, to protect roots of plants with straw and sacking prior to transport
topiaire *a.* topiarian; **art topiaire** topiarian art
topinambour *m.* Jerusalem artichoke (*Helianthus tuberosus*); (*Antilles*) *Maranta* sp. (see **marante**)
torchis *m.* forkful of farmyard manure for lining a hotbed
torchon *m.* dishcloth; **t. végétal** loofah (*Luffa cylindrica*) (= **courge éponge**)
tordeuse *f.* tortrix moth, tortricid, leaf-rolling moth (Tortricidae); **t. à bandes rouges** (*Canada*) red banded leaf roller (*Argyrotaenia velutinana*); **tordeuses des bourgeons** fruit tree tortrix moths (several spp.); **t. brune du fraisier** strawberry tortrix (*Argyroploce lacunana*); **tordeuses des buissons** fruit tree tortrix moths (e.g. *Archips podana, Pandemis ribeana* etc.); **tordeuses de la grappe** *Vit.* grape moths (*Clysiana (Clysia) ambiguella*) (= **cochylis**), (*Lobesia (Polychrosis) botrana*) (= **eudémis**); **t. de la laitue** lettuce tortricid (*Eucosma conterminana*); **t. de l'oeillet** carnation tortrix moth (*Cacoecia pronubana*); **t. orientale du pêcher** oriental fruit moth (*Cydia (Grapholitha) molesta*); **t. de la pelure (des fruits)** summer fruit tortrix moth (*Adoxophyes orana*) (= **t. verte**); **t. du pois** pea moth (*Laspeyresia nigricana*) (= **teigne du pois**); **t. des pruniers** plum tortrix moth (*Hedya pruniana*); **t. rouge (des bourgeons)** (eye-spotted) bud moth (*Spilonota ocellana*); **t. verte** = **t. de la pelure (des fruits)** *q.v.*; rose tortrix moth (*Cacoecia rosana*); **t. verte des bourgeons** fruit tree tortrix moth (*Hedya nubiferana*)
tordu,-e *a.* twisted; **arbre tordu** twisted tree

tordyle *m.* *Tordylium* sp.; **t. d'Apulie** ivory-fruited hartwort (*T. apulum*); **t. élevé** *T. maximum*

torilis (*m.*) **des champs** spreading hedge parsley (*Torilis arvensis*); **t. noueux** knotted hedge parsley (*T. nodosa*)

tormentille *f.* tormentil (*Potentilla erecta* = *P. tormentilla*)

tornélie *f.*, **tornelia** *m.* *Tornelia* sp. (former name for *Monstera* sp.)

toron *m.* 1. strand of rope. 2. wisp of straw

torréfaction *f.* torrefaction, roasting; **comportement à la torréfaction** *Coff.* roasting quality

torréfier *v.t.* to torrefy, to roast

torreya *m.* *Torreya* sp., stinking yew (= **if puant**); **t. de Chine** *T. grandis*; **t. du Japon** kaya (*T. nucifera*)

torsader *v.t.* to twist, to twist together

torsion *f.* torsion, twisting

tortueux,-euse *a.* tortuous, twisted

toruleux,-euse *a.* toruloid, torulose

touffe *f.* tuft, clump, truss (of flowers); *Sug.* stool

touffu,-e *a.* tufted, bushy, leafy, luxuriant

touloucouna *m.* Kunda oil tree (*Carapa procera*)

toupet *m.* tuft

toupie *f.* top (toy); **en toupie** shaped like a top

tourbe *f.* peat, turf; **t. azonale** azonal peat; **t. blonde** (commercial) sphagnum peat, white peat; **t. broyée** milled peat; **t. à brûler** peat fuel; **t. brute** raw peat; **t. à carex** sedge peat; **t. (commerciale) de sphaigne, t. à sphaigne** (commercial) sphagnum peat; **t. extraite mécaniquement** machine peat; **t. horticole** horticultural peat; **t. ligneuse** woody peat; **t. des marais** fen peat; **t. en mottes** handcut peat; **t. de mousse** moss peat; **t. noire** black peat; **t. noire soumise à l'action de la gelée** frozen black peat; **t. à roseau** reed peat

tourbeux,-euse *a.* 1. peaty, boggy. 2. growing in peat bogs

tourbier,-ière *a.* peaty; *n. f.* peat bog; **t. haute** raised bog; **t. de source** spring bog; **t. de surface** blanket bog; **t. de vallée** valley bog; **jardin de tourbière** bog garden

tourmenté,-e *a.* distorted, contorted; **forme tourmentée** distorted shape

tournant,-e *a.* turning; **stade tournant** colour-changing stage of fruit maturity; *n.m.* turning, bend

tourner *v.i.* *Hort.* to ripen, to change colour (of grape etc.), to heart (of lettuce), to bulb (of onion)

tournesol *m.* 1. turnsole, including sunflower (*Helianthus annuus*), heliotrope; **t. (des teinturiers)** dyer's croton (*Chrozophora tinctoria*) (= **tournesolie**). 2. litmus

tournesolie *f.* dyer's croton (*Chrozophora tinctoria*) (= **tournesol des teinturiers**)

tournière *f.* headland

tourniquet *m.* *Hort.* **t. (hydraulique)** rotary sprinkler (= **arroseur rotatif**)

tourteau *m.*, *pl.* **-eaux** oil cake, press cake; **t. moulu** meal (of oil cake); **t. de noix de coco** coconut cake; **tourteaux de filtration** *Sug.* filter cake

toute-bonne *f.*, *pl.* **toutes-bonnes** 1. clary (*Salvia sclarea*) (= **sauge sclarée**), Good King Henry (*Chenopodium bonus-henricus*) (see also **ansérine**). 2. type of pear

toute-épice *f.*, *pl.* **toutes-épices** 1. allspice (*Pimenta officinalis*) (= **poivre de la Jamaïque**). 2. black cumin (*Nigella sativa*) (= **nigelle aromatique**)

toute-saine *f.*, *pl.* **toutes-saines** tutsan (*Hypericum androsaemum*)

toxicité *f.* toxicity; **t. aiguë** acute toxicity; **toxicités comparées** comparative toxicities

toxicodendron *m.* *Toxicodendron* sp. (see also **sumac**)

toxicologie *f.* toxicology

toxine *f.* toxin

toxique *a.* toxic, poisonous; *n. m.* poison

traçage *m.* tracing; **t. isotopique** isotope tracing

traçant,-e *a.* horizontal running (of rhizomes, roots etc.); **racine traçante** creeping root

trace *f.* trace; **t. foliaire** leaf trace; *Bot.* stolon, runner

tracé *m.* lay-out, plan (of plantation etc.)

traceur,-euse *n.* tracing; *n. m.* **traceur (radioactif)** (radio-active) tracer

trachée *f.* *Bot.* tracheid, vessel

trachéide *f.* tracheid

trachélie *f.* *Trachelium* sp., (blue) throat-wort

trachélosperme, trachélospermum *m.* *Trachelospermum* sp., Chinese jasmine, Chinese ivy

trachéogenèse *f.* tracheogenesis

trachéomycose *f.* *PD* trachaeomycosis, wilt disease; **t. du cacaoyer** cocoa die-back (*Calonectria rigidiuscula*); **t. de l'œillet** carnation fusarium wilt (*Fusarium oxysporum* f. *dianthi*); **t. de la tomate** tomato fusarium wilt (*F. oxysporum* f. *lycopersici*)

trachymène *f.* *Trachymene* sp., including *T. caerulea* (= *Didiscus caerulea*), blue lace flower

tracteur *m.* tractor; **t. à chenilles** caterpillar tractor; **t. enjambeur** straddling or high-clearance tractor for high crops; **t. à roues** wheeled tractor

tractoriste *m.* tractor driver

tradescantia *m.*, **tradescantie** *f.* *Tradescantia* sp.; **t. de Virginie** common spiderwort (*T. virginiana*)

tragacanthe *f.* tragacanth (*Astragalus tragacantha*)
traînant,-e *a.* (of plant) trailing
traînasse *f.* 1. stolon, runner. 2. name applied to several trailing plants e.g. knotgrass (*Polygonum aviculare*) (= **renouée traînasse**)
traineau *m., pl.* **-eaux** sledge, picking sledge
traînée *f.* trail (of smoke etc.); *Hort.* row; **semer une trainée de carottes** to sow a row of carrots; *Bot.* runner
traité,-e *a.* treated; **. à la chaleur, t. par la chaleur** heat-treated; **non-traité** untreated (of control plot etc.)
traitement *m.* treatment; **t. aérien** aerial treatment; **t. anti-parasitaire** plant protection treatment; **t. à la chaleur** heat treatment; **t. curatif** curative treatment; **t. dirigé** directed (herbicide etc.) spray; **t. à l'eau chaude** hot water treatment; **t. généralisé** overall treatment; **t. d'induction de la floraison** floral induction treatment; **t. localisé** localized treatment; **t. pendant le repos végétatif** treatment during plant dormancy; **t. de post-émergence** = **t. de post-levée** (*q.v.*); **t. de post-levée** post-emergence treatment; **t. de post-plantation** post-planting treatment; **t. de post-semis** post-sowing treatment; **t. de pré-émergence** = **t. de pré-levée** (*q.v.*); **t. de pré-levée** pre-emergence treatment; **t. de pré-plantation** pre-planting treatment; **t. de pré-semis** pre-sowing treatment; **t. préventif** preventive treatment; **t. de référence** control treatment; **t. des semences** seed treatment; **t. au semis** treatment at sowing; **t. du sol** soil treatment; **t. thermique** heat treatment
traiter *v.t.* to treat, to apply treatment, to dress (seed etc.)
tranchant,-e *a.* cutting, sharp; *n. m.* cutting edge (of tool)
tranchée *f.* small ditch, trench; **t. à ciel ouvert** open trench; **t. de couche** forcing trench; **t. de drainage** draining trench, drain
tranche-gazon *n. m. inv.* turf cutter, grass-edging knife
trancheuse *f.* trenching machine, trencher
transfert *m.* transfer; **t. d'électrons** electron transfer
transformateur *m.* transformer
transformation *f.* transformation
transfusion *f.* transfusion; **tissu de transfusion** *Bot.* transfusion tissue
translocation *f.* translocation
translucide *a.* translucent
transmissible *a.* transmissible; **t. par semis** transmissible by seed, seed-borne
transmission *f.* transmission, drive (of machine); **t. centrale** central drive; **t. latérale** side drive; **t. mécanique** mechanical transmission
transparence *f.* transparency
transpiration *f. Bot.* transpiration
transpirer *v.i. Bot.* to transpire
transplantable *a.* transplantable
transplantation *f.*, **transplantement** *m.* transplantation, transplanting (especially of larger plants) (see also **repiquage**)
transplanté,-e *a. Hort.* transplanted
transplanter *v.t.* to transplant
transplanteur *m.* transplanter, planter; *a. m.* **matériel transplanteur** transplanting material
transplanteuse *f.* transplanting machine
transplantoir *m.* 1. transplanting tool, trowel. 2. transplanting machine
transport *m.* transport; **t. maritime** sea transport
transversal,-e *a., m. pl.* **-aux** transversal
trappe-pince *f., pl.* **trappes-pinces** pincer trap (for voles etc.)
trapu,-e *a.* dumpy, stocky, dwarf(ed) (of plant)
traumatisme *m. Hort.* bruise, injury (on fruit, plant etc.)
travail *m., pl.* **-aux** work, labour; **t. d'entretien** maintenance work; **t. saisonnier** seasonal work
travaillé,-e *a.* worked; *Hort.* cultivated
travée *f.* bay, span (in construction)
traverse *f.* cross bar, traverse beam, girder, rung of ladder
trèfle *m.* 1. clover (*Trifolium* sp.); **t. blanc** white clover (*T. repens*); **t. incarnat** crimson clover (*T. incarnatum*); **t. à quatre feuilles** four-leaved clover; **t. rampant** = **t. blanc** (*q.v.*). 2. Other genera: **t. d'eau** marsh trefoil (*Menyanthes trifoliata*); **t. musqué** blue melilot (*Trigonella caerulea*) (= **mélilot bleu**); **"t. à quatre feuilles porte-bonheur"** "lucky clover" (*Oxalis deppei*)
treillage *m.* trellis, lattice; **t. en fil de fer, t. métallique** wire netting
treillager *v.t.* to cover or furnish with trellis, to trellis, to enclose with wire netting
treillageur, treillagiste *m.* trellis-maker, lattice-maker or seller
treille *f.* vine-trellis, trellised vineyard, trellised vine, espalier vines; **t. en tonnelle** vine arbour
treillis *m.* trellis, lattice; **t. métallique** wire netting
treillissé,-e *a.* trellised, latticed
treillisser *v.t.* to trellis, to lattice, to enclose with wire netting
treillon *m.* small **treille** (*q.v.*)
tremblaie *f.* aspen grove
tremble *m.* aspen (*Populus tremula*) (see also **peuplier**)
trémie *f.* hopper (of implement)
trémière *a. f.* & *n. f.* hollyhock (*Althaea rosea*) (= **rose trémière**)

trempage *m.* dipping, soaking; **t. dans l'eau** dipping, soaking in water
tremper *v.t.* to steep, to soak, to dip; **"trempez au préalable"** "soak to begin with"
treuil *m.* winch
tri *m.* sorting, sorting out; **t. électronique** electronic sorting; **t. manuel** manual sorting; **t. sévère** rigorous selection (see also **triage**)
triage *m.* choosing, sorting, grading; **t. colorimétrique** colorimetric sorting; **triage sélectionné** grading for quality (see also **trier**)
triallate *m.* triallate
triangle *m.* triangle; **en triangle** triangular; **t. liégeux** *PD* corky triangle
triangulaire *a.* triangular; **test triangulaire** triangle test (for taste panels)
tribasique *a.* tribasic
tribu *f.* tribe
tribulus *m.* *Tribulus* sp., caltrop
tricalcique *a.* tricalcic
trichocaule *a.* *Bot.* hairy-stemmed
tricholome *m.* *Tricholoma* sp.; **t. émarginé** *T. sejunctum*; **t. géant (d'Afrique équatoriale)** *T. lobayensis*; **t. géant de Ferney** (*Mauritius*) *T. spectabilis* (= **champignon de Ferney**); **t. petite colombe** dove-coloured tricholoma (*T. columbetta*) (= **colombette**) (see also **chevalier**); **t. de la Saint-Georges** St. George's mushroom (*T. gambosum*) (= **mousseron**)
trichoma, trichome *m.* trichome
trichomane *m.* *Trichomanes* sp., bristle fern
trichosanthes *m.* *Trichosanthes* sp., snake gourd (*T. anguina*) (= **serpent végétal**)
trichotome *a.* trichotomous
tricolor *m.* tricolor amaranth (*Amaranthus gangeticus* var. *tricolor*)
trident *m.* three-pronged pitchfork
tridenté,-e *a.* tridentate
trier *v.t.* to sort, to grade (see also **tri** and **triage**)
trieur,-euse *n.* sorter, grader; **trieur à vibrations** vibrating riddle, vibrating sieve; **trieuse calibreuse** size grader
trifide *a.* trifid
triflore *a.* triflorous
trifluraline *f.* trifluralin
trifolié,-e *a.* trifoliate
trifoliolé,-e *a.* trifoliolate; **oranger trifoliolé** trifoliate orange (*Poncirus trifoliata*)
trifolium *m.* *Trifolium* sp. (see also **trèfle**)
trigame *a.* *Bot.* trigamous
trigemme *a.* three-budded, with three buds
trigone *a.* trigonal, three-cornered; *n.m.* trigone
trigonella, trigonelle *f.* *Trigonella* sp., especially fenugreek (*T. foenum-graecum*)
triguère *f.* *Triguera* sp., including *T. ambrosiaca*
trigyne *a.* *Bot.* trigynous
trille, trillium, trillie *m.* *Trillium* sp., American wood lily
trilobé,-e *a.* trilobate
triloculaire *a.* trilocular
trinervé,-e *a.* trinervate
triparti,-e *a.* tripartite
tripenné,-e *a.* tripinnate
tripétale, tripétalé,-e *a.* tripetalous
triploïde *a.* triploid
tripsacum *m.* *Tripsacum* sp., including *T. laxum*, Guatemala grass
trique-madame *f. inv.* white stonecrop (= **orpin blanc** *q.v.*)
triquètre *a.* triquetrous
trisépale *a.* trisepalous
trisperme *a.* trispermous, three-seeded
trisomie *f.* trisomia
tristeza *f.* *Cit. PD* tristeza virus
tritié,-e *a.* tritiated; **eau tritiée** tritiated water
tritome *m.* red-hot poker (*Kniphofia* sp.)
triumfetta *m.* burweed (*T. cordifolia*), *T. pentandra*
trivalent,-e *a.* trivalent
trivalve *a.* trivalvular
trochet *m.* cluster of fruit, flowers etc.; **en trochet** clustered; **t. de noisettes** cluster of hazel nuts
troène *m.* **t. (commun)** (common) privet (*Ligustrum vulgare*); **t. de Californie** *L. ovalifolium;* **t. de (la) Chine** (1) Chinese privet (*L. sinense*), (2) white wax tree (*L. lucidum*); **t. du Japon** Japanese privet (*L. japonicum*); **t. panaché** golden privet (*L. ovalifolium* var. *aureum*)
trognon *m.* core (of apple, pear etc.), stump (of cabbage etc.)
trolle *m.* *Trollius* sp., globe flower; **t. d'Europe** *T. europaeus* (see also **boule d'or**)
trompe d'éléphant *f.* unicorn plant (*Martynia proboscidea* = *M. louisiana*) (see also **cornaret**)
trompette *f.* *Hort.* trumpet (of narcissus etc.); **t. des morts** *Fung.* horn of plenty (*Craterellus cornucopioides*) (= **craterelle**)
tronc *m.* trunk (of tree etc.)
tronçon *m.* section of cylindrical object, stump; **t. de bois** log
tronconique *a.* in the shape of a truncated cone; **segment tronconique** truncated segment
tronçonnage *m.* cutting trunks etc. into pieces
tronçonné,-e *a.* (of trunks, stems etc.) cut into pieces; **cannes tronçonnées** *Sug.* chopped harvested cane
tronçonner *v.t.* to cut trunks etc. into pieces
tronçonneuse *f.* chain saw, cross-cut saw, (portable) band saw
tronqué,-e *a.* truncated; *Vit.* applied to near-pentagonal leaves
trop *adv.* too, over-; **trop mûr** overripe

tropanique *a.* tropane; **alcaloïdes tropaniques** tropane alkaloids
trophique *a.* trophic, pertaining to nutrition
trophobiose *f.* trophobiosis
tropical,-e *a., m. pl.* **-aux** tropical; **région tropicale humide** humid tropical region
tropisme *m.* tropism
trou *m.* hole; **t. de plantation** planting hole; **t. de sortie** *Ent.* emergence hole
trouaison *f. Plant.* making holes for planting, holing
trouble *m. PD* disorder; **t. physiologique** physiological disorder
truffe *f.* truffle (*Tuber* sp. and other spp.); **t. de Bourgogne** Burgundy truffle (*T. uncinatum*); **t. du Périgord** Perigord truffle (*T. melanosporum*); **t. du Piémont** white Piedmont truffle (*T. magnatum*); **t. d'eau** water chestnut (*Trapa natans*) (= **châtaigne d'eau**)
trufficulteur,-trice *n.* truffle grower
trufficulture *f.* truffle cultivation, truffle growing
truffier,-ère *a.* relating to truffles; **chêne truffier** oak species under which truffles are found (e.g. *Quercus ilex, Q. coccifera*); **région truffière** truffle (-producing) region; **terrain truffier** truffle-producing land
truffière *f.* truffle bed, truffle ground
tsuga *m. Tsuga* sp.; **t. du Canada** eastern hemlock (*Tsuga canadensis*); **t. de l'ouest** western hemlock (*T. heterophylla*)
tube *m.* tube, pipe; **t. fluorescent** fluorescent lamp; *Bot.* tube (of corolla, calyx); **t. calicinal** hypanthium; **t. criblé** sieve tube; **t. pollinique** pollen tube
tubercule *m.* tuber; **tubercules bourgeonnants** sprouting tubers; **t. de pomme de terre** potato tuber; **t. en fuseau** *PD* potato spindle tuber virus
tuberculé,-e *a.* tuberculate
tuberculeux,-euse *a. Bot.* tubercular
tubéreuse *f.* tuberose (*Polianthes tuberosa*); **t. bleue** African lily (*Agapanthus umbellatus*) (see also **agapanthe**)
tubéreux,-euse *a.* tuberous
tubérisation *f.* tuberization
tubérisé,-e *a.* tuberous; **racine tubérisée** tuberous root
tubéroïde *a.* tuberoid
tubérosité *f.* tuberosity
tubulaire *a.* tubular
tubuleux,-euse *a.* tubular, tubulous
tucum *m.* tucum palm (*Astrocaryon* sp., including *A. tucuma*)
tue-chien *m. inv.* meadow saffron (*Colchicum autumnale*); black nightshade (*Solanum nigrum*)
tuf *m.* tuff, tufa; **t. volcanique** volcanic tuff
tulipe *f.* tulip (*Tulipa* sp.); **t. avortée** blind tulip; **t. dragonne** parrot tulip (= **t. perroquet**); **t. de l'Ecluse** lady tulip (*T. clusiana*); **t. des fleuristes, t. de Gesner** garden tulip (*T. gesneriana*); **t. infléchie** tulip affected by topple; **t. des jardins** = **t. des fleuristes** (*q.v.*); **t. odorante** Duc van Thol tulip (*T. suaveolens*); **t. perroquet** parrot tulip (= **t. dragonne**); **t. piquée** tulip affected by fire (*Botrytis tulipae*); **t. radis** = **t. de l'Ecluse** (*q.v.*)
tulipier *m.* **(de Virginie)** 1. tulip tree (*Liriodendron tulipifera*); **t. d'Afrique** African tulip tree (*Spathodea campanulata*). 2. tulip grower
tulipiste *m.* tulip grower
tumeur *f.* tumor; **t. bactérienne du collet et des racines** *PD* crown gall (*Agrobacterium tumefaciens*)
tumoral,-e *a., m. pl.* **-aux** *PD* tumoral, tumorous; **tissus tumoraux** tumorous tissues
tuna *m.* prickly pear (*Opuntia tuna*)
tunique *f. Bot.* tunic, envelope, skin; **t. de bulbe** bulb tunic
tuniqué,-e *a.* tunicate(d)
turban *m. Hort.* turban squash (*Cucurbita maxima*) (= **bonnet turc**)
turbiné,-e *a.* turbinate
turc *m. Hort.* cockchafer larva
turgescence *f.* turgescence, turgidity
turgescent,-e *a.* turgescent
turion *m.* turion; **t. d'asperge** asparagus turion, spear
turnep, turneps *m.* kohl-rabi (*Brassica oleracea* var. *caulorapa*) (= **chou-rave**)
tussilage *m.* coltsfoot (*Tussilago farfara*) (= **pas d'âne**)
tuteur *m. Hort.* prop, stake, stick, pole; **t. en arceau** hooped, curved support; **t. mort** stake, stick; **t. vivant** tree or plant used as support
tuteurage *m. Hort.* staking, propping, supporting
tuteuré,-e *a.* staked, propped, supported
tuteurer *v.t.* to stake, to prop, to support, to tie to supports
tuyau *m., pl.* **-aux** pipe, tube; **t. d'alimentation** feed pipe; **t. d'arrosage** garden hose, watering, irrigation pipe; **t. de chauffage** hot water pipe; **t. de drainage** drainage pipe; **t. d'eau** water pipe
tyndallisation *f.* tyndallization
type *m.* type; **t. de sol** soil type
typha *m. Typha* sp., reed mace
typhlodrome *m.* predacious mite (*Amblyseius* sp.) (see also **acarien**)
tyrosine *f.* tyrosine

U

ubiquiste *a.* ubiquitous
ulex *m.* *Ulex* sp., including *U. europaeus*, furze, gorse, whin
ulluco, ulluque *m.* *Ullucus* sp.; **u. tubéreux** ulluco (*U. tuberosus*)
ulmacé,-e *a.* ulmaceous
ulmaire *f.* meadow sweet (*Filipendula ulmaria*) (= **reine des prés**)
ulmeau *m., pl.* **-eaux** common elm (*Ulmus procera*)
ultrastructural,-e *a., m.pl.* **-aux** ultrastructural; **étude ultrastructurale** ultrastructural study
ultrastructure *f.* ultrastructure
ultra(-)violet,-ette *a.* ultra-violet; *n. m.* ultra-violet
umbilicus (*m.*) **rupestre** pennywort (*Umbilicus rupestris*)
umbo *m. Fung.* umbo
umboné,-e *a. Fung.* umbonate
unicaule *a.* single-stemmed; **taille unicaule** *Coff.* etc. single-stem pruning
unicaulie *f. Coff.* etc. single-stem pruning system (see also **multicaulie**)
unicellulaire *a.* unicellular
unicolore *a.* unicoloured
unifère *a.* flowering once in one year, cropping once (see also **bifère**)
uniflore *a.* uniflorous
unigemme *a.* single-budded
unilabié,-e *a.* unilabiate
unilatéral,-e *a., m.pl.* **-aux** unilateral
uniloculaire *a.* unilocular
uninervé,-e *a.* uninervate (of leaf)
union *f.* union
uniovulé,-e *a.* uniovular, uniovulate
unipare *a. Bot.* uniparous
unisérié,-e *a.* uniseriate
unisexué,-e *a.* unisexual
unité *f.* unit
unitige *a.* single-stemmed
univalent,-e *a.* monovalent
urbec, urebec *m.* hazel leaf, vine leaf roller weevil (*Byctiscus betulae*) (= **cigarier**)
urcéole *m.* 1. urceolus. 2. *Urceolina* sp., urn flower
urcéolé,-e *a.* urceolate, urn-shaped
urebec = **urbec** (*q.v.*)
urée *f.* urea; **u. substituée** substituted urea
urena *m.*, **urène** *f.* urena (*Urena* sp.), including aramina fibre (*U. lobata*)
urginea *m.*, **urginée** *f.* *Urginea* sp.; **u. maritime** sea onion (*U. maritima*) (= **scille maritime**)
urine *f.* urine
urocyste *m. Fung.* *Urocystis* sp.
urticacées *f. pl.* Urticaceae
usage *m.* use
usinable *a. Sug.* etc. millable, capable of being processed
usinage *f. Sug.* etc. processing
usine *f.* factory, mill
usinier,-ère *a.* relating to factories; *n. m.* manufacturer
ustilago *m.* *Ustilago* sp. (see also **charbon**)
usure *f.* wear (and tear)
utriculaire *f.* *Utricularia* sp., including common bladderwort (*U. vulgaris*)
UV = **ultra-violet** (*q.v.*)
uval,-e *a., m.pl.* **-aux** relating to grapes; **cure uvale** grape cure; **station uvale** place where grape cures are carried out
uva-ursi *m.* bearberry (*Arctostaphylos uva-ursi*) (= **busserole, raisin d'ours**)
uvifère *a.* grape-bearing
uviforme *a.* grape-shaped
uvulaire *f.* *Uvularia* sp., bell-wort

V

vachette *f. Coll. Fung. Lactarius volemus* (= **lactaire à lait abondant**)
vaciet *n. Coll.* bilberry (= **airelle** *q.v.*)
vacuole *f.* vacuole; *Bot.* vesicle
vacuome *m.* vacuome
vaginé,-e *a.* vaginate
vain,-e *a.* vain, empty; **graines vaines** empty seeds
vaisseau *m., pl.* **-eaux** *Bot.* vessel; **v. criblé** pitted vessel; **v. ligneux** lignified vessel; **v. ponctué** = **v. criblé** (*q.v.*); **v. spiralé** spiral vessel
valérianacées *n. f. pl.* Valerianaceae
valériane *f.* valerian (*Valeriana* sp.); **grande valériane, v. phu** *V. phu* var. *aurea*; **v. d'Alger** African valerian (*Fedia cornucopiae*); **v. dioïque** lesser valerian (*V. dioica*); **v. officinale** (common) valerian (*V. officinalis*); **v. grecque** Greek valerian (*Polemonium coeruleum*); **v. rouge** red valerian (*Centranthus (Kentranthus) ruber*) (= **barbe de Jupiter**)
valérianelle *f. Valerianella* sp., including lamb's lettuce (*V. locusta*) (= **mâche**)
valeur *f.* value, worth; **mise en valeur** profitable development, rational use; **v. antiparasitaire** pesticidal value; **v. culturale** cultural value; **v. marchande** market value, commercial value; **v. nutritive** food value; **v. résiduelle** residual value; **v. vénale** market value; **v. à la vente** sale value
vallisnérie *f. Vallisneria* sp.; **v. spirale** eel-grass, tape grass (*V. spiralis*)
vallon *m.* small valley, dale
vallonné,-e *a.* cut up by dells, undulating
vallonnement *m. Hort.* laying out of park, garden etc. in dells
vallonner *v.t. Hort.* to lay out park, garden etc. in dells
valvaire *a.* valvate; **préfloraison valvaire** valvate aestivation
valve *f. Bot.* valve
valvule *f.* valvule
vanda *m. Vanda* sp., cowslip-scented orchid
vanguier *m. Vangueria* sp. (see **voavanguier**)
vanille *f.* vanilla; **gousse de vanille** vanilla bean
vanilleraie, vanillerie, vanillière *f.* vanilla plantation
vanillier,-ière *a.* relating to vanilla; **culture vanillière** vanilla growing; *n. m.* vanilla plant (*Vanilla fragrans*)
vanilline *f.* vanillin
vanillisme *m.* vanillism
vanillon *m.* West Indian vanilla (*Vanilla pompona*)
vannage *m.* fanning, winnowing
vanne *f. Irrig.* sluice-gate, water gate
vanner *v.t.* to fan, to winnow
vannerie *f.* 1. basket-making. 2. basket-work, wicker-work
V.A.O.G. see **vignoble**
vapeur *f.* vapour, haze, fumes; **v. (d'eau)** steam, water vapour; **v. de mercure** mercury vapour
vaporisateur *m.* sprayer
vaporisation *f.* vaporization, evaporation
vaquois *m.* screw pine (*Pandanus utilis*)
varaire *m.* or *f. Veratrum* sp., false hellebore (see also **vératre**)
variabilité *f.* variability; **v. génétique** genetical variability
variation *f.* variation
variétal,-e *a., m. pl.* **-aux** varietal; **étude variétale** variety trial
variété *f.* variety; **v. agréée** approved variety; **v. cultivée** cultivated variety; **v. fixée** stable variety; **v. horticole** horticultural variety, cultivar; **v. protégée** protected variety; **v. de référence** control variety; **v. de saison** mid-season variety
variole (de la pomme de terre) *f. PD* black scurf and stem canker of potato (*Corticium (Rhizoctonia) solani*)
varroase *f.* varroasis, the infestation of beehives by the mite *Varroa jacobsoni*
vasculaire *a.* vascular
vascularisation *f.* vascularization, process of becoming vascular
vascularisé,-e *a.* vascularized
vase *m.* 1. vase, pot; **v. à fleurs** flower vase; **v. de pierre** stone urn; *Hort.* **en vases de végétation** in pots, in containers. 2. *Hort.* goblet. 3. *Bot.* vase-shaped corolla
vase *f.* mud, sludge; **v. de filtrage** sewage sludge compost
vaseux,-euse *a.* muddy, slimy
vasistas *m.* opening window frame, ventilator (over door etc.)
vasque *f.* 1. basin (of fountain). 2. shallow basin (topography)

vecteur,-trice *a.* vector; **espèce vectrice** vector species; **insecte vecteur** insect vector; *n. m. PD* vector
végétabilité *f.* vegetability, vegetable nature
végétable *a.* vegetable, vegetating
végétal,-e *a., m. pl.* **-aux** plant, vegetable e.g. in plant life, vegetable kingdom; **cellule végétale** plant cell; **graisse végétale** plant fat; **matériel végétal** plant material; **organes végétaux** plant organs
végétal *m., pl.* **-aux** plant; **végétaux cultivés** cultivated plants; **végétaux supérieurs** higher plants
végétalité *f.* vegetable life, plant life
végétant,-e *a.* vegetating
végétarien,-ienne *a.* & *n.* vegetarian
végétatif,-ive *a.* vegetative
végétation *f.* vegetation; **en végétation** in (active) growth
végéter *v.i.* to vegetate, to grow (of plant)
véhicule *m.* vehicle, carrier
veilleuse, veillotte *f.* meadow saffron (*Colchicum autumnale*) (= **safran des prés, tue-chien, vendangeuse**)
veine *f. Bot.* vein
velamen *m.* velamen
vélanède *f.* acorn cups of valonia oak (used in tanning)
vélani valonia oak (*Quercus aegylops* and allied species)
vélar, vélaret *m.* hedge mustard (*Sisymbrium officinale*) (= **herbe aux chantres**); *Erisymum* sp. (see **érisymum**); **v. d'Orient** hare's ear cabbage (*Conringia orientalis*)
velouté,-e *a.* velvety, downy
velu,-e *a.* hairy, downy, pubescent
velum *m. Fung.* veil, velum
venant,-e *a. Hort.* thriving, growing well (see also **venir**)
vendable *a.* saleable, marketable
vendange *f.* 1. vintage season. 2. grape harvest; harvested grapes
vendangeable *a. Vit.* capable of being harvested
vendangeoir, vendangeret *m. Vit.* grape basket, pannier of grape harvester
vendanger *v.t., v.i. Vit.* to harvest grapes
vendangerot *m.* wicker grape basket
vendangeur,-euse *n. Vit.* grape harvester; *n. f. Bot.* aster (*Aster* sp.), autumn crocus (*Colchicum autumnale*) (= **safran des prés, tue-chien, veilleuse**)
vénéneux,-euse *a.* poisonous
venidium *m. Venidium* sp., including monarch of the veldt (*V. fastuosum*); **v. à fleur de souci** *V. decurrens*
venir *v.i.* to come up, to grow; **plant mal venu** stunted (nursery) plant; **plante qui vient à l'état spontané** self-sown plant (see also **venant**)
vent *m.* wind; **à l'abri du vent** sheltered from the wind; **v. marin** wind from the sea
vente *f.* sale, selling
venté,-e *a.* windy, windswept
venteux,-euse *a.* windy, windswept
ventilateur *m.* 1. ventilator. 2. **v. (rotatif)** fan; **v. (soufflant)** blower
ventilation *f.* ventilation
ventiler *v.t.* to ventilate
ventral,-e *a., m. pl.* **-aux** ventral
ventru,-e *a.* swollen, bulging, stout
ver *m.* 1. worm; **v. (de terre)** earthworm (= **lombric**). 2. larva, grub, caterpillar; **v. blanc** cockchafer grub (*Melolontha* sp.) or similar grub; **v. des caroubes** carob moth (larva) (*Myelois ceratoniae*) (see also **pyrale**); **v. fil de fer** wireworm; **v. des framboises** raspberry beetle (larva) (*Byturus tomentosus*); **v. du fruit** = **v. des pommes** (*q.v.*); **vers de la grappe** *Vit.* = larvae of **tordeuses de la grappe** (*q.v.*); **v. gris** cutworm, noctuid moth caterpillar; **v. des jeunes fruits** fruitlet mining tortrix (larva) (*Pammene rhediella*); **v. des pommes (et des poires)** codling moth (larva) (*Cydia pomonella*) (= **carpocapse**); **v. des prunes** red plum maggot (*Laspeyresia funebrana*); **v. à soie** silkworm
véraison *f. Vit.* veraison, first colour change, (beginning of) ripening in grapes
vératre *m. Veratrum* sp., false hellebore (= **varaire**); **v. blanc** white false hellebore (*V. album*); **v. noir** *V. nigrum*
verbénacé,-e *a.* verbenaceous; *n. f. pl.* Verbenaceae
verdage *m.* green manure crop
verdâtre *a.* greenish
verdeur *f.* 1. greenness. 2. tartness
verdir *v.t.* to make green; *v.i.* to grow, to become green (of plants)
verdissant,-e *a.* growing green, verdant
verdissement *m.* growing green, becoming green
verdoiement *m.* becoming green
verdoyant,-e *a.* verdant, green, greenish (of colour)
verdoyer *v.i.* to become green
verdure *f.* 1. greenness. 2. verdure, greenery; **v. décorative** evergreen plants, evergreens; **tapis de verdure** (*m.*) green sward. 3. greens, pot herbs (in cooking)
véré,-e *a. Vit.* having turned colour (of grapes)
vérer *v.i.* to ripen
véreux,-euse *a.* wormy, maggoty
verge *f.* rod, wand, cane; *Vit.* fruit cane; **v. d'or** golden rod (*Solidago virgaurea*)
verger *m.* orchard; **v. agricole** farm orchard; **v.**

intensif intensive orchard; **v. sur prairie** grass orchard
vergerette *f. Erigeron* sp., fleabane
vergeté,-e *a.* streaky; **v. de** streaked with
vergette *f.* small cane, switch
vergne *m.* alder (*Alnus glutinosa*) (= **aune commun**)
verjus *m.* 1. verjuice. 2. verjuice grape, unripe grape
verjuté,-e *a.* verjuiced, acid, sour
vermicompost *m.* worm compost (= **lombricompost**)
vermiculaire *a.* vermicular (see also **maladie**); *n. f.* **v. (âcre)** yellow stonecrop (*Sedum acris*) (= **orpin**)
vermiculite *f.* vermiculite
vermifuge *a.* vermifuge, anthelmintic
vernal,-e *a., m. pl.* **-aux** vernal
vernalisant,-e *a.* vernalizing
vernalisation *f.* vernalization
vernation *f.* 1. vernation (= **estivation, préfloraison**). 2. spring growth
verne *m.* = **vergne** (*q.v.*)
vernis *m.* varnish, polish; **v. du Japon** 1. lacquer tree (*Rhus verniciflua*), wax tree (*R. succedanea*) (= **laquier**). 2. tree of heaven (*Ailanthus altissima*) (= **ailante, faux vernis du Japon**)
vernonia *m.*, **vernonie** *f.* bitter leaf (*Vernonia* sp., including *V. amygdalina, V. colorata, V. calvoana*), used as pot-herbs
véronique *f. Veronica* sp., speedwell; **v. agreste** field speedwell (*V. agrestis*); **v. des champs** wall speedwell (*V. arvensis*); **v. cressonnée** brooklime (*V. beccabunga*); **v. en épis** *V. spicata*; **v. à feuille de lierre** ivy-leaved speedwell (*V. hederifolia*); **v. germandrée** large speedwell (*V. teucrium*); **v. ligneuse** shrubby veronica (*Hebe* sp.); **v. officinale** common speedwell (*V. officinalis*) (= **thé d'Europe**); **v. de Perse** Persian speedwell (*V. persica*); **v. petit-chêne** germander speedwell (*V. chamaedrys*)
verre *m.* glass, pane of glass; **v. à faible émissivité** low emissivity glass
verrine *f. Hort.* bell glass, cloche
verrue *f.* wart, gall
verruqueux,-euse *a.* verrucose, warted
versadi *m. Vit.* arched long shoot
versant *m.* slope, side, slope of roof
versatile *a. Bot.* versatile
verse *f.* laying, lodging (of crops), shedding (of grain etc.)
verser *v.t.* to pour, to pour into; *v.i.* to become laid, to lodge (of crops)
versicolore *a.* 1. variegated. 2. capable of changing colour
versoir *m.* mouldboard
vert,-e *a.* green; **bois vert** green wood; **café vert** *Coff.* unroasted coffee; **chêne vert** holm oak (*Quercus ilex*); **fruit vert** unripe fruit; **haricot vert** french bean; **légumes verts** green vegetables, greens; **plante verte** ornamental foliage plant; **en vert** *Hort.* during the summer, on the (actively-) growing plant; **greffage en vert** green grafting; *n. m.* green (colour); **v. olive** *a. inv.* olive green; **v. sombre** *a. inv.* dark green; **v. de Chine** Chinese green (*Rhamnus globosus*); **v. de gris** *Mush.* mushroom verdigris disease, mat disease (*Myceliophthora lutea*)
verticalité *f.* verticality, uprightness
verticille *m. Bot.* verticil, whorl
verticillé,-e *a. Bot.* verticillate, whorled
verticilliose *f. Verticillium* disease, wilt; **v. (de la pomme de terre)** potato verticillium wilt; **v. de l'artichaut** globe artichoke wilt; **v. du chrysanthème** chrysanthemum wilt; **v. du concombre** cucumber verticillium wilt (all caused by *Verticillium albo-atrum*); **v. du champignon de couche** *Mush.* dry bubble (*V. malthousei*); **v. de l'oeillet** carnation wilt (*Phialophora cinerescens*)
vertugadin *m. Hort.* lawns at different levels
verveine *f.* vervain, verbena (*Verbena* sp.); **v. hybride** common garden verbena (*V.* × *hybrida*); **v. citron** *Isotoma petraea*; **v. citronnelle** lemon verbena (*V. triphylla* = *Lippia citriodora*); **v. odorante** = **v. citronnelle** (*q.v.*); **v. de Miquelon** *V.* × *aubletia* (= *V. canadensis*); **v. officinale** vervain (*V. officinalis*); **v. rugueuse** *V. rigida* (= *V. venosa*)
vesce *f.* vetch (*Vicia* sp.); **v. cultivée, v. fourragère** common vetch (*V. sativa*); **v. velue** hairy vetch (*V. villosa*); **v. voyageuse** *Vicia peregrina*
vesceron *m.* tufted vetch (*Vicia cracca*)
vésiculo-arbusculaire *a.* vesiculo-arbuscular
vésicule *f.* vesicle; **v. à jus, v. juteuse** *Cit.* juice vesicle
vésiculeux,-euse *a.* vesiculate
vesou *m. Sug.* cane juice (= **jus de canne**)
vespéral,-e *a., m. pl.* **-aux** vespertine, of the evening; **saignée vespérale** *Rubb.* evening tapping
vespère *m. Vit.* vine beetle (*Vesperus xatarti*)
vesse-de-loup *f., pl.* **vesses-de-loup** *Fung.* puffball (*Lycoperdon perlatum*) and related spp.; **v.de-loup géante** giant puffball (*Calvatia gigantea*)
vétiver, vétyver *m. Vetiveria* sp., vetiver
viabilité *f.* viability
viable *a.* viable
vibrateur *m.* vibrator; **v. pour tomates** truss vibrator

vibration *f.* vibration
vibreur *m.* vibrator
vibroculteur *m.* spring-tine cultivator
viburnum *m.* viburnum (see **viorne**)
victoria *f. Victoria* sp.; **v. regia** royal water lily (*V. regia*) (= *V. amazonica*)
vidange *m.* 1. draining, emptying. 2. night-soil
vide *a.* empty; *n. m.* empty space, gap, vacuum; **sous vide** under vacuum
vide-pomme *m., pl.* **vide-pommes** apple-corer
vider *v.t.* to empty, to clear out; **se vider** to become empty, to empty
vie *f.* life; **v. latente** state of dormancy; **v. ralentie** reduced activity (of plant); **v. en vase** vase life (of flowers etc.); **v. végétale** plant life
viellir *v.i.* to grow old, to age
vieillissement *m.* aging; **v. physiologique** physiological aging
vierge *a.* virgin; *Hort.* unpollinated; **arbre vierge** *Rubb.* virgin tree; **canne vierge** *Sug.* plant cane; **forêt vierge** virgin forest
vif, vive *a.* living, alive; **haie vive** quickset hedge; **couleur vive** bright colour; **rose vif** bright pink
vigna *m. Vigna* sp., cowpea (see **niébé**)
vigne *f.* 1. grapevine (*Vitis vinifera*); **v. à bois, v.-mère** vine for rootstock production; **v. américaine** American vine; **v. folle** rank-growing vine with little fruit; **serre à vignes** vinery. 2. vineyard; **planter vigne sur vigne** replanting in a vineyard. 3. vine-like plant; **v. blanche** = **v. de Salomon** (*q.v.*); **v. de Judas, v. de Judée** bittersweet (*Solanum dulcamara*) (= **morelle douce-amère**); **v. russe** kangaroo vine (*Cissus antarctica*); **v. de Salomon** traveller's joy (*Clematis vitalba*); **v. (-)vierge** Virginia creeper (*Parthenocissus quinquefolia*); **v. (-)vierge de Veitch** Boston ivy (*P. tricuspidata*) (= **lierre japonais**)
vigne-mère *f., pl.* **vignes-mères** vine kept to provide cuttings; **v.-mère de greffons** scion-producing vine; **v.-mère de porte-greffes** rootstock-producing vine
vigneron,-onne *n.* vine grower, viticulturist; *a.* **charrue vigneronne** vineyard plough; *n. f.* vineyard plough
vignette *f.* small vine; *Coll.* meadow sweet (*Filipendula ulmaria*), traveller's joy (*Clematis vitalba*), mercury (*Mercurialis* sp.)
vigne-vierge see **vigne** 3
vignoble *m.* vineyard, vines in a vineyard; **le vignoble bordelais** the Bordeaux wine-producing district; **v. d'appellation d'origine guarantie (V.A.O.G.)** vineyard of guaranteed vintage; *a.* **pays vignoble** wine region
vignon *m. Coll.* gorse, furze (*Ulex* sp.)
vigueur *f.* vigour; **manque de vigueur** lack of strength; **v. germinative** germination energy; **v. hybride** hybrid vigour
villarsia *m.,* **villarsie** *f. Villarsia* sp., including floating heart (*Nymphoides (Villarsia) cordata*)
villeux,-euse *a.* villous, hairy
villosité *f.* villosity
vin *m.* wine; **grand vin, v. fin, v. de marque** wine from a famous vineyard, vintage wine; **v. mousseux** sparkling wine; **v. non mousseux** still wine; **v. ordinaire, v. de table** dinner wine, beverage wine; **v. de fruits** fruit wine; **v. de groseilles** currant wine; **v. de palme** palm wine; **v. pétillant** semi-sparkling wine; **v. de prunelles** sloe wine
vinaigre *m.* vinegar
vinaigrier *m.* smooth sumac (*Rhus glabra*), also applied to Sicilian sumach (*R. coriaria*) (see also **sumac**)
vinca *f. Vinca* sp., periwinkle (see also **pervenche**)
vincetoxicum *m.* vincetoxicum (*Cynanchum vincetoxicum*) (= **dompte-venin**)
vinée *f.* 1. fruit branch of a vine. 2. vintage, wine crop
vinette *f.* barberry (*Berberis vulgaris*) (= **épine-vinette**)
vinettier *m.* = **vinette** (*q.v.*)
vineux,-euse *a.* vinous; **année vineuse** good vintage year
vinicole *a.* wine-producing, viticultural
viniculture *f.* viniculture
vinifère *a.* viniferous, wine-producing
vinification *f.* vinification, winemaking
violacé,-e *a.* violaceous; **rouge violacé** violet red; *n. f. pl.* Violaceae
violaxanthine *f.* violaxanthin
violet,-ette *a.* violet, purple; **récolte (de l'asperge) à pointe violette** cutting (asparagus) at the purple-tip stage
violette *f.* violet (*Viola* sp. and other spp.); **v. des Alpes** long-spurred pansy (*V. calcarata*); **v. des bois** woodland violet (*V. sylvestris* = *V. reichenbachiana*); **v. du Cap** African violet (*Saintpaulia ionantha*) (= **saintpaulia**); **v. de la Chandeleur** = **perce-neige** (*q.v.*); **v. cornue** horned violet (*V. cornuta*); **v. du Labrador** Labrador violet (*V. labradorica*); **v. odorante** sweet violet (*V. odorata*); **v. de Parme** Parma violet (*V. alba*); **v. de Rouen** Rouen pansy (*V. rothomagensis* = *V. hispida*)
violier *m* wallflower (*Cheiranthus cheiri*) (= **giroflée jaune**); **v. d'hiver** = **perce-neige** (*q.v.*)
viorne *f.* viburnum (*Viburnum* sp.); **v. cotonneuse** wayfaring tree (*V. lantana*); **v. laurier thym** = **laurier-thym** (*q.v.*); **v. mancienne** =

v. cotonneuse (*q.v.*); **v. obier** guelder rose (*V. opulus*) (see also **boule-de-neige, caillebot**); **v. odorante** *V. farreri*
viorne-tin *f., pl.* **viornes-tin** laurustinus (*Viburnum tinus*) (= **laurier tin**)
vipérine *f. Echium* sp.; **v. (commune) viper's bugloss** (*E. vulgare*)
virage *m. Hort.* colour change
viral,-e *a., m. pl.* **-aux** viral; **infection virale** viral infection
virer *v.i.* (of fruit) to change colour
virescence *f.* virescence (= **chloranthie**)
vireux,-euse *a.* 1. poisonous, noxious. 2. malodorous
viroïde *m.* viroid
virole *f. Virola* sp.; **v. sébifère** *V. sebifera* (= **arbre à suif**) (see also **yayamadou**)
virologie *f.* virology
virose *f.* virus disease; **atteint(e) de virose** virus-infected
virosé,-e *a.* virus-affected, virus-infected
virulent,-e *a.* virulent; **mutant virulent** virulent mutant
virus *m.* virus; **v. filtrant** filterable virus; **v. latent** latent virus; **v. de la mosaïque du concombre** cucumber mosaic virus (CMV) (see also under **mosaïque, panachure**); **v. de rabougrissement** stunt virus; **v. transmis par le sol, v. conservé dans le sol** soil-borne virus
viscosité *f.* viscosity, stickiness
visnage, visnague *m.* bishop's weed (*Ammi visnaga*)
visnée *f. Visnea mocanera*
visqueux,-euse *a.* viscous
vitacées *f. pl.* Vitaceae
vitalité *f.* vitality
vitamine *f.* vitamin
vitaminique *a.* relating to vitamins; **composition vitaminique** vitamin composition
vitesse *f.* speed, promptness; **v. d'avancement** forward speed; **v. de développement** speed of development; **v. germinative** rate of germination
viticole *a.* viticultural; **expérimentation viticole** viticultural research; **sol viticole** vineyard soil
viticulteur *m.* viticulturist, vine-grower
viticulture *f.* viticulture, vine culture
vitifère *a.* vitiferous, vine-producing
viti-viniculture *f.* vitiviniculture
vitrage *m.* glazing; **double vitrage** double glazing
vitré,-e *a.* glazed; **cage vitrée** hand-light; **surface vitrée** glazed surface
vitrer *v.t.* to glaze
vitrescence *f. PD* water core, glassiness (of apple) (= **pommes vitreuses**)
vitreux,-euse *a.* vitreous, glassy; **pommes vitreuses** *PD* water core, glassiness (of apple)
vitroculture *f. in vitro* culture
vitrométhode *f.* method of growing *in vitro*
vitroplant *m.* plant raised *in vitro*
vittigère *a.* vittate, with longitudinal ridges or stripes
vivace *a.* perennial; **pois vivace** everlasting pea (*Lathyrus latifolius*); *n. f. Coll.* perennial plant
vivant,-e *a.* alive, living; **cellules vivantes** living cells
vivipare *a. Bot.* viviparous
vivrier,-ère *a.* food; **culture vivrière** food crop; *n. m.* food crop; plot of food crops
voandzou *m.* Bambara groundnut (*Voandzeia subterranea*) (= **pois de terre**)
voavanguier *m.* voavanga (*Vangueria madagascariensis*), a small tree with edible fruit
vocation *f.* vocation; *Hort., Plant.* suitability (for), potential; **v. cacaoyère d'un sol** suitability of a soil for cacao; **v. culturale** suitability for culture; **v. horticole** suitability for horticulture; **v. oléicole** olive-growing capabilities
voie *f.* 1. way, road; **v. d'accès** access road, track. 2. process, method; **voie humide** *Coff.* wet process
voile *m. Bot.* velamen; *Mush.* veil
volant *m.* reserves required (for continued production)
volatil,-e *a.* volatile; **composés volatils** volatile compounds
volatilisation *f.* volatilization
volatilité *f.* volatility
volée *f.* flight; *Hort.* **à la volée** broadcast; *Mush.* etc. flush
volige *f.* batten; **caisse en voliges** crate
voligeage *m.* battening, cladding
voltage *m.* voltage
volubile *a.* voluble, twining, climbing (of plants)
volubilis *m.* convolvulus; **v. des jardins** morning glory (*Ipomoea purpurea*)
volubilisme *m.* twining property of plants
volume *m.* volume; **v. des pores** pore space
volumineux,-euse *a.* voluminous, large
volumique *a.* voluminal; **masse volumique de sol** soil bulk density
volva *f. Fung.* volva
volvaire *f.* volvaria; **v. comestible, v. cultivée** paddy straw mushroom (*Volvariella volvacea*) (= **champignon des pailles**); **v. gluante** handsome volvaria (*V. speciosa*); **v. soyeuse** silky volvaria (*V. bombycina*)
volve *f. Fung.* = **volva** (*q.v.*)
vomiquier *m.* strychnine tree (*Strychnos nux-vomica*)
vouge *m. Hort.* billhook
voûte *f.* vault, arch; *Arb.* canopy
vrac *m.*, **en vrac** loose, in bulk

vriesea *m.* *Vriesea* sp., including flaming sword (*V. splendens*)

vrille *f.* 1. *Bot.* tendril. 2. *Coll.* bindweed (*Convolvulus arvensis*)

vrillé,-e *a.* *Bot.* with tendrils, tendrilled

vrillée *f.* bindweed (*Convolvulus arvensis*); **v. (bâtarde)** black bindweed (*Polygonum convolvulus*)

vulgarisateur,-trice *n.* *Hort.* advisory worker, extension worker

vulgarisation *f.* *Hort.* advisory work, extension; **service de vulgarisation** advisory service, extension service

vulnéraire *a.* vulnerary, healing; **plante vulnéraire** healing plant; *n. m.* vulnerary; *n. f. Bot.* kidney vetch (*Anthyllis vulneraria*)

vulpin (*m.*) **(des champs)** blackgrass (*Alopecurus myosuroides*); **v. des prés** meadow foxtail (*A. pratensis*)

W

wachendorfie *f. Wachendorfia*, a genus which includes ornamental half-hardy tuberous-rooted plants
wahlenbergie *f. Wahlenbergia* sp., bell-flower
wasabi *m.* wasabi (*Eutrema wasabi*)
washingtonia *m. Washingtonia* sp., palm
wédélie *f. Wedelia* sp., including *W. radiosa*
weigela, weigelia *m. Weigela* sp., Japanese honeysuckle
weinmannia *m.*, **weinmannie** *f. Weinmannia*, a genus containing evergreen ornamental woody plants
wellingtonia *m.* = **sequoia** (*q.v.*)
welwitschie *f.* welwitschia (*Welwitschia mirabilis*)
whitfieldie *f. Whitfieldia* sp., including *W. longifolia*, an African species whose seeds are believed to have magical properties
wigandie *f.*, **wigandia** *m. Wigandia* sp., including *W. viegieri*, an ornamental shrub
wilt *m. PD* wilt; **w. à cochenilles (de l'ananas)** pineapple mealybug wilt (*Dysmicoccus brevipes*)
wistaria, wisteria *m.*, **wistarie** *f. Wistaria* sp., wistaria (see also **glycine**)
witloof *m.* witloof chicory (= **chicorée (à grosse racine) de Bruxelles, endive**)
wulffie *f. Wulffia* sp., including *W. stenoglossa*

X

xanthium *m. Xanthium* sp. (see **lampourde**)
xanthocarpe *a.* xanthocarpous
xanthocéras, xanthocère *m. Xanthoceras* sp., including *X. sorbifolium*, an ornamental tree
xanthophylle *f.* xanthophyll
xanthorrhée *f. Xanthorrhoea* sp., grass tree
xanthorrhize *f.* **(à feuilles de persil)** yellow root (*Xanthorhiza (Zanthoriza) simplicissima* = *X. apiifolia*)
xanthosoma, xanthosome *m. Xanthosoma* sp. (see **chou caraïbe, macabo**)
xanthoxylum, xanthoxyle *m. Zanthoxylum* (*Xanthoxylum*) sp. (= **clavalier**); **x. massue d'Hercule** Hercules' club (*Z. clava-Herculis*) (see also **poivrier du Japon**)
xénie *f.* xenia
xénopollinisé,-e *a.* cross-pollinated
xéranthème *m. Xeranthemum* sp., including *X. annuum* (= **immortelle annuelle, i. de Belleville**); **x. fermé** *X. inapertum*
xérique *a.* xeric
xérophile *a.* xerophilous
xérophyte *a.* xerophytic; *n. f.* xerophyte
ximénie *f. Ximenia* sp.; **x. du Gabon** seaside plum (*X. americana*) (see also **citron de mer, prunier de mer**)
xylébore *m. Xyleborus* sp., shot-hole borer
xylème *m.* xylem
xylémien,-ienne *a.* relating to xylem; **sève xylémienne** xylem sap
xylémique *a.* relating to xylem; **sève xylémique** xylem sap
xylocarpe *a.* xylocarpous
xylogenèse *f.* xylogenesis
xylophage *a.* wood-eating; **insecte xylophage** wood-eating insect
xylopia *m.*, **xylopie** *f. Xylopia* sp. (see also **poivrier de Guinée**)
xyloporose *f. PD* xyloporosis (= **cachexie**)
xyride *f. Xyris* sp., including *X. operculata*

Y

yacon *m.* = **poire de terre** (*q.v.*)
yang-tao *m. inv.* Chinese gooseberry, kiwi fruit (*Actinidia chinensis*) (= **groseille de Chine**)
yayamadou *m.* ucahuba (*Virola surinamensis*) (see also **virole**)
yèble *f.* dwarf elder (*Sambucus ebulus*) (see also **sureau**)
yeuse *f.* holm oak (*Quercus ilex*) (= **chêne vert**)
yeux *m. pl.* (plural of **oeil** (*q.v.*)); **y. latents** latent buds; **y. multiples** multiple buds; **y. d'une plante** eyes, buds of a plant
ylang-ylang ylang-ylang (*Cananga odorata*) (= **ilang-ilang**)
yosta *m.* josta (berry) (blackcurrant × gooseberry hybrid)
yponomeute *f.* ermine moth (*Yponomeuta* sp.) (see also **hyponomeute**)
ypréau *m.* 1. white poplar, abele (*Populus alba*). 2. wych elm (*Ulmus glabra*)
yucca *m.* *Yucca* sp. (see also **papillon**); **y. à feuilles d'aloès** *Y. aloifolia*; **y. superbe** Spanish dagger (*Y. gloriosa*)

Z

zacinthe *f. Zacintha* sp.; **z. verruqueuse** *Z. verrucosa*
zagrinette *f.* beach palm (*Bactris major*)
zaïrois,-e *a.* of Zaïre
zamia *m.*, **zamie** *f. Zamia* sp., zamia (Cycadaceae); **z. de Wallis** *Z. wallisii*
zamier *m.* zamia (plant)
zantedeschia *m. Zantedeschia* sp. (see also **richardia**)
zauschnérie *f. Zauschneria* sp., including Californian fuchsia (*Z. californica*)
zébré,-e *a.* striped
zéro *m.* zero, nought; **z. de germination** germination threshold temperature; **z. végétatif** growth threshold temperature
zeste *m.* 1. peel of citrus fruit, zest. 2. partition quartering the kernel (of walnut)
zeuzère *f.* leopard moth (*Zeuzera pyrina*)
zinc *m.* zinc
zincique *a.* of zinc, zincic; **nutrition zincique** zinc nutrition
zingiber *m. Zingiber* sp. (see **gingembre**)
zinnia *m.* zinnia (*Zinnia* sp.), including *Z. elegans*; **z. du Mexique** *Z. angustifolia* (= *Z. haageana*)
ziziphus *m. Ziziphus* sp. (see also **jujubier**)
zone *f.* zone; **z. aride** arid zone; **z. capillaire** capillary region; **z. de croissance** growth zone; **z. radiculaire (de la tige de la canne à sucre)**; *Sug.* root band (of cane stalk); **zones tropicales humides** humid tropics
zoné,-e *a.* zoned, zonate
zoocécidie *f.* zoocecidium
zoochore *a.* zoochoric; *n. f.* **(plante) zoochore** zoochore
zoospore *f.* zoospore
zornie *f. Zornia* sp., including *Z. diphylla*
zygène *f.* burnet moth (Zygaenidae); **z. des cocotiers** (*Coconut*) *Chalconycles catori*
zygocactus *m. Zygocactus* (*Schlumbergera*) sp., leaf-flowering cactus
zygomorphe *a.* zygomorphic, zygomorphous
zygopétale *m. Zygopetalum* sp. (orchid)
zygophylle *f.*, **zygophyllum** *m. Zygophyllum* sp. (see **fabagelle**)
zygote *m.* zygote
zymase *f.* zymase
zymogène *a.* zymogenous, producing enzymes

ENGLISH–FRENCH INDEX

aberration *n.* **aberration**
abiotic *a.* **abiotique**
abort *v.* **avorter, couler**
aborted *a.* **avorté**
abortion *n.* **avortement, coulure**
abscission *n.* **abscission**; leaf a. **a. foliaire**
absorb *v.* **absorber**
absorbable *a.* **absorbable**
absorbed *a.* **absorbé**
absorbing *a.* **absorbant**
absorption *n.* **absorption, imbibition**
abutilon *n.* **abutilon**
acacia *n.* **acacia**
acanthoid *a.* **acanthoïde**
acanthus *n.* **acanthe**
acaricide *n.* **acaricide, miticide**
acariosis *n.* **acariose**
acarpous *a.* **acarpe**
acauline *a.* **acaule**
acceptor *n.* **accepteur**
access *n.* **abord**
acclimatization *n.* **acclimatation**
acclimatize *v.* **acclimater, naturaliser**
acclimatizer *n.* **acclimateur**
accommodation *n.* **accommodat, accommodation**
accrescent *a.* **accrescent**
accumbent *a.* **accombant**
accumulator *n.* **accumulateur**
acellular *a.* **acellulaire**
achene *n.* **akène**
acicular *a.* **aciculaire**
acid *a.* **acide, aigre, sur**; *n.* **acide**
acidify *v.* **acidifier**
acidity *n.* **acidité**
aconite *n.* **aconit**; winter *a.* **helléborine**
acorn *n.* **gland**
acreage *n.* **acrage**
acrid *a.* **âcre**
acridian *a.* & *n.* **acridien**
acridity *n.* **âcreté**
acrotonic *a.* **acrotone**
activity *n.* **activité**
aculeate *a.* **aculé**
acuminate *a.* **acuminé, alêné**
acylated *a.* **acylé**
adaptable *a.* **complaisant**
additional *a.* **supplémentaire**
additive *a.* **additif**
adhesion *n.* **adhérence**
adhesive *a.* **adhérent, adhésif**; *n.* **adhésif, agglutinant**
adjustable *a.* **réglable**
adjustment *n.* **réglage**
adonis *n.* **adonis**
advancement *n.* **accélération**
adventitious *a.* **adventice, adventif**
aeolian *a.* **éolien**
aerate *v.* **aérer**
aerated *a.* **aéré**
aeration *n.* **aération**
aerial *a.* **aérien**
aerobic *a.* **aérobie**
aeroplane *n.* **avion**
aeroponic *a.* **aéroponique**
aerosol *n.* **aérosol**
aestival *a.* **estival**
aestivation *n.* **estivation, préfleuraison**
aetiology *n.* **étiologie**
affiliation *n.* *PB* **filiation**
African daisy *n.* **dimorphothéca**
African lily *n.* **agapanthe**
African violet *n.* **saintpaulia**
after-ripening *n.* **post-maturation**
after-sale *a.* **après-vente**
age *v.* **vieillir**
agent *n.* **agent**
agricultural *a.* **agricole**
agriculture *n.* **agriculture**
agrimony *n.* **aigremoine**
agroclimatology *n.* **agroclimatologie**
agroecosystem *n.* **agroécosystème**
agroforestry *n.* **agroforesterie**
agrometeorology *n.* **agrométéorologie**
agronomist *n.* **agronome**
agronomy *n.* **agronomie**
air-conditioned *a.* **climatisé**
air-conditioner *n.* **climatiseur**
air conditioning *n.* **climatisation**
air-plant *n.* **aéride**
akee *n.* **fisanier, ris de veau**
albino *a.* **albinos**
albumen *n.* **albumen**
albumin *n.* **albumine**
alburnum *n.* **aubier**
alcohol *n.* **alcool**
alder *n.* **aune**
alder buckthorn *n.* **bourdaine, frangule**
alexanders *n.* **maceron**
alga *n.* **algue**

algicidal *a.* **algicide**
alkaline *a.* **alcalin**
alkalinity *n.* **alcalinité**
alkaloid *a.* & *n.* **alcaloïde**
alkanet *n.* **orcanette**
allele *n. PB* **allèle**
allopolyploid *a.* & *n.* **allopolyploïde**
allspice *n.* **pimenta, toute-épice**
almond *n.* **amande**; a. tree **amandier**; tropical a. tree **a. des Antilles**
aloe *n.* **aloès**
alpine *a.* **alpin**; alpine plant *n.* **alpine**
alternation *n.* **alternance**
alum *n.* **alun**
alveole *n.* **alvéole**
alyssum *n.* **alysse**
Amanita sp. *n.* **amanite**
amaranth *n.* **amarante**
amendment *n.* **amendement**
amino acid *n.* **acide aminé, amino-acide**
ammonia *n.* **ammoniaque**
amorphous *a.* **amorphe**
amortization *n.* **amortissement**
amount *n.* **teneur**
amphibious *a.* **amphibie**
amputate *v.* **amputer**
anaerobic *a.* **anaérobie**
anaerobiosis *n.* **anaérobiose**
analyser *n.* **analyseur**
analysis *n.* **analyse**
anchorage *n.* **ancrage**
anchored *a.* **ancré**
androgenetic *a.* **androgénétique**
andromonoecious *a.* **andromonoïque**
anemone *n.* **anémone**; wood a. **sylvie**
anhydrous *a.* **anhydre**
aniseed *n.* **anis**; aniseed-flavoured *a.* **anisé**
annatto *n.* **rocouyer**
annona *n.* **anone**
annual *a.* **annuel**
ant *n.* **fourmi**
anther *n.* **anthère**
antheridium *n.* **anthéridie**
anthesis *n.* **anthèse**
anthocyanin *n.* **anthocyane**
anthracnose *n.* **anthracnose**
antidote *n.* **contre-poison**
antifeedant *n.* **anorexigène, antiappétant**
anti-hail *a.* **paragrêle**
anti-leprosy *a.* **antilépreux**
anti-shock *a.* **antichoc**
antitranspirant *a.* & *n.* **antitranspirant**
anti-tumour *a.* **antitumoral**
apera, silky *n.* **jouet-du-vent**
aphid *n.* **aphidien, puceron**
apiarian *a.* **abeiller**
apiary *n.* **rucher**
apogamy *n.* **apogamie**
apomictic *a.* **apomictique**
apoplexy *n.* **apoplexie**
apparatus *n.* **appareil**
apple *n.* **pomme**; a. tree **pommier**
appraisement *n.* **appréciation**
apricot *n.* **abricot**; a. tree **abricotier**; clingstone a. **alberge**; clingstone a. tree **albergier**
apricot-flavoured *a.* **abricoté**
arabis *n.* **arabette**
arable *a.* **cultivable, labourable**
arboreal *a.* **arboricole**
arboriculture *n.* **arboriculture**
arbor-vitae *n.* **thuya**
arbour *n.* **charmille, gloriette, tonnelle**
arched *a.* **arqué**; a. shoot **arceau**
archegonium *n.* **archégone**
arcure *n.* **arcure**
area *n.* **aire**
areca palm *n.* **arec, aréquier**
argan tree *n.* **arganier**
aril *n.* **arille**
aristate *a.* **aristé**
arolla pine *n.* **arole**
arrangement *n.* **disposition**
arranging *n.* **aménagement**
arrowhead *n.* **sagittaire**
arrowing *n. Sug.* **fléchage**
artichoke *n.* Chinese a. **crosne**; globe a. **artichaut**; Jerusalem a. **topinambour**
asbestos *n.* **amiante**
asepalous *a.* **asépale**
asexual *a.* **asexuel**
ash *n.* **cendre**
ash (tree) *n.* **frêne**
asparagus *n.* **asperge**
aspect *n.* **exposition, orientation**
aspen *n.* **tremble**
asphyxia *n.* **asphyxie**
assimilable *a.* **assimilable**
associated *a.* **associé**
aster, China *n.* **reine-marguerite**
astringency *n.* **astringence**
atomize *v.* **atomiser**
attractant *n.* **attractif**
aubrietia *n.* **aubriétie**
auricle *n.* **auricule**
auricula *n.* **auricule**
autoanalyser *n.* **autoanalyseur**
autoclaving *n.* **appertisation**
autograft *n.* **autogreffe**
availability *n.* **disponibilité**
avens *n.* **benoîte**
average *n.* see **moyen**
avocado (fruit) *n.* **avocat**; a. tree **avocatier**
awn *n.* **arête**
axe *n.* **hache**
axenic *a.* **axénique**
axil *n.* **aisselle**

axillary *a.* **axillaire**
azalea *n.* **azalée**
azarole (fruit) *n.* **azerole**; a. tree **azerolier**

bacterial disease *n.* **bactériose**
bacterium *n.* **bactérie**
bag *n.* **sac**
bag *v.* **ensacher**
bagger *n.* **ensacheur**
bait *n.* **appât**
bait-trap *n.* **piège-appât**
balance *n.* **équilibre**; b. (sheet) **bilan**
balanced *a.* **équilibré**
balance sheet *n.* **bilan**
bale *n.* **balle, ballot**
ball up (*v.*) plant **emmotter**
balm *n.* **citragon**
balsam *n.* **balsamine, baume**
balsam apple *n.* **momordique**
balsam tree *n.* **balsamier**
bamboo *n.* **bambou**
banana *n.* (fruit) **banane**; b. plant **bananier**
band *n.* **bande**; sticky band **bande engluée**
band (*v.*) trees **engluer**
banger *n.* **pétard**
bank *n.* **talus**; b. of manure etc. **accot**; data b. **banque de données**; gene b. **b. de gènes**
bank-bed *n.* **ados**
bank up *v.* **rechausser**
baobab *n.* **baobab**
bar *n.* **barre**
Barbados cherry *n.* **acérolier**
barbed *a.* **barbelé**
barberry *n.* **épine-vinette**
bare *a.* **nu**
bare *v.* **déchausser**
bark (of plant) *n.* **écorce**
bark (*v.*) tree **écorcer**
bark-dwelling *a.* **corticole**
barnyard grass *n.* **panisse**
barrel *n.* **tonneau**
barren *a.* **infécond**
barrenness *n.* **infécondité**
base *n.* **base**; data b. **b. de données**
Basella sp. *n.* **baselle**
basket *n.* **corbeille, hotte, panier**
basic slag *n.* **laitier basique**
basil *n.* **basilic**
basin *n.* **bassin, cuvette**
basipetal *a.* **basipète**
basitonic *a.* **basitone**
bat (Chiroptera) *n.* **chauve-souris**; fruit b. **roussette**
batten *n.* **volige**
bay *n.* **daphné, laurier**
bead tree *n.* **adénanthère; margousier**
bean *n.* **haricot, fève**
bean caper *n.* **fabagelle**
bearberry *n.* **busserole**
bearded *a.* **barbu**
beard-tongue *n.* **galane**
bear's breech *n.* **branche-ursine**
bed *n.* **banc, côtière, couche, planche**
bed *v.* **asseoir**
bedeguar *n.* **bédégar**
bee *n.* **abeille**
beech *n.* **hêtre**
beechnut *n.* **faîne**
beehive *n.* **ruche**
beet *n.* **bette, betterave**
beetle *n.* **coleoptère**; bark b. **scolyte**; ground b. **carabe**
begonia *n.* **bégonia**
behaviour *n.* **comportement**
belladonna lily *n.* **amaryllis**
bells of Ireland *n.* **clochettes d'Irlande**
bellflower *n.* **campanule**
bell-wort *n.* **uvulaire**
belt *n.* **ceinture**
bench *n.* **banc**
bending *n.* **flexion**
bent-grass *n.* **agrostide**
benzoin *n.* **benjoin**
bergamot (fruit) *n.* **bergamote**; b. tree **bergamotier**
Bermuda grass *n.* **chiendent**
berry *n.* **baie**
betel *n.* **bétel**
betony *n.* **bétoine**
bevel *n.* **biseau**
biennial *a.* **bisannuel**; b. bearing **alternant**
bifoliate *a.* **bifolié**
bigarreau (fruit) *n.* **bigarreau**; b. tree **bigarreautier**
bilberry *n.* **airelle, myrtille**; b. plant **myrtillier**
billhook *n.* **croissant, gouet, serpe, serpette**
bilobed *a.* **dilobé**
bindweed *n.* **campanelle, liseron**
bioassay *n.* **bioessai**
biochemical *a.* **biochimique**
biochemistry *n.* **biochimie**
biocoenosis *n.* **biocénose**
biogas *n.* **biogaz, gaz de fumier**
biogenesis *n.* **biogenèse**
bioindicator *n.* **bioindicateur**
biomass *n.* **biomasse**
biosphere *n.* **biosphère**
birch *n.* **bouleau**
bird's foot trefoil *n.* **lotier**
biriba *n.* **abriba, fausse chérimole**
birthwort *n.* **aristoloche**
bitter *a.* **amer, âpre**
bitterness *n.* **amertume**
bitter pit *n.* **tache amère**
bittersweet *n.* **douce-amère**

black *a.* **noir**
blackberry *n.* **mûre**
blackbird *n.* **merle**
black-currant *n.* **cassis**; b. bush **cassissier**
blackgrass *n.* see **vulpin**
black leg *n. PD* **pied-noir**
bladder cherry *n.* **coqueret**
bladder nut *n.* **staphylier**
bladder senna *n.* **faux séné**
bladderwort *n.* **utriculaire**
blanch *v.* **blanchir**
blast *n. PD* **dessèchement**
bleeding heart *n.* **coeur-de-Jeannette**
blewit *n. Fung.* **pied-bleu**
blight *n.* **cloque**
blighted *a.* **cloqué**
blind (of plant) *a.* **borgne**
blind (of window) *n.* **store**
block *n.* **bloc**
block-making machine *n.* **presse-mottes**
blood-root *n.* **sanguinaire**
bloom *n.* **fleur**; (on fruit etc.) **efflorescence, pruine**
blossom *n.* **fleur**
blossom *v.* **éclore, fleurir**
blow *n.* **coup**
blower *n.* **souffleur**
bluebell *n.* **jacinthe**; Californian b. **némophile**
blueberry *n.* (*Canada*) **bleuet, bleuetier**
blue-eyed Mary *n.* **petite bourrache**
blunt *a.* **obtus**
blusher *n. Fung.* **golmotte**
boiler *n.* **chaudière**
boiling *n.* **ébullition**
bole *n.* **fût**
boletus *n. Fung.* **bolet**
bolt *v.* **monter**
bolter *a.* & *n.* **accéléré**
bolting *n.* **montaison, montée**
bombyx moth *n.* **bombyx**
bone *n.* **os**
boom *n.* **rampe**; b. carrier **porte-rampe**
borage *n.* **bourrache**
Bordeaux mixture *n.* **bouillie bordelaise**
border *n.* **marge**; b. effect **effet de bordure**
bore *v.* **sonder**
borehole *n.* **forage, sondage**
borer *n.* **borer, foreur**
bottleneck *n.* **goulet**
bough *n.* **rameau**
bourse *n.* **bourse**
bowl *n.* **godet**
box *n.* **boîte, caissette**
box (tree) *n.* **buis**
box elder *n.* **négondo**
box thorn *n.* **lyciet**
brackish *a.* **saumâtre**
bract *n.* **bractée**
brake *v.* **freiner**
bramble *n.* **ronce**
bran *n.* **son**
branch *n.* **branche, charpentière, ramure**
branched *a.* **branchu**
branching *n.* **branchaison, embranchement**
break *n.* **brisure, cassure**
break *v.* **concasser, écabosser, émotter**
breakdown *n.* **décomposition**; (of machine) **panne**
breaking *n.* **cassage, casse, cassement**
breed *v.* **sélectionner**
breeder *n.* **obtenteur, sélectionneur**
brenner *n.* **rougeot**
brewing *a.* **brassicole**
brindle *n.* **brindille**
brittle *a.* **cassant**
broad-leaved tree *n.* **feuillu**
broccoli *n.* **brocoli**
bromeliad *n.* **bromélie**
bromine *n.* **brome**
broom *n.* **balai**
broom (plant) *n.* **genêt**
broomrape *n.* **orobanche**
brown *a.* **brun**
brown *v.* **brunir, roussir**
browning *n.* **brunissement, roussissement**
browsed *a.* **abrouti**
bruchid beetle *n.* **bruche**
bruise *n.* **meurtrissure, talure**
brush *n.* **brosse**
brush *v.* **brosser**
bryony *n.* black b. **tamier**; white b. **bryone**
buckle *v.* **boucler**
buckthorn *n.* **nerprun**
bud *n.* **bourgeon, bouton, gemme, œil, prompt-bourgeon**; basal b. **sous-oeil**; b. (for budding) **écusson**
bud *v.* **gemmer**; **écussonner**
budding knife *n.* **écussonnoir**
budwood *n.* **bois de greffe**
buffer *n.* **tampon**
bug *n.* **punaise**
bulb *n.* **bulbe, oignon**
bulbil *n.* **bulbille**
bulbing *n.* **bulbaison**
bulbous plant *n.* **bulbeuse**
bullfinch *n.* **bouvreuil**
bumble-bee *n.* **bourdon**
bunch *n.* **grappe, régime**; (of flowers) **bouquet**
bunching machine *n.* **botteleuse**
bund *n.* **diguette**
bundle *n.* **botte, faisceau**
burdock *n.* **bardane**
burn *n.* **brûlure**
burn *v.* **brûler**
burr-knot *n.* **broussin**
burrowing *a.* **mineur**

burst *n.* **éclatement**; bud b. **bourgeonnement, boutonnement, éclatement**
burst *v.* **éclater**
bury *v.* **enfouir, enterrer**
brush *n.* **arbuste, buisson**
buttercup *n.* **bouton d'or, renoncule**
buttress *n.* **contrefort**

cabbage *n.* **chou**; c. white butterfly *n.* **piéride**
cacao *n.* **cacao**; c. tree **cacaoyer**
calabash *n.* **calebasse**; c. tree **calebassier**; c. nutmeg tree *n.* **faux muscadier**
caladium *n.* **caladion**
calamint *n.* **calament**
calcareous *a.* **calcaire**
calcic *a.* **calcique**
calcicolous *a.* **calcicole**
calcifugous *a.* **calcifuge**
calendula *n.* **calendule**
calibrate *v.* **calibrer, étalonner**
callus *n.* **cal, callus, callosité**; callus-forming **callogène**
calyx *n.* **calice**
camellia *n.* **camélia**
campanula *n.* **campanule**
camphor *n.* **camphre**; c. tree **camphrier**
campion *n.* **coquelourde, lychnide**
Canary grass *n.* **alpiste**
candytuft *n.* **corbeille**
canebrake *n.* **cannaie**
cane rat *n.* **agouti, aulacode**
canker *n.* **chancre**
canna lily *n.* **balisier, canna**
cannery *n.* **conserverie**
canopy *n.* **canopée**
cap *n.* **capuchon**
capacity *n.* **capacité**
caper *n.* **câpre**, c. bush **câprier**
capillarity *n.* **capillarité**
capillary *a.* **capillaire**
capping (soil) *a.* **battant, glaçant**
carambola *n.* **carambole**; c. tree **carambolier**
caraway *n.* **carvi**
carbohydrate *n.* **glucide**
carbon *n.* **carbone**; c. balance **bilan carbonique**
carboxylation *n.* **carboxylation**
carcinogen *n.* **cancerigène**
card *n.* **carte**; punched c. **c. perforée**
cardamom *n.* **cardamome**
cardoon *n.* **cardon**
carnation *n.* **oeillet**
carob (bean) *n.* **caroube**; c. tree **caroubier**
carpel *n.* **carpelle**
carpet *n.* **tapis**
carrot *n.* **carotte**
carrying capacity *n.* *AH* **charge**
cart *n.* **charrette**
caruncle *n.* **caroncule**
case *n.* **boîtier, caisse**
case *v.* *Mush.* **gobeter**
cashew tree *n.* **anacardier, cajou**
cask *n.* **tonne**
cassava *n.* **manioc**
castrate *v.* **châtrer**
casuarina *n.* **filao**
catalyst *n.* **catalyseur**
catchfly *n.* **silène**
catechu *n.* **cachou**
caterpillar *n.* **chenille**
catmint *n.* **cataire**
cauliflower *n.* **chou-fleur**
cavity *n.* **cavité**
cedar *n.* **cèdre**
celandine *n.* **chélidoine, éclaire, ficaire**
celeriac *n.* **céleri-rave**
celery *n.* **céleri**
cell *n.* **cellule**
cep *n.* **cèpe**
Cercospora disease *n.* **cercosporiose**
cespitose *a.* **cespiteux**
chain *n.* **chaîne**
chalcid *n.* **chalcis**
chalk *n.* **craie**
chalky *a.* **calcaire**
chamber *n.* **chambre**
chamomile *n.* **camomille, matricaire**
change colour *v.* **virer**
channel *n.* **buse, rigole**
chard *n.* **carde**
chaste tree *n.* **gattilier**
chayote *n.* **chayote, christophine**
check *n.* **contrôle**
chelate *n.* **chélate**
chemosterilant *n.* **chimiostérilant**
chemotaxonomy *n.* **chimiotaxonomie**
chemotropism *n.* **chimiotropisme**
chemotype *n.* **chiomiotype**
cherimoya (fruit) *n.* **chérimole**; c. tree **chérimolier**
cherry *n.* **cerise**; c. tree **cerisier**
cherry-laurel *n.* **laurier-cerise**
cherry pie *n.* **cloucourde, héliotrope du Pérou**
chervil *n.* **cerfeuil**
chestnut *n.* **châtaigne**. c. tree **châtaignier**; chestnut growing **castanéiculture**
chick pea *n.* see **chiche**
chicory *n.* **chicorée**
chilling requirement *n.* **exigence en froid**
chimera *n.* **chimère**
chimney *n.* **cheminée**
chinquapin *n.* **chincapin**
chives *n.* **ciboulette**
chloride *n.* **chlorure**
chlorine *n.* **chlore**
chlorophyll *n.* **chlorophylle**

chloroplast *n.* **chloroplaste**
chlorosis *n.* **chlorose, jaunisse**
chorology *n.* **chorologie**
chromatography *n.* **chromatographie**
chrysanthemum *n.* **chrysanthème**
cider *n.* **cidre**
ciliate *a.* **cilié**
cineraria *n.* **cinéraire**
cinnamon (bark) *n.* **cannelle**; c. tree **cannellier**
cinquefoil *n.* **quintefeuille**
circle *n.* **cercle, rond**
cistern *n.* **citerne**
citron (fruit) *n.* **cédrat**; c. tree **cédratier**
citronella *n.* **citronnelle**
citrus *n.* **agrume**
cladding *n.* **revêtement**
clary *n.* **orvale, sauge, sclarée**
clay *n.* **argile, glaise**
clean *v.* **curer, décaper, émonder, nettoyer**
clear *v.* **déblayer, débroussailler, débrousser, défricher**; c. mud *v.* **débourber**
clearing *n.* **clairière**
clematis *n.* **clématite**
clementine (fruit) *n.* **clémentine**; c. tree **clémentinier**
click-beetle *n.* **taupin**
climacteric *a.* & *n.* **climatérique**
climate *n.* **climat**
climbing *a.* **grimpant, volubile**
cloche *n.* **cloche**
clod *n.* **motte**
clone *n.* **clone**; meristem-culture c. **mériclone**; *v.* **cloner**
close *a.* **serré**
close-cropped *a.* **ras**
closing *n.* **fermeture**
clove *n.* **girofle**; c. tree **giroflier**
clover *n.* **trèfle**
clubroot *n.* **hernie**
clump *n.* **massif**
coarse *a.* **grossier**
coat *v.* **enduire, enrober**
cobweb disease *n.* *Mush.* **toile**
coca *n.* **coca**; c. shrub **cocaïer**
cockchafer *n.* **hanneton**
cockscomb *n.* **célosie, crête-de-coq**
cocksfoot *n.* **dactyle**
cocoa *n.* **cacao**; cacoa growing **cacaoculture**
coconut *n.* **coco**; c. palm **cocotier**
cocoplum *n.* **chrysobalanier**
cocoyam *n.* **colocase, macabo**
codling moth *n.* **carpocapse**
cofactor *n.* **cofacteur**
coffee *n.* **café**; c. bush **caféier**
co-kriging *n.* *Stat.* **cokrigeage**
collar *n.* **collet, collier**
collect *v.* **ramasser**
Colorado beetle *n.* **doryphore**
colouring *n.* **coloration**
colourless *a.* **incolore**
coltsfoot *n.* **tussilage**
columbine *n.* **ancolie**
column *n.* **colonne**
coma *n.* *Bot.* **chevelure**
community *n.* **communauté**
compaction *n.* **compactage**
compatibility *n.* **compatibilité**
compete *v.* **concurrencer**
composite *n.* **composé**
compost *n.* **compost, terreau**
compound *n.* **composé**
computer *n.* **ordinateur**
concrete *n.* **béton**
conditioner *n.* **conditionneur**
conductivity *n.* **conductance, conductibilité, conductivité**
conidium *n.* **conidie**
connective *a.* **connectif**
constituent *a.* & *n.* **constituant**
consumer *n.* **consommateur**
container *n.* **conteneur, récipient**
contaminate *v.* **contaminer**
content *n.* **teneur**
contour *n.* see **courbe**
control *n.* *PC* **lutte**; *Stat.* **témoin**; *v.* **lutter (contre)**
convolvulus *n.* **volubilis**
cool *v.* **refroidir**
cooperative *a.* **coopératif**; *n.* **coopérative**
copaiba (tree) *n.* **copaïer**
coppice *n.* **cépée**
copse *n.* **bois, boqueteau, taillis**
cordgrass *n.* **spartine**
cordon *n.* **cordon**
core *n.* **trognon**
coriander *n.* **coriandre**
cork *n.* **bouchon (de liège)**; **liège**
corky bark *n.* *PD* **écorce liégeuse**
cormlets *n.* *pl.* **kralen**
cornelian cherry *n.* **cornouiller**
cornflower *n.* **bleuet, centaurée**
corolla *n.* **corolle**
corymb *n.* **corymbe**
cost *n.* **coût**
costmary *n.* **balsamite, menthe-coq**
cotoneaster *n.* **cotonéastre**
couch *n.* **chiendent**
coulter *n.* **coutre**
course (of rotation) *n.* **sole**
cover *v.* **couvrir**; c. (again) **recouvrir**
cowpea *n.* **niébé**
crabgrass *n.* **panic sanguin**
crab stock *n.* **égrain**
crack *n.* **craquelure, crevasse, fente, gerçure**
cranberry *n.* **airelle, canneberge, myrtille**
crape myrtle *n.* **lagerstroemie**

creasing (of citrus) *n.* *PD* **gaufrage**
creeper *n.* **liane**
creeping *a.* **rampant**
cress *n.* **cresson**; winter c. **barbarée**
crinkle (virus) *n.* **enroulage, frisolée, frisure**
crocus *n.* **crocus**
crop *n.* **culture**
cropping plan *n.* **assolement**
cropping twice *a.* **bifère**
cross *n.* **croisement**
cross-pollinated *a.* **allogame**
cross vine *n.* **bignone**
crotch *n.* **empattement, enfourchure**
crown *n.* **couronne**
crumbly *a.* **friable**
crumbly berry *n.* *PD* **grenaille**
crush *v.t.* **broyer, concasser, écraser**
crust *n.* **croûte**
crusting *n.* **encroûtement**
cucumber *n.* **concombre**
culling *n.* **élimination sélective**
cultivate *v.* **cultiver**
cultivation *n.* **culture, façon, intervention**
cultural *a.* **cultural**
culture *n.* **culture**; c. *in vitro* **vitroculture**
cup *n.* **tasse**
Cupid's dart *n.* **cupidone**
cupric *a.* **cuivrique**
cure *n.* **guérison**
currant *n.* **cassis, groseille**; c. bush **cassissier, groseillier**, (*Canada*) **gadelier**
current *n.* **courant**
cush-cush yam *n.* **couche-couche**
cut *v.* **couper, encocher**; c. back **rabattre, ravaler, recéper**; c. up **morceler**
cuticle *n.* **cuticule**
cutting *n.* **bouture, éclat**
cyanide *n.* **cyanure**
cyclamen *n.* **cyclamen**
cycle *n.* **cycle**
cyme *n.* **cyme**
cypress *n.* **cyprès**
cyst *n.* **kyste**
cytokinin *n.* **cytokinine**

daffodil *n.* **narcisse**
dahlia *n.* **dahlia**
daisy *n.* **marguerite, pâquerette**
dale *n.* **vallon**
damage *n.* **dégat, endommagement**
damping-off *n.* **fonte**
damson *n.* **damas**
dandelion *n.* **pissenlit**
dard *n.* **dard**
data *n. pl.* **données**
date (palm) *n.* **(palmier-) dattier**; d. growing **phéniculture**
daub *v.* **badigeonner**
dawn *n.* **aube**
day *n.* **jour, journée**
dead-nettle *n.* **lamier**
decaffeinated *a.* **décaféiné**
deciduous *a.* **caduc**
decline *n.* **déclin, décroissance, dépérissement**
decompose *v.* **décomposer**
decorticate *v.* **décortiquer**
deep *a.* **profond**
deficiency *n.* **carence**
deficient *a.* **carencé**
deficit *n.* **déficit**
defoliant *n.* **défoliant**
defoliate *v.* **défeuiller, effeuiller**
deforestation *n.* **déboisement, déforestation**
deform *v.* **déformer**
deformation *n.* **déformation, malformation**; d. by browsing **abroutissement**
defrosting *n.* **dégivrage**
degeneration *n.* **abâtardissement, dégénérescence**
degree *n.* **degré**
degree-day *n.* **degré-jour**
degreen *v.t.* **déverdir, éverdumer**
dehydration *n.* **déshydratation**
delay *n.* **retard, retardation**
dendroclimatology *n.* **dendroclimatologie**
dendrometer *n.* **dendromètre**
denitrify *v.* **dénitrifier**
dense *a.* **dru**
density *n.* **densité**
dentate *a.* **denté**
denticulate *a.* **découpé**
denuded *a.* **dénudé**
depleted *a.* **dégarni**
deposit *n.* **dépôt**
depreciation *n.* **tare**
depressed *a.* **déprimé**
depression *n.* **baissière, bas-fond, cuvette, dépression**
deproteinized *a.* **déprotéiné**
depth *n.* **profondeur**
depurative *a.* & *n.* **dépuratif**
derivation *n.* **origine**
desaturated *a.* **désaturé**
desert date *n.* **soump, balanites**
desertification *n.* **désertification**
desiccation *n.* **dessiccation**
destem *v.* **égrapper**
detergent *n.* **détergent**
deterioration *n.* **altération, détérioration**
determinate *a.* **déterminé**
deterrent *n.* **dissuadant**
deviation *n.* **écart**
dew *n.* **rosée**
dewlap *n.* *Sug.* **ocréa**
diallel *a.* **diallèle**

dialysis *n.* **dialyse**
dibble *n.* **plantoir**
dicotyledon *n.* **dicotylédone**
differentiation *n.* **différenciation**
diffused *a.* **diffus**
dig *v.* **creuser**; d. with spade **bêcher**
digitate *a.* **digité**
dilacerated *a.* **dilacéré**
dilute *v.* **diluer**
dioxide *n.* **dioxyde**; carbon dioxide **d. de carbone**
dip (*v.*) in slurry **praliner**
dipetalous *a.* **dipétale**
Diptera *n. pl.* see **diptère**
disbud *v.* **éborgner, éboutonner, ébouturer, épincer, escionner**
disc *n.* **disque**; discing **discage**
discharge *v.* **déverser**
discolour *v.* **décolorer**
disease *n.* **affection, maladie**
disinfect *v.* **désinfecter**
disintegrate *v.* **désagréger**
disorder *n.* **affection, désordre**
display unit *n.* **présentoir**
dissolve *v.* **dissoudre**
distichous *a.* **distique**
distribution *n.* **distribution, répartition, partage**
ditch *n.* **fossé**
dittany *n.* **dictame**
diuretic *a.* & *n.m.* **diurétique**
divide *v.* **diviser**
dock *n.* **rumex**
dodder *n.* **cuscute**
donor *n.* **donneur**
dormancy *n.* **dormance**
dormouse *n.* **lérot, loir**
dose *n.* **dose**
dotted *a.* **pointillé, ponctué**
double-graft *v.* **contre-greffer, surgreffer**
double-work *v.* see double-graft
downy *a.* **duveteux**
drain *v.* **drainer, égoutter**
dredge *v.* **draguer, dévaser**
dressing (of manure) *n.* **apport**
drill (sowing) *n.* **semoir**
drinkable *a.* **buvable**
dripper *n. Irrig.* **goutteur**
drooping *a.* **retombant**
drop (fall) *n.* **baisse, chute**
drop (of liquid) *n.* **goutte**
drought *n.* **sécheresse**
drug *n.* **drogue**
drupe *n.* **drupe**
dry *a.* **sec**; medium-dry **demi-sec**
dry *v.* **assécher, faner, ressuyer, sécher**
dum palm *n.* **doum**
dumpy *a.* **trapu**
dung *n.* **fumier**
duration *n.* **durée**
durian *n.* **durio**
dust *v.* **empoussiérer, poudrer, saupoudrer, soufrer**
Dutch light *n.* **châssis**
dwarf *a.* **nain**
dwarfing *a.* **nanisant**; dwarfing compound *n.* **nanifiant**
dyer's rocket *n.* **gaude**

early *a.* **hâtif, précoce**
earth *n.* **sol, terre**
earth up *v.* **chausser, enchausser, terrer**
earthworm *n.* **lombric, ver de terre**
eatable *a.* **consommable**
edaphic *a.* **édaphique**
edible *a.* **édule, mangeable**
EEC CEE
eelworm *n.* **anguillule, nématode**
effect *n.* **effet**; residual e. **arrière-effet**
efficacity *n.* **efficacité**
eggplant *n.* **aubergine**
elder *n.* **sureau**
electrofocusing *n.* **électrofocalisation**
elm *n.* **orme**
embryo *n.* **embryon**
embryoid *n.* **embryoïde**
emergence (of plant) *n.* **levée**
emit *v.* **émettre**
empty *v.* **vider**
emulsible *a.* **émulsionnable**
emulsion *n.* **émulsion**
enclosure *n.* **clos, clôture, enceinte, parc**
endive *n.* **endive**
endomycorrhiza *n.* **endomycorhize**
endosperm *n.* **endosperme**
endosymbiont *n.* **endosymbiote**
ensile *v.* **ensiler**
entomophilous *a.* **entomogame, entomophile**
envelope *n.* **enveloppe**
epicotyl *n.* **épicotyle**
epidermis *n.* **épiderme**
epigeal *a.* **épigé**
epigynous *a.* **épigyne**
epiphyllous *a.* **épiphylle**
equalize *v.* **égaliser**
equipment *n.* **équipement, outillage**
eradicate *v.* **extirper**
erect *a.* **dressé, érigé**
ergot *n.* **ergot**
ermine moth *n.* **hyponomeute, yponomeute**
erode *v.* **éroder**
espalier *n.* **espalier**
esparto grass *n.* **alfa**
establishment *n.* **établissement**
etiolated *a.* **étiolé**
Etrog citron *n.* **citron-cédrat**
eutrophic *a.* **eutrope**

evaporimeter *n.* **évaporimètre**
evening primrose *n.* **oenothère**
everbearing *a.* **remontant**
everlasting flower *n.* **immortelle**
excess *n.* **excès**
exchange *n.* **échange**; base e. **é. de bases**; cation e. **é. cationique**
excise *v.* **exciser**
excision *n.* **ablation**
excoriosis *n.* **excoriose**
excrescence *n.* **excroissance**
excretion *n.* **excrétion**
exfoliate *v.* **exfolier**
exhausting *a.* **épuisant**
exogenous *a.* **exogène**
expansion *n.* **dilatation**
experiment *v.* **expérimenter**
experiment *n.* **expérience**
extensive *a.* **extensif**
extract *v.* **extraire**
extremity *n.* **bout, extrémité**
extricate from mud *v.* **désembourber**
exudate *n.* **exsudat**
eye (of plant) *n.* **oeil, oeilleton**

factor *n.* **facteur**
factory *n.* **usine**
fade *v.* **flétrir**
faggot *n.* **fagot, fascine**
fairy ring champignon *n.* **faux mousseron**
fall (of river) *n.* **décrue**
fallow *n.* **friche, jachère**
false acacia *n.* **robinier**
family *n.* **famille**
fan *n.* **évantail, queue-de-paon**; fan-shaped *a.* **en palmette**
fanging (of roots) *n.* **éclatement**
fat *a.* **gras**
fat hen *n.* **chénopode**
feather *n.* (*Hort.*) **anticipé, rameau**
febrifuge *a.* & *n.* **fébrifuge**
feeding *n.* **alimentation**
feijoa *n.* **féijoa**
fell *v.* **abattre**
fennel *n.* **fenouil**
fenugreek *n.* **fenugrec, sénegré**
ferment *v.* **fermenter**
fern *n.* **fougère**
fertigation *n.* **fertirrigation**
fertility *n.* **fertilité**
fertilize *v.* **fertiliser, fumer**
fertilizer *n.* **engrais**; slow-release f. **e. à libération lente**
fescue *n.* **fétuque**
fibre *n.* **fibre**
fibrous *a.* **fibreux**
field *n.* **champ**
fig *n.* **figue**; f. tree **figuier**
fig marigold *n.* **ficoïde**
fig wasp *n.* **blastophage**
fill (up) *v.* **combler, reboucher, remplir**
filler (tree) *n.* **temporaire**
film-coating (of seed) *n.* **pelliculage**
filter *v.* **filtrer**
financial year *n.* **exercice**
fir *n.* **abies, sapin**
fire *n.* **feu**
firebreak *n.* **pare-feu**
firm (soil) *v.* **borner**
fish farming *n.* **pisciculture**
fish leaf (*Tea*) *n.* **kepel**
fissured *a.* **fissuré**
flame gun *n.* **agriflamme**
flamingo flower *n.* **anthure**
flap *n.* **lambeau**
flat *a.* **plat**
flavedo *n.* **flavedo**
flavour *n.* **goût, saveur**; flavourless *a.* **fade**
flea beetle *n.* **altise**
fleeting *a.* **fugace**
flesh *n.* **chair**
fleshy *a.* **charnu**
floating *a.* **flottant, nageant**
flood *n.* **crue, inondation**
flood *v.* **inonder**
floral *a.* **floral**
floret *n.* **fleuron**
floriculture *n.* **floriculture**
flower *n.* **fleur**
flower *v.* **fleurir**; f. (again) **refleurir, remonter**
flower bed *n.* **corbeille, plate-bande, parterre**
flower holder *n.* **pique-fleurs**
flowering *n.* **floraison**
fluorescence *n.* **fluorescence**
fluorodensitometry *n.* **fluorodensitométrie**
fluted *a.* **cannelé**
fly *n.* **mouche**
foam *n.* **écume, mousse**
focus (of disease) *n.* **foyer**
foliage *n.* **feuillage**
foliar *a.* **foliaire**
foliate *a.* **folié**
foliation *n.* **feuillaison, frondaison**
follicle *n.* **follicule**
food *n.* **nourriture**
footbath *n.* **pédiluve**
forage (of bees) *v.* **butiner**
force *v.* **forcer**
forcing *n.* **forçage**
forecasting *n.* **prévision**
forestry *n.* **foresterie**
forget-me-not *n.* **myosotis**
fork *n.* **fourche, griffe**; f. (in tree) **enfourchure**
forked *a.* **bifurqué, fourchu**
form *n.* **forme**

formula *n.* **formule**
fountain *n.* **fontaine**
foxglove *n.* **digitale**
foxtail grass *n.* **alopécure, vulpin**
fragrant *a.* **odorant, parfumé**
frame *n.* **bâche, châssis**
framework *n.* **charpente, encadrement, ossature**
frangipani *n.* **frangipanier**
fraxinella *n.* **fraxinelle**
free stock *n.* **franc**
freeze *v.* **geler, congeler**
freeze dry *v.* **lyophiliser**
fresh *a.* **frais**
friable *a.* **ameubli, meuble**
fringe *n.* **frange**
fritillary *n.* **fritillaire**
frond *n.* **fronde**; (of palm) **palme**
frost *n.* **gel, gelée**
frost-tender *a.* **délicat**
froth *n.* **écume**
fructiferous *a.* **fructifère**
fruit *v.* **fructifier, fruiter**
fruit *n.* **fruit**
fruity *a.* **fruité**
fuel *n.* **combustible**
fuel oil *n.* **fuel**
full *a.* **plein**
full-bodied (of wine) *a.* **corsé**
fumigate *v.* **fumiger**
fumitory *n.* **fumeterre**
fungicide *n.* **fongicide**
fungitoxic *a.* **fongitoxique**
fungus *n.* **champignon**
funnel *n.* **entonnoir**
furrow *n.* **dérayure, interbillon, sillon**
Fusarium disease *n.* **fusariose**

gaillardia *n.* **gaillarde**
galangal *n.* **galanga**
galenic *a.* **galénique**
gall *n.* **galle, intumescence**; g.-forming **cécidogène, gallicole**; g.-midge **cécidomyie**
gallant soldier *n.* **galinsoga**
gallery *n.* **galerie**
gamboge *n.* **guttier**
game (wild animals) *n.* **gibier**
gang (of labourers) *n.* **équipe**
gap *n.* **intervalle**
gap-up *v.* **remplacer**
garden *n.* **jardin**
gardener *n.* **jardinier, maraîcher, paysagiste**
gardener's garters *n.* **alpiste panaché, ruban de bergère**
gardenia *n.* **gardénia**
gari *n.* **attiéké, garri**
garlic *n.* **ail**
garou bush *n.* **garou**
gaseous *a.* **gazeux**
gauge *n.* **jauge, gabarit**; rain g. **pluviomètre**
gean *n.* **guigne**; g. tree **guignier**
gene *n.* **gène**
gentian *n.* **gentiane**
genus *n.* **genre**
geometrid moth *n.* **géomètre, phalène**
geotropism *n.* **géotropisme**
geranium *n.* **géranium, pélargonium**
germander *n.* **germandrée**
germinability *n.* **germinabilité**
germinate *v.* **germer**
germination *n.* **germination, levée**
germinator *n.* **germoir**
gherkin *n.* **cornichon**
gill *n.* *Mush.* **lame**
ginger *n.* **gingembre**
girdling tool *n.* **inciseur**
glabrous *a.* **glabre**
gladiolus *n.* **glaïeul**
gland *n.* **glande**
glass *n.* **verre**
glasshouse *n.* **serre**
glaucous *a.* **glauque**
glaze *v.* **vitrer**
glean *v.* **glaner**
globe amaranth *n.* **amarantoïde, gomphrène**
globe flower *n.* **boule d'or, trolle**
gloriosa lily *n.* **glorieuse**
glucan *n.* **glucane**
goblet *n.* **gobelet**
gold of pleasure *n.* **caméline**
golden shower *n.* **averse dorée**
Good King Henry *n.* **bon-henri**
gooseberry *n.* **groseille**; g. bush **groseillier**; Cape g. **coqueret**; Chinese g. **actinidia, kiwi, yang-tao**
goosefoot *n.* **ansérine**
goosegrass *n.* **gaillet-gratteron**
gorse *n.* **ajonc**
gourd *n.* **courge, gourde**
graft *n.* **ente, greffe**
graft *v.* **enter, greffer**
grafting knife *n.* **entoir, greffoir**
grafting wax *n.* **mastic**
grain *n.* **grain**
grains of paradise *n.* **maniguette**
granadilla *n.* **barbadine, grenadille, passiflore**
granary *n.* **grenier**
granulated *a.* **granulé**
granulosis *n.* **granulose**
grape *n.* **raisin**
grapefruit *n.* **grapefruit**
grapevine *n.* **vigne**
grass *n.* **graminée, herbe**
grass down *v.* **enherber**
grasshopper *n.* **acridien, sauterelle**
gravel *n.* **gravier**

grease *n.* **graisse**
green *v.* **verdir, verdoyer**; g. (again) **reverdir**
greenback (of tomato) *n. PD* **collet vert (de la tomate)**
greengage *n.* **reine-claude**; g. tree **reine-claudier**
greenhouse *n.* **serre**
ground *n.* **terrain, sol**
ground elder *n.* **égopode, podagraire**
groundnut, Bambara *n.* **voandzou**
groundsel *n.* **seneçon**
grove *n.* **bosquet**
grow *v.* **cultiver; pousser, venir**; g. (again) **repousser, reprendre**
grower *n.* **cultivateur**
growing *n.* **culture**; fruit g. **fructiculture**; glasshouse g. **serriculture**; g. under plastic **plasticulture**
growth *n.* **croissance; pousse**
growth chamber *n.* **cabinet de croissance; chambre climatisée**
growth regulator, g. substance *n.* **phytohormone, phytorégulateur**
guava *n.* **goyave**; g. tree **goyavier**
guelder rose *n.* **aubour, obier**
gullying *n.* **ravinement**
gum *n.* **gomme**
gum tree *n.* **gommier**
gummosis *n.* **gommose**
guy (rope) *n.* **hauban**
gynaecium *n.* **gynécée**
gynodioecy *f.* **gynodioécie**
gynogenetic *a.* **gynogénétique**

habit *n.* **comportement, port, tenue**
haemagglutinin *n.* **hémaglutinine**
hail *n.* **grêle**
hairy *a.* **pileux, poilu**
hammer mill *n.* **broyeur à marteaux**
hamper *n.* **cageot**
hand *n.* **main**
handle *n.* **anse**
handling *n.* **manutention**
haploid *a.* & *n.m.* **haploïde**
harden (off) *v.* **durcir, endurcir**
hardness *n.* **dureté**
hardy *a.* **rustique**
hare *n.* **lièvre**
hare's tail *n.* **queue-de-lièvre**
harmful *a.* **nuisible**
harrow *n.* **herse**; disc h. **pulvérisateur**
hart's tongue *n.* **scolopendre**
harvest *n.* **moisson, récolte**; grape h. **vendange**
harvester (machine) *n.* **moissonneuse, récolteuse**
hastate *a.* **hasté**
hatch *v.* **éclore**
hawfinch *n.* **gros-bec**
hawkweed *n.* **épervière**; mouse-ear h. **piloselle**
hawthorn *n.* **aubépine**
hazel nut *n.* **noisette**; h. tree **noisetier**
head *n.* **tête**
head *v.* **étêter**
headland *n.* **chaintre**
heal *v.* **cicatriser**
healthy *a.* **sain**
heap *n.* **tas**
heap *v.* **entasser**
heart *n.* **coeur**
heart *v.* **pommer**
heartwood *n.* **duramen**
heat *n.* **chaleur**; heat resistant *a.* **thermorésistant**; heat-treated *a.* **thermotraité**
heat *v.* **chauffer**
heather *n.* **bruyère**
heathland *n.* **lande**
heavy *a.* **lourd**; (of soil) **fort**
hedge *n.* **haie**
heel in *v.* **enjauger, mettre en jauge**
height *n.* **hauteur**
heliotrope *n.* **héliotrope**
hellebore *n.* **ellébore**
hemlock *n.* **cigüe**
hemp *n.* **chanvre**
henbane *n.* **jusquiame**
hepatica *n.* **hépatique**
herb *n.* **herbe**; culinary herbs **fines herbes**
herbicide *n.* **désherbant, herbicide**
herborize *v.* **herboriser**
hereditary *a.* **héréditaire**
Hibiscus sp. *n.* **ketmie**
hillock *n.* **monticule, tertre**
hilum *n.* **hile, nombril**
histogenesis *n.* **histogenèse**
histological *a.* **histologique**
hoar-frost *n.* **givre**
hoe *v.* **biner, sarcler**
hoe *n.* **binette, houe**; h. and fork **serfouette**
hogweed *n.* **berce**
hold (of ship) *n.* **cale**
holding *n.* **exploitation**
hole *n.* **trou**; planting h. **poquet**
hollow *a.* **creux, évidé**
holly *n.* **houx**
hollyhock *n.* **rose trémière**
honesty *n.* **lunaire**
honey *n.* **miel**
honey fungus *n.* **armillaire**
honeysuckle *n.* **chèvrefeuille**
hook *n.* **croc, crochet**
hop *n.* **houblon**
hopper (of implement) *n.* **trémie**
hormone *n.* **hormone**; plant h. **phytohormone**
horn *n.* **corne**
hornbeam *n.* **charme**
horse chestnut (tree) *n.* **marronnier**
horseradish *n.* **raifort**

horticultural *a.* **horticole, horticultural**
horticulture *n.* **horticulture**
horticulturist *n.* **horticulteur**
host *n.* **hôte**
hotbed *n.* **couche, meule**
houseleek *n.* **joubarbe**
hover-fly *n.* **syrphe**
humid *a.* **humide**
humus *n.* **humus**
hurdle *n.* **claie, clayette**
husk *n.* **brou, coque, cosse**
hyacinth *n.* **jacinthe**; grape h. **muscari**; tassel h. **m. à toupet, poireau roux**
hybrid *n.* **hybride, métis**
hydrangea *n.* **hortensia**
hydrocarbon *n.* **hydrocarbure**
hydrocooling *n.* **réfrigération par eau glacée**
hydrogen *n.* **hydrogène**
hydrolysis *n.* **hydrolyse**
hydroponic *a.* **hydroponique**
hypertrophy *n.* **hypertrophie**
hypha *n.* **hyphe**
hypocotyl *n.* **collet, hypocotyle**
hypogeal *a.* **hypogé**
hypovirulent *a.* **hypoagressif, hypovirulent**
hyssop *n.* **hysope**

ice *n.* **glace**
ice-nucleating *a.* **glaçogène**
ice plant *n.* **glaciale**
imbalance *n.* **déséquilibre**
immersed *a.* **immergé**
immunoelectrophoresis *n.* **immunoélectrophorèse**
immunoenzymatic *a.* **immunoenzymatique**
impervious *a.* **impénétrable, imperméable**
implanted *a.* **implanté**
imprint *n.* **empreinte**
improve *v.* **améliorer, amender**
impurity *n.* **impureté**
inactivated *a.* **inactivé**
incised *a.* **incisé**
incompatibility *n.* **incompatibilité**
incorporated *a.* **incorporé**
indeterminate *a.* **indéterminé**
index *n.* **index, indice**
indigenous *a.* **autochtone, indigène**
indigo (plant) *n.* **indigotier**
indole *a.* **indolique**
induced *a.* **induit**
ineradicable *a.* **indéracinable**
inermous *a.* **inerme**
infected *a.* **infecté**
infectivity *n.* **infectivité**
infestation *n.* **infestation**
inflatable *a.* **gonflable**
inhibit *v.* **inhiber**
inject *v.* **injecter**
ink cap *n.* **coprin**
inoculum *n.* **inoculat**
input *n.* **intrant**
insect *n.* **insecte**; i. control **désinsectisation**; i. pest **i. nuisible**; i. repellent **insectifuge**
insecticide *n.* **insecticide**
insolation *n.* **ensoleillement**
insoluble *a.* **insoluble**
inspected *a.* **contrôlé**
insufficiency *n.* **insuffisance**
insulation *n.* **isolation**
intensity *n.* **intensité**
interchangeable *a.* **amovible**
internode *n.* **mérithalle**
interplanted *a.* **intercalaire**
interplanting *n.* **interplantation, contreplantation**
interrow *n.* **entreligne, interligne**
interstock *n.* **(sujet) intermédiaire, mésobiote**
intracellular *a.* **intracellulaire**
introgression *n.* *PB* **introgression**
invasion *n.* **envahissement**
invasive *a.* **envahissant**
iodine *n.* **iode**
ionization *n.* **ionisation**
iridoid *n.* **iridoïde**
iris *n.* **iris**
iron *n.* **fer**
iron impoverishment (of soil) *n.* **déferrification**
ironstone cap *n.* **cuirasse**
irrigate *v.* **irriguer**
isobar *n.* **isobare**
isohyet *n.* **isohyète**
isolation *n.* **isolement**
isomer *n.* **isomère**
isotherm *n.* **isotherme**
isovalue *n.* *Stat.* **isovaleur**
ivy *n.* **lierre**

jackfruit *n.* **jaque**; j. (tree) **jaquier**
jam *n.* **confiture**
jasmine *n.* **jasmin**
jassid *n.* **jasside**
Johnson grass *n.* **sorgo d'Alep**
join *v.* **souder**
jonquil *n.* **jonquille**
Judas tree *n.* **cercis, gainier**
juice *n.* **jus**
juiciness *n.* **jutosité**
jujube (tree) *n.* **jujubier**
juniper *n.* **genévrier**
Jupiter's beard *n.* **barbe de Jupiter**
juvenility *n.* **juvénilité**

kapok tree *n.* **capoquier**

karengiya *n.* **cram-cram**
karyology *n.* **caryologie**
kikuyu grass *n.* **kikuyu**
keeled *a.* **caréné**
kerria *n.* **corète**
ketonic *a.* **cétonique**
knapweed *n.* **jacée**
knife *n.* **couteau**
knot *n.* **noeud**
knotgrass *n.* **renouée**
kohl-rabi *n.* **chou-rave**
kola tree *n.* **kolatier**

label *n.* **étiquette, fiche**
labour *n.* **main-d'oeuvre**
laburnum *n.* **cytise**
lack *n.* **manque**
lacquer tree *n.* **laquier**
ladder *n.* **échelle**
ladybird *n.* **coccinelle**
lady's mantle *n.* **alchémille**
lady's slipper *n.* **cypripède**
lalang *n.* **impérata**
lamb's lettuce *n.* **boursette, mâche**
lamina *n.* **lame**
lamp *n.* **lampe, irradiateur**
landscape *n.* **paysage**
lane *n.* **allée**
larch *n.* **mélèze**
larkspur *n.* **éperonnière, pied-d'alouette**
larva *n.* **larve**
lasting *a.* **durable**
late *a.* **tardif**
laticifer *n.* **laticifère**
laurel *n.* **laurier**
laurustinus *n.* **laurier-tin**
lavandin *n.* **lavandin**
lavender *n.* **lavande**
lavender cotton *n.* **santoline**
lawn *n.* **pelouse**
layer *n.* **couche, gisement**; *Hort.* **marcotte**
layer *v.* **marcotter**
layer-bed *n.* **marcottière**
leach *v.* **lessiver**
lead *n.* **plomb**
leaf *n.* **feuille**; leaf curl *PD* **bouclage**
leafhopper *n.* **cicadelle**
leaflet *n.* **foliole**
lease *n.* **bail**
leasure *n.* **loisir**
leek *n.* **poireau**
legume *n.* **légumineuse**
lemon *n.* **citron**; l. tree **citronnier**
length *n.* **longueur**
lenticel *n.* **lenticelle**
lentil *n.* **lentille**
lettuce *n.* **laitue**
level *n.* **niveau**
level *v.* **égaliser, niveler**
licencing *n.* **homologation**
lift *v.* **arracher, déplanter, soulever**; lift turf **déplaquer**
light *n.* **lumière**
lighten *v.* **alléger**
lighting *n.* **éclairage, irradiation**
lightning *n.* **foudre**
lignan *n.* **lignane**
lignified *a.* **aoûté, lignifié**
lignin *n.* **lignine**
lignoid *n.* **lignoïde**
lilac *n.* **lilas**
lily *n.* **lis**; **amaryllis, hémérocalle**
lily-of-the-valley *n.* **muguet**
lime *n.* **chaux**; *v.* **chauler**
lime (citrus) tree *n.* **citron vert, limettier, limonier**; l. (linden) tree **tilleul**
lime-hating *a.* **calcifuge**
lime impoverishment *n.* **décalcarisation**
lime-loving *a.* **basiphile, calcicole**
line *n.* **ligne**; l. up *v.* **aligner**; l. out **contre-planter**
linkage *n.* *PB* **liaison**
lipophilic *a.* **lipophile**
liqueur-like *a.* **liquoreux**
liquorice *n.* **réglisse**
litchi *n.* **litchi**
liverwort *n.* **hépatique**
load *v.* **charger**
loam *n.* **limon**; see **meuble**
lobelia *n.* **lobélia**
lobed *a.* **lobé**
loculus *n.* **loge**
locus *n.*, *pl.* loci *PB* **locus**
locust *n.* **criquet, locuste, sauterelle**
locust bean *n.* **nété**
lodging *n.* **verse**
lodging control, see **antiverse**
London pride *n.* **désespoir du peintre**
looper (caterpillar) *n.* **arpenteur**
loosen (soil) *v.* **décompacter**
loosestrife *n.* **lysimachie, salicaire**
lop *v.* (branches) **ébrancher**
loquat (fruit) *n.* **bibasse**; l. (tree) **bibassier**
loss *n.* **perte**
louse *n.* **pou**
lovage *n.* **livèche**
love-lies-bleeding *n.* **amarante**
lowering *n.* **abaissement**
lucerne *n.* **luzerne**
luffa *n.* **loofa**
lungwort *n.* **pulmonaire**
lupin *n.* **lupin**
lysimeter *n.* **case lysimétrique, lysimètre**

macadamia tree *n.* **macadamier**
mace *n.* **macis**
madder *n.* **garance**
maggot *n.* **asticot**
magnesia *n.* **magnésie**
magnolia (tree) *n.* **magnolier**
mahaleb cherry *n.* **mahaleb, sainte-lucie**
maiden *n.* *Hort.* **scion**
maidenhair fern *n.* **adiante, cheveux de Vénus**
maintenance *n.* **entretien**
maize *n.* **maïs**
maladjustment *n.* **inadaptation**
male sterile *a.* **androstérile, mâle-stérile**
mal secco *n.* *PD* **mal sec**
mallow *n.* **mauve, guimauve**
mammey *n.* **mammea**
management *n.* **conduite, gestion, régie**
manchineel (tree) *n.* **mancenillier**
mandarin *n.* **mandarine**; m. tree **mandarinier**
man-day *n.* **homme-jour**
mandrake *n.* **mandragore**
mango (fruit) *n.* **mangue**; m. tree **manguier**
mangosteen (fruit) *n.* **mangoustan**; m. (tree) **mangoustanier**
mangrove *n.* **manglier, palétuvier**
Manila hemp *n.* **abaca**
manure *n.* **engrais, fumure**; farmyard manure **fumier**
manure *v.* **fumer**
map *n.* **carte**
maple *n.* **érable**
marcotting *n.* **marcottage**
marjoram *n.* **marjolaine, origan**
marker *n.* **marqueur**
market *n.* **marché**
marketable *a.* **commercialisable, vendable**
market gardener *n.* **hortillonneur, maraîcher**
market gardening *n.* **maraîchage**
marigold *n.* **souci**; marsh m. **populage**
marking *n.* **marquage**
marl *n.* **marne**
marrow (cucurbit) *n.* **courge**; m. (pith) **moelle**
marsh *n.* **marais**
marvel of Peru *n.* **belle-de-nuit**
masterwort *n.* **impératoire**
masticatory *a.* & *n.* **masticatoire**
mat *n.* **nappe, paillasson**; heating m. **n. chauffante**; irrigation m. **n. d'irrigation**
material *n.* **matériel**; m. (for building etc.) **matériau**
mating disruption *n.* *PC* **confusion sexuelle**
matter *n.* **matière**
mattock *n.* **pioche**
mattress *n.* **matelas**
maturity *n.* **maturité**
meadow grass *n.* **pâturin**
meadow sweet *n.* **ulmaire**
meatus *n.* **méat**
mechanize *v.* **mécaniser**
medium *a.* **moyen**; *n.* **milieu**
medlar *n.* **nèfle**; m. (tree) **néflier**
medullary *a.* **médullaire**
meiotic *a.* **méiotique**
melick *n.* **mélique**
melilot *n.* **mélilot**
melissa *n.* **mélisse**
melliferous *a.* **mellifère**
mellow (*v.*) soil **ameublir**
melon *n.* **melon**
melting *a.* **fondant**
mercury (plant) *n.* **mercuriale**
meristem *n.* **méristème**
mesophyll *n.* **mésophylle**
metabolism *n.* **métabolisme**
microanalysis *n.* **microanalyse**
microflora *n.* **microflore**
micrografting *n.* **microgreffage**
microorganism *n.* **microorganisme**
microprobe *n.* **microsonde**
micropropagation *n.* **microbouturage**
microtapping *n.* *Rubb.* **microsaignée**
midge *n.* **moucheron**
midrib *n.* **côte**
mignonette *n.* **réséda**
mildew *n.* **mildiou, moisissure, oïdium**
milfoil *n.* **achillée, millefeuille**
milk cap *n.* *Fung.* **lactaire**
mill *n.* **moulin**
millable *a.* **usinable**
millipede *n.* **iule, mille-pattes**
millstone *n.* **meule**
miniblock *n.* **minimotte**
mint *n.* **menthe**
mirabelle plum *n.* **mirabelle**; m. plum (tree) **mirabellier**
mirid *n.* **miride**
missing *a.* **manquant**
mist *n.* **brouillard, brume**
mist blower *n.* **nébulisateur**
mist chamber *n.* **miste**
misting *n.* **brumisation**
mite *n.* **acarien, érinose**; predacious m. **acarien prédateur, typhlodrome**
mix *v.* **mélanger, mêler**
mixture *n.* **mélange**; *PC* **bouillie**
modelling *n.* **modélisation**
moisten *v.* **humecter, mouiller**
moisture *n.* **humidité**; moisture meter **humidimètre**
molasses grass *n.* **mélinis**
mole *n.* **taupe**
mole-cricket *n.* **courtilière**
monilia disease *n.* **moniliose**
monkshood *n.* **capuchon-de-moine, napel**
monocotyledon *n.* **monocotylédone**
monovalent *a.* **univalent**

monsoon *n.* **mousson**
moonflower *n.* **belle-de-nuit**
morel *n.* **morille**
morning glory *n.* **belle-de-jour, ipomée**
morphogenic *a.* **morphogène**
mosaic *n.* **bigarrure, mosaïque**
moss *n.* **mousse**
moth *n.* **lépidoptère, papillon; abraxas, mite, noctuelle, zeuzère**
mother-plant *n.* **pied-mère**
motherwort *n.* **agripaume, léonure**
motor *n.* **moteur**; motor hoe **motobineur, motohoue**; m. mower **motofaucheuse**
motorized *a.* **motorisé**
mottle *n. PD* **marbrure**
mould *n.* (vegetable) **terreau**; (fungus) **moisissure**
mound *n.* **motte**
mouse *n.* **souris**; field m. **mulot**
mouse-ear *n.* **céraiste**
mouse-tail *n.* **queue-de-souris**
mow *v.* **faucher, tondre**
mower *n.* **faucheur, tondeuse**
mud *n.* **boue, vase**
mugwort *n.* **armoise**
mulberry *n.* **mûre**; m. (tree) **mûrier**; Indian m. **morinde**
mulching *n.* **paillage**; *v.* **pailler**; radiant m. **p. radiant**
mullein *n.* **molène**; moth m. **blattaire**
multiflorous *a.* **pluriflore**
mummified *a.* **momifié**
mushroom *n.* **champignon, mousseron**; oyster m. **pleurote**
musk mallow *n.* **abelmosch, ambrette**
muskrat *n.* **ondatra, rat musqué**
musky *a.* **musqué**
must *n.* **moût**
mustard *n.* **moutarde**; Indian m. **m. brune**
mutagenic *a.* **mutagène**
mutation *n.* **mutation**
mycelial *a.* **mycélien**
mycelium *n.* **mycélium**
mycoplasma *n.* **mycoplasme**
mycorrhization *n.* **mycorhization**
mycosis *n.* **mycose**
myrobalan *n.* **myrobolan**; emblic m. **groseillier de Ceylan**
myrrh *n.* **myrrhe**
myrtle *n.* **myrte**

narcissus *n.* **narcisse**
named place *n.* **lieu(-)dit**
nard *n.* **nard**
nasturtium *n.* **capucine**
Natal plum *n.* **carisse**
necrosis *n.* **nécrose**
nectar flow *n.* **miellée**
nectarine *n.* (fruit) **brugnon, nectarine**; n. (tree) **brugnonier**
need *n.* **besoin**
needle *n. Bot.* **aiguille**
nematode *n.* **nématode**
nerine *n.* **nérine**
nervature *n.* **nervation**
nest *n.* **nid**
nesting box *n.* **nichoir**
net *n.* **filet**
nettle *n.* **ortie**
nettle-tree *n.* **micocoulier**
network *n.* **réseau**
nick *v.* **entailler**
night period *n.* **nyctipériode**
nightshade *n.* **belladone, morelle**
night soil *n.* **gadoue, vidange**
nitrate *n.* **azotate, nitrate**
nitrogen *n.* **azote**
nitrogenous *a.* **azoté**
nitrous *a.* **nitreux**
noctiflorous *a.* **noctiflore**
node *n.* **noeud**
nodule *n.* **nodosité**
non-cultivation *n.* **inculture**
non-run (of fabric) *a.* **indémaillable**
non-skidding *a.* **antidérapant**
non-target *a. PC* **non-cible**
notch *n.* **cran, échancrure**
notch *v.* **entailler**
nucellus *n.* **nucelle**
nucleic *a.* **nucléique**
nucleus *n.* **noyau**
nudiflorous *a.* **nudiflore**
nursery *n.* **pépinière**
nut *n.* **noix**; nut grower **nuciculteur**; n. tree **noyer**
nutmeg *n.* **muscade**; n. tree **muscadier**
nutrition *n.* **alimentation, nutrition**
nymph *n.* **nymphe**

oak *n.* **chêne**; **rouvre**
oasthouse *n.* **séchoir**
oat *n.* **avoine**
offset bulb *n.* **caïeu**
offshoot *n.* **rejet**
oil *n.* **huile**; essential o. **essence**; fuel o. **fuel, mazout**
oil cake *n.* **tourteau**
oil plants *n.pl.* **oléagineux**
okra *n.* **gombo**
old line *n. Cit.* **vielle lignée**
oleander *n.* **laurier-rose**
oleaster *n.* **chalef**
olive *n.* **olive**; o. tree **olivier**
onion *n.* **oignon**; Welsh o. **ciboule**

ontogenesis *n.* **ontogénèse**
open *v.* (of flower) **épanouir**
opening *n.* **ouverture**
opposite *a.* **opposé**; opposite-leaved **oppositifolié**
orache *n.* **arroche**
orange *n.* **orange**; o. tree **oranger**
orchard *n.* **verger**
orchid *n.* **orchidée**
organ *n.* **organe**
organic *a.* **organique**
organochlorine compound *n.* **organochloré**
organogenesis *n.* **organogénèse**
oriole *n.* **loriot**
ornamental *a.* **ornemental**
orthostichy *n.* **orthostique**
osmosis *n.* **osmose**
Otaheite gooseberry *n.* **chérimbélier, surette**
outflow *n.* **écoulement**
outlet *n.* **débouché**; *Irrig.* **colature**
out-of-season *a.* **hors saison**
output *n.* **débit**
ovary *n.* **ovaire**
oven *n.* **four**
overdose *n.* **surdosage**
overlapping *a.* **chevauchant**
overliming *n.* **surchaulage**
over-ripe *a.* **blet**
over-ripening *n.* **surmaturation**
overwintered *a.* **hiverné**
ovicidal *a.* **ovicide**
ovipositor *n.* **ovipositeur**
ovule *n.* **ovule**
ovum *n.* **oeuf**
own-rooted *a.* **sur ses propres racines**
oxidation *n.* **oxydation**
oyster cap *n.* *Fung.* **pleurote**

pack *v.* **emballer, empaqueter**; p. in straw **empailler**
pad *n.* **bourrelet, coussinet**
padding *n.* **rembourrage**
paint *n.* **peinture**
pairing *n.* **appariement**; chromosome p. **a. chromosomique**
palisade *n.* **palissade**
pallet *n.* **palette**
palm (tree) *n.* **palmier**; oil palm **p. à huile**; palm cabbage **chou-palmiste**; palm grove **palmeraie**; palm kernel **palmiste**
palmette *n.* **palmette**
panel *n.* **panneau**
panicle *n.* **panicule**
panmictic *a.* *PB* **panmictique**
panmixia *n.* *PB* **panmixie**
pansy *n.* **pensée**
papaya *n.* see pawpaw
papeda, Mauritius *n.* **combara**
paradise apple *n.* **paradis**
parasite *n.* **parasite**
parasol *n.* **parasol**; parasol mushroom **coulemelle, lépiote**
parenchyma *n.* **parenchyme**
parent (of plant) *n.* **géniteur**
parent rock *n.* **roche-mère**
park *n.* **parc**
parsley *n.* **persil**; fool's p. **ache des chiens**
parsley piert *n.* **alchémille**
parsnip *n.* **panais**
part *n.* **partie**
partition *n.* **cloison, paroi**
pasque flower *n.* **pulsatille**
passion flower *n.* **grenadille, passiflore**
paste *n.* **enduit**
pasture *n.* **pâture**
path *n.* **chemin, sentier**
pathology *n.* **pathologie**; plant p. **phytiatrie**
pathotype *n.* **pathotype**
paving *n.* **dallage**
pawpaw (fruit) *n.* **papaye**; p. tree **papayer**
pea *n.* **pois**
peach *n.* **pêche**; p. tree **pêcher**
pear *n.* **poire**; oriental p. **nashi, poire-pomme**; p. tree **poirier**
peat *n.* **tourbe**
pecan (nut) *n.* **pacane**; p. tree **pacanier**
pedicel *n.* **pédicelle**
pedogenesis *n.* **pédogénèse**
peduncle *n.* **pédoncule**
peel *n.* **pelure, zeste**
peel *v.* **éplucher, peler**
peg *n.* **fiche, piquet**
peg out *v.* **piqueter**
pelargonium *n.* **pélargonium**
pellitory *n.* **pariétaire**
peony *n.* **pivoine**
pepper (capsicum) *n.* **chilli, piment, poivron**
pepper (*Piper*) *n.* **poivre**; p. vine **poivrier**
pepperwort *n.* **passerage**
perennial *a.* **pérenne, vivace**
perfume *n.* **parfum**
perianth *n.* **périanthe**
pericarp *n.* **péricarpe**
period *n.* **époque, période**
perishable *a.* **périssable**
periwinkle *n.* **pervenche**
permeability *n.* **perméabilité**
perry *n.* **poiré**
Persian wheel *n.* **noria**
persimmon *n.* **kaki, plaqueminier**
persistence *n.* **rémanence**
pest *n.* **déprédateur, ravageur**
pesticide *n.* **pesticide** (*m.*); p. science **phytopharmacie**
petal *n.* **pétale**
petunia *n.* **pétunia**

pharmacology *n.* **pharmacologie**
pharmacopoeia *n.* **pharmacopée**
phase *n.* **phase**
phenology *n.* **phénologie**
phenomenon *n.* **phénomène**
phenophase *n.* **phénophase**
phloem *n.* **liber, phloème**
phlox *n.* **phlox**
Pholiota sp. n. **pholiote**
phoretic *a.* **phorétique**
phosphate *n.* **phosphate**
photoperiod *n.* **photopériode**
photosensitive *a.* **photosensible**
photosynthesis *n.* **photosynthèse**
phyllody *n.* **phyllodie**
phylloxera *n.* **phylloxéra**
physic nut *n.* **médicinier**
phytopathogenic *a.* **phytopathogène**
phytotherapy *n.* **phytothérapie**
phytotoxicity *n.* **phytotoxicité**
pick (up) *v.* **ramasser**; p. over **entrecueillir, éplucher**
pickaxe *n.* **pic, pioche**
picker *n.* **cueille-fleurs, cueille-fruits, cueilleur, cueilloir**
pigeon pea *n.* **ambrevade**
pinch (off) *v.* **pincer**
pine *n.* **pin**
pineapple *n.* **ananas**
pink *n.* **mignardise, oeillet**
pinnate *a.* **penné**
pip *n.* **pépin**
pipe *n.* **buse, tuyau**; delivery p. *Irrig.* **ajutage**
pistachio (nut) *n.* **pistache**; p. tree **pistachier**
pistil *n.* **pistil**
pit *n.* **fosse**
Pitanga cherry *n.* **cerisier de Cayenne**
pitch *n.* **brai, poix**
placement (of fertilizer) *n.* **localisation**
placing *n.* **mise**
plane tree *n.* **plane, platane**
plant *n.* **plante, végétal**; *in vitro* plant **vitroplant**; nursery p. **plant**; oil plants **oléagineux**; protein plants **protéagineux**
plant *v.* **planter**
plant community *n.* **phytocénose**
plant parasite *n.* **phytoparasite**
plant sociology *n.* **phytosociologie**
plantain *n.* **plantain**
plantation *n.* **plantation**
planter *n.* **planteur**
planting *n.* **plantation**; p. machine **planteuse**
plasmolysis *n.* **plasmolyse**
plastic *a.* & *n.* **plastique**; p. greenhouse **abri-serre**; p. mat **plastisson**; p. material (not woven) **agrotextile**; p. mulching **bâchage**
plateau *n.* **plateau**
playing field *n.* **terrain de sport**
pleiomerous *a.* **pléiomère**
pleophagous *a.* **pléophage**
plot *n.* **parcelle**
plough *n.* **charrue**; **brabant**; heavy p. **défonceuse**
plough *v.* **labourer**; p. deeply **défoncer**
ploughing *n.* **labour, labourage**
plugging index *n.* *Rubb.* **index de plugging, indice d'obstruction**
plum *n.* **prune**; p. tree **prunier**
plumbago *n.* **dentelaire**
plume *n.* **plumet**
pocket *n.* **poche**
pod *n.* **cosse, gousse, cabosse**
poinsettia *n.* **poinsettia, étoile de Noël**
poisoning *n.* **empoisonnement, intoxication**
pole *n.* **gaule, perche**
polish *v.* **lustrer**
pollard *v.* **étêter, étronçonner**
pollinate *v.* **polliniser**
polluted *a.* **pollué**
pomaceous *a.* **pomacé**
pomegranate (fruit) *n.* **grenade**; p. plant **grenadier**
Pomoideae *n. pl.* **pomoïdées**
pomology *n.* **pomologie**
pond *n.* **mare**; pond apple **corossol**
poplar *n.* **peuplier**
poppy *n.* **coquelicot, pavot**
porcupine *n.* **porc-épic**
porous *a.* **poreux**
post *n.* **poteau**
postdormancy *n.* **postdormance**
post-driver *n.* **enfonce-pieux**
post-emergence *n.* **post-levée**
post-sowing *n.* **post-semis**
pot (herb etc.) *a.* **potager**; *n.* **pot**
pot *v.* **empoter**; **relever**; p. (out) **dépoter**
pot-herb *n.* **brède**
potash *n.* **potasse**
potful *n.* **potée**
pour *v.* **déverser, verser**
powder *n.* **poudre**
pox (plum) *n.* *PD* **sharka**
pozzolana *n.* **pouzzolane**
predatory *a.* **prédateur**; *n.* **prédateur, contre-parasite**
pre-emergence *n.* **pré-levée**
premature *a.* **prématuré**
pre-nursery *n.* **pré-pépinière**
pre-packaged *a.* **pré-emballé**
pre-ripening *n.* **prématuration**
preservative *a.* & *n.m.* **préservatif**
preserve *n.* (food) **conserve**
pre-sowing *n.* **pré-semis**
press *n.* **pressoir**
pressure *n.* **pression**
pre-treatment *n.* **pré-traitement**
preventive *a.* **préventif**

prey *n.* **proie**
price *n.* **prix**
prick *n.* **piqûre**
prickle *n.* **aiguillon**
primordium *n.* **ébauche**
primrose *n.* **primevère**
primula *n.* **primevère**
privet *n.* **troène**
probe *n.* **sonde**
process *n.* **procédé, processus**
producer *n.* **producteur**
product *n.* **produit**
production *n.* **production**; **obtention**
profitable *a.* **payant, rentable**
progeny *n.* **descendance**
programming *n.* **programmation**
proliferation *n.* **prolifération**
prolongation *n.* **prolongement**
promising *n.* **prometteur**
prop *v.* **échalasser, tuteurer**; p. up **béquiller, étayer**
propagate *v.* **multiplier, oeilletonner, propager**
propagator *n.* **multiplicateur, propagateur**
prophyll *n.* **préfeuille**
protandrous *a.* **protandre**
protein *n.* **protéine**
protogynous *a.* **protogyne**
protoplast *n.* **protoplaste**
prune *n.* **pruneau**
prune *v.* **égauler, élaguer, émonder, tailler**; **rafraîchir**
pseudo-bulb *n.* **pseudo-bulbe**
pseudocarp *n.* **pseudocarpe**
pseudostem *n.* **faux tronc**
psorosis *n.* **psorose**
psylla *n.* **psylle**
pterocarpous *a.* **ptérocarpe**
puddle *v.* **praliner**
puffiness *n.* **boursouflement**
puffy (of soil) *a.* **soufflé**
pulp *v.* **dépulper**
pulse *n.* **légume à gousse, légumineuse sèche**
pulse (CO_2) *n.* **choc (CO_2)**
pulverize *v.* **pulvériser**
pummelo (fruit) *n.* **pamplemousse**; p. tree **pamplemoussier**
pump *n.* **pompe**; electric p. **électropompe**
pumpkin *n.* **citrouille**
punnet *n.* **flein**
pupa *n.* **chrysalide, pupe**
purity *n.* **pureté**
purslane *n.* **pourpier**; winter p. **claytone**
pyrethrum *n.* **pyrèthre**

quaking grass *n.* **amourette, brize**
quality *n.* **qualité**
quarantine *n.* **quarantaine**
quelea finch *n.* **mange-mil, quéléa**
quince *n.* **coing**; q. tree **cognassier**
quincunx *n.* **quinconce**
quinoa *n.* **quinoa**

rabbit *n.* **lapin**
raceme *n.* **racème**
radiation *n.* **radiation, rayonnement**
radicular *a.* **radiculaire**
radioactive *a.* **radioactif**; r. tracer **radiotraceur**
radish *n.* **radis**; Japanese r. **daïkon**; wild r. **ravenelle**
rake *n.* **râteau**
rake *v.* **racler, ratisser**
rain *n.* **pluie**
rainfall *n.* **pluviosité**
rainy *a.* **pluvieux**; r. season (tropics) **hivernage**
ram (*v.*) soil **plomber**
rambutan *n.* **ramboutan**
ramify *v.* **ramifier**
rampion *n.* **raiponce**
range *n.* **gamme**
ranunculus *n.* **renoncule**
raspberry *n.* **framboise**; r. plant **framboisier**
rat *n.* **rat, surmulot**
rate *n.* **dose, taux**
ratio *n.* **rapport**
ratoon *n.* **rejet, rejeton, repousse**
raw *a.* **brut, cru**
ray *n.* **rayon**
reabsorb *v.* **résorber**
reactive *a.* **réactif**
reagent *n.* **réactif**
receiver *n.* **receveur**
receptacle *n.* **réceptacle**
recessive *a.* **récessif**
reclaim (land) *v.* **assainir**
recovery *n.* **guérison, rétablissement**
redden *v.* **rougir**
redox *n.* **oxydoréduction**
reduce *v.* **réduire**
reed *n.* **roseau**
reed grass *n.* **baldingère**
refinery *n.* **raffinerie**
refrigerate *v.* **réfrigérer**
refuse *n.* **détritus**
regenerate *v.* **régénérer**
regimen *n.* **régime**
regraft *v.* **regreffer**
regrowth *n.* **recrû, repousse**
reheat *v.* **réchauffer**
rejuvenate *v.* **rajeunir**
rejuvenation *n.* **réjuvénilisation**
release *v.* **lâcher**
remontant *a.* **reflorescent**
remote sensing *n.* **télédétection**
removal *n.* **suppression**

remove *v.* **supprimer**; r. moss **émousser**; r. plastic cover **débâcher**; r. seeds **épépiner**; r. stalks **équeuter**; r. stones **épierrer**; r. stumps **dessoucher, essoucher**; r. stalks **dérafler**; r. stolons **déstolloner**; r. suckers **égourmander**; r. thistles **échardonner**; r. tie **déligaturer**
remunerative *a.* **rémunérateur**
renewal *n.* **rénovation**
renovate *v.* **rénover**
repellent *n.* **répulsif**
replacement *n.* **remplacement**
replant *v.* **replanter**
report *n.* **rapport**
repot *v.* **rempoter**
reproduction *n.* **reproduction**
requirement *n.* **exigence**
research *n.* **recherche**; r. worker **chercheur, expérimentateur**
residual *a.* **rémanent, résiduel**
residue *n.* **résidu**
resin *n.* **résine**
resistant *a.* **résistant**
resow *v.* **resemer, réensemencer**
respiration *n.* **respiration**
rest *n.* **repos**
restore *v.* **restaurer**
result *n.* **résultat**
retardant (of growth) *n.* **retardateur de croissance**
reticulated *a.* **réticulé**
retractable *a.* **escamotable**
re-turf *v.* **regazonner**
return (profit) *n.* **rapport**
reversible *a.* **réversible**
rewetting *n.* **réhumectation**
rhizoctonia *n.* **rhizoctone**
rhizome *n.* **rhizome**
rhizomorph *n.* **rhizomorphe**
rhododendron *n.* **rhododendron**
rhubarb *n.* **rhubarbe**
rhythm *n.* **rythme**
rib *n.* **côte**
ribbon *n.* **ruban**
ribosome *n.* **ribosome**
riddle *n.* **crible, tamis**; *v.* **cribler**
riddled *a.* **criblé, tamisé**
ridge *n.* **faîte**; **billon**
ridge *v.* **billonner, butter**
ring *n.* **anneau, bague, rond**; annual r. **cerne**
ringing *n.* **ceinturage**; r. knife **bagueur**; r. tool **pince-sève**
rinse *v.* **rincer**
ripen *v.* **mûrir**
roast *v.* **torréfier**
rock *n.* **roche**
rockery *n.* **rocaille**
rocket *n.* **julienne, roquette**
rock wool *n.* **laine de roche**
rod *n.* **baguette, verge**
rodent *n.* **rongeur**
rodenticide *n.* **raticide**
rogue *a.* **illégitime**
roguing *n.* **élimination**
roll *v.* **rouler**; r. (soil) **plomber**; r. up **enrouler**
roller *n.* **rouleau**; leaf r. *Ent.* **enrouleuse**
roof *n.* **toit**
rook *n.* **freux**
roost (of birds) *n.* **dortoir**
root *n.* **racine**; r. bunch **griffe**
root *v.* **enraciner, implanter, prendre, raciner**; scion r. **s'affranchir**
root-forming *a.* **radicigène**
rootlet *n.* **radicelle**
rootstock *n.* **porte-greffe, sujet**
rose *n.* **rose**; r. bush **rosier**; r. garden **roseraie**
rose apple *n.* **jambose**; r. a. tree **jambosier**
rosemary *n.* **romarin**
rot *n.* **pourridié, pourriture, rot**
rot *v.* **pourrir**
rotary tiller *n.* **motobineuse**
rotation *n.* **assolement, rotation**
rotavate *v.* **fraiser**
rot-proof *a.* **imputrescible**
rotten *a.* **pourri**
rough *a.* **rude**
round(ed) *a.* **arrondi**
row *n.* **alignée, ligne, raie, rang**; r. (of fruit trees) **contre-espalier**
royal agaric *n.* **oronge**
rubber *n.* **caoutchouc**; r. tree **hévéa**; r. growing **hévéaculture**
ruderal *a.* **rudéral**
rue *n.* **rue**
runner *n.* **coulant, stolon, trace**
run-off *n.* **ruissellement**
rush *n.* **jonc**; flowering r. **butôme**
russet *a.* **roux**
russeting *n.* **rugosité**
russula *n.* **russule**
rust *n.* **rouille**
rusty *a.* **rouillé**
rye *n.* **seigle**
rye-grass *n.* **ray-grass**

sackful *n.* **sachée**
saffron *n.* **safran**
sage *n.* **sauge**
sago *n.* **sagou**; s. palm **sagoutier**
Sahelian *a.* **sahélien**
sainfoin *n.* **sainfoin**
St. John's wort *n.* **millepertuis**
salad *n.* **salade**
sale *n.* **vente**
salinity *n.* **salinité**
salsify *n.* **salsifis**

salt *n.* **sel**; salt out *v.* *Chem.* **relarguer**
salting *n.* **salage**
saltpetre *n.* **salpêtre**
salvia *n.* **sauge**
samara *n.* **samare**
sample *n.* **échantillon, prélèvement**
sample *v.* **échantillonner, sonder**
sand *n.* **sable**
sand-leek *n.* **rocambole**
sanitary *a.* **sanitaire**
sansevieria *n.* **sansevière**
santonica *n.* **semen-contra**
sap *n.* **sève**; s.-drawer **appel-sève**
sapling *n.* **baliveau, plantard**
sapodilla (plum) *n.* **sapote, sapotille**; s. tree **sapotier**
sappy *a.* **séreux**
sarcocolla (tree) *n.* **sarcocollier**
sarmentous *a.* **sarmenteux**
sarsaparilla *n.* **salsepareille**
saturated *a.* **saturé**
sausage tree *n.* **saucissonnier**
savory *n.* **sarriette**
saw *n.* **scie**
sawfly *n.* **hoplocampe, tenthrède**
saxifrage *n.* **saxifrage**
scab *n.* **gale, tavelure**
scabious *n.* **scabieuse**
scald *v.* **ébouillanter, échauder**
scale (range) *n.* **gamme**
scale *n.* **écaille**; s. insect **cochenille, kermès**
scammony *n.* **scammonée**
scape *n.* **hampe**
scar *n.* **cicatrice**
scarify *v.* **scarifier**
scent *n.* **odeur, senteur**
scentless *a.* **inodore**
schistous *a.* **schisteux**
scion *n.* **ente, greffon**
sclerenchyma *n.* **sclérenchyme**
sclerophyll *n.* **sclérophylle**
sclerosed *a.* **sclérifié**
sclerotinia disease *n.* **sclérotiniose**
sclerotium *n.* **sclérote**
scorching *n.* **grillage**
scorzonera *n.* **scorsonère**
scraping *n.* **grattage**
scratch *n.* **strie**
screen *n.* **écran, rideau**; thermal s. **é. thermique**
screening *n.* **criblage**
scrub *n.* **broussaille**; s. killer **débroussaillant**
scythe *n.* **faux**
sea buckthorn *n.* **argousier**
sea-grape *n.* **raisinier**
seakale *n.* **crambé**
season *n.* **saison**; **campagne**
season *v.* **assaisonner**
seasonal *a.* **saisonnier**
seaweed *n.* **goémon**
secateur *n.* **sécateur**
secretory *a.* **sécréteur, sécrétoire**
sectorial *a.* **sectoriel**
sedge *n.* **carex, laîche, souchet**
seed *n.* **graine, semence**; s. bed **lit de semences**; s. hole **poquet**; s. plant **porte-graine**
seedless *a.* **apyrène, asperme**
seed tray *n.* **bac à semis**
selaginella *n.* **sélaginelle**
select *v.* **sélectionner**
selectivity *n.* **sélectivité**
self *v.* **autoféconder**
self-cleaning *a.* **autonettoyant**
self-compatibility *n.* *PB* **autocompatibilité**
self-draining *a.* **autodrainant**
self-fertile *a.* **autofécond, autofertile, autogame**
self-pollinated *a.* **autopollinisé**
self-propelled *a.* **automoteur**
self-regulating *a.* **autorégulant**
self-sown *a.* **spontané**
self-sterile *a.* **autostérile**
semi-determinate *a.* **semi-déterminé**
semi-early *a.* **demi-hâtif**
seminiferous *a.* **séminifère**
semi-sparkling (of wine) *a.* **pétillant**
semi-variogram *n.* **demi-variogramme**
senescent *a.* **sénescent**
senile *a.* **sénile**
sensitization *n.* **sensibilisation**
separate *v.* **séparer**
septoria disease *n.* **septoriose**
serradella *n.* **serradelle**
service tree *n.* **cormier, sorbier**
set *v.* (of fruit) **nouer**
sett *n.* **bouture, éclat**
settled (of soil) *a.* **rappuyé**
sex determination *n.* **déterminisme du sexe**
shadbush *n.* **amélanchier**
shade *n.* **ombrage, ombre**; shade-loving *a.* **sciaphile**
shadoof *n.* **chadouf**
shake *v.* **secouer**
shallot *n.* **échalote**
shape *n.* **forme**
shard *n.* **tesson**
shea-nut tree *n.* **karité**
shears *n.* **cisaille, ciseaux**
sheath *n.* **gaine**
sheave *v.* **gerber**
shed *n.* **remise**; fruit s. **fruitier**
sheet *n.* **nappe**
shell *n.* **coque, coquille**
shell *v.* **écaler, écosser, égrener**
shelter *v.* **abriter**
shelterbelt *n.* **brise-vent**
shield *n.* **bouclier**; s. for budding **écusson**
shining *a.* **luisant**

shoeflower *n.* **rose de Chine**
shoot *n.* **brin, pousse, rameau**
shoot *v.* (of plant) **lever, pousser**
shorten *v.* **raccourcir**
shot-hole disease *n.* **criblure**
shovel *n.* **pelle**
shower *n.* **averse, giboulée**
shred *v.* **déchiqueter**
shrink *v.* **rétrécir**
shrub *n.* **arbrisseau, arbuste**
shrubby *a.* **arbustif**
sickle *n.* **faucille**
sickly *a.* **débile, malingre**
side *n.* **côté**
sieve *n.* **tamis**; *v.* **tamiser**
siftings *n.* **criblure**
sigatoka disease *n.* **cercosporiose (du bananier)**
silage *n.* **ensilage**
silica *n.* **silice**
silk cotton tree *n.* **kapokier**
silt *n.* **limon**
silvering (tomato) *n. PD* **argenture**
silver leaf *n. PD* **plomb**
singe *v.* **couliner**
sinuate *a.* **sinué**
sinus, grapevine leaf petiolar, see **accolade**
size *n.* **grosseur**
skeletonize *v.* **squelettiser**
skin *n.* **peau, pellicule**
skirret *n.* **berle, chervi**
slag *n.* **scorie**
sledge *n.* **traîneau**
sleeve *n.* **manche, manchon**
slender *a.* **grêle, effilé**
slip *n.* **bion, bouture, caïeu, éclat**
slipperwort *n.* **calcéolaire**
sloe *n.* **prunelle**; s. (bush) **prunellier**
slope *n.* **pente, versant**
sludge *n.* **gadoue**
slug *n.* **limace**
sluice *n.* **partiteur**; s.-gate *n.* **vanne**
slurry *n.* **pralin**
smoked *a.* **fumé**
smoke tree *n.* **arbre à perruque, coquecigrue**
smut *n. PD* **charbon**
snag *v.* **désongletter**
snail *n.* **escargot**
snapdragon *n.* **gueule-de-lion, muflier**
sneezewort *n.* **bouton d'argent**
snowberry *n.* **chiocoque**
snowdrop *n.* **perce-neige**
snowflake *n.* **nivéole**
soak *v.* **tremper**
soapberry *n.* **savonnier**
soapwort *n.* **saponaire**
soft *a.* **mou**
soften *v.* **ramollir**
soil *n.* **sol, terre**; s. climate **pédoclimat**; s. core **carotte**; s. cover (plant) **couvre-sol**; s. science **pédologie**
soilless *a.* **hors-sol**; s. culture **culture h.-s.**
sole (plough) *n.* **semelle, sep**
Solomon's seal *n.* **sceau de Salomon**
solvent *n.* **solvant**
soncoya *n.* **corossol**
sooty mould *n.* **fumagine**
sorghum *n.* **sorgo**
sorrel *n.* **oseille**; red s. *Afr.* **o. de Guinée, roselle**
sort *v.* **trier**
sour *a.* **aigre, amer**
source *n.* **provenance, source**
sour cherry *n.* **griotte**; s. c. tree **griottier**
sour orange *n.* **bigarade**; s. o. tree **bigaradier**
soursop *n.* **corossol**; s. tree **corossolier**
southern pea *n.* **niébé**
southernwood *n.* **aurone**
sow *v.* **semer, ensemencer**; s. again **resemer, sursemer**
sowing *n.* **semis**
sowthistle *n.* **laiteron**
soyabean *n.* **soya**
space (out) *v.* **distancer, échelonner, écarter, espacer**
spade *n.* **bêche**
spadix *n.* **spadice**
span (of greenhouse) *n.* **chapelle**
Spanish oyster plant *n.* **cardouille**
Spanish plum *n.* **mombin**
sparkling (of wine) *a.* **mousseux**
sparse *a.* **clairsemé**
spathe *n.* **spathe**
spatulate *a.* **spatulé**
spawn *v. Mush.* **larder**
species *n.* **espèce, essence**
speck *n.* **moucheture**
spectrography *n.* **spectrographie**
spectrometry *n.* **spectrométrie**
spectroscopy *n.* **spectroscopie**
speed *n.* **vitesse**
sphinx moth *n.* **sphinx**
spice *n.* **aromate, épice**
spider *n.* **araignée**; s. mite **araignée, tisserand**
spider flower *n.* **cléome**
spiderwort *n.* **éphémère**
spike (of flower) *n.* **épi**
spikelet *n.* **épillet**
spinach *n.* **épinard**; New Zealand s. **tétragone**
spindle *n.* **fuseau**; s. tree **fusain**
spine *n.* **épine**
spinney *n.* **bois, boqueteau**
spiraea *n.* **spirée**
spiroplasma *n.* **spiroplasme**
splash *n.* **éclaboussure**
splinter *n.* **éclat**
split *a.* **fendu**
split (*v.*) ridges **débutter**

splitting up *n.* **fractionnement**
spoilt *a.* **gâté, taré**
spongy *a.* **spongieux**
sporangium *n.* **sporange**
sporogenesis *n.* **sporogenèse**
sport *n. PB* **sport**
sports field *n.* **terrain de sport**
spot *n.* **macule, moucheture, tache**
spray *v.* **pulvériser**; **seringuer**
sprayer *n.* **atomiseur, pulvérisateur**
spread *v.* **épandre**; (of plant) **coloniser**
spreader *n.* **épandeur**
sprig *n.* **brindille**
spring (water) *n.* **fontaine, source**
sprinkler *n.* **arroseur, asperseur**
sprinkling *n.* **aspergement, bassinage**
sprout *n.* **germe, pousse**
sprouting *n.* **départ, poussée**
spruce *n.* **épicéa, sapinette**
spur *n.* **ergot, éperon**; (on fruit tree) **courson, lambourde**
spurge *n.* **épurge, euphorbe**
spurrey *n.* **spergulaire, spergule**
spurt *n.* **saillie**
squall *n.* **rafale**
squamous *a.* **squameux**
squash *n.* **courge, pâtisson**; turban s. **turban**
squill *n.* **scille**
stack *v.* **empiler**
stack *n.* **meule**
staff, managerial *n.* **encadrement**
staff, surveyor's *n.* **jalon**
stage *n.* **stade**
staggering *n.* (of dates etc) **décalage**
staghorn fern *n.* **corne d'élan**
stain *n.* **tache**
stake *n.* **échalas, pal, pieu, rame, tuteur**
stale, become *v.* **rassir**
stalk *n.* **tige**; **queue, rafle**
stamen *n.* **étamine**
stand (of plants) *n.* **peuplement**
standard *a.* **classique**; s. tree *n.* **haute-tige**
standardize *v.* **étalonner, normaliser**
staple *v.* **agrafer**
star anise *n.* **badiane**; s. a. tree **badianier**
star apple *n.* **caïmite**; s. a. tree **caïmitier**
star of Bethlehem *n.* **belle-d'onze-heures**
starchy *a.* **féculent**
start *v.* **démarrer**
statocyst *n.* **statocyte**
stavisacre *n.* **staphisaigre**
stele *n.* **stèle**
stem *n.* **tige**
stembuilder *n.* **intermédiaire**
stemless *a.* **acaule**
stem pitting *n.* **cannelure**
step (of scion etc.) *n.* **épaulement**
sterilize *v.* **stériliser**
sticker *n. PC* **adhésif**
stickiness *n. PC* **ténacité, viscosité**
sticky *a.* **collant, gluant**
stigma *n.* **stigmate**
stipe *n.* **stipe**
stock *n.* **giroflée**; annual s. **quarantaine**
stock (of plant) *n.* **pied, souche, sujet**
stolon *n.* **stolon**; s. formation **stolonnage**
stoma *n.* **stomate**
stone (of fruit) *n.* **noyau**
stone (*v.*) fruit **dénoyauter, énoyauter, énucléer**
stonecrop *n.* **orpin, sédum**
stop *v.* **étêter, épointer**
storage *n.* **entreposage, stockage**; cold. s. **e. frigorifique**
store *n.* **dépôt, entrepôt, resserre**
store *v.* **emmagasiner, entreposer**
storksbill *n.* **bec-de-grue**
stove *n.* **étuve, réchaud**
stowing *n.* **arrimage**
strain *n. PB* **lignée, race, souche**
strainer (wire) *n.* **raidisseur, tendeur**
stratified *a.* **stratifié**
stratum *n.* **strate**
straw *n.* **paille**
strawberry *n.* **fraise**; s. plant **fraisier**; s. tree **arbousier**
straw mat *n.* **paillasson**
streaky *a.* **vergeté**
strew *v.* **parsemer**
striated *a.* **strié**
strike (of cutting) *n.* **reprise**
string *n.* **ficelle**
strip *n.* **bande**; grass s. **b. enherbée**; plant s. **b. végétale**
strip (*v.*) flowers **déflorer**; s. bark **écorcer**
stripe *n.* **rayure**
striped *a.* **rayé, tigré, zébré**
stub *n.* **argot**
stubble *n.* **chaume**
stump *n.* **chicot, moignon, tronçon**; s. removal **dessouchage**
stunt *v.* **rabougrir**
subalpine *a.* **subalpin**
subculturing *n.* **repiquage**
sub-deficiency *n.* **sub-carence**
suberose *a.* **subéreux**
submerged *a.* **submergé**
sub-shrub *n.* **sous-arbrisseau**
subsidence *n.* **affaissement**
subsoil *v.* **sous-soler**
sub-species *n.* **sous-espèce**
substrate *n.* **substrat**
subulate *a.* **subulé**
sucker *n.* **accru, drageon, gourmand, rejet, rejeton, surgeon**
sucker *v.* **drageonner, rejeter, surgeonner**
sucking *a.* **suceur**

sucrose *n.* **saccharose**
suction *n.* **aspiration**
sugar *n.* **sucre**
sulphate *n.* **sulfate**
sulphur *n.* **soufre**; s. dusting **soufrage**
sumach *n.* **sumac, vinaigrier**
sun *n.* **soleil**; s. scald **insolation**; s. scorch **échaudage, grillage**
sun-crack *n.* **fente d'insolation**
sundew *n.* **droséra, rossolis**
sundial *n.* **cadran solaire**
sunflower *n.* **soleil, tournesol**
sunny *a.* **ensoleillé**
sunset tree *n.* **boumou**
sunshine recorder *n.* **héliographe**
superior *a. Bot.* **supère**
supernatant *a.* & *n.* **surnageant**
supplementary *a.* **supplémentaire**
supplier *n.* **fournisseur**
support (*v.*) plants **ramer**
suppress *v.* **supprimer**
surface *n.* **aire, superficie, surface**
surface active *a.* **tensio-actif**
surfactant *n.* **agent de surface, surfactant**
surge *n.* **afflux**
survey *n.* **enquête**
surveying, land *n.* **arpentage**
survival *n.* **survie**
susceptibility *n.* **sensibilité, susceptibilité**
suslik *n.* **spermophile**
suspension *n.* **suspension**
swamp *n.* **marécage**
sward *n.* **gazon**
swede *n.* **rutabaga**
sweepings *f. pl.* **balayures**
sweet cicely *n.* **cerfeuil odorant**
sweetener *n.* **édulcorant**
sweet gale *n.* **galé**
sweet potato *n.* **patate**
sweetsop *n.* **attier, corossol**
sweet sultan *n.* **centaurée musquée**
sweet william *n.* **jalousie**
swell *v.* **gonfler, renfler**
swelling *a.* **gonflant**
Swiss chard *n.* **poirée**
sword bean *n.* **pois sabre**
sycamore *n.* **sycomore**
syconium *n.* **sycone**
sylvan *a.* **sylvestre**
symbiont *n.* **symbiote**
symmetrical *a.* **symétrique**
symplast *n.* **symplaste**
symptom *n.* **symptôme**
syncarp *n.* **syncarpe**
synergy *n.* **synergie**
synthesis *n.* **synthèse**
syringa *n.* **seringa**
syringe *n.* **seringue**
syrupy *a.* **sirupeux**
system *n.* **système**
systemic *a.* **systémique**

table *n.* **table, tableau**
tachinid fly *n.* **tachinaire**
take (of graft etc.) *n.* **reprise**
tallow *n.* **suif**
tamarind *n.* **tamarinier**
tank *n.* **bac**; evaporation t. **b. à évaporation**
tannia *n.* **macabo**
tansy *n.* **tanaisie**
tap *n.* **robinet**
tap *v. Rubb.* **saigner**
taproot *n.* **pivot**
tar *n.* **goudron**
target *n.* **cible**
taro *n.* **taro**
tarragon *n.* **estragon**
tart *a.* **acerbe, aigrelet**
taste *n.* **goût**
taster *n.* **dégustateur**
taxon *n., pl.* taxa **taxon**
tea *n.* **thé**; t. bush **théier**
tear *v.* **déchirer**
teasel *n.* **cardère**
tegument *n.* **tégument**
telluric *a.* **tellurique**
tender *a.* **frileux, gélif**
tenderness *n.* **tendreté**
tendril *n.* **crampon, griffe, vrille**
tepal *n.* **tépale**
terebinth *n.* **térébinthe**
terrace *n.* **terrasse**; bench t. **banquette**
test *n.* **épreuve, essai**
test-tube *n.* **éprouvette**
thallus *n.* **thalle**
thaw *v.* **dégeler**
therapeutic *a.* **thérapeutique**
thermofogging *n.* **thermonébulisation**
thermometer *n.* **thermomètre**
thermoperiod *n.* **thermopériode**
thermophile *a.* **thermophile**
therophyte *n.* **thérophyte**
thicket *n.* **fourré**; spiny t. **épinaie**
thin *v.* **démarier, éclaircir**
thistle *n.* **chardon**; globe t. **échinope**; t. hook **échardonnet**
thong *n.* **lanière**
thorn *n.* **épine**
thornbush *n.* **épine**
thorny *a.* **épineux**
thread *n.* **fil**
thresher *n.* **batteuse**
threshold *n.* **seuil**
thrips *n.* **thrips**
throat *n.* **gorge**

throw *v.* **jeter**
thyme *n.* **thym**
tie *n.* **accolure, accouple, attache, lien**
tie *v.* **accoler, attacher, nouer, relever**
tier *n.* **étage**
tiger flower *n.* **tigridie**
tight *a.* **serré**
tiller *v.* **taller**
tillering *n.* **tallage**
tilling *n.* **labour, labourage**
tineid moth *n.* **teigne**
tip *v.* *Hort.* **rabattre**
tipburn *n.* *PD* **brunissement marginal, rand**
tissue *n.* **tissu**
tobacco plant *n.* **tabac**
tolerance *n.* **accoutumance**
tomato *n.* **tomate**
tomentose *a.* **tomenteux**
tool *n.* **outil**
toolbar *n.* **porte-outil**
tooth *n.* **dent**
top *n.* **cime, sommet**
top *v.* **écimer, éhouper, étêter, décolleter**
topple *n.* *PD* **huilage**
topsoil *n.* **couche arable**
tormentil *n.* **tormentille**
tough *a.* **coriace**
tow *n.* **étoupe**
toxicity *n.* **toxicité**
tracheid *n.* **trachéide**
tracing *m.* **traçage**
track *n.* **chemin**
tractor *n.* **tracteur**; t. driver **tractoriste**; horticultural t. **micro-tracteur**; walking t. **motoculteur**
trade *n.* **commerce**
trailer *n.* **remorque**
trailing *a.* **traînant**
training *n.* **formation, palissage**
trampling *n.* **piétinement**
transpire *v.* **transpirer**
transplant *v.* **repiquer, transplanter**
trap *n.* **piège**; pincer t. **trappe-pince**
traveller's tree *n.* **ravenala**
tray *n.* **maniveau, plateau**
treat *v.* **traiter**
treatment *n.* **traitement**
tree *n.* **arbre**; fruit t. **fruitier**; t. guard **corset**
tree of heaven *n.* **ailante**
trellis *n.* **treille, treillis**
trench *n.* **jauge, fossé, tranchée**
trenching machine *n.* **trancheuse**
trial *n.* **épreuve, essai**
trickler *n.* *Irrig.* **goutteur, juteur**
trifoliate *a.* **trifolié**
trim *v.* **ébarber, parer, rogner**
triploid *a.* **triploïde**
tritiated *a.* **tritié**
trophic *a.* **trophique**
trough *n.* **auge**
trowel, garden *n.* **déplantoir**
truck *n.* **chariot**
truffle *n.* **truffe**
trumpet creeper *n.* **bignone**
truncated *a.* **tronqué**
trunk *n.* **tronc**
truss *n.* **bras**
tub *n.* **bac, cuve**
tuber *n.* **tubercule**
tuberose *n.* **tubéreuse**
tuberous *a.* **tubéreux**
tuff *n.* **tuf**
tuft *n.* **touffe**
tulip *n.* **tulipe**
tulip tree *n.* **tulipier**
tumor *n.* **tumeur**
tung tree *n.* **aleurite**
tunic *n.* **tunique**
turf *v.* **engazonner, enherber, gazonner**
turgidity *n.* **turgescence**
turion *n.* **turion**
turmeric *n.* **curcuma**
turnip *n.* **navet**
turn over *v.* **retourner**
tutsan *n.* **toute-saine**
twig *n.* **brindille**
twisted *a.* **tordu**
tying *n.* **accolage, attachage, liage**
tyre *n.* **pneu**

udo *n.* **oudo**
umbilicate *a.* **ombiliqué**
unbreakable *a.* **incassable**
uncultivated *a.* **inculte**
undamaged *a.* **indemne**
underground *a.* **souterrain**
undergrowth *n.* **sous-bois**
underlying *a.* **sous-jacent**
undermining *n.* **affouillement**
understorey *n.* **sous-étage**
undeteriorating *a.* **inaltérable**
undifferentiated *a.* **indifférencié**
undivided *a.* **indivis**
undrinkable *a.* **inbuvable**
undulating *a.* **ondulé**
uneatable *a.* **immangeable**
uneven *a.* **accidenté**
unfavourable *a.* **défavorable**
unfruitful *a.* **infructueux**
unicellular *a.* **unicellulaire**
unicoloured *a.* **unicolore**
uniflorous *a.* **uniflore**
uninervate *a.* **uninervé**
uniovular *a.* **uniovulé**
unit *n.* **unité**

unite *v.* **souder**
unproductive *a.* **improductif**
unsaleable *a.* **invendable**
unsaturated *a.* **insaturé**
unstick *v.* **décoller**
untie *v.* **déligaturer**
unusual *a.* **inhabituel**
unwinding *n.* **déroulement**
upkeep *n.* **entretien**
upright *a.* **dressé, redressé**
uproot *v.* **arracher, déraciner**
urea *n.* **urée**
use *n.* **emploi, usage**

valerian *n.* **valériane**
valonia oak *n.* **vélani**
value *n.* **prix, valeur**
valve *n.* **soupape**
vanilla *n.* **vanille**; v. plant **vanillier**
vapour *n.* **buée, vapeur**
vapourer moth *n.* **orgyie**
variability *n.* **variabilité**
variegated *a.* **diversicolore, versicolore**
variegation *n.* **panachure**
variety *n.* **variété**
vascular *a.* **vasculaire**
vase *n.* **vase**
vault *n.* **voûte**
vector *a.* & *n.* **vecteur**
vegetable *n.* **légume**
vegetarian *a.* & *n.* **végétarien**
vegetate *v.* **végéter**
veil *n.* *Mush.* **voile**
vein *n.* **nervure, veine**
velamen *n.* **voile**
velvety *a.* **velouté**
ventilate *v.* **ventiler**
Venus' fly-trap *n.* **dionée**
vermicompost *n.* **lombricompost**
vermiculture *n.* **lombriculture**
vernal *a.* **printanier, vernal**
vernalization *n.* **printanisation, vernalisation**
veronica *n.* **véronique**
versatility *n.* **polyvalence**
verticillium disease *n.* **verticilliose**
vervain *n.* **verveine**
vesicle *n.* **vacuole, vesicule**
vesiculo-arbuscular *a.* **vésiculo-arbusculaire**
vessel *n.* **vaisseau**
vetch *n.* **gesse, vesce**
viable (of seed) *a.* **capable de germer**
vibrator *n.* **vibrateur, vibreur**
viburnum *n.* **viorne**
vigour *n.* **vigueur**
villous *a.* **villeux**
vine *n.* **vigne, cep**; v. harvest **vendange**; v. science **ampélographie**; v. scientist **ampélographe**
vinegar *n.* **vinaigre**
vineyard *n.* **vignoble**
vinous *a.* **vineux**
violet *n.* **violette**
virgin *a.* **vierge**
Virginia creeper *n.* see **vigne-vierge**
viroid *n.* **viroïde**
virulent *a.* *PD* **agressif**
virus *n.* **virus**; plant v. **phytovirus**; p. virus disease **phytovirose**
viticultural *a.* **vinicole, viticole**
viticulture *n.* **viticulture**
viticulturist *n.* **vigneron, viticulteur**
volatile *a.* **volatil**
vole *n.* **campagnol**
voluble *a.* **volubile**
voluminous *a.* **volumineux**
volvaria *n.* **volvaire**

wadi *n.* **oued**
wages *n.* **salaire**
wall *n.* **mur**
wallflower *n.* **giroflée**
walnut *n.* **noix**; w. tree **noyer**
warning *n.* see **avertissement**
warp (land) *v.* **colmater**
wart *n.* **verrue**; w. disease **galle verruqueuse**
washing *n.* **lavage**
wasp *n.* **guêpe**
waste *n.* **déchet**; *v.* **gaspiller**
water *n.* **eau**; w. core *PD* **vitrescence**; w. culture **aquaculture**; w. lily **nénufar**; w. table **nappe phréatique**
water *v.* **abreuver, arroser, bassiner, irriguer**
watering can *n.* **arrosoir**
water leaf *n.* **grassé**
waterlogging *n.* **engorgement, ennoyage**
watermelon *n.* **pastèque**
waterproof *a.* **imperméable**
water-retaining product *n.* **hydrorétenteur, rétenteur d'eau**
wattle *n.* **claie**
wattle (tree) *n.* **acacia**
wax *n.* **cire**
wax gourd *n.* **bénincase**
way *n.* **chemin, voie**
wayfaring tree *n.* **mancienne**
weak *a.* **faible, chétif**
wear *n.* **usure**
weather *v.* **désagréger**
weathering *n.* **altération, désagrégation**
weathercock *n.* **girouette**
web *n.* **toile**
wedge *n.* **coin, éclisse**
weed *n.* **adventice, mauvaise herbe**; w. control

désherbage; w. invasion **salissement**; w. science **malherbologie**
weed *v.* **désherber, sarcler**
weeping *a.* **pleureur**
weevil *n.* **charançon**
weighing *n.* **pesée**
weight *n.* **poids**
well *n.* **puits**
wet *a.* **humide, mouillé**
wettability *n.* **mouillabilité**
wettable *a.* **mouillable**
wetter *n.* **mouillant**
wheel *n.* **roue**
wheelbarrow *n.* **brouette**
whitebeam *n.* **alisier, allouchier**
whitefly *n.* **aleurode**
whole *a.* **entier**
wholesaler *n.* **grossiste**
width *n.* **largeur**
wild cherry *n.* **merise**; w. c. tree **merisier**
wilding *n.* **sauvageon**
wild oat *n.* **folle avoine**
wild pomegranate *n.* **balauste**; w. p. bush **balaustier**
wild rose *n.* **églantine**; w. r. bush **églantier**
willow *n.* **saule**
wilt disease *n.* **trachéomycose**
wilt *v.* **flétrir, faner**
winch *n.* **treuil**
wind *n.* **vent**
windbreak *n.* **abrivent, brise-vent**
windmill *n.* **moulin à vent**
windrow *v.* **andainer**
wine *n.* **vin**
wing *n.* **aile**
winnow *v.* **vanner**
winter *n.* **hiver**
winter daffodil *n.* **amaryllis**
winter-flowering *a.* **nivéal**
wintering *a.* **hivernant**
winter moth *n.* **chéimatobie, hibernie**
winter squash *n.* **potiron**
winter sweet *n.* **chimonanthe**
wistaria *n.* **glycine**
witch-hazel *n.* **hamamélis**
wither *v.* **dépérir, flétrir**
witloof chicory *n.* **endive, witloof**
wood *n.* **bois, futaie**
wooded *a.* **boisé**
woodruff *n.* **petit muguet**
woody *a.* **ligneux**
woody ornamental *n.* **ligneux d'ornement**
wool *n.* **laine**; glass w. **l. de verre**; rock w. **l. de roche**
woolly *a.* **laineux**
worm *n.* **ver**
worm compost *n.* **lombricompost**
worm culture *n.* **lombriculture**
worm-killer *n.* **lombricide**
wormwood *n.* **absinthe**
wound *n.* **blessure, plaie**
wrap *v.* **emballer**; w. in paper **empapilloter**
wrinkle *v.* **rider**
wych elm *n.* **ypréau**

xeric *a.* **xérique**
xerophilous *a.* **xérophile**
xerophyte *a.* & *n.* **xérophyte**
xylem *n.* **bois**

yellow *a.* **jaune**
yellowing *n.* **jaunisse, jaunissement**
yew *n.* **if**
yield *n.* **rendement**; *v.* **rendre**

Zanthoxylum n. **xanthoxylum**
zymogenous *a.* **zymogène**